An Introduction to Materials Informatics

Tongyi Zhang

An Introduction to Materials Informatics

The Elements of Machine Learning

Tongyi Zhang
Shanghai University
Shanghai, China

ISBN 978-981-99-7994-3 ISBN 978-981-99-7992-9 (eBook)
https://doi.org/10.1007/978-981-99-7992-9

Jointly published with Science Press
The print edition is not for sale in China mainland. Customers from China mainland please order the print book from: Science Press.
ISBN of the Co-Publisher's edition: 978-7-03-072898-2

This Springer imprint is published by the registered company Springer Nature Singapore Pte Ltd.
The registered company address is: 152 Beach Road, #21-01/04 Gateway East, Singapore 189721, Singapore

Preface

Materials informatics is a fast-developing interdisciplinary science and technology field, as evidenced by Fig. 1, which was retrieved on October 4, 2021, from the Web of Science under the search topic "materials informatics." The publication and citation numbers were both negligible in the last century and increased exponentially in the first two decades of this century, reaching over 450 and around 10,000, respectively, in 2020. Although many books and monographs are available on statistical learning, machine learning, and data mining, few are on materials informatics.

In 2011, the author participated in the S14 Xiangshan Science Conference: The Systematic Engineering in Materials Science (Functional and Structural Materials Genome Engineering) [材料科学系统工程(功能、结构材料基因组工程)] in Beijing, China, right after the announcement of Materials Genome Initiative (MGI) in the USA (https://www.mgi.gov). Since then, the author has devoted himself to the development of materials genome engineering (MGE) and materials informatics. The Materials Genome Institute of Shanghai University was established in 2014 with considerable support from Shanghai University and the great endeavors of all MGI colleagues. To promote MGE in China, the Chinese Government launched the MGE research and development scheme in 2016, and many local governments of China have financially supported tremendous numbers of MGE research and development projects. To educate the contemporary working force and the next generation of materials workers, engineers, and researchers with the knowledge of MGE and materials informatics, Shanghai University has developed a novel undergraduate program, namely Materials Design Science and Engineering, and takes statistical learning, machine learning, and data mining as required courses. Obviously, textbooks on materials informatics are urgently needed for this education innovation to meet the need of fast development of such an interdisciplinary science and technology field, and this motivates the author to write textbooks on materials informatics. We have a strong background on materials science and engineering and solid mechanics and truly believe that this background will help us significantly reach the goal that readers with limited knowledge of statistics, probability, and optimization can still learn materials informatics from the textbooks without difficulty. To reach this goal, we strive to write the textbooks in a comprehensible way with as many detailed

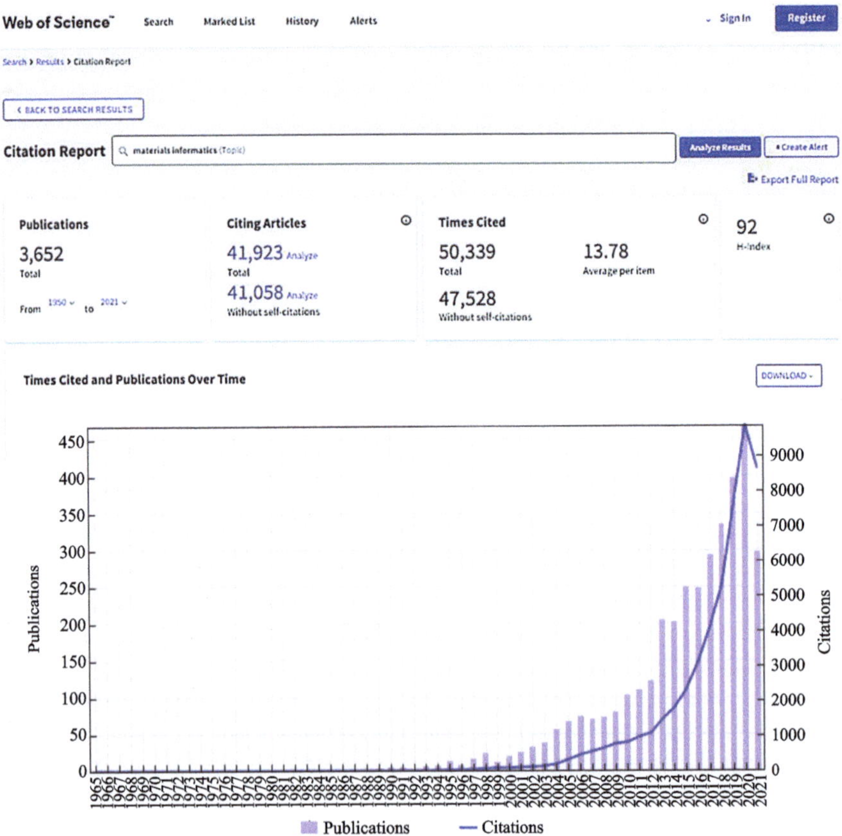

Fig. 1 Searching result of the topic "materials informatics" on October 4, 2021, from the Web of Science

examples as possible. The main theme is manifested in the following two volumes: *An Introduction to Materials Informatics* (I): *The Elements of Machine Learning* and *An Introduction to Materials Informatics* (II): *Advances.* The first volume focuses on essential and fundamental AI and ML knowledge and algorithms; the second volume covers recent advances in materials informatics, introducing advanced AI and ML algorithms and overviewing materials informatics, current status, challenges, and prospects. The textbooks are designed for researchers and students, whose majors are not in mathematics, computer science and engineering, and automation and control, but in other broad fields, such as materials science and engineering, mechanics science and engineering, physics, and chemistry.

The first volume consists of 12 chapters. During the writing, the author's former and present students, Dr. Xiong Jie, Mr. Wei Qinghua, Mr. Cao Bin, Mr. Wang Yuanhao, Mr. Xu Sheng, Mr. Zhang Shouyang, and Mr. Ma Jiaxuan provide full support in preparing examples, exercise problems and appendixes, and check all formats of texts, figures, tables, and references. The author's colleagues, Prof. Sun

Sheng, Prof. Qian Quan, Prof. Ouyang Runhai, Prof. Xie Wei, Prof. Huang Jiaqiang, and Prof. Wang Wei, read the draft version of the textbook and provide many constructive comments and suggestions.

In the course of writing this textbook, the author received heartful concern and encouragement from many friends and colleagues, particularly colleagues working in the fields of MGE and materials informatics. The author would like to thank Prof. Xie Jianxin for his leadership in promoting MGE in China. During the publication process, the author works very closely with the Science Press via Mrs. Zhang Shuxiao, who gives the author numerous encouragement, which makes the writing an enjoyable journey.

Shanghai, China
February 2022

Tongyi Zhang

The original version of the book has been revised. A correction to this book can be found at https://doi.org/10.1007/978-981-99-7992-9_13

Contents

Symbols and Notations

x	Input variable (feature)
$\boldsymbol{x}$	Feature vector
$\boldsymbol{X}$	In Chap. 10: Observed variable sequence In other chapters: Feature matrix of input data
$\boldsymbol{X}_i$	The i-th sample vector in feature matrix
x_{ij}	Value of the j-th component of feature in the i-th sample
x_j	The j-th component of feature vector
σ_{x_j}	The standard deviation of the j-th component of feature vector
$\tilde{x}$	Standardized feature data
$\mathring{x}$	Normalized and centralized feature data
$\check{x}$	Centralized and scaled feature data
y	Target or response variable (data label in classification)
$\boldsymbol{y}$	Target vector
$\boldsymbol{Y}$	Target matrix
$\hat{y}$	Predicted value of target
$\boldsymbol{Y}_i$	Target value in the i-th sample
$\bar{y}$	Mean value of y
$\hat{y}_i$	Predicted value of target for the i-th sample
$y_i^{(k)}$	Value of the k-th component of target in the i-th sample
w_0	Intercept
w	Coefficients
$\boldsymbol{W}$	Coefficient vector (coefficient matrix)
$\bar{\boldsymbol{W}}$	Mean of $\boldsymbol{W}$
$\Vert\boldsymbol{W}\Vert$	Norm of $\boldsymbol{W}$
b	Bias (intercept)

$\hat{b}$	Estimated value of bias
$\boldsymbol{b}$	Bias vector
$\boldsymbol{\mu}$	Mean vector (expectation vector)
M	Integer, Variable In Chap. 4, Margin
λ	In Chap. 4: Lagrange multipliers In Chap. 6: Preset constant In Chap. 10: Intrinsic parameter set of an HMC (see later) In other chapters: Penalty parameter
L_{D}	Dual objective function
RSS	Residual Sum-of-Squares
$\boldsymbol{I}$	Unit matrix
$\boldsymbol{\Sigma}$	Covariance matrix
$\boldsymbol{v}_j$	Normalized and orthogonal basis set
$\boldsymbol{V}$	Normalized and orthogonal eigenvector matrix
$\boldsymbol{D}^2$	Diagonal eigenvalue matrix
$\boldsymbol{D}$	Diagonal matrix
$\boldsymbol{U}$	Semi-orthogonal matrix
$\boldsymbol{p}_j$	The j-th principal component
$\boldsymbol{P}$	Principal component (PC) matrix
$\mathcal{L}$	Loss function
ϕ	High-dimensional space mapping
$K(\boldsymbol{X}_i, \boldsymbol{X}_j)$	Kernel function
ξ	Slack variable
ε	Error
Ω	Penalty term
$\Re$	Generalized Rayleigh quotient
$\boldsymbol{S}$	Symmetric between-class scatter matrix
$\boldsymbol{H}$	Symmetric centering matrix
R_k	Data region
n_{R_k}	Number of data in region R_k
T	Decision tree
$\mid T \mid$	Number of terminal nodes of tree T
G	Ensemble learner
g	Base learner
ε_j	Error rate of base learner g_j
α_j	Weight of base learner g_j
β_j	Parameters involved in base learner g_j
Z	In Chap. 6: Normalizing constant In Chap. 7: Latent variables
γ	In Chap. 6: Preset constant In Chap. 7: Half width at half peak strength

μ_i	Expectation of the i-th Gaussian distribution
σ_i	Variance of the i-th Gaussian distribution
w_i	Weight of the i-th Gaussian distribution
z_i	The i-th Gaussian distribution
$p_M(\cdot)$	Distribution probability density of a mixture
$L(\cdot)$	Likelihood function
$\mathrm{LL}(\cdot)$	Log-likelihood function
$\mathrm{LLL}(\cdot)$	Lagrange function of the log-likelihood function
I_j	Frequency or intensity at x_j
R_{wp}	Weighted profile factor
R_{p}	Profile factor
α_{ji}	Contribution weight of distribution i to $p_M(x_j \| Z)$
$g_l(g_l = 1, 2, \cdots, G_l)$	Neurons in layer l
$r_{g_l}^{(l)}$	Received latent variable by neuron g_l in layer l
$z_{g_l}^{(l)}$	Output latent variable from neuron g_l in layer l
$Z_{g_l}^{(l)}$	Output latent variable vector from layer l $Z_{g_l}^{(l)} = \left(z_1^{(l)}, z_2^{(l)}, \cdots, z_{G_l}^{(l)}\right)$
$w_{g_{l-1}\to g_l}^{(l)}$	The weight connected neuron g_{l-1} and neuron g_l
$\boldsymbol{W}_{g_{l-1}\to g_l}^{(l)}$	Weight vector $\boldsymbol{W}_{g_{l-1}\to g_l}^{(l)} = \left(w_{1\to g_l}^{(l)}, w_{2\to g_l}^{(l)}, \cdots, w_{G_{l-1}\to g_l}^{(l)}\right)^{\mathrm{T}}$
$b_{g_l}^{(l)}$	Bias generated by neuron g_l in layer l
$\varphi_{g_l}^{(l)}$	Activation function acting on neuron g_l in layer l
s_k	The k-th state
$S = \{s_k\}\ (k = 1, 2, \cdots, q)$	State set
$\pi_i^{(k)} = P(s_k \in S)$	s_k occupation probability at event i
π_0	Initial state occupation probability
T_{kl}	Transition probability from state k to l
$\boldsymbol{T} = \{T_{kl}\}$	Transition probability matrix
$E_{P(\theta)}(f(\theta))$	Expectation of $f(\theta)$ over probability distribution $P(\theta)$
$\alpha_{i\to i+1}$	Acceptance probability of candidates
o_j	The j-th observation basis
$\boldsymbol{O} = \{o_j\}\ (j = 1, 2, \cdots, m)$	Observation (emission) basis set

$\boldsymbol{B} = \{B_{kj}\} = \left\{b_k\left(x_i^{(j)} = o_j\right)\right\}$	Observation (emission) matrix
$\lambda = (\boldsymbol{T}, \boldsymbol{B}, \pi_0)$	The intrinsic parameter set
$\alpha_i(k) = P\left(x_{i-1}, x_i, \pi_i^{(k)} \mid \lambda\right)$	The forward probability at event i
$\beta_i(k) = P\left(x_n, x_{n+1}, \cdots, x_{i+1} \mid \lambda, \pi_i^{(k)}\right)$	The backward probability
$\gamma_i(k)$	The probability of a state sequence
$\xi_i(k, l) = P\left(\pi_i^{(k)}, \pi_{i+1}^{(l)} \mid \boldsymbol{X}, \lambda\right)$	The state joint probability
$\mathrm{cov}(\cdot, \cdot)$	Covariance
$R(\cdot, \cdot)$	Pearson correlation coefficient
$E[\cdot]$	Expectation
$I(\cdot, \cdot)$	The mutual information
$I^*(\cdot, \cdot)$	The normalized maximum mutual information
$P(\cdot)$	Probability
$p(\cdot)$	Probability density
$H(\cdot)$	Information entropy
$\mathrm{MIC}(\cdot)$	Maximal Information Coefficient
$B^i\ (i = 1, 2, \cdots)$	Interval middles between two adjacent values
$\mathbf{S}_M$	Original feature set
$\mathbf{S}_m$	Feature subset ($m \leq M$)
$\hat{\mathbf{S}}_m$	Feature subset with the maximum dependency
$H_q\ (1, \cdots, Q)$	The q-th nearest neighbor among the Q nearest neighbors in the same class (class k)
$c_k, c_{\hat{k}}(\hat{k} \neq k)$	Class k, class $\hat{k}$
$M_q^{c_{\hat{k}}}(q = 1, 2, \cdots, Q)$	Q nearest neighbors in class $c_{\hat{k}}$
$W_{x_j}^i$	Weight of feature x_j in sample i
$\mathrm{Imp}_\varphi(\cdot)$	Mean decrease impurity in a decision tree φ
ϕ_i	SHAP value of feature x_i
$\phi_{i,j}$	Interaction SHAP value between feature x_i and feature x_j
$\phi_{i,i}$	Pure SHAP value of feature x_i

Chapter 1
Introduction

Materials informatics is an emerging and rapidly developing field, particularly after the launch of Materials Genome Initiative (MGI) in 2011 (https://www.mgi.gov). Ramakrishna et al. (2019) defined materials informatics as that "the materials informatics employs techniques, tools, and theories drawn from the emerging fields such as data science, internet, computer science and engineering, and digital technologies to the materials science and engineering to accelerate materials, products and manufacturing innovations". Agrawal and Choudhary (2016) proposed that materials informatics has become the "fourth paradigm" in the research and development of materials. Following the guidance of MGI, many free-access and open-to-public material databases have been built up, and materials data are exponentially growing thanks to the development of high-throughput experiments and high-throughput calculation techniques, as well as data sharing among the materials communities. A list of available materials databases can be found in the review paper of Ramakrishna et al. (2019). Materials informatics integrates materials science and engineering and the sciences and technologies from artificial intelligence (AI), database, and machine learning (ML) to accelerate the innovation in the whole materials development continuum from discovery, development, property optimization, system design and integration, manufacturing to deployment, and to speed up the process from data to knowledge in order to understand and master the relationship between material micro-structures and macro-properties based on material composition, processing, and performance. Materials informatics adds the novel tools of AI and ML into the toolbox of materials science and engineering, which will definitely strengthen and enhance the power of the methodologies in materials research and development. Materials informatics utilizes AI and ML to analyze a large ensemble of materials data from experiments, computations, manufactures, industries, and daily life, etc., efficiently and cost-effectively, and to deliver materials knowledge and technology in user-friendly ways to designers, scientists, engineers, and manufacturers of materials and products. ML gives computers the ability to learn from data and make predictions based on data (Samuel, 1967). Mitchell (1997) defined ML as follows:

T. Zhang, *An Introduction to Materials Informatics*,
https://doi.org/10.1007/978-981-99-7992-9_1

"A computer program is said to learn from experience E with respect to some class of tasks T and performance measure P, if its performance at tasks in T, as measured by P, improves with experience E." In materials informatics, ML algorithms learn from existing material data with some properties or/and performance of interest, in order to improve targeted properties and performance of existing materials or/and to design and discover new materials with desired properties and performance. Since material data are usually small, adaptive design with feedback from experiments or/and computations has been proposed to enhance the ML ability. Figure 1.1 shows the adaptive learning and design in materials informatics. Initially successful and failed data, with material properties and performance of interest and material composition, processing, testing conditions, or/and service environments, etc. of adjustable parameters, are generated from experiments, calculations, or/and manufacturing. A dataset is usually built up by one's own produced data or/and by collecting data from various sources. The material properties and performance of interest are usually called output variables, and the parameters of material composition, processing, or/and testing conditions are normally called input variables. In addition, many molecular, atomic and electronic parameters, thermodynamic properties, kinetic parameters, crystalline and amorphous information, etc. are often employed as input or/and output variables. ML is conducted on an initial dataset. The learning results will be evaluated and interpreted automatically or/and by domain experts, who are authorities in a particular area or topic, and clearly, domain experts in materials informatics are materials experts and scientists. The adaptive learning process is an iterative and interactive loop. The learning results are adopted to design and guide next experiments, calculations, or/and manufacture, and the new results will validate the ML prediction and simultaneously be fed back to the dataset for the next cycle of iterations in the adaptive learning loop, which means that the dataset will grow after each iteration. Obviously, the next round of adaptive learning may give a different result because of more data added, which provides modified guidance on experiments, calculations, or/and manufacture. This means that each step in the adaptive design loop will refine the inputs and outputs more or less during the cycle of iterations, and the iterations will go continuously until the designed goal of material properties is reached. Lookman et al. (2019) called the adaptive design as active learning on adaptive sampling considering uncertainties for targeted design. In materials informatics, the knowledge of materials science and engineering is the hub because without the domain knowledge it will be hard to understand and analyze materials data. High-level domain knowledge guides each step in the process and reduces the gap between information and knowledge. A trustful discovery of new promising materials with identification of anomalies and scientific advancement depends on the knowledge retrieval, acquisition, representation, management, and knowledge level modeling, etc., in the adaptive learning and design. If the goal of designed material properties is reached, the iteration process might be ended for the time being, while it will be resumed when a new goal is planned. In principle, adaptive learning and design never end, and the accumulated data are always valuable.

ML converts data to theories/models, viz., knowledge, which is able to understand why things happened, viz., explanation; to predict what will happen, viz., prediction;

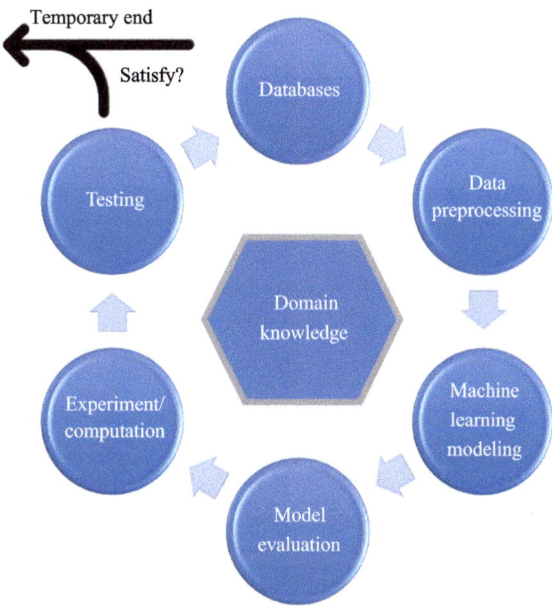

Fig. 1.1 Adaptive learning and design in materials informatics, where the domain knowledge is materials science and engineering

and to manage things that are happening, viz., control. When theories/models in materials science and engineering are represented by mathematical formulas, the theories/models represent materials behavior and performance in a most straightforward, clear, concise, and accurate manner, thereby playing vital roles in the knowledge gain and in the development of advanced materials and technologies. Before ML is available, theories/models were developed by human beings through observation of nature and materials behaviors with various instruments, together with curiosity, intuition, derivation, and/or summarization from the accumulation of experience and knowledge. The development of natural science is actually a continuous and endless process that human beings observe and explore nature and materials behaviors to generate data, learn and analyze the generated data to develop theories/models, and validate the developed theories/models with new observations/experiments. When differences appear between the available theories/models and new observations/experiments, the available theories/models will be revised or completely replaced by new theories/models, which implies that theories/models are valid only within certain conditions. The development of knowledge is, in general, in such a spiral rise mode with cycles of revising theories/models, and each cycle increases the gaining of knowledge. The birth of a completely new theory/model indicates a great breakthrough and a quantum jump in knowledge.

A few typical examples are illustrated here to demonstrate the data-driven discovery of theories/models. The first example is the Hume-Rothery rules, which are the classical work of data-driven formula about the substitutional solid solution (Cottrell, 1975; Darken et al., 1953; Hume-Rothery et al., 1934; Pauling, 1939, 1960). The Hume-Rothery rules state that three factors, called features nowadays in AI and ML, determine whether an alloy will be the substitutional solid solution: (1) The atomic size factor states that if the relative difference in atomic diameter between solute and solvent is less than 14%, the substitutional solid solution will be favorable; (2) The electrochemical factor declares that the probability of forming a substitutional solid solution will be much larger when the difference in electronegativity between two elements is less than 0.4; (3) The relative valency factor expresses that a metallic element of lower valency is more likely to dissolve itself into another of higher valency than vice versa. The second example is the Paris law for fatigue crack growth, which was discovered completely from experimental data (Paris et al., 1961). The Paris law states that the fatigue crack growth rate is a power function of the applied range of stress intensity factors, and the value of the exponent is determined purely by data. The Paris law is widely used in industries and academia because fatigue is the major mode of failure of materials and structures. The third example is the Hall–Petch equation for grain-size-dependent strength of polycrystalline metals, which initially is completely developed based on experimental data (Hall, 1951; Petch, 1953). The Hall–Petch equation states that the yield strength of a polycrystalline metal is inversely proportional to the square root of grain size. The Hall–Petch equation is extremely important to the entire metal industry because most mechanical properties, except for a few, e.g., creep resistance, of metals and alloys will be enhanced as the grain size gets smaller and smaller, down to sub-micrometer scale (The reverse Hall–Petch relationship has been observed at the nanometer scale). It is interesting to notice that the Hall–Petch equation was later derived from at least two different models (Cottrell, 1958; Li, 1963).

The human being learning paradigm to develop theories/models, however, requires long time, huge efforts, and in particular, genius of human beings. Nowadays, the rapid development of AI techniques provides an opportunity for automatic formula construction. The new paradigm of data-driven science and technology prevails rapidly in the field of materials science and engineering by using AI and ML techniques to discover knowledge and science, to design advanced materials, to uncover equations of state, to automatically manage manufacture, etc., where data are generated from experiments, computational simulations, and industrial practice. The data-driven paradigm combined with domain knowledge provides state-of-the-art methodologies for the development of theories/models. A recent milestone in materials informatics is that Burger et al. (2020) developed a mobile robot chemist to search for improved photocatalysts for hydrogen production from water. Equipped with an AI/ML algorithm, viz., the Bayesian optimization algorithm, the mobile robot chemist does the adaptive learning and design by itself, working autonomously over eight days, performing 688 experiments within a ten-variable experimental space, resulting in photocatalyst mixtures that are six times more active than the initial formulations.

In natural sciences, there are a lot of accepted laws, principles, and equations, such as Newton's laws, Maxwell equations, thermodynamic laws, the Schrödinger equation, etc., which are valid in certain circumstances. Theoretically, AI and ML are based on statistics and probability without utilizing any accepted laws, principles, and equations. It is usually taken for granted in AI and ML that a true solution can

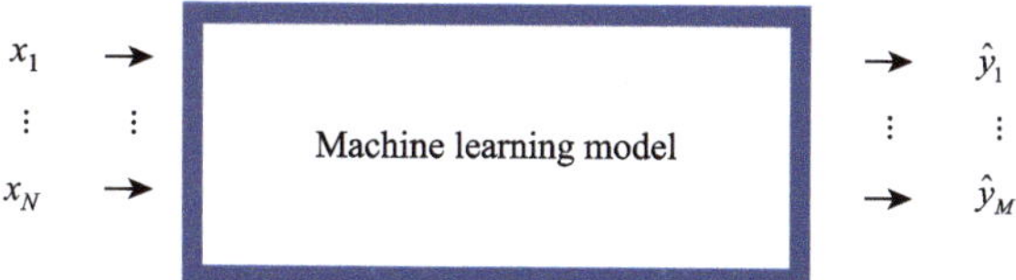

Fig. 1.2 Schematic illustration of a black-box ML model

only be approximately estimated and the task of ML is to find the best estimate based on a given set of data. Thus, it is very general that an inexplicit AI and ML-based model functions as a "black box". For example, a deep learning neural network is a black box, and usually involves many flexible mathematical functions with a lot of parameters to be trained; thus big data are required to train the deep learning neural network, viz., to determine the values of the ML model's internal parameters by fitting the model with data. Prediction and control are the major tasks for such "black box" models. Figure 1.2 schematically shows the working principle of ML, where data contain input variables $(x_1, x_2, \ldots, x_N)$ and output variables $(y_1, y_2, \ldots, y_M)$ (the latter are called responses as well). The ML task is to build up a correlation model between the input variables and the responses, viz., an ML model gives responses $(\hat{y}_1, \hat{y}_2, \ldots, \hat{y}_M)$, which should be as accurate as possible in comparison with the available data, and its application window, viz. the capacity of generalization, is as large as possible.

However, data commonly do not contain sufficiently enough information about the correlation between input variables and responses, and there are a number of hidden variables. The lack of information or the existence of hidden variables introduces an intrinsic error. The financial information of a company in the stock market, as a typical example, is so lacking, where hidden variables cause a high intrinsic error, therefore, it is hard for immature stock investors to make money from the stock market. In scientific research and technology development, however, researchers and engineers do their best to reduce experimental or/and computational errors, and try to obtain sufficiently enough information about the correlation between input variables and output responses by well designing and seriously implementing experiments and computations so that the intrinsic errors are getting smaller and smaller along with the improved methodologies and quality-enhanced devices, instrument, hardware, and software. In addition, various errors, conventionally called noises, appear randomly in experimental tests. Noise can be analyzed statistically if the data are big enough, which will be discussed later. For simplicity, input variables are treated as features for the time being, although features could be combinations of fundamental input variables. Unlike in purely mathematic analysis, there might be some correlations between different features. For example, if x is an independent variable, x and x^2 are not independent variables in purely mathematic analysis, but both x and x^2 are features in ML. A set of features behave like a basis in algebra. Features and feature space are two essential and crucial issues in ML.

As mentioned above, there are a lot of well-developed laws, principles, equations, and formulas in natural sciences. Based on available and widely recognized scientific laws, principles, equations, and formulas, researchers and engineers are developing new theories and formulas, and this deterministic paradigm of knowledge development is called the expert deduction for knowledge. On the one hand, in the expert deduction for knowledge, the deduced equations and formulas are highly biased by these people who believe them as the true solutions for the relationships between outputs and inputs. On the other hand, the data-driven development paradigm is called the data induction for knowledge, where the true solutions for outputs versus inputs are never achieved; instead, only the attainment of optimal approximate solutions is obtained for outputs versus inputs.

Besides the intrinsic error induced by the hidden variables, there are two types of errors in ML: bias and variance. Bias is defined as the squared error of an optimized ML model with respect to the true solution, and variance is defined as the squared error of all possible ML models with respect to the optimized ML model. In ML and statistical learning, there exists a bias-variance dilemma, meaning that a high bias corresponds to a low variance and vice versa. Thus, the total error of an ML model is the sum of intrinsic error, noise, bias, and variance. For an ML task, the sum of bias and variance can be minimized by selecting appropriate features and feature space that is often called search space as well. That is why features and feature space are two essential and crucial issues in ML. Feature selection selects the most influential features to responses, and the most influential features can be initial features and/or their combinations. Feature selection is one of the most crucial and challenging steps in ML because the selected features consequently generate a feature space or search space within which the optimal approximate solution is searched for.

Currently, materials computational data are relatively big, whereas materials experimental data are small. The properties of a material depend on its chemical composition, microstructure, material fabrication process condition, service environment, etc., meaning that there are a large number of features that determine the material performance. In addition, some experimental data scatter greatly, and experiments are typically expensive and time-consuming, which make it hard to generate big data. Generally speaking, there are two kinds of small data. The first kind of small data means that the number of repeated tests is not enough due to great scattering in test results. The second kind of small data means highly sparse data, but the data (or the means of data) are reliable. The second kind of small data has been widely studied in the AI and ML fields, leading to many effective algorithms and methods to handle them, e.g., transfer learning and data augmentation. On the one hand, if repeated experimental data are big, the data will exhibit the probability distribution of experimental results under the same testing conditions on the same material. On the other hand, if repeated experimental data are small with great scattering, the first kind of small data makes the data themselves unreliable, because the scattering in experimental results is caused by unknown parameters.

For instance, brittle and quasi-brittle fracture generally shows considerably high scattering, as illustrated by the strengths measured by three-point bending tests on piezoelectric ceramics, where the tests are repeated over 50 times under each testing condition (Fu & Zhang, 1998). This is because brittle and quasi-brittle fracture is extremely sensitive to defects and other detrimental factors. In many cases, it is very hard to obtain big experimental data from time-consuming or/and high-cost tests. Obviously, small data of repeated tests cannot provide sufficient information about the probability distribution of experimental results. The arithmetic mean of such small data about a material property may greatly deviate from the expectation of the studied property, especially if the estimated variance of the small data is large. Furthermore, small data of brittle and quasi-brittle fracture do not provide any information about the probability of failure, while reliable safety design requires a high survival probability. In this circumstance, domain knowledge of fracture plays a critical role in the extraction of useful information on strength and fracture toughness from small data. As an attempt, Wang et al. (2021) conducted statistical learning of small repeated test data with domain knowledge in the investigation of sample size- and pre-notch length- dependent strength of concrete. Based on the knowledge of strength and fracture domain, they proposed four hypotheses: (1) The larger the concrete sample size is, the lower the strength will be. This is because in a certain material, there exists the largest defect, and the probability of the largest defect existing in a larger sample is higher than that in a smaller sample. (2) The larger the concrete sample size is, the smaller the variance will be. The physical argument is the same as that for the first hypothesis. (3) The deeper the pre-notch length is, the lower the strength will be. This is because a deep pre-notch is the artificial defect larger than any defects inside a tested sample, and causes high-stress concentration at the pre-notch tip. (4) The degree of data scattering should be decreased with the increase of pre-notch length due to the same reason in the third hypothesis. Subsequently, they proposed a size-dependent normal distribution model and a size-and pre-notch length-dependent normal distribution model to statistically analyze the small data of sample size-and pre-notch length-dependent strength of concrete. The statistical learning indicates that the two normal distribution models reduce, to a large extent, the number of parameters involved and makes the results much more reliable than those just based on simple averages. Finally, the domain knowledge of fracture is utilized to reliably determine the fracture toughness of the concrete and the size of the fracture process zone. More endeavors are needed to tightly integrate the domain knowledge with AI and ML to handle the challenge of small data with big noise. Nevertheless, data in most cases in the current AI and ML are regarded to be reliable.

To obtain an optimal approximate solution, a dataset is divided into three subsets, namely a training subset, a validating subset, and a testing subset. For small data, it is very often to divide a dataset into a training subset and a validating subset. During training an ML model, as the ML model's complexity increases, the error of training will monotonically decrease until reaching a limiting value, which could be zero as an extreme. In the determinative paradigm of knowledge development, if one completely biases a formula, he/she will fit data with the biased formula as accurately as possible. In the data-driven paradigm of knowledge development, however, the goal is to find the best optimal solution by balancing model accuracy and generalizability. The validating and testing data subsets assess the model's generalizability. As the ML model's complexity increases, the error of validating (and testing) will decrease first, reach a minimum, and then increase, as illustrated in Fig. 1.3. The error of validating determines the performance of an ML model. It is called the just-right fitting at the minimum testing error, and the regions before and beyond the just-right fitting are

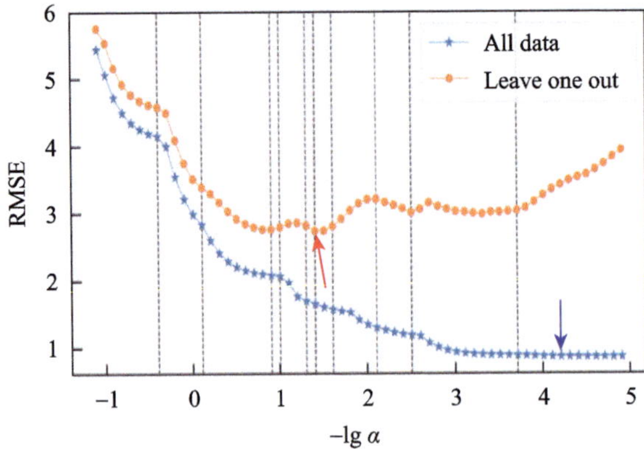

Fig. 1.3 The ML model accuracy represented by the root of mean squared error (RMSE) versus the model complexity represented by $-\lg \alpha$ in the least absolute shrinkage and selection operator, where the red arrow indicates the minimum RMSE, and the blue arrow points out the critical value α_0 and associated RMSE (Wang et al., 2019). The leave-one-out cross validation and the least absolute shrinkage and selection operator with penalty parameter α will be introduced in Chap. 2

the underfitting and overfitting regions, respectively. Data-driven theories/models are not unique, and different theories/models can fit available data with accepted accuracy, because the information provided by the data with noise is not sufficiently adequate, as schematically shown in Fig. 1.4(a). It is worth mentioning that the same dataset is also able to fit two diffident mathematic formulas which are derived from different physical arguments and bases. For example, Fig. 1.4(b) shows the same experimentally measured data of indentation crack length c and indentation load p, where the data are fitted by the formula of $\frac{p}{c^{3/2}} = Ac^{1/2} + B$ (Marshall & Lawn, 1977) with A and B being fitting constants, and another formula $\frac{p}{c^{3/2}} = \frac{\gamma - \delta\sigma_r t}{c^{1/2}} + \frac{K_{IC}}{\chi}$ (Zhang et al., 1999) with t being the film thickness, σ_r the residual film stress, K_{IC} the fracture toughness of the film, γ, δ and χ fitting constants. Clearly, the fitting accuracies with the two formulas are more or less the same. Figure 1.4(c) is another example (Xu & Zhang, 2004), indicating that the same nanoindentation data can be fitted by two formulas with about the same fitting accuracies. The two examples imply that due to experimental errors, the solutions cannot be uniquely determined, and different formulas can fit available data with accepted accuracy, in some way similar to ML models.

Although such "black-box" ML models are widely applicable, they are not the best choices if the goal is to pursue the understanding of underlying mechanisms between inputs and outputs. From data to knowledge necessitates theories/models with simple explicit mathematical expressions. Symbolic regression is one of the most popular AI and ML methods to give exact or approximate analytic and explicit descriptions and formulas of studied phenomena from data in order to explore mechanisms, and thus can be widely applied by materials scientists and engineers to gain knowledge from data (Sun et al., 2019). In addition, great and heroic efforts are continuously being devoted to the AI and ML fields to make "black-box" models more transparent

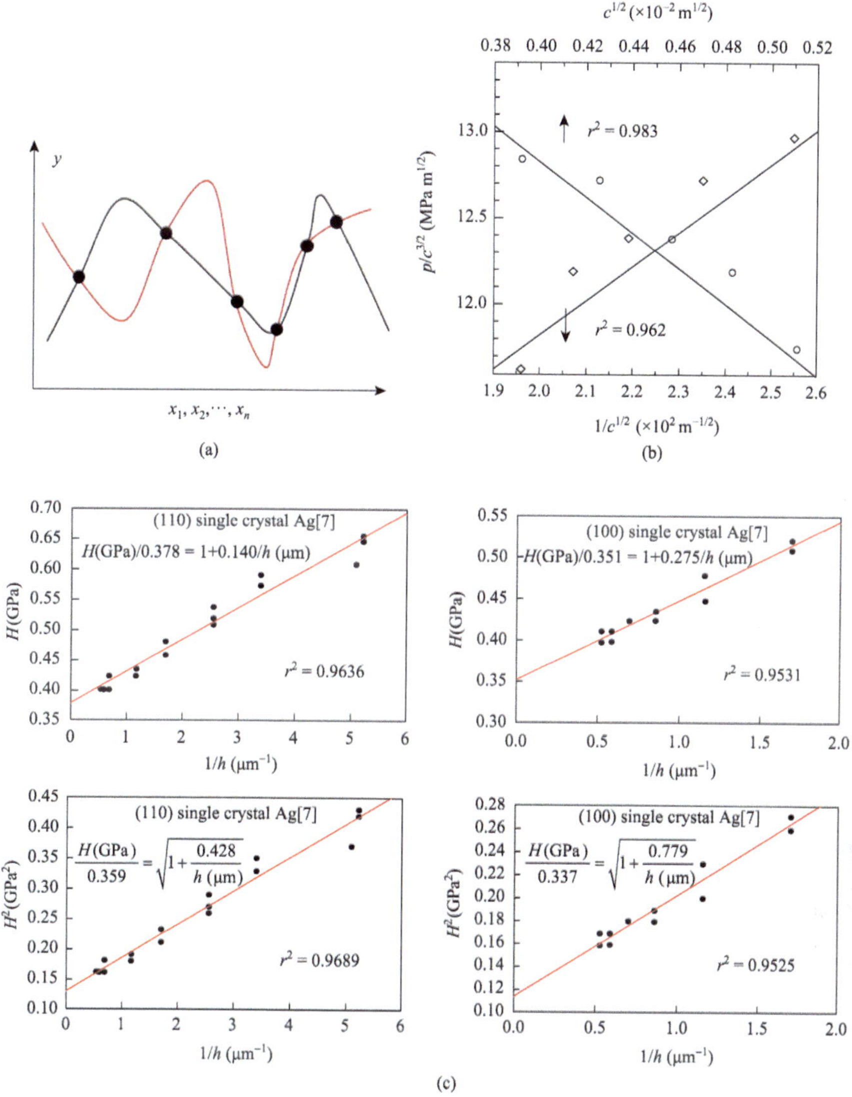

Fig. 1.4 **(a)** A schematic of data fitted by two different ML models with accepted accuracy. **(b)** the same experimentally measured data of indentation crack length c and indentation load p fitted by $\frac{p}{c^{3/2}}=Ac^{1/2}+B$ (Marshall & Lawn, 1977) with A and B being fitting constants, and by $\frac{p}{c^{3/2}}=\frac{\gamma-\delta\sigma_r t}{c^{1/2}}+\frac{K_{IC}}{\chi}$ (Zhang et al., 1999) with t being the film thickness, σ_r the residual film stress, K_{IC} the fracture toughness of the film, γ, δ and χ fitting constants. **(c)** indentation depth-dependent hardness fitted by $H=H_0\sqrt{1+\frac{h^*}{h}}$, where H_0 is the hardness for sufficiently depth of nanoindentation, h is the indentation depth, and h^* is a constant (Nix & Gao, 1998), and by $H=H_0+g\frac{f}{h}$, where g is the tip geometry-related constant and f is the apparent surface stress (Zhang & Xu, 2002)

and interpretative, where again the integration of domain knowledge and AI and ML are definitely a must.

To increase the readability and help students and researchers in materials science and engineering more easily grasp the essential and fundamental knowledge of AI and ML, linear regression forms the next chapter, because least squares linear regression is widely and frequently used in the materials community. A penalty term is added to increase the generalization power of a linear ML model, which results in the Least Absolute Shrinkage and Selection Operator (LASSO, L_1), Ridge Regression (L_2), and elastic net algorithms. After Chap. 2, Chap. 3 covers linear classification, where the focus is on "logistic classification", which was known as logistic regression historically. Once readers are familiar with linear regression and classification, it will be easy for readers to understand support vector machines, including support vector classifier and regressor, in Chap. 4. More importantly, the kernel trick will be first introduced in Chap. 4. Chapter 5 lectures decision trees and K-nearest-neighbors, which can perform classification and regression as well. The scientific and fundamental concept of information entropy will be defined during the learning of classification trees. Tree shape is of the basic expression mode in AI and ML, and a decision tree is used as a base learner in ensemble learning, which is described in Chap. 6. Many base learners can be assembled in parallel or serial manners, leading to the popular tree-based ensemble learning algorithms of Random Forest and eXtreme Gradient Boosting (XGBoost), respectively. Probability is the foundation of statistic learning, and Bayesian Theorem plays a critical role in materials informatics. Chapter 7 introduces Bayesian Theorem and likelihood briefly. Combining Bayesian Theorem and the maximization of likelihood yields the Expectation–Maximization (EM) Algorithm. Global optimization is the core problem in materials informatics, and in AI and ML as well. It is very hard or even impossible to find a global optimum for concave problems, although global optimization can be done on convex problems. Instead, a maximum (or minimum) among many local maximums (or minimums) can be found by following the Darwin's theory of evolution in natural selection. Chapter 8 "Symbolic Regression" describes the most used heuristic algorithms. Chapter 9 explains Artificial Neural Networks (ANN), which mimic the biological neural networks and become the foundation of deep learning. The performance of a material depends highly on its microstructure that may vary along with service time. The microstructure of a material might be regarded as the state in Markov chains and the change in microstructure can be described by the variation in state. Chapter 10 briefly introduces hidden Markov chains, which are the foundation of the powerful Markov decision process algorithm in reinforcement learning. As a whole, Chaps. 2–10 include the most basic and simplest ML algorithms widely used in materials informatics. As mentioned above, materials data are small in size, high in dimension, and big in noise; especially, some experimental data reported in the literature might be inconsistent. Data preprocessing must be conducted in materials informatics, and some algorithms are introduced in the first section of Chap. 11 for data preprocessing. The two crucial issues in materials informatics are features and feature space. The number of initially proposed features is usually very large. Therefore, feature ranking and selection, described in the rest sections of Chap. 11, must be carried out to select the most important features to the outputs studied. The last chapter in the first part, Chap. 12, describes two interpretive feature methods, SHapley Additive exPlanation (SHAP) and Partial Dependency Plot (PDP). Basic statistical and probability and matrix operation are concisely introduced in the Appendix part for the convenience of readers.

References

Agrawal, A., & Choudhary, A. (2016). Perspective: Materials informatics and big data: Realization of the "fourth paradigm" of science in materials science. *APL Materials, 4*(5), 053208.

Burger, B., Maffettone, P. M., Gusev, V. V., et al. (2020). A mobile robotic chemist. *Nature, 583*(7815), 237–241.

Cottrell, A. H. (1958). Theory of brittle fracture in steel and similar metals. *Metallurgical Society of AIME, 1*, 212.

Cottrell, A. H. (1975). *An introduction to metallurgy*. Edward Arnold.

Darken, L. S., Gurry, R. W., & Bever, M. B. (1953). *Physical chemistry of metals*. McGraw-Hill Book Company.

Fu, R., & Zhang, T. Y. (1998). Effect of an applied electric field on the modulus of rupture of poled lead zirconate titanate ceramics. *Journal of the American Ceramic Society, 81*(4), 1058–1060.

Hall, E. O. (1951). The deformation and ageing of mild steel: III discussion of results. *Proceedings of the Physical Society, 643*(9), 747–752.

Hume-Rothery, W., Mabbott, G. W., & Channel-Evans, K. M. (1934). The freezing points, melting points, and solid solubility limits of the alloys of sliver and copper with the elements of the sub-groups. *Philosophical Transactions of the Royal Society of London, 233*(721–730), 1–97.

Li, J. C. (1963). Petch relation and grain boundary sources. *Transactions of the Metallurgical Society of AIME, 227*(1), 239.

Lookman, T., Balachandran, P. V., Xue, D., et al. (2019). Active learning in materials science with emphasis on adaptive sampling using uncertainties for targeted design. *NPJ Computational Materials, 5*(1), 1–17.

Marshall, D. B., & Lawn, B. R. (1977). An indentation technique for measuring stresses in tempered glass surfaces. *Journal of the American Ceramic Society, 60*(1–2), 86–87.

Mitchell, T. (1997). Machine learning meets natural language (abstract). In E. Coasta & A. Cardoso (Eds.), *Progress in artificial intelligence*. Berlin: Springer.

Nix, W. D., & Gao, H. (1998). Indentation size effects in crystalline materials: A law for strain gradient plasticity. *Journal of the Mechanics & Physics of Solids, 46*(3), 411–425.

Paris, P. C., Gomez, M. P., & Anderson, W. E. (1961). *A rational analytical theory of fatigue*. University of Washington Press.

Pauling, L. (1939). Recent work on the configuration and electronic structure of molecules; with some applications to natural products. In: L. Zechmeister (Ed.), *Fortschritte der Chemie Organischer Naturstoffe. Fortschritte der Chemie Organischer Naturstoffe* (Vol. 3, pp. 203–235). Vienna: Springer.

Pauling, L. (1960). *The nature of the chemical bond* (3rd ed.). Cornell University Press.

Petch, N. J. (1953). The cleavage strength of polycrystals. *Journal of the Iron and Steel Institute, 174*, 25–28.

Ramakrishna, S., Zhang, T. Y., Lu, W. C., et al. (2019). Materials informatics. *Journal of Intelligent Manufacturing, 30*(6), 2307–2326.

Samuel, A. L. (1967). Some studies in machine learning using the game of checkers. II—Recent progress. *IBM Journal of Research and Development, 11*(6), 601–617.

Sun, S., Ouyang, R., Zhang, B., et al. (2019). Data-driven discovery of formulas by symbolic regression. *MRS Bulletin, 7*, 559–564.

Wang, J. H., Jia, J. N., Sun, S., et al. (2021). Statistical learning of small repeated test data with domain knowledge—Sample size-and pre-notch length-dependent strength of concrete. *Engineering Fracture Mechanics, 259*, 108160.

Wang, J. H., Sun, S., Yanlin, H. E., et al. (2019). Machine learning prediction of the hardness of tool and mold steels. *Scientia Sinica Technologica, 49*(10), 1148–1158.

Xu, W. H., & Zhang, T. Y. (2004). Surface effect for different types of materials in nanoindentation. *Key Engineering Materials, 261–263*, 1587–1592.

Zhang, T. Y., Chen, L. Q., & Fu, R. (1999). Measurements of residual stresses in thin films deposited on silicon wafers by indentation fracture. *Acta Materialia, 47*(14), 3869–3878.

Zhang, T. Y., & Xu, W. H. (2002). Surface effects on nanoindentation. *Journal of Materials Research, 17*(7), 1715–1720.

Chapter 2
Linear Regression

2.1 Least Squares Linear Regression

Linear regression uses a linear function to express the relationship between responses and input variables. There may be m input variables of $(x_1,\ x_2,\ \cdots,\ x_m)$ and p responses of $(y_1,\ y_2,\ \cdots,\ y_p)$. The simplest case is only one input and one response, and the linear function can be expressed by $y = w_0 + wx$, where w_0 and w are the intercept and coefficient (slope), respectively, and both values are determined from least squares linear regression (LSLR). In many cases, to have the linear relationship, inputs and responses are converted with simple mathematic operators from directly measured or/and calculated data. For example, the temperature dependence of diffusion coefficient $D(T)$ is usually described by the Arrhenius equation, $D = D_0\exp\left(-\frac{Q}{RT}\right)$, where R is the gas constant, T is the absolute temperature, D_0 is the pre-factor, and Q is called the activation energy (enthalpy) of diffusion. Diffusion coefficients at various temperatures are experimentally measured; thus one has the original dataset of $(T_i,\ D_i)$, $i = 1, 2, 3, \cdots, n$. The Arrhenius equation of diffusion is rewritten as $\ln D = \ln D_0 - \frac{Q}{RT}$, which is a linear equation of $y = w_0 + wx$ with $y = \ln D$, $w_0 = \ln D_0$, $w = -Q/R$ and $x = 1/T$. In this way, the original dataset is converted to (x_i, y_i), $i = 1, 2, 3, \cdots, n$. With the experimental measured dataset of (x_i, y_i), the values of w_0 and w are determined by minimizing Residual Sum-of-Squares (RSS), which is defined by

$$\mathrm{RSS}(w_0, w) = \sum_{i=1}^{n} (y_i - w_0 - wx_i)^2. \tag{2.1}$$

The RSS minimization requires

$$\frac{\partial \mathrm{RSS}}{\partial w_0} = -2\sum_{i=1}^{n} (y_i - w_0 - wx_i) = 0, \tag{2.2a}$$

T. Zhang, *An Introduction to Materials Informatics*,
https://doi.org/10.1007/978-981-99-7992-9_2

$$\frac{\partial \text{RSS}}{\partial w} = -2\sum_{i=1}^{n} x_i(y_i - w_0 - wx_i) = 0. \tag{2.2b}$$

The linear equation set of Eqs. (2.2a) and (2.2b) can be rewritten in the matrix form

$$\begin{bmatrix} n & \sum_{i=1}^{n} x_i \\ \sum_{i=1}^{n} x_i & \sum_{i=1}^{n} x_i^2 \end{bmatrix} \begin{bmatrix} w_0 \\ w \end{bmatrix} = \begin{bmatrix} \sum_{i=1}^{n} y_i \\ \sum_{i=1}^{n} x_i y_i \end{bmatrix}. \tag{2.2c}$$

If matrix $\begin{bmatrix} n & \sum_{i=1}^{n} x_i \\ \sum_{i=1}^{n} x_i & \sum_{i=1}^{n} x_i^2 \end{bmatrix}$ is inversible, the values of w_0 and w are determined by

$$\begin{bmatrix} w_0 \\ w \end{bmatrix} = \begin{bmatrix} n & \sum_{i=1}^{n} x_i \\ \sum_{i=1}^{n} x_i & \sum_{i=1}^{n} x_i^2 \end{bmatrix}^{-1} \begin{bmatrix} \sum_{i=1}^{n} y_i \\ \sum_{i=1}^{n} x_i y_i \end{bmatrix}. \tag{2.3}$$

The determined values of w_0 and w depend on the data of inputs and outputs. Usually, data contain noise, and thus different values of w_0 and w can be obtained from different datasets, meaning that the values of w_0 and w are the best estimates for the given dataset, rather than true values. To differentiate true and estimated values, the estimated values are denoted by $\hat{w}_0$ and $\hat{w}$, and hence the prediction is expressed by $\hat{y} = \hat{w}_0 + \hat{w}x$. Since the regression accuracy is gauged by RSS, this linear regression is called LSLR. LSLR is a widely used method in the determination of parameters involved in linear equations. In this simple example, there is only one input variable (temperature) and one response (diffusion coefficient). As described in Chap. 1, responses are usually material properties or other parameters of interest. In the diffusion case, the relationship between input temperature and response diffusion coefficient is well established via the Arrhenius equation. With such a well-established physical equation, the values of parameters involved in the equation will be determined as accurately as possible, which indicates that using a preset equation to fit data biases the equation greatly.

Example 2.1 Table 2.1 lists the experimental results of hydrogen diffusion coefficients in commercially pure iron at temperatures of 25, 50, 70, and 80 °C, and Fig. 2.1 shows the LSLR of the data, which gives the activation energies of diffusion of 7560 J·mol^{-1} and 7110 J·mol^{-1} for the Group I and Group II samples, respectively (Zhang & Zheng, 1998).

Equations (2.1)–(2.3) can be expressed in a much more condensed manner with matrix. Considering m input features $(x_1, x_2, \cdots, x_m)$ and only one response y, we

Table. 2.1 Hydrogen diffusion coefficients in pure iron ($\times 10^{-5}$cm^2·s^{-1})

Groups	25 °C	50 °C	70 °C	80 °C
Group I	4.11	4.92	6.10	6.66
Group II	4.27	4.92	6.14	6.62

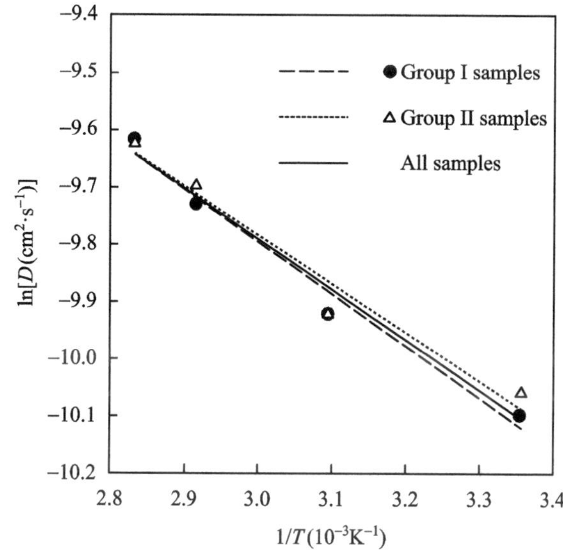

Fig. 2.1 The LSLR of ln *D* versus 1/*T*

have the linear correlation between the response and the inputs as

$$y = w_0 + \sum_{j=1}^{m} w_j x_j, \tag{2.4a}$$

where w_j ($j = 1, 2, 3, \cdots, m$) are the coefficients, and w_0 is the intercept. If a dull input feature of one is introduced, Eq. (2.4a) is rewritten as

$$y = \sum_{j=0}^{m} w_j x_j, \tag{2.4b}$$

or

$$y = \boldsymbol{x}\boldsymbol{W}, \tag{2.4c}$$

with $\boldsymbol{x} = (1, x_1, \cdots, x_m)$ being a $1 \times (m + 1)$ row vector and $\boldsymbol{W} = (w_0, w_1, \cdots, w_m)^{\mathrm{T}}$ being a $(m + 1) \times 1$ column vector. In this case, the RSS is given by

$$\mathrm{RSS}(\boldsymbol{W}) = \sum_{i=1}^{n} \left(y_i - \sum_{j=0}^{m} w_j x_{ij} \right)^2, \tag{2.5a}$$

where y_i and $x_{ij} = (x_{i0} \equiv 1, x_{i1}, \cdots, x_{im})$ $(i = 1, 2, \cdots, n)$ represent the response value and the input feature values in the ith datum, respectively. Equation (2.5a) can be rewritten in the matrix form of

$$\mathrm{RSS}(\boldsymbol{W}) = (\boldsymbol{y} - \boldsymbol{X}\boldsymbol{W})^{\mathrm{T}}(\boldsymbol{y} - \boldsymbol{X}\boldsymbol{W}), \tag{2.5b}$$

where $\boldsymbol{y}$ is a $n \times 1$ output vector $[y_1, y_2, \cdots, y_n]^{\mathrm{T}}$ and $\boldsymbol{X}$ is a $n \times (m+1)$ input matrix $\begin{bmatrix} x_{10} & \cdots & x_{1m} \\ \vdots & & \vdots \\ x_{n0} & \cdots & x_{nm} \end{bmatrix}$. RSS is one widely used loss function in machine learning. The values of $\boldsymbol{W}$ are determined by the minimum of RSS, i.e.,

$$\hat{\boldsymbol{W}} = \arg\min_{\boldsymbol{W}} \mathrm{RSS}(\boldsymbol{W}) = \underset{\boldsymbol{W}}{\mathrm{argmin}}[(\boldsymbol{y} - \boldsymbol{X}\boldsymbol{W})^{\mathrm{T}}(\boldsymbol{y} - \boldsymbol{X}\boldsymbol{W})]. \tag{2.5c}$$

The RSS minimization requires

$$\frac{\partial \mathrm{RSS}(\boldsymbol{W})}{\partial w_j} = \sum_{i=1}^{n} \left[-2 \left(y_i - \sum_{j=0}^{m} w_j x_{ij} \right) x_{ij} \right] = 0, \quad j = 0, 1, \cdots, m. \tag{2.6a}$$

Equation (2.6a) represents an equation set of $(m+1)$ equations, which can be expressed in a compact form of

$$\frac{\partial \mathrm{RSS}(\boldsymbol{W})}{\partial \boldsymbol{W}} = -2\boldsymbol{X}^{\mathrm{T}}(\boldsymbol{y} - \boldsymbol{X}\boldsymbol{W}) = 0. \tag{2.6b}$$

Obviously, if the sampling number n is smaller than the coefficient number $(m+1)$, one will not be able to determine the $(m+1)$ values uniquely. When n is bigger than $(m+1)$ and $\boldsymbol{X}^{\mathrm{T}}\boldsymbol{X}$ has the rank of $(m+1)$, we have, from Eq. (2.6b), $\boldsymbol{X}^{\mathrm{T}}\boldsymbol{X}\boldsymbol{W} = \boldsymbol{X}^{\mathrm{T}}\boldsymbol{y}$ and

$$\hat{\boldsymbol{W}} = (\boldsymbol{X}^{\mathrm{T}}\boldsymbol{X})^{-1}\boldsymbol{X}^{\mathrm{T}}\boldsymbol{y}, \tag{2.6c}$$

where the inverse matrix of $(\boldsymbol{X}^{\mathrm{T}}\boldsymbol{X})^{-1}$ exists. The prediction is then given by

$$\hat{\boldsymbol{y}} = \boldsymbol{X}\hat{\boldsymbol{W}}. \tag{2.6d}$$

If the matrix $\boldsymbol{X}^{\mathrm{T}}\boldsymbol{X}$ cannot be inversed, all the values of $\boldsymbol{W}$ components cannot be uniquely determined. In this case, pseudo-inverse of matrix $\boldsymbol{X}^{\mathrm{T}}\boldsymbol{X}$ gives the best solution. In addition, three approaches are strongly suggested to solve the problem that the sampling number n is smaller than the coefficient number $(m+1)$. The first and direct approach is to obtain more data, e.g., doing more tests, to make $n \geqslant m+1$. The second approach is to reduce the number of input variables (features) by recoding and/or dropping redundant features to meet $m+1 \leqslant n$. The third approach is to replace $(\boldsymbol{X}\boldsymbol{X})$ by $(\boldsymbol{X}\boldsymbol{X}+\lambda\boldsymbol{I})$, where $\boldsymbol{I}$ is the unit $(m+1)\times(m+1)$ matrix and λ is a positive constant. The matrix $(\boldsymbol{X}^{\mathrm{T}}\boldsymbol{X}+\lambda\boldsymbol{I})$ is invertible and Eq. (2.6d) is revised to

$$\hat{\boldsymbol{W}} = (\boldsymbol{X}^{\mathrm{T}}\boldsymbol{X}+\lambda\boldsymbol{I})^{-1}\boldsymbol{X}^{\mathrm{T}}\boldsymbol{y}. \tag{2.6e}$$

The constant λ is called the regularization coefficient. How to add the regularization term will be introduced in detail in the ridge regression section.

There are parameters evaluating the goodness of fitting. The widely used ones include the Mean Squared Error (MSE),

$$\mathrm{MSE} = \frac{1}{n}\sum_{i=1}^{n}(\hat{y}_i - y_i)^2, \tag{2.7a}$$

the Root of MSE (RMSE)

$$\mathrm{RMSE} = \sqrt{\frac{\sum_{i=1}^{n}(\hat{y}_i - y_i)^2}{n}}, \tag{2.7b}$$

the Relative RMSE (RRMSE)

$$\mathrm{RRMSE} = \sqrt{\frac{1}{n}\sum_{i=1}^{n}\left(\frac{y_i - \hat{y}_i}{y_i}\right)^2}, \tag{2.7c}$$

the Mean Absolute Error (MAE)

$$\mathrm{MAE} = \frac{1}{n}\sum_{i=1}^{n}\left|y_i - \hat{y}_i\right|, \tag{2.7d}$$

the Mean Percent Error (MPE)

$$\mathrm{MPE} = \frac{1}{n}\sum_{i=1}^{n}\left|\frac{y_i - \hat{y}_i}{y_i}\right|, \tag{2.7e}$$

the coefficient of determination (R^2)

$$R^2 = 1 - \frac{\sum_{i=1}^{n} (y_i - \hat{y}_i)^2}{\sum_{i=1}^{n} (y_i - \overline{y})^2}, \tag{2.7f}$$

and the Pearson correlation coefficient (R)

$$R = \frac{\sum_{i=1}^{n} (y_i - \overline{y})(\hat{y}_i - \overline{\hat{y}})}{\sqrt{\sum_{i=1}^{n} (y_i - \overline{y})^2 \sum_{i=1}^{n} (\hat{y}_i - \overline{\hat{y}})^2}}, \tag{2.7g}$$

where n is the number of data, y_i is the true value, $\hat{y}_i$ is the prediction, $\overline{y}$ is the mean of y_i and $\overline{\hat{y}}$ is the mean of $\hat{y}_i$.

The coefficient of determination (R^2) is widely used in ML for regression. In the best case, the predicted values exactly match the observed values, which results in $R^2 = 1$, while the baseline model always predicts the average value $\hat{y}_i = \overline{y}$ and have $R^2 = 0$. Models that have worse predictions than this baseline will have a negative R^2. The Pearson correlation coefficient $R = 1$ represents perfectly positive linear correlation between two variables, while $R = -1$ represents perfectly negative linear correlation. The square of Pearson correlation coefficient R^2 is also often used in the assessment of fitting goodness, which ranges from 0 to 1, and 1 means that the prediction is completely linear correlated with the observation.

Example 2.2 Table 2.2 lists the experimental results of plastic strain rates from the stress relaxation tests on coarse-grained Cu at temperatures of ~22, 30, 40, 50 and 75 °C (Yang et al., 2016). The plastic strain rate $\dot{\varepsilon}_p(\sigma, T)$ depends on temperature T and applied stress σ. Based on domain knowledge, Yang et al. proposed that the plastic strain rate follows the equation of $\dot{\varepsilon}_p(\sigma, T) = \dot{\varepsilon}_{00}\left(\frac{\sigma}{\sigma_0}\right)^n \exp\left(-\frac{\Delta G_0}{kT} + \frac{\sigma V}{kT\sqrt{3}}\right)$,

Table. 2.2 Plastic strain rates from the stress relaxation tests on coarse-grained Cu at various stresses and temperatures

$1/kT$ (atom·J^{-1})	Strain rates at different stress levels for coarse-grained Cu @ 1st cycle (s^{-1})							
	113 MPa	112 MPa	111 MPa	110 MPa	109 MPa	108 MPa	107 MPa	106 MPa
2.46×10^{20}	3.95×10^{-5}	2.51×10^{-5}	1.42×10^{-5}	7.79×10^{-6}	4.59×10^{-6}	2.65×10^{-6}	1.67×10^{-6}	1.06×10^{-6}
2.39×10^{20}	4.35×10^{-5}	2.83×10^{-5}	1.74×10^{-5}	9.66×10^{-6}	5.06×10^{-6}	3.11×10^{-6}	2.07×10^{-6}	1.50×10^{-6}
2.31×10^{20}	4.95×10^{-5}	3.65×10^{-5}	2.51×10^{-5}	1.54×10^{-5}	9.99×10^{-6}	5.97×10^{-6}	4.35×10^{-6}	3.11×10^{-6}
2.24×10^{20}	5.65×10^{-5}	4.05×10^{-5}	2.89×10^{-5}	1.76×10^{-5}	1.19×10^{-5}	8.23×10^{-6}	5.47×10^{-6}	4.12×10^{-6}
2.08×10^{20}	6.67×10^{-5}	5.17×10^{-5}	3.90×10^{-5}	2.79×10^{-5}	1.91×10^{-5}	1.35×10^{-5}	9.65×10^{-6}	8.09×10^{-6}

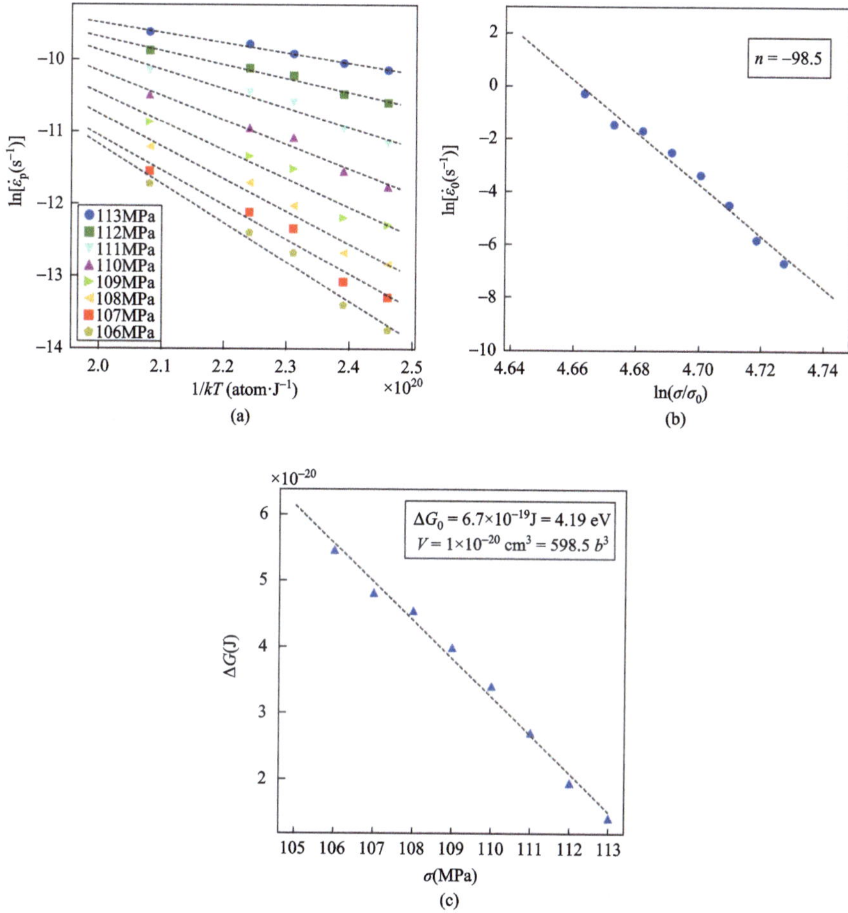

Fig. 2.2 Plots of three LSLRs

where $\dot{\varepsilon}_{00}$ is a constant, $\sigma_0 = 1$MPa is a reference stress, n is the stress exponent, k is the Boltzmann constant, ΔG_0 is a stress-independent activation energy, and V is the nominal activation volume conjugated to shear stress. Yang et al. (2016) used LSLR three times to estimate the values of n, ΔG_0, and V. They firstly estimated $\dot{\varepsilon}_0(\sigma) = \dot{\varepsilon}_{00}\left(\frac{\sigma}{\sigma_0}\right)^n$ and $\Delta G = \Delta G_0 - \sigma V/\sqrt{3}$ by LSLR of $\ln\dot{\varepsilon}_p$ versus $1/(kT)$, and then estimated the n value by LSLR of $\ln\dot{\varepsilon}_0$ versus $\ln\sigma$, and the values of ΔG_0 and V by LSLR of ΔG versus σ. Their results are $\Delta G_0 = 3.97$ eV, $V = 607.3\ b^3$, and $n = -99.9$, where $b = 0.256$ nm is the magnitude of a perfect $\frac{1}{2}\langle\bar{1}\ 1\ 0\rangle\{1\ 1\ 1\}$ dislocation in face-centered cubic Cu. The triple LSLRs can be done simultaneously. Letting $y = \ln\dot{\varepsilon}_p$, $x_0 \equiv 1$, $x_1 = \ln\left(\frac{\sigma}{\sigma_0}\right)$, $x_2 = \frac{1}{kT}$, and $x_3 = \frac{\sigma}{\sqrt{3}kT}$, we have $y = \sum_{j=0}^{3} w_j x_j$. With the data listed in Table 2.2, the coefficients are estimated to be $n = -98.75$, $\Delta G_0 = 4.19$ eV, and $V = 598.5\ b^3$. Figure 2.2 replots the three LSLRs in the original publication.

The above described LSLR are easily extended to multiple outputs, $(y^{(1)}, y^{(2)}, \cdots, y^{(p)})$. For multiple LSLR, Eqs. (2.4b) and (2.4c) are extended to

$$y^{(k)} = \sum_{j=0}^{m} w_j^{(k)} x_j \quad \text{for } k = 1, 2, \cdots, p \tag{2.8a}$$

or in a compact form

$$\boldsymbol{y} = \boldsymbol{x}\boldsymbol{W}, \tag{2.8b}$$

where $\boldsymbol{y} = (y^{(1)}, y^{(2)}, \cdots, y^{(p)})$ is a $1 \times p$ row vector, $\boldsymbol{x} = (1, x_1, x_2, \cdots, x_m)$ is a $1 \times (m+1)$ row vector, and $\boldsymbol{W}$ is a $(m+1) \times p$ matrix. With n data, Eq. (2.8b) is extended to

$$\boldsymbol{Y} = \boldsymbol{X}\boldsymbol{W}, \tag{2.8c}$$

where $\boldsymbol{Y}$ is a $n \times p$ output matrix and $\boldsymbol{X}$ is a $n \times (m+1)$ input matrix. The loss function of RSS for the multiple outputs takes the form

$$\mathrm{RSS}(\boldsymbol{W}) = \sum_{k=1}^{p} \sum_{i=1}^{n} \left(y_i^{(k)} - \sum_{j=0}^{m} w_j^{(k)} x_{ij} \right)^2 = \mathrm{tr}[(\boldsymbol{Y} - \boldsymbol{X}\boldsymbol{W})^{\mathrm{T}} (\boldsymbol{Y} - \boldsymbol{X}\boldsymbol{W})], \tag{2.9a}$$

where $\mathrm{tr}[\cdot]$ denotes the trace of a matrix. The values of coefficient matrix are determined by minimizing the loss function, i.e.,

$$\hat{\boldsymbol{W}} = \underset{\boldsymbol{W}}{\mathrm{argmin}} \left\{ \mathrm{tr}[(\boldsymbol{Y} - \boldsymbol{X}\boldsymbol{W})^{\mathrm{T}} (\boldsymbol{Y} - \boldsymbol{X}\boldsymbol{W})] \right\}. \tag{2.9b}$$

When the inverse matrix of $(\boldsymbol{X}^{\mathrm{T}}\boldsymbol{X})^{-1}$ exists, solving Eq. (2.9b) gives the results of

$$\hat{\boldsymbol{W}} = (\boldsymbol{X}^{\mathrm{T}}\boldsymbol{X})^{-1}\boldsymbol{X}^{\mathrm{T}}\boldsymbol{Y}, \tag{2.9c}$$

$$\hat{\boldsymbol{Y}} = \boldsymbol{X}\hat{\boldsymbol{W}}. \tag{2.9d}$$

Since $\boldsymbol{Y} = (\boldsymbol{Y}^{(1)}, \boldsymbol{Y}^{(2)}, \cdots, \boldsymbol{Y}^{(p)})$ and $\hat{\boldsymbol{W}} = \begin{bmatrix} \hat{w}_0^{(1)} & \cdots & \hat{w}_0^{(p)} \\ \vdots & & \vdots \\ \hat{w}_m^{(1)} & \cdots & \hat{w}_m^{(p)} \end{bmatrix} = (\hat{\boldsymbol{W}}^{(1)} \; \hat{\boldsymbol{W}}^{(2)} \; \cdots \; \hat{\boldsymbol{W}}^{(p)})$, we have $\hat{\boldsymbol{W}}^{(k)} = (\boldsymbol{X}^{\mathrm{T}}\boldsymbol{X})^{-1}\boldsymbol{X}^{\mathrm{T}}\boldsymbol{Y}^{(k)}$ and $\hat{\boldsymbol{Y}}^{(k)} = \boldsymbol{X}\hat{\boldsymbol{W}}^{(k)}$ $(k = 1, 2, \cdots, p)$, meaning that when using the same input dataset, multiple outputs are the set of all single output LSLRs' estimates.

Example 2.3 Multi-objective machine learning of four mechanical properties of steels was conducted by MultiTaskLASSO (Wei et al., 2020), based on 360 data samples from the Japan National Institute for Materials Science (NIMS) database. The four mechanical properties are fatigue strength, ultimate tensile strength, true fracture strength, and hardness. Tables 2.3 and 2.4 list the used 16 features and 20 data of steels, respectively.

With the 20 NIMS data, we have

$$\boldsymbol{X} = \begin{bmatrix} 1 & x_{1,1} & \cdots & x_{1,16} \\ \vdots & \vdots & & \vdots \\ 1 & x_{20,1} & \cdots & x_{20,16} \end{bmatrix}_{20\times 17} = \begin{bmatrix} 1 & 900 & \cdots & 0.03 \\ \vdots & \vdots & & \vdots \\ 1 & 845 & \cdots & 0 \end{bmatrix}_{20\times 17} \text{ and}$$

$$\boldsymbol{Y} = \begin{bmatrix} y_1^{(1)} & \ldots & y_1^{(4)} \\ \vdots & & \vdots \\ y_{20}^{(1)} & \ldots & y_{20}^{(4)} \end{bmatrix}_{20\times 4} = \begin{bmatrix} 510 & \ldots & 279 \\ \vdots & & \vdots \\ 513 & \ldots & 297 \end{bmatrix}_{20\times 4}.$$

Then, we obtain the transposed matrix of $\boldsymbol{X}$, $\boldsymbol{X}^{\mathrm{T}} = \begin{bmatrix} 1 & \cdots & 1 \\ 900 & \cdots & 845 \\ \vdots & & \vdots \\ 0.03 & \cdots & 0 \end{bmatrix}_{17\times 20}$ and the determinant of matrix $\boldsymbol{X}^{\mathrm{T}}\boldsymbol{X}$, $\left|\boldsymbol{X}^{\mathrm{T}}\boldsymbol{X}\right| = 5.24 \times 10^{-8}$. The determinant of matrix

Table. 2.3 Feature description of the 20 NIMS data

Features	Descriptions
NT	Normalizing temperature (°C)
QT	Quenching temperature (°C)
TT	Tempering temperature (°C)
C	wt.% of carbon
Si	wt.% of silicon
Mn	wt.% of manganese
P	wt.% of phosphorus
S	wt.% of sulphur
Ni	wt.% of nickel
Cr	wt.% of chromium
Cu	wt.% of copper
Mo	wt.% of molybdenum
RR	Reduction ratio (%)
dA	Fraction of plastic work-inclusions (%)
dB	Fraction of discontinuous array-inclusions (%)
dC	Fraction of isolated inclusions (%)

Table. 2.4 Twenty NIMS data

Sample number	NT	QT	TT	C	Si	Mn	P	S	Ni	Cr	Cu	Mo	RR	dA	dB	dC	*P*1	*P*2	*P*3	*P*4
1	900	845	600	0.32	0.24	0.58	0.009	0.01	2.65	0.71	0.06	0	700	0.01	0.01	0.03	510	886	1714	279
2	870	855	550	0.41	0.25	0.74	0.013	0.019	0.05	0.99	0.08	0.16	700	0.03	0	0	566	1150	1898	361
3	870	855	600	0.33	0.27	0.69	0.015	0.008	0.04	0.99	0.05	0.18	640	0.03	0	0	535	973	1786	315
4	870	845	630	0.4	0.23	0.7	0.011	0.01	1.82	0.78	0.09	0.18	740	0.05	0	0.01	543	984	1733	307
5	865	865	650	0.35	0.25	0.82	0.016	0.023	0.03	0.01	0.01	0	825	0.07	0.04	0	341	631	1395	201
6	870	845	550	0.38	0.21	1.59	0.017	0.015	0.03	0.2	0.05	0	480	0.02	0	0	432	804	1569	258
7	870	855	650	0.4	0.22	0.7	0.01	0.004	0.04	0.94	0.12	0.17	820	0.02	0	0.02	517	933	1709	300
8	845	845	550	0.45	0.24	0.71	0.01	0.009	0.12	0.06	0.13	0	610	0.06	0	0.01	465	874	1510	289
9	870	855	550	0.4	0.26	0.77	0.018	0.005	0.05	0.98	0.13	0.17	610	0.05	0	0.01	562	1160	1821	371
10	870	845	550	0.4	0.22	1.56	0.011	0.019	0.06	0.09	0.05	0	530	0.02	0	0.01	451	854	1584	281
11	825	825	650	0.52	0.21	0.8	0.021	0.021	0.02	0.1	0.02	0	1740	0.09	0	0	410	742	1464	249
12	825	825	650	0.52	0.29	0.83	0.016	0.025	0.04	0.04	0.02	0	825	0.12	0.02	0	402	746	1446	235
13	870	855	600	0.37	0.29	0.76	0.017	0.009	0.12	1.01	0.12	0.16	500	0.05	0	0.02	542	996	1759	322
14	900	845	550	0.32	0.3	0.52	0.013	0.009	2.67	0.77	0.06	0	1740	0.03	0	0	570	1010	1715	316
15	870	855	600	0.41	0.25	0.74	0.013	0.019	0.05	0.99	0.08	0.16	700	0.03	0	0	527	1002	1758	326
16	870	855	550	0.4	0.25	0.87	0.011	0.015	0.07	1.06	0.11	0.18	700	0.07	0.01	0	603	1146	1873	367
17	870	845	680	0.4	0.23	0.7	0.011	0.01	1.82	0.78	0.09	0.18	740	0.05	0	0.01	482	868	1687	276
18	870	845	550	0.41	0.24	1.51	0.02	0.008	0.04	0.22	0.08	0	610	0.02	0.01	0.02	425	890	1552	289
19	870	855	600	0.38	0.27	0.8	0.015	0.012	0.07	1.07	0.12	0.16	1120	0.03	0	0.02	592	1067	1725	339
20	845	845	550	0.45	0.25	0.79	0.018	0.016	0.02	0.13	0.02	0	1740	0.07	0	0	513	923	1617	297

Note *P*1, *P*2, *P*3, and *P*4 denote fatigue strength (MPa), ultimate tensile strength (MPa), true fracture strength (MPa), and hardness (HV), respectively

$\boldsymbol{X}^{\mathrm{T}}\boldsymbol{X}$ isn't zero, which indicates the inverse matrix of $\boldsymbol{X}^{\mathrm{T}}\boldsymbol{X}$ exists, and the inverse matrix of $\boldsymbol{X}^{\mathrm{T}}\boldsymbol{X}$ is $(\boldsymbol{X}^{\mathrm{T}}\boldsymbol{X})^{-1} = \begin{bmatrix} 0.04 \cdots & 0.27 \\ \vdots & \vdots \\ 0.27 \cdots & 16791.47 \end{bmatrix}_{17\times 17}$.

Finally, with Eq. (2.9c) we have the coefficient matrix

$$\hat{\boldsymbol{W}} = (\boldsymbol{X}^{\mathrm{T}}\boldsymbol{X})^{-1}\boldsymbol{X}^{\mathrm{T}}\boldsymbol{Y} = \begin{bmatrix} w_0^{(1)} & \cdots & w_0^{(4)} \\ w_1^{(1)} & \ldots & w_1^{(4)} \\ \vdots & & \vdots \\ w_{16}^{(1)} & \cdots & w_{16}^{(4)} \end{bmatrix}_{17\times 4} = \begin{bmatrix} 1847.43 & \cdots & 919.71 \\ -2.31 & \ldots & 2.49 \\ \vdots & & \vdots \\ -1130.20 & \cdots & 243.40 \end{bmatrix}_{17\times 4}. \tag{2.10}$$

With the values of coefficient matrix $\hat{\boldsymbol{W}}$, multiple LSLRs give the following mathematical expressions for fatigue strength (FaS, MPa), ultimate tensile strength (TS, MPa), true fracture strength (FrS, MPa), and hardness (H, HV).

$$\begin{aligned} \mathrm{FaS} = {} & 1847.43 - 2.31\mathrm{NT} + 1.17\mathrm{QT} - 0.82\mathrm{TT} - 190.16\mathrm{C} + 285.64\mathrm{Si} + 62.82\mathrm{Mn} \\ & - 5385.72\mathrm{P} - 127.37\mathrm{S} + 24.02\mathrm{Ni} + 138.55\mathrm{Cr} - 101.31\mathrm{Cu} + 218.67\mathrm{Mo} \\ & + 0.06\mathrm{RR} + 322.88\mathrm{dA} - 618.62\mathrm{dB} - 1130.20\mathrm{dC}, \end{aligned} \tag{2.11a}$$

$$\begin{aligned} \mathrm{TS} = {} & 3092.58 + 5.43\mathrm{NT} - 6.99\mathrm{QT} - 2.42\mathrm{TT} + 576.76\mathrm{C} + 488.93\mathrm{Si} \\ & - 135.67\mathrm{Mn} + 1658.90\mathrm{P} + 3014.20\mathrm{S} - 42.94\mathrm{Ni} - 157.19\mathrm{Cr} \\ & + 623.68\mathrm{Cu} + 2154.18\mathrm{Mo} + 0.10\mathrm{RR} - 507.93\mathrm{dA} + 2497.92\mathrm{dB} + 1234.61\mathrm{dC}, \end{aligned} \tag{2.11b}$$

$$\begin{aligned} \mathrm{FrS} = {} & -591.34 + 1.83\mathrm{NT} + 2.06\mathrm{QT} - 1.79\mathrm{TT} + 803.97\mathrm{C} \\ & - 1143.07\mathrm{Si} - 77.82\mathrm{Mn} - 3523.24\mathrm{P} - 1602.52\mathrm{S} \\ & + 15.94\mathrm{Ni} + 360.97\mathrm{Cr} - 2331.49\mathrm{Cu} \\ & + 124.03\mathrm{Mo} - 0.08\mathrm{RR} + 1734.00\mathrm{dA} - 2178.35\mathrm{dB} + 4328.09\mathrm{dC}, \end{aligned} \tag{2.11c}$$

$$\begin{aligned} \mathrm{H} = {} & 919.71 + 2.49\mathrm{NT} - 2.91\mathrm{QT} - 0.71\mathrm{TT} + 121.59\mathrm{C} \\ & + 130.42\mathrm{Si} - 56.96\mathrm{Mn} + 809.06\mathrm{P} + 1181.72\mathrm{S} - 27.75\mathrm{Ni} - 83.99\mathrm{Cr} \\ & + 321.24\mathrm{Cu} + 755.14\mathrm{Mo} + 0.03\mathrm{RR} - 47.17\mathrm{dA} + 374.24\mathrm{dB} + 243.40\mathrm{dC}. \end{aligned} \tag{2.11d}$$

The equations have strong predictive power for the 20 samples, and the correlation coefficient R values of Eqs. (2.11a–d) are 0.995, 0.995, 0.979 and 0.988, respectively. Figure 2.3 shows the measured values against the predicted values of the four properties.

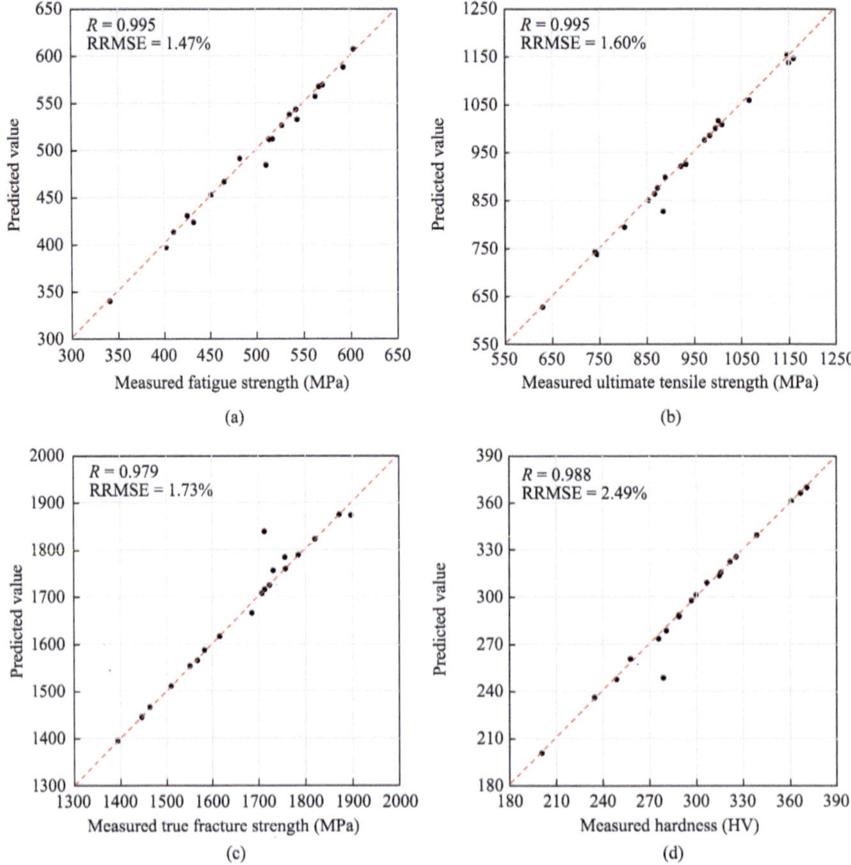

Fig. 2.3 Predicted values of multiple LSLR models against the measured values for **(a)** fatigue strength, **(b)** ultimate tensile strength, **(c)** true fracture strength, and **(d)** hardness

---

2.2 Principal Component Analysis and Principal Component Regression

Principal Component Regression (PCR) is based on Principal Component Analysis (PCA) (Whitlark & Dunteman, 1990). PCA is to find a basis set, and along the directions of basis unit vectors, input data should spread out as largely as possible. PCA is an important and essential ML method in classifying and clustering input data, and is widely used in digital image processing. If the feature input data are expressed by the $n \times m$ matrix $\boldsymbol{X}$ with n and m being the numbers of data and features, respectively, the $1 \times m$ means that vector $\overline{\boldsymbol{X}}$ can be expressed by

$$\overline{X} = (\overline{x}_1, \overline{x}_2, \cdots, \overline{x}_m) = \frac{1}{n}(1 \quad 1 \quad \cdots \quad 1)_{1\times n}\begin{bmatrix} x_{11} & \cdots & x_{1m} \\ \vdots & & \vdots \\ x_{n1} & \cdots & x_{nm} \end{bmatrix} = \frac{1}{n}\mathbf{1}_n^{\mathrm{T}}X, \quad (2.12a)$$

where $\overline{x}_j = \frac{1}{n}\sum_{i=1}^{n} x_{ij}, j = 1, 2, \cdots, m$ and $\mathbf{1}_n$ is a $(n \times 1)$ column vector $\mathbf{1}_n = \begin{pmatrix} 1 \\ \vdots \\ 1 \end{pmatrix}_{n\times 1}$. The $m \times m$ covariance matrix $\boldsymbol{\Sigma}$ of feature input data is calculated by

$$\boldsymbol{\Sigma} = (X - \mathbf{1}_n\overline{X})^{\mathrm{T}}(X - \mathbf{1}_n\overline{X}) = X^{\mathrm{T}}H^{\mathrm{T}}HX, \quad (2.12b)$$

where $H = \left(I - \frac{1}{n}\mathbf{1}_n\mathbf{1}_n^{\mathrm{T}}\right)$ is an $n \times n$ matrix with I of the $n \times n$ unit matrix and has the idempotent property of $H^a = H$ with a being an arbitrary integer. Equation (2.12b) indicates

$$X - \mathbf{1}_n\overline{X} = X - \frac{1}{n}\mathbf{1}_n\mathbf{1}_n^{\mathrm{T}}X = \left(I - \frac{1}{n}\mathbf{1}_n\mathbf{1}_n^{\mathrm{T}}\right)X = HX.$$

The covariance matrix $\boldsymbol{\Sigma}$ is real and symmetric. Considering a normalized and orthogonal basis set $v_j (j = 1, 2, \cdots, m)$, the variance of feature input data on the projected line v_j is calculated by

$$\mathrm{var}(X_{v_j}) = v_j^{\mathrm{T}}\boldsymbol{\Sigma}v_j. \quad (2.13a)$$

The maximization of the variance, subject to the constraint of the normalized and orthogonal basis set $v_j^{\mathrm{T}}v_j = 1$ and in the Lagrange form, determines the basis set, i.e.,

$$v_j = \underset{v_j}{\mathrm{argmax}}\left(v_j^{\mathrm{T}}\boldsymbol{\Sigma}v_j - d_j^2\left(v_j^{\mathrm{T}}v_j - 1\right)\right), \quad (2.13b)$$

where d_j^2 is a parameter, often called the Lagrange multiplier. The above equation yields

$$\boldsymbol{\Sigma}v_j = d_j^2 v_j, \quad (2.13c)$$

and

$$d_j^2 = v_j^{\mathrm{T}}\boldsymbol{\Sigma}v_j. \quad (2.13d)$$

Equations (2.13c) and (2.13d) indicate that v_j is the normalized eigenvector and d_j^2 is the eigenvalue associated with $v_j (j = 1, 2, \cdots, m)$ of the covariance matrix

$\boldsymbol{\Sigma}$. Equation (2.13d) can be expressed in the matrix form as

$$\boldsymbol{D}^2 = \begin{bmatrix} d_1^2 & \cdots & 0 \\ \vdots & & \vdots \\ 0 & \cdots & d_m^2 \end{bmatrix} = \boldsymbol{V}^{-1}\boldsymbol{V}, \tag{2.14a}$$

where $\boldsymbol{D}^2$ is the diagonal eigenvalue matrix, and the eigenvalues are ranked by $d_1^2 \geqslant d_2^2 \geqslant \cdots \geqslant d_m^2 \geqslant 0$; $\boldsymbol{V} = \begin{bmatrix} v_{11} & \cdots & v_{1m} \\ \vdots & & \vdots \\ v_{m1} & \cdots & v_{mm} \end{bmatrix}$ is the $m \times m$ normalized and orthogonal eigenvector matrix, and has the property of $\boldsymbol{V}^{-1} = \boldsymbol{V}^{\mathrm{T}}$. Introducing a centered $n \times m$ feature input matrix $\hat{\boldsymbol{X}} = \boldsymbol{X} - \mathbf{1}_n\overline{\boldsymbol{X}}$, we have $\boldsymbol{\Sigma} = \hat{\boldsymbol{X}}^{\mathrm{T}}\hat{\boldsymbol{X}}$. The centered input matrix $\hat{\boldsymbol{X}}$ can be decomposed by Singular Value Decomposition (SVD):

$$\hat{\boldsymbol{X}} = \boldsymbol{U}\boldsymbol{D}\boldsymbol{V}^{\mathrm{T}}, \tag{2.14b}$$

where $\boldsymbol{U}$ is an $n \times m$ semi-orthogonal matrix satisfying $\boldsymbol{U}^{\mathrm{T}}\boldsymbol{U} = \boldsymbol{I}_{m\times m}$, and $\boldsymbol{D}$ is a diagonal matrix with diagonal entries $d_1 \geqslant d_2 \geqslant \cdots \geqslant d_m \geqslant 0$. In case that there are some zero eigenvalues, the eigenvectors associated with zero eigenvalues are dropped, and the decomposition is called tight SVD, which happens if the rank of covariance matrix $\boldsymbol{\Sigma} = \hat{\boldsymbol{X}}^{\mathrm{T}}\hat{\boldsymbol{X}}$ is not full, i.e., lower than m. The eigenvectors $(\boldsymbol{v}_1, \boldsymbol{v}_2, \cdots, \boldsymbol{v}_m)^{\mathrm{T}}$ form the basis set for principal components (PCs). Projecting the centered input matrix $\hat{\boldsymbol{X}}$ on the eigenvector basis set gives the PC matrix $\boldsymbol{P} \triangleq \hat{\boldsymbol{X}}\boldsymbol{V}$ and from Eq. (2.14b), we have

$$\boldsymbol{P} \triangleq \hat{\boldsymbol{X}}\boldsymbol{V} = \boldsymbol{U}\boldsymbol{D}. \tag{2.14c}$$

The $n \times 1$ vector $\boldsymbol{p}_j \equiv \hat{\boldsymbol{X}}\boldsymbol{v}_j$ $(j = 1, 2, \cdots, m)$ is termed as the jth principal component of $\hat{\boldsymbol{X}}$, where $\boldsymbol{v}_j$ is the jth $m \times 1$ eigenvector. Along the jth principal component direction, $\boldsymbol{p}_j \equiv \hat{\boldsymbol{X}}\boldsymbol{v}_j = \boldsymbol{u}_j d_j$, and the $n \times 1$ vector $\boldsymbol{u}_j$ is called the normalized jth principal component because of $\boldsymbol{U}^{\mathrm{T}}\boldsymbol{U} = \boldsymbol{I}_{m\times m}$. Thus, the jth principal component has the sample variance

$$\mathrm{var}(\boldsymbol{p}_j) \equiv \mathrm{var}(\hat{\boldsymbol{X}}\boldsymbol{v}_j) = \mathrm{var}(\boldsymbol{u}_j d_j) = \frac{d_j^2}{n} \quad (j = 1, 2, \cdots, m). \tag{2.14d}$$

Clearly, the first principal component has the largest sample variance due to $d_1^2 \geqslant d_2^2 \geqslant \cdots \geqslant d_m^2 \geqslant 0$.

It is taken for granted in the above analysis that all features have the same physical units. In materials science and engineering, however, features have various physical dimensions or/and scales. How to analyze data with different physical dimensions or/and scales is an essential and fundamental aspect in materials informatics. Some researchers (e.g. Li, 2019) prefer to use the standardized input data in PCA. The standardized data are defined by

$$\tilde{x}_{ij} = \frac{x_{ij} - \overline{x}_j}{\sqrt{\text{var}(x_j)}}, \quad i = 1, 2, \cdots, n; \ \ j = 1, 2, \cdots, m, \tag{2.15a}$$

where $\overline{x}_j$ is the mean of feature j's inputs, and the variance of feature j's inputs is calculated by

$$\text{var}(x_j) = \frac{1}{n}\sum_{i=1}^{n}(x_{ij} - \overline{x}_j)^2, \tag{2.15b}$$

which is adopted in the machine learning package of Python, scikit-learn (Kramer, 2016). Very often, the unbiased variance (Li, 2019) is used here

$$\text{var}(x_j) = \frac{1}{n-1}\sum_{i=1}^{n}(x_{ij} - \overline{x}_j)^2. \tag{2.15c}$$

Obviously, the standardized input data are dimensionless, and have a zero mean and a unit variance for all features' inputs.

Alternately, original data for the inputs of each feature can be first normalized into the (0, 1) range by $\check{x}_{ij} = \frac{x_{ij} - x_{\min,j}}{x_{\max,j} - x_{\min,j}}, \ \ i = 1, 2, \cdots, n; \ \ j = 1, 2, \cdots, m$, where $x_{\max,j}$ and $x_{\min,j}$ are the maximum and minimum of feature j inputs, respectively. Obviously, such normalized data are dimensionless. The normalized data are then centered by $\hat{\check{x}}_{ij} = \check{x}_{ij} - \overline{\check{x}}_j, \ i = 1, 2, \cdots, n; \ \ j = 1, 2, \cdots, m$, where $\overline{\check{x}}_j = \frac{1}{n}\sum_{i=1}^{n}\check{x}_{ij}$ is the mean of normalized data. After that, PCA is conducted in the same way as described above. Similarly, original data for each-feature inputs can be centered first and then normalized by $\check{\hat{x}}_{ij} = \frac{\hat{x}_{ij}}{|\hat{x}_{ij}|_{\max}}, i = 1, 2, \cdots, n; \ \ j = 1, 2, \cdots, m$, where $|\hat{x}_{ij}|_{\max}$ is the maximum of each-feature absolute centered inputs. All these three data pre-treatments implement centralization, which makes the calculation of covariance matrix easier. Example 2.4 is going to show the effect of these dimensionless data treatments on PCA.

PCR uses PCs as input variables so that Eq. (2.4b) is rewritten, with the original y_i for the response, as

$$y_i = \sum_{j=1}^{m^{\text{pc}}} w_j^{\text{pc}} p_{ij} \quad (i = 1, 2, \cdots, n), \tag{2.16a}$$

where $m^{\text{pc}} \leqslant m$. $\boldsymbol{P} = (\boldsymbol{p}_{i1}, \boldsymbol{p}_{i2}, \cdots, \boldsymbol{p}_{im})$ $(i = 1, 2, \cdots, n)$ is an $n \times m$ input matrix, and $\boldsymbol{W}^{\text{pc}} = \left(w_1^{\text{pc}}, w_2^{\text{pc}}, \cdots, w_m^{\text{pc}}\right)^{\text{T}}$ is the column coefficient vector. Thus, $\boldsymbol{y} = \boldsymbol{P}\boldsymbol{W}^{\text{pc}}$ and

$$\hat{\boldsymbol{W}}^{\text{pc}} = \underset{\boldsymbol{W}^{\text{pc}}}{\operatorname{argmin}} \operatorname{RSS}(\boldsymbol{W}^{\text{pc}}) = \underset{\boldsymbol{W}^{\text{pc}}}{\operatorname{argmin}}[(\boldsymbol{y} - \boldsymbol{P}\boldsymbol{W}^{\text{pc}})^{\text{T}}(\boldsymbol{y} - \boldsymbol{P}\boldsymbol{W}^{\text{pc}})], \tag{2.16b}$$

which yields

$$\hat{\boldsymbol{W}}^{\text{pc}} = (\boldsymbol{P}^{\text{T}}\boldsymbol{P})^{-1}\boldsymbol{P}^{\text{T}}\boldsymbol{y}, \tag{2.16c}$$

since $(\boldsymbol{P}^{\text{T}}\boldsymbol{P})^{-1} = (\boldsymbol{D}^2)^{-1} = \begin{bmatrix} 1/d_1^2 & \cdots & 0 \\ \vdots & & \vdots \\ 0 & \cdots & 1/d_m^2 \end{bmatrix}$, Eq. (2.16c) is reduced to

$$\hat{\boldsymbol{W}}^{\text{pc}} = (\boldsymbol{D}^2)^{-1}\boldsymbol{P}^{\text{T}}\boldsymbol{y}. \tag{2.16d}$$

The prediction is then given by

$$\hat{\boldsymbol{y}} = \boldsymbol{P}\hat{\boldsymbol{W}}^{\text{pc}} = \boldsymbol{P}(\boldsymbol{D}^2)^{-1}\boldsymbol{P}^{\text{T}}\boldsymbol{y}. \tag{2.16e}$$

The above analysis indicates that raw data are expressed in the original basis set of features. PCA takes the eigenvectors as another basis set to express the feature data. Since the eigenvector basis set is obtained from the linear transformation in the feature space, each of the basis units in PCA is a linear combination of the original basis units. As an example, Fig. 2.4 illustrates five data in a two-dimensional feature space, where the original basis set is (x_1, x_2) and the basis set in PCA is (v_1, v_2).

Example 2.4 Table 2.5 lists five experimental data about F-M steels (Ferritic-Martensitic steels) exposed to the supercritical water environment. The data include two features of Cr element content (wt.%) and exposure time (h), and one response of weight gain (mg·cm^{-2}). Conduct PCA and Principle Component Regression (PCR) on the dataset.

The raw feature input matrix is

$$\boldsymbol{X}_{5\times 2} \triangleq [\boldsymbol{x}_1, \boldsymbol{x}_2] = \begin{bmatrix} 8.895 & 40 \\ 8.960 & 40 \\ 8.630 & 110 \\ 9.500 & 100 \\ 8.370 & 505 \end{bmatrix}.$$

Firstly, we perform PCA on the raw data without normalization and show the result in Fig. 2.4(a). Centralizing input data only shifts the coordinate system with the mean vector as the origin so that the centralized data locate near the origin. The

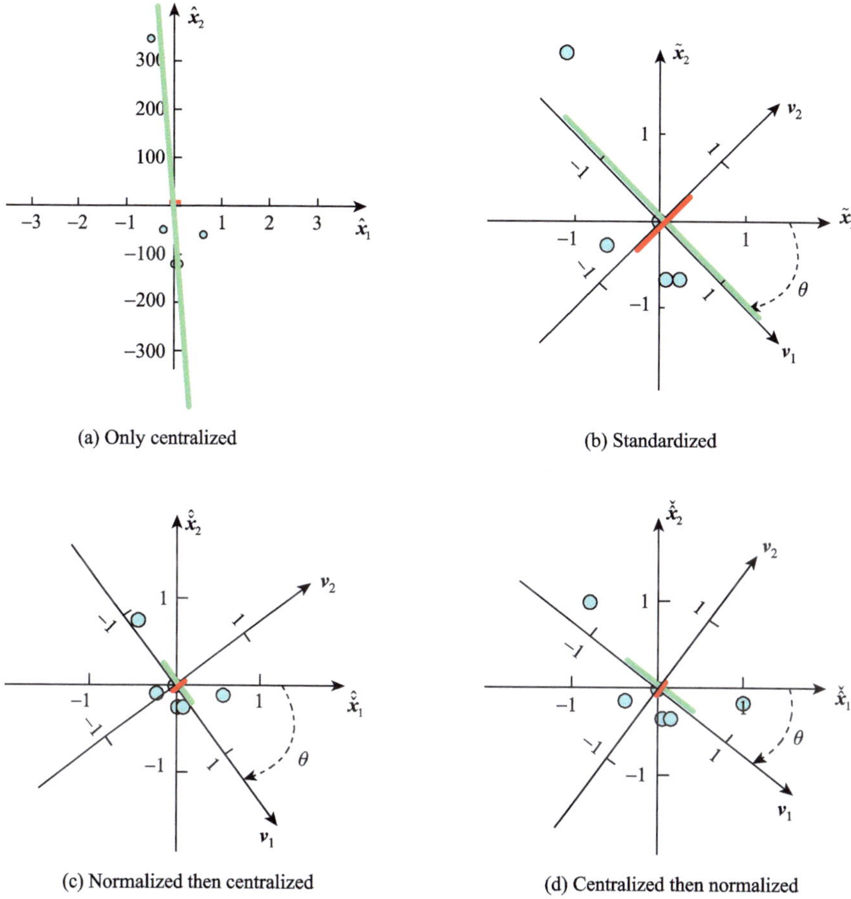

Fig. 2.4 **(a)** The data in the centralized feature coordinate system, where the scales in $\hat{x}_1$ and $\hat{x}_2$ axes have different physical dimensions and magnitudes. **(b)** the data in the standardized feature coordinate system. **(c)** the data in the normalized and then centralized feature coordinate system. **(d)** the data in the centralized and then scaled feature coordinate system. *Note* The half lengths of green and red lines show the data variances on PC axes, respectively

Table. 2.5 Weight gain of different F-M steels in supercritical water

Cr (wt.%)	Exposure time (h)	Weight gain (mg·cm^{-2})
8.895	40	0.340
8.960	40	0.500
8.630	110	1.202
9.500	100	1.184
8.370	505	1.870

input matrix after the centralization is

$$\hat{X}_{5\times2} = X_{5\times2} - \frac{1}{n}\mathbf{1}_n\mathbf{1}_n^{\mathrm{T}}X = \begin{bmatrix} 0.024 & -119 \\ 0.089 & -119 \\ -0.241 & -49 \\ 0.629 & -59 \\ -0.501 & 346 \end{bmatrix}.$$

SVD gives the decomposition of $\hat{X}_{5\times2} = U_{5\times2}D_{2\times2}V_{2\times2}^{\mathrm{T}}$ with

$$U_{5\times2} = \begin{bmatrix} 0.3033 & -0.2157 \\ 0.3033 & -0.1156 \\ 0.1249 & -0.4755 \\ 0.1504 & 0.8442 \\ -0.8819 & -0.0373 \end{bmatrix}, \quad D_{2\times2} = \begin{bmatrix} 392.3268 & 0 \\ 0 & 0.6488 \end{bmatrix}, \text{ and}$$

$$V_{2\times2}^{\mathrm{T}} = \begin{bmatrix} 0.0014 & -0.9999 \\ 0.9999 & 0.0014 \end{bmatrix}.$$

The normalized eigenvectors are $v_1 = \begin{bmatrix} 0.0014 \\ -0.9999 \end{bmatrix}$ and $v_2 = \begin{bmatrix} 0.9999 \\ 0.0014 \end{bmatrix}$, which form the PC basis unit vectors. The PC inputs are given by

$$P_{5\times2} \triangleq [p_1, p_2] = \hat{X}_{5\times2} \times V_{2\times2} = \begin{bmatrix} 118.9881 & -0.1426 \\ 118.9882 & -0.0776 \\ 48.9948 & -0.3096 \\ 58.9950 & 0.5463 \\ -345.9661 & -0.0165 \end{bmatrix}.$$

The information ratio of the jth ($j = 1, 2$) principal component is calculated by $I_j = \frac{d_j^2}{\sum_{j=1}^{m} d_j^2}$, which are $I_1 \approx 1$ and $I_2 \approx 0$. The first principal component is almost parallel to the axis of exposure time. With the centralized input data, a linear regression is conducted as $\hat{y} = 0.3497 \times \hat{x}_1 + 0.0032 \times \hat{x}_2 - 1.02$ ($R^2 = 0.784$), while with the PC input data, the linear regression function is $\hat{y} = -0.00271 \times v_1 + 0.3498 \times v_2 + 1.0192$ ($R^2 = 0.784$). The same value of $R^2 = 0.784$ indicates that the predicted outputs are independent of the basis unit vectors, whereas the coefficient values vary with the change in the basis unit vectors.

Secondly, we calculate PCA on the standardized input data and show the result in Fig. 2.4(b). The standardized input data are

$$\tilde{X}_{5\times2} \triangleq [\tilde{x}_1, \tilde{x}_2] = \begin{bmatrix} 0.063545 & -0.67824 \\ 0.235648 & -0.67824 \\ -0.6381 & -0.27928 \\ 1.66542 & -0.33627 \\ -1.32651 & 1.97203 \end{bmatrix}.$$

SVD gives $\tilde{X}_{5\times2} = U_{5\times2} D_{2\times2} V^{\mathrm{T}}_{2\times2}$ with

$$U_{5\times2} = \begin{bmatrix} 0.1832 & -0.3240 \\ 0.2257 & -0.2333 \\ -0.0886 & -0.4836 \\ 0.4943 & 0.7007 \\ -0.8145 & 0.3403 \end{bmatrix}, D_{2\times2} = \begin{bmatrix} 2.8637 & 0 \\ 0 & 1.3414 \end{bmatrix}, \text{and}$$

$$V^{\mathrm{T}}_{2\times2} = \begin{bmatrix} 0.7071 & -0.7071 \\ 0.7071 & 0.7071 \end{bmatrix}.$$

A rotation matrix $M(\theta) = \begin{bmatrix} \cos\theta & -\sin\theta \\ \sin\theta & \cos\theta \end{bmatrix}$ is an operator that rotates the coordinator system counterclockwise θ degree around the origin. Obviously $M(\theta)$ is an orthogonal matrix defined in V of PCA. The normalized eigenvectors are $v_1 = \begin{bmatrix} 0.7071 \\ -0.7071 \end{bmatrix}$ and $v_2 = \begin{bmatrix} 0.7071 \\ 0.7071 \end{bmatrix}$, which form the PC basis unit vectors. It is interesting to notice that the PC coordinate system rotates $\theta = -\frac{\pi}{4}\left(\frac{7\pi}{4}\right)$ with respect to the original coordinate system of the centralized input data. This result might be expected because both variances along the standardized $\tilde{x}_1$ axis and $\tilde{x}_2$ axis have the same value of one, and the maximum variance might be along the direction of $-\frac{\pi}{4}$ with respect to the standardized $\tilde{x}_1$ axis. The PC input matrix is

$$P_{5\times2} \triangleq [p_1, p_2] = \tilde{X}_{5\times2} \times V_{2\times2} = \begin{bmatrix} 0.52451617 & -0.43465083 \\ 0.64621020 & -0.31295680 \\ -0.25372162 & -0.64867940 \\ 1.41540963 & 0.93984196 \\ -2.33242114 & 0.45644719 \end{bmatrix}.$$

The information ratio of the jth ($j = 1, 2$) principal component is calculated as $I_1 = 0.820$ and $I_2 = 0.180$. Being Compared with the information ratio with the only centralized data, standardization makes the first (second) PC information ratio lower (higher). Varying the method in data treatment changes the values of coefficients. With the standardized feature inputs, the linear regression function is $\hat{y} = 0.132 \times \tilde{x}_1 + 0.561 \times \tilde{x}_2 + 1.02$ ($R^2 = 0.784$), while in the PC coordinate system, the linear regression function is $\hat{y} = -0.304 \times v_1 + 0.490 \times v_2 + 1.02$ ($R^2 = 0.784$).

Thirdly, PCA is conducted on normalized and then centralized data, and the result is shown in Fig. 2.4(c). The normalized and centralized inputs are

$$\hat{\check{\boldsymbol{X}}}_{5\times 2} \triangleq [\hat{\check{\boldsymbol{x}}}_1, \hat{\check{\boldsymbol{x}}}_2] = \begin{bmatrix} 0.021239 & -0.255914 \\ 0.078761 & -0.255914 \\ -0.213274 & -0.105376 \\ 0.556637 & -0.126882 \\ -0.443363 & 0.744086 \end{bmatrix}.$$

SVD gives $\hat{\check{\boldsymbol{X}}}_{5\times 2} = \boldsymbol{U}_{5\times 2}\boldsymbol{D}_{2\times 2}\boldsymbol{V}^{\mathrm{T}}_{2\times 2}$ with

$$\boldsymbol{U}_{5\times 2} = \begin{bmatrix} 0.2060 & -0.3100 \\ 0.2418 & -0.2165 \\ -0.0536 & -0.4887 \\ 0.4426 & 0.7344 \\ -0.8368 & 0.2809 \end{bmatrix}, \boldsymbol{D}_{2\times 2} = \begin{bmatrix} 1.0228 & 0 \\ 0 & 0.4737 \end{bmatrix}, \text{ and}$$

$$\boldsymbol{V}^{\mathrm{T}}_{2\times 2} = \begin{bmatrix} 0.6377 & -0.7702 \\ 0.7702 & 0.6377 \end{bmatrix}.$$

The normalized eigenvectors are $\boldsymbol{v}_1 = \begin{bmatrix} 0.6377 \\ -0.7702 \end{bmatrix}$ and $\boldsymbol{v}_2 = \begin{bmatrix} 0.7702 \\ 0.6377 \end{bmatrix}$, which form the PC basis unit vectors. Then, the PC inputs are

$$\boldsymbol{P}_{5\times 2} \triangleq [\boldsymbol{p}_1, \boldsymbol{p}_2] = \hat{\check{\boldsymbol{X}}}_{5\times 2} \times \boldsymbol{V}_{2\times 2} = \begin{bmatrix} 0.2106 & -0.1468 \\ 0.2473 & -0.1025 \\ -0.0548 & -0.2315 \\ 0.4527 & 0.3478 \\ -0.8558 & 0.1330 \end{bmatrix}.$$

The information ratio of the jth ($j = 1, 2$) principal component is calculated as $I_1 = 0.823$ and $I_2 = 0.177$. With normalized and then centralized inputs, the linear regression function is $\hat{y} = 0.395 \times \hat{\check{\boldsymbol{x}}}_1 + 1.488 \times \hat{\check{\boldsymbol{x}}}_2 + 1.019$ ($R^2 = 0.784$), while with the PC inputs, the linear regression function is $\hat{y} = -0.894 \times \boldsymbol{v}_1 + 1.253 \times \boldsymbol{v}_2 + 1.019$ ($R^2 = 0.784$).

Fourthly and finally, PCA is conducted on centralized and then normalized data, and the result is shown in Fig. 2.4(d). The centralized and then scaled inputs are

$$\check{\hat{\boldsymbol{X}}}_{5\times 2} \triangleq [\check{\hat{\boldsymbol{x}}}_1, \check{\hat{\boldsymbol{x}}}_2] = \begin{bmatrix} 0.038156 & -0.343931 \\ 0.141494 & -0.343931 \\ -0.383148 & -0.141618 \\ 1.00000 & -0.170520 \\ -0.796502 & 1.00000 \end{bmatrix}.$$

SVD gives $\check{\hat{\boldsymbol{X}}}_{5\times 2} = \boldsymbol{U}_{5\times 2}\boldsymbol{D}_{2\times 2}\boldsymbol{V}^{\mathrm{T}}_{2\times 2}$ with

$$U_{5\times 2} = \begin{bmatrix} 0.1502 & -0.3406 \\ 0.2015 & -0.2545 \\ -0.1360 & -0.4724 \\ 0.5612 & 0.6483 \\ -0.7768 & 0.4192 \end{bmatrix}, D_{2\times 2} = \begin{bmatrix} 1.5976 & 0 \\ 0 & 0.7321 \end{bmatrix}, \text{ and}$$

$$V_{2\times 2}^{\mathrm{T}} = \begin{bmatrix} 0.7926 & -0.6098 \\ 0.6098 & 0.7926 \end{bmatrix}.$$

The normalized eigenvectors are $v_1 = \begin{bmatrix} 0.7926 \\ -0.6098 \end{bmatrix}$ and $v_2 = \begin{bmatrix} 0.6098 \\ 0.7926 \end{bmatrix}$, which form the PC basis unit vectors. Then, the PC inputs are

$$P_{5\times 2} \triangleq [p_1, p_2] = \check{\hat{X}}_{5\times 2} \times V_{2\times 2} = \begin{bmatrix} 0.2400 & -0.2493 \\ 0.3219 & -0.1863 \\ -0.2173 & -0.3459 \\ 0.8966 & 0.4746 \\ -1.2411 & 0.3069 \end{bmatrix}.$$

The information ratio of the jth ($j = 1, 2$) principal component is calculated as $I_1 = 0.826$ and $I_2 = 0.174$. With the centralized and then scaled inputs, the linear regression function is $\hat{y} = 0.220 \times \check{\hat{x}}_1 + 1.107 \times \check{\hat{x}}_2 + 1.019$ ($R^2 = 0.784$), while in the PC coordinate system, the linear regression function is $\hat{y} = -0.5007 \times v_1 + 1.0116 \times v_2 + 1.019$ ($R^2 = 0.784$).

Figure 2.4 demonstrates the centralized input data in the four coordinate systems and associated PC coordinate systems, and Table 2.6 lists the PC eigenvectors in the four coordinate systems. In the only centralized coordinate system, the axes have different basis units in the physical dimension and in scale as well. It is therefore suggested to normalize the centralized input data. Standardization uses the variance of each-feature inputs so that in the standardized coordinate system, the basis unit of an axis has the magnitude of the feature variance. Thus, the PC in the two-feature space has a great possibility along the diagonal direction of the basis unit square. Since the commonly-used normalization utilizes the maximum and minimum of inputs to normalize original inputs into the range of (0, 1) and then centralizes the normalized data, the absolute values of data along the positive and negative directions of an axis will be the same, and the variance along any directions, including the PC

Table. 2.6 The eigenvectors of feature centralized input data matrix

Treatments	v_1^{T}	v_2^{T}
Only centralized	(0.0014, −0.9999)	(0.9999, 0.0014)
Standardized	(0.7071, −0.7071)	(0.7071, 0.7071)
Normalized then centralized	(0.6377, −0.7702)	(0.7702, 0.6377)
Centralized then normalized	(0.7926, −0.6098)	(0.6098, 0.7926)

directions, in the normalized-centralized feature coordinate system will be smaller than one. The result in the centralized-scaled feature coordinate system is similar to that in the normalized-centralized feature coordinate system, except that the absolute value of data along the positive and genitive diction of an axis will be different.

LSLR is the fundamental and essential method in machine learning. Many advanced ML algorithms are developed on the basis of LSLR, which will be described in the following chapters. LSLR regresses input and output variables via a linear relationship, no matter whether inputs are really single variables, combinations of single variables, or functions of single variables. There are two challenges faced by LSLR. The first challenge is that when there are too many input variables (features), LSLR brings in a great computational burden and requires big data to obtain a satisfactory LSLR solution with statistic reliability. Therefore, feature selection becomes a crucial step to select appropriate features from a huge number (e.g., $> 10^9$) of feature candidates. There are many feature selection methods, a few of which are based on LSLR. The second challenge is that if linear models are not built on solid scientific foundations, linear regressions will just be a game of purely linear data fitting.

As described in the Introduction, the bias-variance dilemma manifests that an LSLR model with the lowest RSS, equivalent to the highest fitting accuracy, might be overfitted and has a very low generalization power. To avoid overfitting, a penalty term is added to the objective function of LSLR. The penalty term is also called the regularization term, and LSLR with the regularization term is called regularized LSLR. Furthermore, K-fold cross-validation is widely used to avoid overfitting, where the whole data are randomly and equally split into K folds; K-1 folds form the training set and one fold forms the testing set. An ML model is trained on the training set and tested on the testing set. This process is repeated K times, and the validation performance is obtained by the mean of the K testing results (Pedregosa et al., 2011). The extreme case in K-fold cross-validation is the leave-one-out cross-validation (LOOCV) method, where a machine learning model is trained on $n-1$ data of a dataset with n data and validated by one datum, and cycles n times to let every single datum to be the validation one. The LOOCV method might be the best cross-validation one if the computation power is large enough and therefore suitable for small data. The validation performance can be regarded as an estimation of the generalization ability (Arlot & Celisse, 2010).

2.3 Least Absolute Shrinkage and Selection Operator (L_1)

The least absolute shrinkage and selection operator (LASSO (Tibshirani, 2011)) is based on LSLR by adding a penalty constraint on the regression coefficients that the sum of absolute values of w_i should be smaller than a preset positive value t, i.e., $\sum_{j=1}^{m}|w_j| \leqslant t$. LASSO estimates the values of $\hat{w}_j$ $(j = 0, 1, \cdots, m)$ from the criterion

$$\hat{w}^{\text{LASSO}} = \underset{w}{\text{argmin}} \sum_{i=1}^{n} \left(y_i - w_0 - \sum_{j=1}^{m} w_j x_{ij} \right)^2, \quad \text{subject to} \sum_{j=1}^{m} |w_j| \leqslant t. \tag{2.17a}$$

The LASSO problem can also be written in the equivalent Lagrange form

$$\hat{w}^{\text{ LASSO}} = \underset{w}{\text{argmin}} \left[\sum_{i=1}^{n} \left(y_i - w_0 - \sum_{j=1}^{m} w_j x_{ij} \right)^2 + \lambda \sum_{j=1}^{m} |w_j| \right] \quad (\lambda \geqslant 0). \tag{2.17b}$$

Subsequently, the LASSO predicted response is given by

$$\hat{y}_i = \hat{w}_0 + \sum\nolimits_{j=1}^{m} \hat{w}_j x_{ij}. \tag{2.17c}$$

For a given value of t, there is a unique value of λ, viz., the parameter λ corresponds one-to-one to parameter t. Obviously, LASSO is reduced to LSLR if $\lambda = 0$ or $t \geqslant t_0$, where $t_0 = \sum_{j=1}^{m} |w_j|$ is the sum over $|w_j|$ calculated by LSLR. When $\lambda > 0$, a sufficiently large value of λ may lead some of the regression coefficients $\hat{w}_j$ to be zero, which makes LASSO capable of feature selection. When using LASSO in feature selection, the value of λ is monotonically decreased from a sufficiently large number. Along with the decrease, the coefficients $\hat{w}_j$ emerge one by one, and the values of the coefficients depend on the value of λ. According to the appearance order, the first one, two, three and so on emerged coefficients are called the LASSO one-, two-, three-dimensional coefficients and so on, respectively. The right number of features is determined by the minimum error in validation (or testing). The feature selection via LASSO, on the hardness of tool and mold steels, is taken here as an example to illustrate the function of LASSO, where the LOOCV method and RMSE are used to evaluate the regression accuracy. The LOOCV RMSE represents the validation fitness.

Example 2.5 LASSO was conducted to investigate the contributions of alloying elements and select the most important alloying elements (Wang et al., 2019). Table 2.7 lists the hardness data of 18 tool and mold steels, which are obtained from hierarchical clustering of the original data (GB/T1299-2014, Tool and Mould Steels). The steels contain 12 alloying elements, which are taken as initial features plus another initial feature of square root of carbon content.

To show the Fig. 2.5(a) clearly, the estimated values of coefficients are enlarged (or reduced) by 4 times for Cr, double for Mo, triple for Si, 1/5 times for $\sqrt{\text{C}}$, double for Ni, 4 times for W, 4 times for Co, 1/3 times for Cu, 1/13 times for P, and 1/ 22 times for S. The inset shows the real values. In Fig. 2.5(b), the RMSE versus $-\lg \lambda$. The red arrow indicates the minimum RMSE and the blue arrow points out the critical value λ_0 and associated RMSE.

Table. 2.7 Hardness of 18 tool and mold steels (Unit: wt.%)

Samples	Cr	C	Mo	Si	V	$\sqrt{C}$	Ni	Mn	W	Co	Cu	P	S	H
1	16	0.17	0	0.5	0	0.41	2	0.75	0	0	0	0	0	49
2	18	0.95	0	0.4	0	0.97	0.3	0.4	0	0	0	0	0	55
3	17	0.36	1.15	0.45	0.25	0.60	0.8	0.7	0	0	0	0.013	0.003	50
4	16.5	0.39	1.05	0.5	0	0.62	0.5	0.75	0	0	0	0	0	46
5	13	0.2	0	0.5	0	0.44	0.3	0.5	0	0	0	0	0	45
6	13.5	0.4	0	1.05	0.3	0.63	0.225	0.55	0	0	0	0.005	0.002	50
7	18	0.9	1.15	0.4	0.095	0.94	0.3	0.4	0	0	0	0	0	55
8	13	0.4	0.3	0.3	0	0.63	0	0.4	0	0	0	0	0	50
9	12	1.5	0.95	0.3	0.8	1.22	0	0.3	0	0.5	0	0	0	59
10	12	1.5	0.95	0.3	0.8	1.22	0.1	0.3	0	0.5	0.15	0.015	0.015	59
11	11.75	1.575	0.5	0.2	0.225	1.25	0	0.2	0	0	0	0	0	58
12	11.75	1.575	0.5	0.2	0.225	1.25	0.1	0.2	0	0.5	0.15	0.015	0.015	58
13	8.5	0.505	1.475	0.9	0.425	0.71	0	0.425	0	0	0	0.015	0.008	59
14	12.25	2.15	0	0.2	0	1.46	0	0.2	0	0	0	0	0	60
15	7	0.73	2.1	0.95	2	0.85	0	0.2	0	0	0	0	0	60
16	12	2.15	0	0.25	0	1.46	0	0.45	0.7	0	0	0	0	60
17	8	1.75	0	0.4	0	1.32	0	0.4	0	0	0	0	0	63
18	8.05	0.99	2.4	1	0.325	0.99	0	0.35	0	0	0	0	0	62

Note "H" represents Rockwell hardness

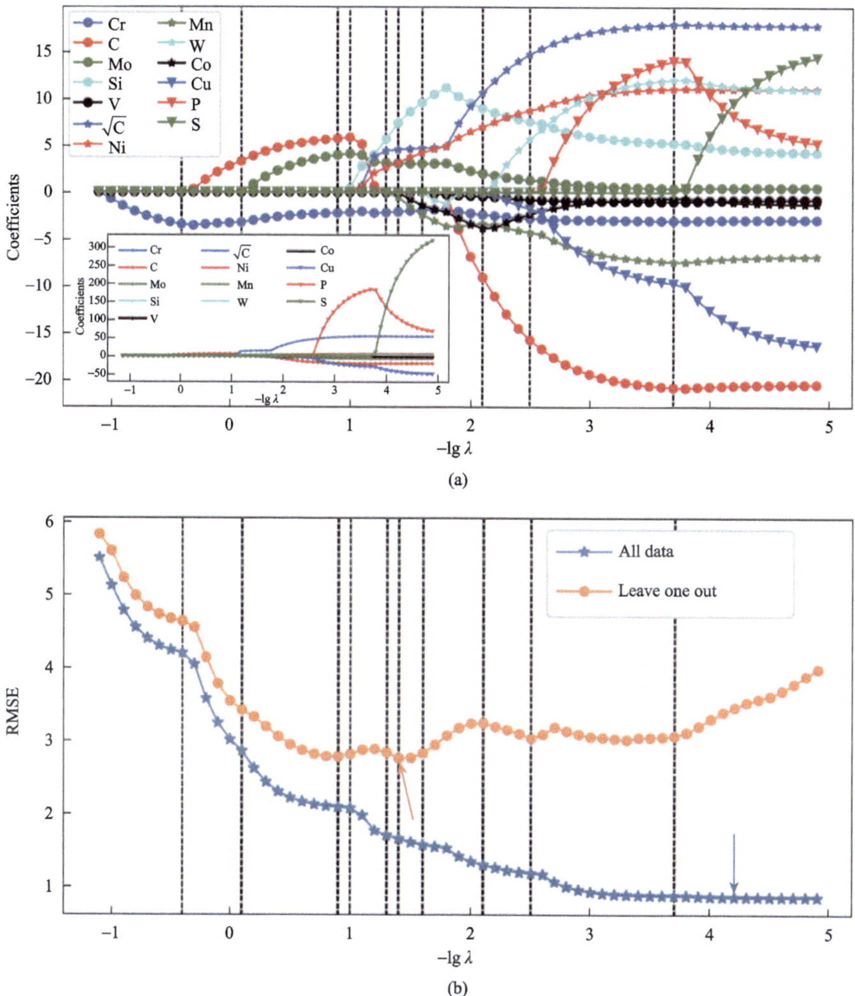

Fig. 2.5 **(a)** The estimated values of coefficients versus $-\lg \lambda$. **(b)** the estimated values of RMSE versus $-\lg \lambda$

With the criterion of minimal LOOCV RMSE, the LASSO determines the critical value of λ_0, selects 6 features from the original 13 features, and yields the hardness formula

$$\begin{aligned} \text{H (HRC)} = -0.53\ \text{Cr} + 1.53\ \text{Mo} + 2.45\ \text{Si} + 13.53\sqrt{\text{C}} + 1.56\ \text{Ni} - 0.52\ \text{Mn} + 46.87 \\ \text{with RMSE} = 1.6615\ \text{HRC}, \end{aligned} \tag{2.18a}$$

where Fe will balance the rest. LASSO determines the number of features, rather than providing the best regression result. With the 6 features determined by LASSO, LSLR gives the formula of

$$\begin{aligned}\mathrm{H\,(HRC)} &= -0.40\ \mathrm{Cr} + 1.36\ \mathrm{Mo} + 4.96\ \mathrm{Si} + 14.69\sqrt{\mathrm{C}} + 2.88\ \mathrm{Ni} - 5.09\ \mathrm{Mn} + 44.60 \\ &\text{with RMSE} = 1.5466\ \text{HRC}.\end{aligned} \tag{2.18b}$$

Obviously, the fitness is improved by LSLR.

LASSO is widely used in feature selection, and the purpose of feature selection is to find the most influential features for the modeling of targeted properties without redundancy. Ghiringhelli et al. (2015) proposed four important properties of a material feature, called descriptor by them: (1) A descriptor uniquely characterizes a material as well as property-relevant elementary processes; (2) Materials that are very different (similar) should be characterized by very different (similar) descriptor values; (3) The determination of descriptors must not involve calculations as intensively as those needed for the evaluation of the property to be predicted; (4) The number of descriptors should be as few as possible (for a certain accuracy request). To demonstrate how meaningful descriptors can be found systematically by LASSO, they took the energy difference of zinc blende or wurtzite and rock-salt semiconductors from 82 octet binary materials as targeted property. About 4500 feature candidates were formed by combining 23 initial features with simple math operators (+, −, × , /, absolute, etc.) in a heterogeneous manner. Then, LASSO was employed to select features from the about 4500 candidates, leading to the best 1D, 2D and 3D features. Obviously, LASSO increases the generalization by selecting the most important features.

2.4 Ridge Regression (L_2)

Like LASSO, ridge regression (Frank & Fridman, 1993) is also based on LSLR by adding a penalty constraint on the regression coefficients that the sum of w_i square must be smaller than a preset positive value t, i.e., $\sum_{j=1}^{m} w_j^2 \leqslant t$. Mathematically, ridge regression requires

$$\begin{aligned}\hat{w}^{\mathrm{ridge}} &= \underset{w}{\operatorname{argmin}} \sum_{i=1}^{n} \left(y_i - \sum_{j=1}^{m} w_j x_{ij} \right)^2, \\ &\text{subject to} \sum_{j=1}^{m} w_j^2 \leqslant t,\end{aligned} \tag{2.19a}$$

or in the equivalent Lagrange form

$$\hat{w}^{\text{ridge}} = \underset{w}{\operatorname{argmin}}\left[\sum_{i=1}^{n}\left(y_i - \sum_{j=1}^{m} w_j x_{ij}\right)^2 + \lambda \sum_{j=1}^{m} w_j^2\right]. \tag{2.19b}$$

As described above, the input $\boldsymbol{X}$ and coefficient $\boldsymbol{W}$ become $n \times m$ matrix and $m \times 1$ vector, respectively, with centered inputs. The RSS of Eq. (2.19b) can be expressed in the compact matrix form

$$\text{RSS}(\boldsymbol{W}, \lambda) = (\boldsymbol{y} - \boldsymbol{X}\boldsymbol{W})^{\mathrm{T}}(\boldsymbol{y} - \boldsymbol{X}\boldsymbol{W}) + \lambda \boldsymbol{W}^{\mathrm{T}}\boldsymbol{W}. \tag{2.20a}$$

The solution of ridge regression is given by

$$\hat{\boldsymbol{W}}^{\text{ridge}} = (\boldsymbol{X}^{\mathrm{T}}\boldsymbol{X} + \lambda \boldsymbol{I})^{-1}\boldsymbol{X}^{\mathrm{T}}\boldsymbol{y}, \tag{2.20b}$$

which yields the prediction of ridge regression

$$\hat{\boldsymbol{y}}^{\text{ridge}} = \boldsymbol{X}(\boldsymbol{X}^{\mathrm{T}}\boldsymbol{X} + \lambda \boldsymbol{I})^{-1}\boldsymbol{X}^{\mathrm{T}}\boldsymbol{y}. \tag{2.20c}$$

Equations (2.20b) and (2.20c) hold for standardized feature input matrix $\boldsymbol{X}$, and here the standardized feature input matrix is also expressed by $\boldsymbol{X}$ rather than $\hat{\boldsymbol{X}}$ without causing any confusion. As described in PCR, the standardized feature input matrix $\boldsymbol{X}$ can be replaced by the PC input matrix $\boldsymbol{P}$, and the prediction is

$$\hat{\boldsymbol{y}} = \boldsymbol{P}\hat{\boldsymbol{W}}^{\text{pc}}. \tag{2.21a}$$

The coefficient vector $\hat{\boldsymbol{W}}^{\text{pc}}$ is given in Eq. (2.16d), which is rewritten as

$$\hat{\boldsymbol{W}}^{\text{pc}} = (\boldsymbol{D}^2)^{-1}\boldsymbol{P}^{\mathrm{T}}\boldsymbol{y} = \begin{bmatrix} \frac{1}{d_1^2} & \cdots & 0 \\ \vdots & & \vdots \\ 0 & \cdots & \frac{1}{d_m^2} \end{bmatrix}\begin{bmatrix} q_1 \\ \vdots \\ q_m \end{bmatrix} \quad \text{with} \quad \begin{bmatrix} q_1 \\ \vdots \\ q_m \end{bmatrix} = \boldsymbol{P}^{\mathrm{T}}\boldsymbol{y}. \tag{2.21b}$$

In the PC coordinates, the ridge regression L_2 takes the form of

$$\hat{\boldsymbol{y}}^{\text{pc,ridge}} = \boldsymbol{P}\hat{\boldsymbol{W}}^{\text{pc,ridge}} = \boldsymbol{P}(\boldsymbol{P}^{\mathrm{T}}\boldsymbol{P} + \lambda \boldsymbol{I})^{-1}\boldsymbol{P}^{\mathrm{T}}\boldsymbol{y} = \boldsymbol{P}(\boldsymbol{D}^2 + \lambda \boldsymbol{I})^{-1}\boldsymbol{P}^{\mathrm{T}}\boldsymbol{y} \tag{2.21c}$$

and

$$\hat{\boldsymbol{W}}^{\text{pc,ridge}} = \begin{bmatrix} \frac{1}{d_1^2+\lambda} & \cdots & 0 \\ \vdots & & \vdots \\ 0 & \cdots & \frac{1}{d_m^2+\lambda} \end{bmatrix}\begin{bmatrix} q_1 \\ \vdots \\ q_m \end{bmatrix}. \tag{2.21d}$$

Comparing Eq. (2.21d) with Eq. (2.21b) indicates that in the PC coordinates, each coefficient of $\hat{\boldsymbol{W}}^{\text{pc,ridge}}$ with L_2 regularization is shrunk by a factor $\frac{d_j^2}{d_j^2+\lambda}$ ($j = 1, 2, \cdots, m$), because λ is a positive number and hence $0 < \frac{d_j^2}{d_j^2+\lambda} \leqslant 1$.

Example 2.6 Take the 20 NIMS fatigue data in Table 2.4 as an example to illustrate the influence of the penalty factor λ on the ridge regression coefficients $\hat{w}^{\text{ridge}}$ and the ridge regression prediction $\hat{\boldsymbol{y}}^{\text{ridge}}$, and take only fatigue strength as the regression target. The ridge regression coefficients approach 0 as λ approaches infinity. Figure 2.6 shows the variation of ridge regression coefficients $\hat{w}^{\text{ridge}}$ when the λ decreases.

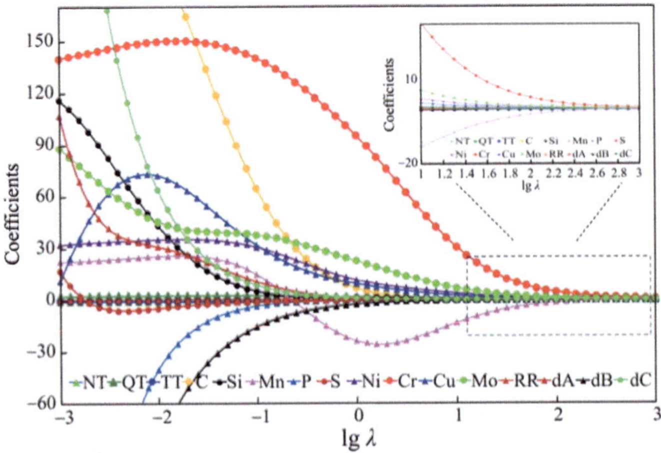

Fig. 2.6 Ridge track plot of ridge regression coefficients versus lgλ

Unlike in LASSO, the coefficients in ridge repression appear simultaneously. As shown in Fig. 2.6, ridge track plot was proposed by Hoerl and Kennard (1970) to eliminate multicollinearity. For multicollinearity problems, the coefficients tend to become too large in absolute value, and it is possible that some will even have the wrong sign in LSLR. In the ridge track plot, the λ is the best when the change of the coefficients tends to be flat. However, there is no uniform standard to determine when the curves "tend to be flat" for this method. The value λ should be determined by cross-validation. Figure 2.7 shows the cross validation (CV = 5) coefficient of determination R^2 value versus lgλ, where $\lambda = 0.025$ reaches the maximum $R^2 = 0.80$. Compared with cross validation, the ridge track plot might be too vague to determine the λ value.

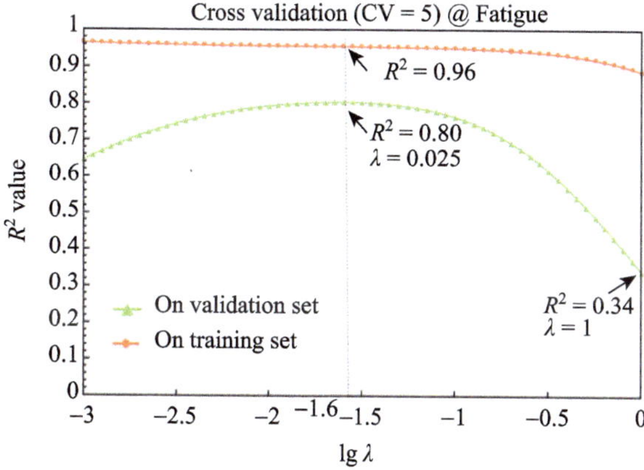

Fig. 2.7 The R^2 value versus lg λ in fivefold cross-validation

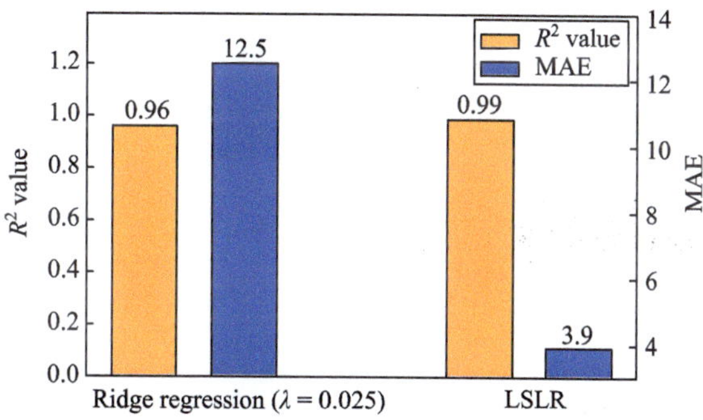

Fig. 2.8 Comparison of ridge regression with $\lambda = 0.025$ and LSLR on the training set

Figure 2.8 shows the comparison of ridge regression and LSLR in terms of R^2 and MAE. The results indicate that the regression accuracy of ridge regression does not significantly decrease, while the robustness of the model will be improved, especially when the matrix $\boldsymbol{X}^{\mathrm{T}}\boldsymbol{X}$ cannot be inverted. As Eqs. (2.21a–d) indicated, the generalization power in ridge regression is increased by reducing the absolute magnitude of fitted coefficients, while LASSO increases the generalization power via the reduction in the feature number.

The regression criterion for LSLR with a general penalty constraint is given by Hastie et al. (2009) as

$$\hat{w}^q = \arg\min_w \left[\sum_{i=1}^{n} \left(y_i - \sum_{j=0}^{m} w_j x_{ij} \right)^2 + \lambda \sum_{j=1}^{m} |w_j|^q \right] \quad \text{for} \quad q \geqslant 0. \tag{2.22}$$

Obviously, $q = 1$ is for LASSO and $q = 2$ for ridge regression, which are called the L_1 and L_2 regularizations, respectively, or the L_1 norm and the L_2 norm, respectively. In addition, if the coefficient vector of w is very sparse, the L_0 regularization will be used to put a penalty on the number of non-zero coefficients. Usually, k-sparse means that there are k components in the m coefficients which are non-zero. Generally, the penalty form in Eq. (2.22) with an arbitrary value of q is called the L_q norm. When $q \geqslant 1$, the constraint region is convex; otherwise, it is non-convex and the optimization becomes more difficult.

As described above, RSS is a widely-used loss function for regression. There are many kinds of loss functions for regression and classification. In general, consider an arbitrary loss function $\mathcal{L}(f(X_i), y_i)$, where f or $f(X_i)$ denotes an ML model. Without any penalty term, the parameters involved in the ML model are determined from the minimization of the loss function

$$\min_f \left[\frac{1}{n} \sum_{i=1}^{n} \mathcal{L}(f(X_i), y_i) \right]. \tag{2.23a}$$

Equation (2.23a) is called the empirical risk minimization, because it favours the model and often causes overfitting with low generalization prediction power. When a penalty term $\lambda\Omega(f)$ is added into Eq. (2.23a), the parameters involved in the ML model will be determined from the minimization of the regularized loss function

$$\min_f \left[\lambda\Omega(f) + \frac{1}{n} \sum_{i=1}^{n} \mathcal{L}(f(X_i), y_i) \right], \tag{2.23b}$$

where λ is a preset positive constant and called "cost" parameter, penalty parameter, hyperparameter or regularization parameter. Equation (2.23b) represents a common regularized problem and is termed as the structural risk minimization, where the value of regularization parameter λ controls the tradeoff between the model prediction accuracy and the model complexity (or generalization power), as illustrated in the Examples 2.5. and 2.6.

2.5 Elastic Net Regression

Elastic net regression is a combination of LASSO and ridge regression (Zou & Hastie, 2005), and thus requires

$$\hat{w}^{\text{elastic net}} = \underset{w}{\text{argmin}} \sum_{i=1}^{n} \left(y_i - w_0 - \sum_{j=1}^{m} w_j x_{ij} \right)^2,$$

$$\text{subject to } \alpha \sum_{j=1}^{m} |w_j| + (1-\alpha) \sum_{j=1}^{m} w_j^2 \leqslant t, \tag{2.24a}$$

or in the equivalent Lagrange form

$$\hat{w}^{\text{elastic net}} = \underset{w}{\text{argmin}} \left\{ \sum_{i=1}^{n} \left\{ y_i - w_0 - \sum_{j=1}^{m} w_j x_{ij} \right\}^2 + \lambda \left[\alpha \sum_{j=1}^{m} |w_j| + (1-\alpha) \sum_{j=1}^{m} w_j^2 \right] \right\}, \tag{2.24b}$$

where α is the weight of LASSO. Obviously, when $\alpha = 1$, elastic net regression is reduced to LASSO, whereas $\alpha = 0$ reduces elastic net regression to ridge regression. Equation (2.24b) can be rewritten as

$$\hat{w}^{\text{elastic net}} = \underset{w}{\text{argmin}} \left\{ \sum_{i=1}^{n} \left\{ y_i - w_0 - \sum_{j=1}^{m} w_j x_{ij} \right\}^2 + \lambda_1 \sum_{j=1}^{m} |w_j| + \lambda_2 \sum_{j=1}^{m} w_j^2 \right\}. \tag{2.24c}$$

Clearly, there are two regularization parameters of λ and α, which are called hyperparameters in ML, because their appropriate values should be determined by cross validation, as done above for one hyperparameter in LASSO or ridge regression. To determine the appropriate values of two (or multiple) hyperparameters is a challenging task, because it demands finding the global minimum of a loss function or the global maximum of fitting goodness.

Example 2.7 The 360 data of steel fatigue strength are collected from the NIMS database, which can be found in https://github.com/George-JieXIONG/Materials-Dataset/blob/main/Chapter4/NIMS-Fatigue.csv. Table 2.8 lists the initial 17 features. The elastic net regression with tenfold cross validation is employed to determine the values of the initial features.

Table. 2.8 The features used in this example

Features	Descriptions
NT	Normalizing temperature (°C)
QT	Quenching temperature (°C)
TT	Tempering temperature (°C)
C	wt.% of carbon
$\sqrt{C}$	squared wt.% of carbon
Si	wt.% of silicon
Mn	wt.% of manganese
P	wt.% of phosphorus
S	wt.% of sulphur
Ni	wt.% of nickel
Cr	wt.% of chromium
Cu	wt.% of copper
Mo	wt.% of molybdenum
RR	Reduction ratio
dA	Plastic work-inclusions (%)
dB	Discontinuous array-inclusions (%)
dC	Isolated inclusions (%)

The 2D search space of hyperparameters α and λ is set to be within the region of $\alpha \in (0, 1)$ and $\lambda \in (0.00001, 0.1)$ and is discretized equally with the grid length 1% for α and 10^{-5} for λ, giving about 10^6 discrete points with each discrete point being a combination of α and λ values. The search for optimal values of hyperparameters α and λ is conducted by randomly sampling 0.01% (i.e., 10^2 points), 0.1% (i.e., 10^3 points), 1% (i.e., 10^4 points), 10% (i.e., 10^5 points) of all candidates, and repeats the sampling without retuning three times at each sampling number. The greedy search is also conducted on the entire 10^6 candidates. The search results are tabulated in Table 2.9.

The search results are very interesting in that when randomly sampling the higher and equal to 1% candidates, the optimized α value is always 1, meaning that the optimized elastic net is actually LASSO, the tenfold CV-R^2 (coefficient of determination) is about 0.9267, and 11 features are selected. Two cases in randomly sampling 0.1% candidates also give the optimized α value of 1 and the same 11 features, except for slightly lower values of the tenfold CV-R^2. When randomly sampling 0.01% candidates and one case in 0.1% sampling, the optimized α value is lower than 1, meaning that the elastic net actively works without selecting features.

Data in Table 2.7 are also used to test elastic net regression. A greedy search associated with cross-validation is conducted to find the best combination of λ and α from the search space of λ ranging from 0.00001 to 10, and α varying from 0 to 1. RMSE of LOOCV is used to evaluate the prediction performance. The greedy search

Table. 2.9 The search results of elastic net with tenfold cross validation

Search methods	α	λ	Tenfold CV-R^2	Selected features
Greedy search (100%)	1	0.07452	0.926734	NT, QT, TT, C, $\sqrt{\text{C}}$, Mn, Ni, Cr, Cu, Mo, RR
Randomly sampling (10%)	1	0.07463	0.926730	
	1	0.07466	0.926729	
	1	0.07461	0.926731	
Randomly sampling (1%)	1	0.07566	0.926693	
	1	0.07464	0.926730	
	1	0.07521	0.926709	
Randomly sampling (0.1%)	1	0.08096	0.924370	
	1	0.06984	0.923352	
	0.2323	0.00026	0.921060	NT, QT, TT, C, $\sqrt{\text{C}}$, Si, Mn, P, S, Ni, Cr, Cu, Mo, RR, dA, dB, dC
Randomly sampling(0.01%)	0.4848	0.00028	0.920959	
	0.8081	0.00117	0.921039	
	0.9091	0.00288	0.920947	

gives $\lambda = 0.037$ and $\alpha = 1$. Again, for this case, the elastic net is reduced to LASSO and yields the same regression results, as described in Example 2.5.

Example 2.8 Conduct elastic net regression of data listed in Table 2.10, including three features of x_1, x_2, and x_3, and one target of y. RMSE of LOOCV is used to evaluate the prediction performance. The greedy search gives $\lambda = 1 \times 10^{-5}$ and $\alpha = 0$. In this case, the elastic net is reduced to ridge regression and the regression gives

$$y = -5.93x_1 + 121.88x_2 - 0.00856x_3 - 0.08$$

with RMSE = 1.5794.

Examples 2.7 and 2.8 might give a clue that when the feature number is large, elastic net regression will be reduced to LASSO, while elastic net regression will be reduced to ridge regression when the feature number is small. It is under study whether this reduction is sharp or diffuse, and whether there is a critical number of features in this reduction.

Table. 2.10 Dataset

x_1	x_2	x_3	y	x_1	x_2	x_3	y
0.4	0.6324555	1.5	51	0.85	0.9219544	0.5	62
0.27	0.5196152	1.5	48	0.8	0.8944272	1.5	62
0.36	0.6	1.3	48	1.025	1.0124228	0.2	62
0.86	0.9273619	0.35	64	0.575	0.7582875	0.8	58
0.34	0.5830952	0.8	52	0.795	0.8916277	0.2	62
1	1	0.5	60	0.9	0.9486833	1.05	62
1	1	0.5	60	0.975	0.9874209	1.2	62
0.9	0.9486833	1.85	62	0.895	0.9460444	0.2	62
0.7	0.83666	2.15	61	0.9	0.9486833	1.05	62
0.9	0.9486833	1.85	62	0.995	0.9974969	0.2	62
0.7	0.83666	0.85	60	1.095	1.0464225	0.2	62
0.86	0.9273619	0.325	64	1.195	1.0931606	0.2	62
0.6	0.7745967	0.3	58	1.375	1.1726039	0.2	64
0.52	0.7211103	0.95	60	1.3	1.1401754	0.2	62
0.5	0.7071068	0.55	58	1.375	1.1726039	0.2	64
0.55	0.7416198	1	62	0.975	0.9874209	0.95	62
0.7	0.83666	0.85	60	1.15	1.0723805	0.2	62
0.85	0.9219544	0.4	64	0.4	0.6324555	0.2	53
0.9	0.9486833	0.325	64	1.15	1.0723805	0.2	62
0.875	0.9354143	0.2	62	0.5	0.7071068	0.2	55
0.8	0.8944272	0.95	60	0.4	0.6324555	0.2	53
0.52	0.7211103	0.95	60	0.5	0.7071068	0.2	55
0.9	0.9486833	0.275	64	1.185	1.0885771	0.2	60
0.8	0.8944272	0.95	60	0.6	0.7745967	0.2	57
0.9	0.9486833	0.45	62	1.185	1.0885771	0.2	60
1.025	1.0124228	0.2	62	0.6	0.7745967	0.2	57
0.695	0.8336666	0.2	62	0.65	0.8062258	0.2	60
0.575	0.7582875	0.8	58	0.65	0.8062258	0.2	60

2.6 Multiply Task LASSO (MultiTaskLASSO)

Equation (2.8c) of $\boldsymbol{Y} = \boldsymbol{X}\boldsymbol{W}$ gives the general expression of multiply task linear regression (Argyriou et al., 2006). If the number of data is n, $\boldsymbol{Y}$ is $n \times p$ matrix, $\boldsymbol{X}$ is an $n \times (m + 1)$ matrix, and $\boldsymbol{W}$ is an $(m + 1) \times p$ matrix. MultiTaskLASSO introduces the L_1L_2 regularization, which is defined by $\|\boldsymbol{W}\|_{21} = \sum_{j=1}^{m}\sqrt{\sum_{k=1}^{p}\left(w_j^{(k)}\right)^2}$ and modifies Eq. (2.8b) as

$$\hat{\boldsymbol{W}} = \underset{\boldsymbol{W}}{\mathrm{argmin}}\left\{\frac{1}{2}\mathrm{tr}(\boldsymbol{Y} - \boldsymbol{X}\boldsymbol{W})^{\mathrm{T}}(\boldsymbol{Y} - \boldsymbol{X}\boldsymbol{W}) + \lambda\|\boldsymbol{W}\|_{21}\right\}, \tag{2.25}$$

where λ is a penalty parameter. Obviously, MultiTaskLASSO is reduced to the normal multiple task regression if $\lambda = 0$. When $\lambda > 0$, a sufficiently large value of λ will make all regression coefficients $\hat{w}_j^{(k)}$ zero, thereby rendering MultiTaskLASSO the capability of feature selection. When using MultiTaskLASSO in feature selection, the value of λ is monotonically decreased from a sufficiently large number. Along with the decrease, the coefficients $\hat{w}_j^{(k)}$ emerge one by one, and the values of the coefficients depend on the value of λ. According to the appearance order, the first, second, third, etc. emerged coefficients are called the MultiTaskLASSO one-, two-, three-dimensional coefficients and so on. The right number of features is determined by the minimum error in validation (or testing).

Example 2.9 Take the 20 NIMS fatigue data in Table 2.4 to conduct MultiTaskLASSO. The four properties are considered with equal weight in the MultiTaskLASSO example. Figure 2.9 shows the variations of feature coefficients with the λ decrease, indicating that the coefficients $\hat{w}_j^{(k)}$ emerge one by one. The number of selected features and the associated λ values are determined by the minimal LOOCV error, as shown in Fig. 2.10. In the single-task LASSO approach, each task selects a feature set and a λ value, giving four λ values, as shown in Fig. 2.11a. Figure 2.11b1 shows the feature set selected by MultiTaskLASSO, while Fig. 2.11b2 shows the feature sets selected by single-task LASSO.

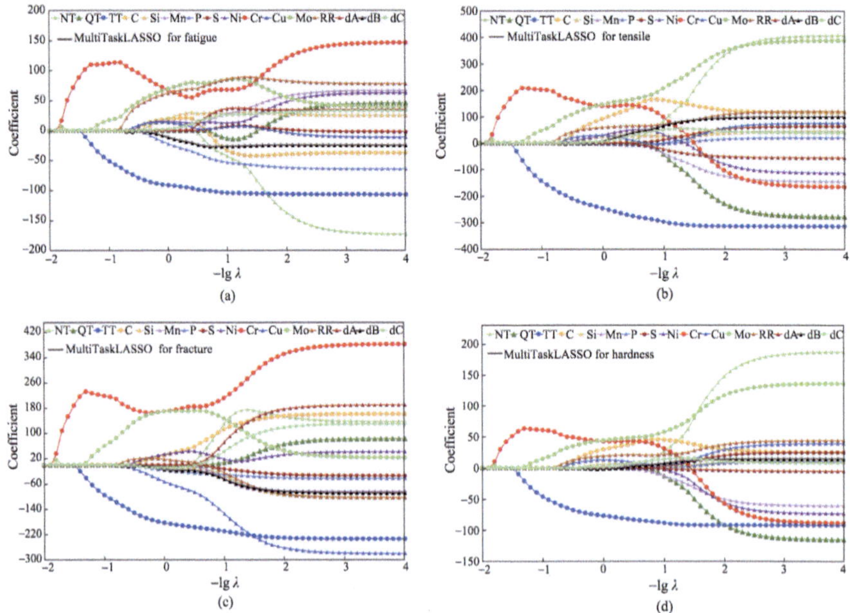

Fig. 2.9 The feature coefficients of four properties varying with the value of λ in MultiTaskLASSO

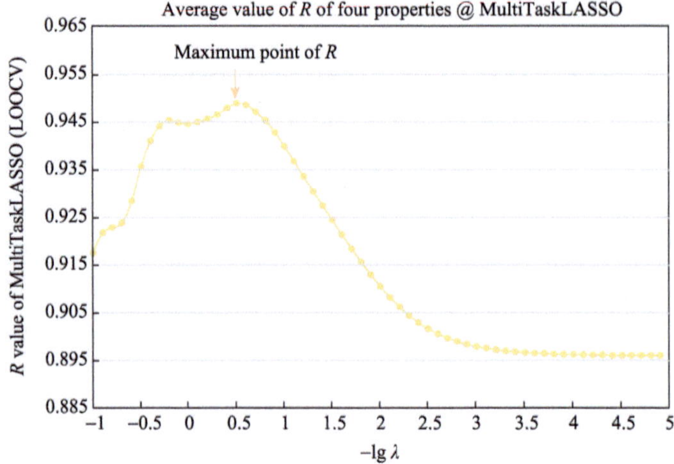

Fig. 2.10 When using LOOCV, the average of correlation coefficient R value of the four properties of MultiTaskLASSO varies with the value of λ

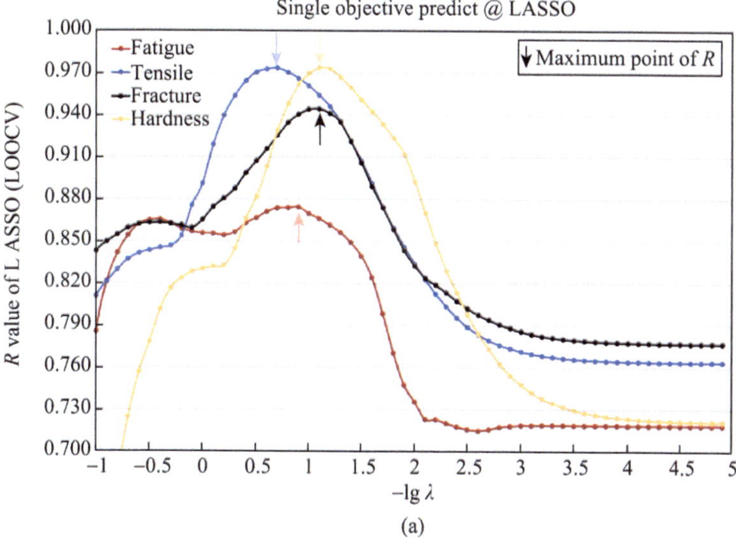

(a)

Coefficients non-zero location @ Optimal λ

	Fatigue	Tensile	Fracture	Hardness
−lg λ	0.5			
NT	−1.49	−2.88	−1.17	−0.79
QT	0.63	0.78	0.13	0.04
TT	−0.81	−2.00	−1.36	−0.63
C	−1.32	29.20	12.90	8.48
Si	0	0	0	0
Mn	8.13	12.07	14.07	2.55
P	0	0	0	0
S	0	0	0	0
Ni	25.54	35.30	25.55	5.99
Cr	137.55	257.97	238.98	76.03
Cu	64.75	106.39	-217.90	54.24
Mo	155.06	341.96	559.51	119.46
RR	0.03	0.02	0.00	0.01
dA	0	0	0	0
dB	0	0	0	0
dC	0	0	0	0

(b1) Multi-objective predict @ MultiTaskLASSO

	Fatigue	Tensile	Fracture	Hardness
−lg λ	0.9	0.7	1.1	1.1
NT	0.76	1.63	2.30	0.52
QT	0.03	0.04	0.01	0.01
TT	−0.67	−1.78	−1.35	−0.55
C	284.52	927.72	768.73	300.85
Si	0	58.32	16.59	0
Mn	−20.23	−26.85	−6.86	−10.48
P	0	0	0	0
S	0	0	0	0
Ni	4.99	3.58	7.30	−2.28
Cr	94.61	185.93	139.92	63.63
Cu	27.33	0	−558.10	8.55
Mo	239.94	530.37	1036.58	122.03
RR	0.04	0.03	0.02	0.01
dA	0	0	0	0
dB	0	0	0	0
dC	0	0	0	0

(b2) Single-objective predict @ LASSO

Fig. 2.11 **(a)** The correlation coefficient R value (LOOCV) of the four properties of LASSO varies with the value of λ. Comparison of coefficients non-zero location between **(b1)** multiTaskLASSO and **(b2)** LASSO with the right values of λ

Homework

1. Derive Eqs. (2.9c) and (2.20c) in LSLR and ridge regression, and understand what the regularization coefficient λ does on linear regression. (Refer to Appendix 1. Matrix Calculation).
2. Please use matrix $\boldsymbol{A} = \begin{bmatrix} 0 & 1 \\ 1 & 1 \\ 1 & 0 \end{bmatrix}$ to conduct a full SVD decompose.
3. Please use all the data in Table 2.4 (from the NIMS database) to conduct PCA linear regression and evaluate its fitting effect by LOOCV.
4. Please use LASSO and cross validation (CV = 5, 10 and LOOCV) to screen out the most important feature subset of data in Table 2.4 by referring to Example 2.5. And give the more concise linear equations to predict the four mechanical properties of steel.
5. Please give the fitting effect, (R, R^2 and MSE) of LSLR, LASSO and ridge regression on the training set of Table 2.7, where the λ in LASSO and ridge regression can be determined by cross validation, respectively. Then, use LOOCV to reevaluate the three models.
6. Please reproduce Example 2.8 and compare the difference between elastic net and LASSO.
7. Only focus on fatigue strength and ultimate tensile strength of steel on Table 2.4, and please conduct MultiTaskLASSO and find out the best common feature subset of them.

References

Argyriou, A., Evgeniou, T. & Pontil, M. (2006). *Multi-task feature learning: Advances in neural information processing systems 19*. Paper presented at the Advances in Neural Information Processing Systems 19 (NIPS 2006). Vancouver, British Columbia, Canada.

Arlot, S., & Celisse, A. (2010). A survey of cross-validation procedures for model selection. *Statistics Surveys, 4*, 40–79.

Frank, I. E., & Friedman, J. H. (1993). A statistical view of some chemometrics regression tools (with discussion). *Technometrics, 35*(2), 109–135.

Ghiringhelli, L. M., Vybiral, J., Levchenko, S. V., et al. (2015). Big data of materials science—critical role of the descriptor. *Physical Review Letters, 114*(10), 105503.

Hastie, T., Tibshirani, R., Friedman, J. H., et al. (2009). *The elements of statistical learning: Data mining, inference, and prediction*. Springer.

Hoerl, A. E., & Kennard, R. W. (1970). Ridge regression: Biased estimation for nonorthogonal problems. *Technometrics, 12*(1), 55–67.

Kramer, O. (2016). Scikit-learn. *His machine learning for evolution strategies* (pp. 45–53). Springer.

Li, H. (2019). *Statistical earning method (统计学习方法)*. Tsinghua University Press.

Pedregosa, F., Varoquaux, G., Gramfort, A., et al. (2011). Scikit-learn: Machine learning in Python. *The Journal of Machine Learning Research, 12*, 2825–2830.

Tibshirani, R. (2011). Regression shrinkage and selection via the LASSO: A retrospective. *Journal of the Royal Statistical Society, 73*(3), 267–288.

Wang, J. H., He, Y. L., Zhang, T. J., et al. (2019). Machine learning prediction of the hardness of tool and mold steels. *Scientia Sinica Technologica, 49*(10), 1148–1158.
Wei, Q. H., Xiong, J., Sun, S., et al. (2020). Multi-objective machine learning of four mechanical properties of steels. *Science China Technological Sciences, 51*, 722–736.
Whitlark, D., & Dunteman, G. H. (1990). Principal components analysis. *Journal of Marketing Research, 27*(2), 243.
Yang, X. S., Dai, L. H., & Zhai, H. R. (2016). Time, stress, and temperature-dependent deformation in nanostructured copper: Stress relaxation tests and simulations. *Acta Materialia, 108*, 252–263.
Zhang, T. Y. & Zheng, Y. R. (1998). Effects of absorption and desorption on hydrogen permeation—I. theoretical modeling and room temperature verification. *Acta Materialia, 46* (14), 5023–5033.
Zou, H., & Hastie, T. (2005). Regularization and variable selection via the elastic net. *Journal of the Royal Statistical Society, 67*(2), 301–320.

Chapter 3
Linear Classification

Regression predicts quantitative outputs, while classification predicts qualitative outputs. Classification divides data into various classes different in quality. For example, electric conductors and electric insulators are classified based on electric conductivity; amorphous materials and crystalline materials are separated by the features in atom/molecule arrangement of the materials. Classification finds the boundary between different classes.

The performance of an ML classifier is evaluated by a confusion matrix. As an example, a confusion matrix for binary classifiers is shown below (Table 3.1).

Table 3.1 Confusion matrix

Real label	Predicted label	
	Positive	Negative
Positive	True positive (TP)	False negative (FN)
Negative	False positive (FP)	True negative (TN)

The confusion matrix divides all data into four categories according to the combination of the real label and the predicted label: True Positive (TP) and False Negative (FN) denote that the real positive is predicted as positive and negative, respectively; False Positive (FP) and True Negative (TN) denote that the real negative is predicted as positive and negative, respectively. The confusion matrix can be used to define the following binary classification judgment indicators.

Accuracy (ACC) is the global judgment index, which measures the proportion of the number of all correctly classified data over the total number of entire data, as given by

$$\mathrm{ACC} = \frac{\mathrm{TP} + \mathrm{TN}}{\mathrm{TP} + \mathrm{TN} + \mathrm{FN} + \mathrm{FP}}. \tag{3.1a}$$

T. Zhang, *An Introduction to Materials Informatics*,
https://doi.org/10.1007/978-981-99-7992-9_3

True negative rate (TNR) measures the proportion of the correctly classified number of real negative data over the number of all real negative data, which is defined by

$$\text{TNR} = \frac{\text{TN}}{\text{FP} + \text{TN}}. \tag{3.1b}$$

False positive rate (FPR) indicates the proportion of the misclassified number of real negative data over the number of all real negative data, which is defined by

$$\text{FPR} = \frac{\text{FP}}{\text{FP} + \text{TN}}. \tag{3.1c}$$

Obviously, we have FPR $= 1 - $TNR. True positive rate (TPR) or recall rate indicates the proportion of the correctly classified number of real positive data over the number of all real positive data, which is defined by

$$\text{TPR} = \frac{\text{TP}}{\text{TP} + \text{FN}}. \tag{3.1d}$$

Receiver operating characteristic (ROC) curve (Spackman, 1989) is the curve of the true positive rate (y) versus the false positive rate (x). If it is known in advance that there are N_p positive and N_n negative data in a dataset, a random guess will give the TPR $= \frac{N_\text{p}}{N_\text{p}+N_\text{n}}$ and the TNR $= \frac{N_\text{n}}{N_\text{p}+N_\text{n}}$. Since FPR $= 1 -$ TNR $= \frac{N_\text{p}}{N_\text{p}+N_\text{n}} =$ TPR, the random guess will give the diagonal line from (0, 0) to (1, 1) in the ROC plot. A good classifier should yield a high true positive rate with a low false positive rate, and its ROC curve should be above the diagonal line from (0, 0) to (1, 1). Extremely, the ROC curve of an ideal perfect classifier will go through the point (0, 1). The area under the Curve (AUC) evaluates the classifier performance. The larger the AUC is, the better the classifier performance will be. Figure 3.1a shows a smooth ROC curve with a large number of data, and Fig. 3.1b shows a jagged ROC curve with a small number of data.

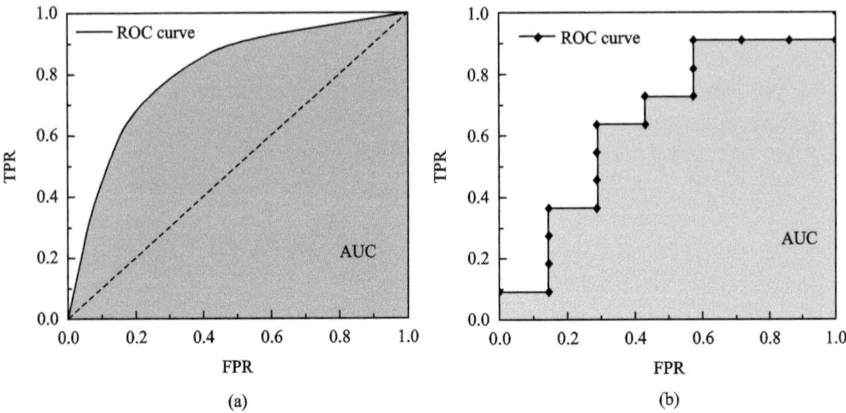

Fig. 3.1 The ROC curve and AUC based on **(a)** a large number of data and **(b)** a small number of data

3.1 Perceptron

Consider a dataset $D = \{(X_1, y_1), (X_2, y_1), \cdots, (X_n, y_n)\}$ of n separable data, where $X_i = (x_{i1}, x_{i2}, \cdots, x_{im})(i = 1, 2, \cdots, n) \in \mathbf{R}^m$ denotes an $1 \times m$ input vector, and $y \in \{-1, +1\}$ is the label of the data. The perceptron algorithm (Rosenblatt & Papert, 1957) seeks for a separating hyperplane with a normal vector $W = (w_1, w_2, \cdots, w_m)^{\mathrm{T}}$ and an intercept b separates the training data, meaning that $y_i = +1$ corresponds to $X_i W + b > 0$, and $y_i = -1$ corresponds to $X_i W + b < 0$. Thus, the misclassification leads to $-y_i(X_i W + b) > 0$, and hence a loss function is given by

$$\mathcal{L}(W, b) = -\sum_{X_i \in M} y_i(X_i W + b), \tag{3.2a}$$

where M denotes the misclassified dataset. The values of (W, b) are determined by the minimization of loss function, i.e.,

$$(W, b) = \underset{W,b}{\operatorname{argmin}}\left[-\sum_{X_i \in M} y_i(X_i W + b)\right]. \tag{3.2b}$$

The gradient of loss function of

$$\frac{\partial \mathcal{L}(W, b)}{\partial W} = -\sum_{X_i \in M} y_i X_i, \tag{3.3a}$$

$$\frac{\partial \mathcal{L}(W, b)}{\partial b} = -\sum_{X_i \in M} y_i, \tag{3.3b}$$

will be used in iteration to find a separating hyperplane. If all misclassified data are used in each iteration to update the parameter W, the iteration approach is called the whole batch method. Alternately, the Rosenblatt's Perceptron Learning Algorithm (Rosenblatt, 1958) uses the stochastic gradient descent in iteration, meaning that rather than computing the sum of the gradient contributions of all misclassified data in one iteration step, it randomly selects the data (X_i, y_i) from the misclassified dataset M to update the parameter W, which can be mathematically expressed by

$$W^{\mathrm{T}} = W^{\mathrm{T}} + \eta y_i X_i, \quad (X_i, y_i) \in M, \tag{3.4a}$$

$$b = b + \eta y_i, \quad (X_i, y_i) \in M, \tag{3.4b}$$

where η is the learning rate, and usually the value of $0 < \eta \leq 1$ is used; $(\boldsymbol{W}, b)$ in the left and right sites of Eqs. (3.4a) and (3.4b) take the values of iteration t and $t-1$, respectively. In the classic gradient optimization method, the learning rate η is also a parameter to be optimized by exactly line search or Amijo Rule (Boyd et al., 2004).

Perceptron is the fundamental method in Support Vector Machine and Neural Networks, which are described in Chaps. 4 and 9, respectively.

Example 3.1 Table 3.2 shows 4 alloys, including 2 crystalline alloys (CRA) and 2 bulk metallic glasses (BMG), collected from the reference of Xiong et al. (2020). The enthalpy of mixing (H_{mix}) and entropy of mixing (S_{mix}) (Takeuchi & Inoue, 2005) are the only two features considered in the example, which are calculated respectively by $S_{\text{mix}} = -R\sum_{i=1}^{N} a_i \ln a_i$, $H_{\text{mix}} = 4\sum_{j=i}^{N}\sum_{i=1}^{N} \Delta H_{ij} a_i a_j$, where R is the gas constant, a_i is the atomic fraction of the i-th constituent, and ΔH_{ij} is the molar mixing enthalpy for binary liquid alloys. With H_{mix} and S_{mix}, classify CRA and BMG by the perceptron algorithm and show the result in confusion matrix.

Table 3.2 Formed phases of 4 alloys

Data	Alloys	H_{mix}	S_{mix}	Phases
$\boldsymbol{X}_1$	$Ag_{10.5}Cu_{80.5}P_9$	−5.09	5.22	CRA
$\boldsymbol{X}_2$	$Ag_{10}Ce_5Cu_{85}$	−3.49	4.31	CRA
$\boldsymbol{X}_3$	$Cu_{34}Zr_{50}Ag_8Al_8$	−25.87	9.29	BMG
$\boldsymbol{X}_4$	$(Zr_{0.55}Al_{0.20}Co_{0.20}Cu_{0.05})_{95}Ag_5$	−40.10	10.52	BMG

Let CRA be the positive class as mentioned above (labeled $y = +1$) and BMG be the negative class (labeled $y = -1$). The correct decision boundary satisfies the following conditions: $y = +1$ corresponds to $\boldsymbol{X}_i\hat{\boldsymbol{W}} + \hat{b} > 0$ and $y_i = -1$ corresponds to $\boldsymbol{X}_i\hat{\boldsymbol{W}} + \hat{b} < 0$. The following steps are adopted to solve $\hat{\boldsymbol{W}}$ and $\hat{b}$.

Initialize $\boldsymbol{W}^0 = (0, 0)^{\mathrm{T}}$, $b^0 = 0$, and set the learning rate $\eta = 1$. The initial value of $Z^0 = \{\boldsymbol{W}^0,\ b^0\}$ denotes a reduced decision boundary $\boldsymbol{X}\boldsymbol{W}^0 + b^0 = 0$, which is the origin, as shown in Fig. 3.2a. With the datum of alloy $Ag_{10.5}Cu_{80.5}P_9$ [CRA, $\boldsymbol{X}_1 = (-5.09,\ 5.22)$ and $y_1 = +1$], we get $\boldsymbol{X}_1\boldsymbol{W}^0 + b^0 = (-5.09) \times 0 + 5.22 \times 0 + 0 = 0$. This is a wrong classification. We update parameters according to Eqs. (3.4a) and (3.4b),

$$\boldsymbol{W}^1 = \boldsymbol{W}^0 + \eta \times y_1 \times \boldsymbol{X}_1^{\mathrm{T}} = (0,\ 0)^{\mathrm{T}} + 1 \times 1 \times (-5.09,\ 5.22)^{\mathrm{T}} = (-5.09,\ 5.22)^{\mathrm{T}},$$

$$b^1 = b^0 + \eta \times y_1 = 0 + 1 \times 1 = 1.$$

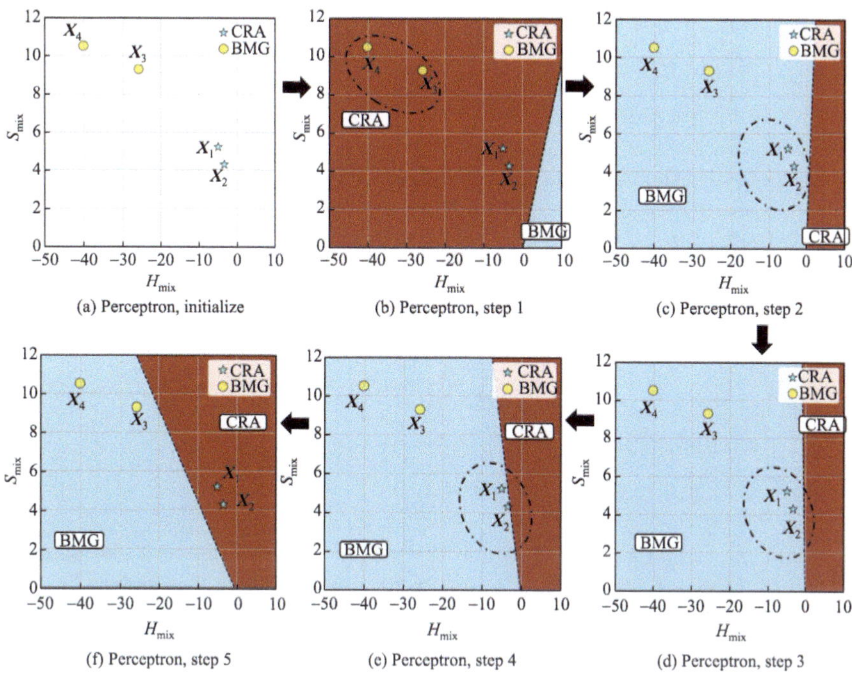

Fig. 3.2 The boundaries calculated by perceptron in the $(H_{\text{mix}}, S_{\text{mix}})$ feature space. *Note* Cyan and yellow dots represent the CRA and BMG data, respectively; the dashed lines are the separating hyperplanes calculated by perceptron; data inside the dashed ellipses are wrongly classified

The decision boundary is moved to $\boldsymbol{X}\boldsymbol{W}^1 + b^1 = 0$ and simply expressed by $Z^1 = \{\boldsymbol{W}^1, b^1\}$, as shown in Fig. 3.2b. With Z^1, the datum of alloy $Ag_{10.5}Cu_{80.5}P_9$ [CRA, $\boldsymbol{X}_1 = (-5.09,\ 5.22)$ and $y_1 = +1$] is correctly classified by $\boldsymbol{X}_1\boldsymbol{W}^1 + b^1 = -5.09 \times (-5.09) + 5.22 \times 5.22 + 1 = 54.16 > 0$.

With Z^1, the datum of alloy $Ag_{10}Ce_5Cu_{85}$ [CRA, $\boldsymbol{X}_2 = (-3.49,\ 4.31)$ and $y_2 = +1$] is also correctly classified due to $\boldsymbol{X}_2\boldsymbol{W}^1 + b^1 = -5.09 \times (-3.49) + 5.22 \times 4.31 + 1 = 41.26 > 0$, while the datum of alloy $Cu_{34}Zr_{50}Ag_8Al_8$ [BMG, $\boldsymbol{X}_3 = (-25.87,\ 9.29)$ and $y_3 = -1$] is wrongly classified because of $\boldsymbol{X}_3\boldsymbol{W}^1 + b^1 = -5.09 \times (-25.87) + 5.22 \times 9.29 + 1 = 181.17 > 0$. Due to this wrong classification, we update Z^1 to $Z^2 = \{\boldsymbol{W}^2, b^2\}$ with Eqs. (3.4a) and (3.4b) and datum $\boldsymbol{X}_3$:

$$\begin{aligned} \boldsymbol{W}^2 &= \boldsymbol{W}^1 + \eta \times y_3 \times \boldsymbol{X}_3^{\mathrm{T}} = (-5.09,\ 5.22)^{\mathrm{T}} + 1 \\ &\quad \times (-1) \times (-25.87,\ 9.29)^{\mathrm{T}} = (20.78,\ -4.07)^{\mathrm{T}}, \\ b^2 &= b^1 + \eta \times y_3 = 1 - 1 \times 1 = 0. \end{aligned}$$

The updated decision boundary $Z^2 = \{\boldsymbol{W}^2,\ b^2\}$ is shown in Fig. 3.2c. With Z^2, the first datum of alloy $Ag_{10.5}Cu_{80.5}P_9$ [CRA, $\boldsymbol{X}_1 = (-5.09,\ 5.22)$ and $y_1 = +1$] is wrongly classified due to $\boldsymbol{X}_1\boldsymbol{W}^2 + b^2 = 20.78 \times (-5.09) - 4.07 \times 5.22 + 0 = -127.0 < 0$. Then, Z^2 is revised to Z^3 with $\boldsymbol{X}_1$, as shown in Fig. 3.2d.

$$\begin{aligned}\boldsymbol{W}^3 &= \boldsymbol{W}^2 + \eta \times y_1 \times \boldsymbol{X}_1^{\mathrm{T}} = (20.78,\ -4.07)^{\mathrm{T}} + 1 \times 1 \times (-5.09,\ 5.22)^{\mathrm{T}} \\ &= (15.69,\ 1.15)^{\mathrm{T}}, \\ b^3 &= b^2 + \eta \times y_1 = 0 + 1 \times 1 = 1.\end{aligned}$$

Check firstly whether $\boldsymbol{X}_1$ is correctly classified with $Z^3 = \{\boldsymbol{W}^3,\ b^3\}$, we have

$$\boldsymbol{X}_1\boldsymbol{W}^3 + b^3 = 15.69 \times (-5.09) + 1.15 \times 5.22 + 1 = -72.86 < 0,$$

which is wrongly classified. Then, Z^3 is revised to Z^4 with $\boldsymbol{X}_1$ again, as shown in Fig. 3.2e.

$$\begin{aligned}\boldsymbol{W}^4 &= \boldsymbol{W}^3 + \eta \times y_1 \times \boldsymbol{X}_1^{\mathrm{T}} = (15.69,\ 1.15)^{\mathrm{T}} + 1 \times 1 \times (-5.09,\ 5.22)^{\mathrm{T}} \\ &= (10.6,\ 6.37)^{\mathrm{T}}, \\ b^4 &= b^3 + \eta \times y_1 = 1 + 1 \times 1 = 2.\end{aligned}$$

Again, $\boldsymbol{X}_1$ is firstly checked, which is wrongly classified due to $\boldsymbol{X}_1\boldsymbol{W}^4 + b^4 = 10.6 \times (-5.09) + 6.37 \times 5.22 + 2 = -18.70 < 0$. With $\boldsymbol{X}_1$, Z^4 is revised by

$$\begin{aligned}\boldsymbol{W}^5 &= \boldsymbol{W}^4 + \eta \times y_1 \times \boldsymbol{X}_1^{\mathrm{T}} = (10.6,\ 6.37)^{\mathrm{T}} + 1 \times 1 \times (-5.09,\ 5.22)^{\mathrm{T}} \\ &= (5.51,\ 11.59)^{\mathrm{T}}, \\ b^5 &= b^4 + \eta \times y_1 = 2 + 1 \times 1 = 3.\end{aligned}$$

The decision boundary of $Z^5 = \{\boldsymbol{W}^5, b^5\}$ is shown in Fig. 3.2f. Again, $\boldsymbol{X}_1$ is firstly checked, which is correctly classified because of $\boldsymbol{X}_1\boldsymbol{W}^5 + b^5 = 5.51 \times (-5.09) + 11.59 \times 5.22 + 3 = 35.5 > 0$. Then, $\boldsymbol{X}_2$ is checked $\boldsymbol{X}_2\boldsymbol{W}^5 + b^5 = 5.51 \times (-3.49) + 11.59 \times 4.31 + 3 = 33.7 > 0$, which is correctly classified. $\boldsymbol{X}_3$ is checked $\boldsymbol{X}_3\boldsymbol{W}^5 + b^5 = 5.51 \times (-25.87) + 11.59 \times 9.29 + 3 = -31.9 < 0$, which is also correctly classified. Finally, $\boldsymbol{X}_4$ is checked $\boldsymbol{X}_4\boldsymbol{W}^5 + b^5 = 5.51 \times (-40.10) + 11.59 \times 10.52 + 3 = -96 < 0$, which is also correctly classified. At this stage, all data are correctly classified with the decision boundary of $Z^5 = \{\boldsymbol{W}^5,\ b^5\}$, as shown in Fig. 3.2f.

The confusion matrix of final result is given by Table 3.3.

Table 3.3 Confusion matrix of final result

Real label	Predicted label	
	CRA	BMG
CRA	2	0
BMG	0	2

3.2 Logistic Regression

As described above, in binary classification, one class is labeled by one when $z > 0$, denoted by $y = 1|_{z>0}$, and the other by zero when $z < 0$, expressed by $y = 0|_{z<0}$. The binary classification can be described by the unit step function with the boundary at $z = 0$, which means that as variable z increases from negative to positive, output y jumps at $z = 0$ from zero to one. Mathematically, the value of 0.5 is assigned to y at $z = 0$, denoted by $y = 0.5|_{z=0}$. The unit step function separates two classes with a sharp boundary. Alternately, a diffuse boundary is also able to separate two classes. For example, a logistic function

$$y = \frac{1}{1 + e^{-z}}, \tag{3.5a}$$

separates two classes with a diffuse boundary with the decision boundary at $z = 0$. Figure 3.3 shows the unit step function and the logistic function. With a diffusion boundary, two thresholds for the two classes or a threshold for the boundary thickness should be defined to separate the two classes. In Fig. 3.3, for the boundary thickness of 10, and the logistic function gives the thresholds of $y > 0.993 \approx 1|_{z>5}$ for class 1 and $y < 0.0067 \approx 0|_{z<-5}$ for class 0. If $z = w_0 + wx$, Eq. (3.5a) is rewritten as

$$y = \frac{1}{1 + e^{-(w_0 + wx)}}. \tag{3.5b}$$

Equation (3.5b) indicates that the linear function moves the decision boundary to $\hat{x} = -\frac{w_0}{w}$ and shrinks the diffuse boundary's width by $1/w$ times, meaning that a diffuse boundary will gradually evolve to a sharp boundary as the absolute value of w increases. When there are multiple inputs and one output, the linear equation

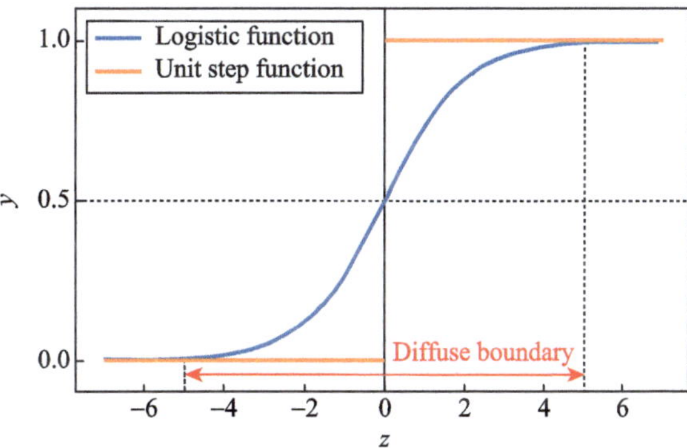

Fig. 3.3 Schematic plot of Eq. (3.5a)

is given by $z = \boldsymbol{XW}$ with $\boldsymbol{X} = (1, x_1, \cdots, x_m)$ being a row vector, and $\boldsymbol{W} = (w_0, w_1, \cdots, w_m)^{\mathrm{T}}$ being a column vector. In this case, the logistic function (Trevor et al., 2005) takes the form of

$$y = \frac{1}{1 + \mathrm{e}^{-\boldsymbol{XW}}}. \tag{3.5c}$$

Equation (3.5c) shows that most values of y will approximately be one or zero when $\boldsymbol{XW}$ varies from negative infinity to positive infinity except for the diffuse boundary region. Statistically, the probability of $\boldsymbol{X}$ belonging to class 1 can be defined by

$$P_1(y = 1|\boldsymbol{X}) = \frac{1}{1 + \mathrm{e}^{-\boldsymbol{XW}}}. \tag{3.6a}$$

Then the probability of $\boldsymbol{X}$ belonging to class 0 is given by

$$P_0(y = 0|\boldsymbol{X}) = 1 - P_1(y = 1|\boldsymbol{X}) = \frac{1}{1 + \mathrm{e}^{\boldsymbol{XW}}}. \tag{3.6b}$$

Introduce the ratio of $P_1(y = 1|\boldsymbol{X}) \big/ P_0(y = 0|\boldsymbol{X})$, which is termed as odds, measuring the relative probability that $\boldsymbol{X}$ belonging to class 1 with respect to that belonging to class 0. The logit is defined by ln(odds), i.e.,

$$\ln\left(\frac{P_1(y = 1|\boldsymbol{X})}{P_0(y = 0|\boldsymbol{X})}\right) = \ln\left(\frac{y}{1 - y}\right) = \boldsymbol{XW}. \tag{3.6c}$$

Obviously, $\boldsymbol{XW} = 0$ is a hyperplane in the m-dimensional space, and determines the location of the decision boundary, at which the probability of $\boldsymbol{X}$ belonging to class 1 equals to the probability of $\boldsymbol{X}$ belonging to class 0. The maximum likelihood method is used to estimate the values of $\boldsymbol{W}$ with available n data of $(1, x_{i1}, \cdots, x_{im})$ and $y_i \in \{0,\ 1\}$, $i = 1, 2, \cdots, n$. If each of the n data (observations) is independent, the likelihood is given by

$$L = \prod_{i=1}^{n} P(y_i|X_i;\ \boldsymbol{W}). \tag{3.7a}$$

The log-likelihood reads as

$$LL = \ln(L) = \sum_{i=1}^{n} \ln(P(y_i|\boldsymbol{X}_i;\ \boldsymbol{W})). \tag{3.7b}$$

Considering $y_i \in \{0,\ 1\}$, the log-probability of each data belonging to class 1 or class 0 can be expressed by

$$\begin{aligned} \ln(P(y_i|\boldsymbol{X}_i;\ \boldsymbol{W})) &= y_i \ln(P_1(\boldsymbol{X}_i;\ \boldsymbol{W})) + (1 - y_i)\ln(P_0(\boldsymbol{X}_i;\ \boldsymbol{W})) \\ &= y_i \boldsymbol{X}_i \boldsymbol{W} - \ln(1 + \mathrm{e}^{\boldsymbol{X}_i \boldsymbol{W}}). \end{aligned} \tag{3.7c}$$

The right-side term in Eq. (3.7c) is also called the cross entropy (Li, 2012), which measures the distance between two probabilities, and is often used in deep learning. Substituting Eq. (3.7c) into Eq. (3.7b) yields

$$LL(\boldsymbol{W}) = \sum_{i=1}^{n} \ln(P(y_i|\boldsymbol{X}_i; \boldsymbol{W})) = \sum_{i=1}^{n} [y_i \boldsymbol{X}_i \boldsymbol{W} - \ln(1 + e^{\boldsymbol{X}_i \boldsymbol{W}})]. \tag{3.7d}$$

The values of $\boldsymbol{W}$ are determined from the maximum of the log-likelihood

$$\hat{\boldsymbol{W}} = \underset{\boldsymbol{W}}{\operatorname{argmax}}\, LL(\boldsymbol{W}). \tag{3.8a}$$

Equivalently, a loss function is introduced as the negative log-likelihood and thus

$$\hat{\boldsymbol{W}} = \underset{\boldsymbol{W}}{\operatorname{argmin}}\, \mathcal{L}(\boldsymbol{W}). \tag{3.8b}$$

Both Eqs. (3.8a) and (3.8b) require

$$\frac{\partial \mathcal{L}(\boldsymbol{W})}{\partial \boldsymbol{W}} = -\frac{\partial LL(\boldsymbol{W})}{\partial \boldsymbol{W}} = \sum_{i=1}^{n} \boldsymbol{X}_i \left[y_i - \frac{e^{\boldsymbol{X}_i \boldsymbol{W}}}{1 + e^{\boldsymbol{X}_i \boldsymbol{W}}} \right] = 0. \tag{3.8c}$$

Since $\boldsymbol{W} = (w_0, w_1, \cdots, w_m)^{\mathrm{T}}$ is a $m + 1$ vector, Eq. (3.8c) represents a set of $m + 1$ equations. Equation (3.8b) is numerically solved by the coordinate descent method (Nocedal & Wright, 2006). The coordinate descent method first signs $\boldsymbol{W}$ initial values randomly, and then calculate $\boldsymbol{W}$ values by iterations. In each iteration step, the j-th component W_j is allowed to change under the fixed values of other components of $\boldsymbol{W}$ so that W_j value is updated by the gradient descent method, i.e.,

$$w_j^{t+1} = w_j^t + r \times X_{ij} \left[y_i - \frac{e^{\boldsymbol{X}_i \boldsymbol{W}^t}}{1 + e^{\boldsymbol{X}_i \boldsymbol{W}^t}} \right] \text{ with a preset value of learning rate } r. \tag{3.8d}$$

During each iteration, once the j-th component is updated to $w_j^{(t+1)}$, this new value immediately replaces the old w_j^t and is then used in the subsequent updates of the remaining components of **W**. Proceeding component-wise, all elements of **W** are updated recursively. The iteration goes on until convergence, and thus the $\boldsymbol{W}$ values are determined.

To make the classification model more generalized, we can add L_1 or L_2 regularization term to the loss function. With L_1 regularization, the loss function is expressed by

$$\mathcal{L}(\boldsymbol{W}) = C \sum_{i=1}^{n} \ln(P(y_i|\boldsymbol{X}_i; \boldsymbol{W})) + L_1$$

$$= C\sum_{i=1}^{n}[y_i X_i W - \ln(1 + e^{X_i W})] + \sum_{j=1}^{m}|w_j|, \tag{3.9a}$$

or equivalently,

$$\mathcal{L}(W) = \sum_{i=1}^{n} \ln(P(y_i|X_i; W)) + L_1$$
$$= \sum_{i=1}^{n}[y_i X_i W - \ln(1 + e^{X_i W})] + \lambda\sum_{j=1}^{m}|w_j|. \tag{3.9b}$$

Similarly, with L_2 regularization, the loss function is expressed by

$$\mathcal{L}(W) = C\sum_{i=1}^{n} \ln(P(y_i|X_i; W)) + L_2$$
$$= C\sum_{i=1}^{n}[y_i X_i W - \ln(1 + e^{X_i W})] + \frac{1}{2}\sum_{j=1}^{m} w_j^2, \tag{3.9c}$$

or equivalently,

$$\mathcal{L}(W) = \sum_{i=1}^{n} \ln(P(y_i|X_i; W)) + L_2$$
$$= \sum_{i=1}^{n}[y_i X_i W - \ln(1 + e^{X_i W})] + \frac{\lambda}{2}\sum_{j=1}^{m} w_j^2. \tag{3.9d}$$

The C or λ is the penalty parameter, with C decreases or λ increases, the proportion of the regularization term increases, and the performance of the model on the training set will deteriorate, but the model robustness will increase. As described in Chap. 2, cross validation should be conducted to optimize the value of C or λ.

Example 3.2 With the data in Table 3.2, we calculate the decision boundary of logistic regression. For the sake of simplicity, we shift the classification hyperplane to pass through the origin by setting $b = 0$ to classify the data of $X = \begin{bmatrix} X_{11} & X_{12} \\ X_{21} & X_{22} \\ X_{31} & X_{32} \\ X_{41} & X_{42} \end{bmatrix} =$

$\begin{bmatrix} -5.09 & 5.22 \\ -3.49 & 4.31 \\ -25.87 & 9.29 \\ -40.10 & 10.52 \end{bmatrix}$. The CRA and BMG are labeled by one and zero, respectively, and the label matrix is $\boldsymbol{Y} = \begin{bmatrix} y_1 \\ y_2 \\ y_3 \\ y_4 \end{bmatrix} = \begin{bmatrix} 1 \\ 1 \\ 0 \\ 0 \end{bmatrix}$.

Take nonzero $\boldsymbol{W}^0 \neq (0, 0)$ as the initial value of $\boldsymbol{W}^0 = \left(w_1^0, w_2^0\right)^{\mathrm{T}} = (5.55,\ 0.5)^{\mathrm{T}}$, and set the learning rate $r = 1$, as shown in Fig. 3.4a. With the datum $\boldsymbol{X}_1 = (X_{11},\ X_{12}) =(-5.09,\ 5.22)$ and $y_1 = 1$, the probability of $\boldsymbol{X}_1$ belonging to CRA ($y_1 = 1$) is calculated with Eq. (3.2a) to be $P_1(y = 1|\boldsymbol{X}_1) = \frac{1}{1+\mathrm{e}^{-\boldsymbol{X}_1 \boldsymbol{W}^0}} = \frac{1}{1+\mathrm{e}^{-(-5.09\times5.55+5.22\times0.5)}} =\approx 0 < \frac{1}{2}$, which is wrong classification. The coefficient value is iterated with Eqs. (3.8c) and (3.8d).

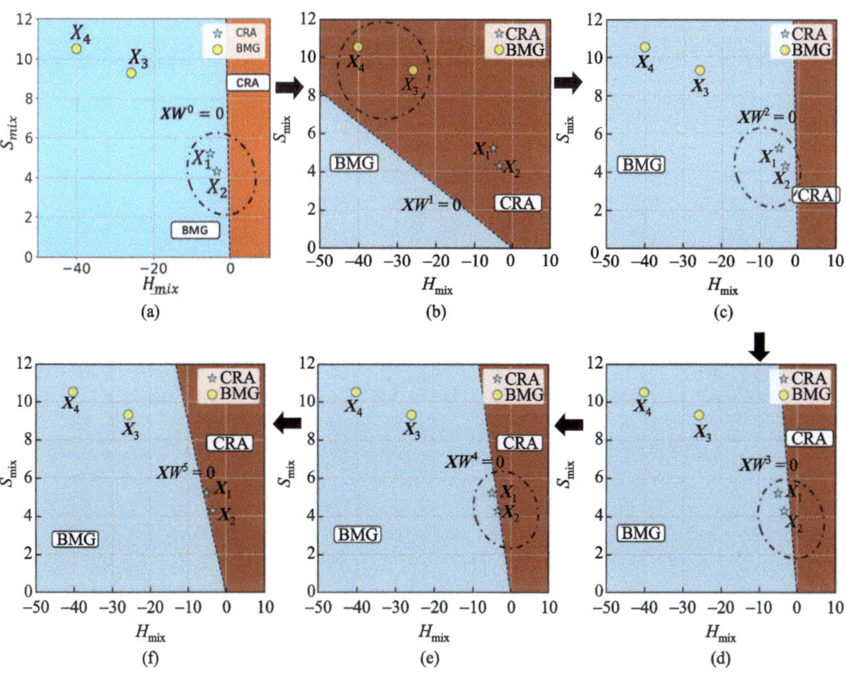

Fig. 3.4 The boundaries calculated by logistic regression in the (H_{mix}, S_{mix}) feature space. *Note* Cyan and yellow dots represent the CRA and BMG data, respectively; the dashed lines are the separating hyperplanes calculated by logistic regression; data inside the dashed ellipses are wrongly classified

$$w_1^1 = w_1^0 + r \times X_{11}\left[y_1 - \frac{\mathrm{e}^{X_1 W^0}}{1+\mathrm{e}^{X_1 W^0}}\right]$$
$$= 5.55 + 1 \times (-5.09) \times \left[1 - \frac{\mathrm{e}^{-5.09\times 5.55+5.22\times 0.5}}{1+\mathrm{e}^{-5.09\times 5.55+5.22\times 0.5}}\right] = 0.46,$$
$$w_2^1 = w_2^0 + r \times X_{12}\left[y_1 - \frac{\mathrm{e}^{X_1 W^0}}{1+\mathrm{e}^{X_1 W^0}}\right]$$
$$= 0.5 + 1 \times 5.22 \times \left[1 - \frac{\mathrm{e}^{-5.09\times 0.46+5.22\times 0.5}}{1+\mathrm{e}^{-5.09\times 0.46+5.22\times 0.5}}\right] = 2.8.$$

With $\boldsymbol{W}^1 = \left(w_1^1, w_2^1\right)^{\mathrm{T}} = (0.46,\ 2.8)^{\mathrm{T}}$, the decision boundary is $\boldsymbol{X}\boldsymbol{W}^1 = 0$, as shown in Fig. 3.4b.

With the datum $\boldsymbol{X}_2 = (X_{21},\ X_{22}) = (-3.49,\ 4.31)$ and $y_2 = 1$, the probability of $\boldsymbol{X}_2$ belonging to CRA ($y_2 = 1$) is calculated to be $P_2(y = 1|\boldsymbol{X}_2) = \frac{1}{1+\mathrm{e}^{-X_2 W^1}} = \frac{1}{1+\mathrm{e}^{-(-3.49\times 0.46+4.31\times 2.8)}} \approx 1 > \frac{1}{2}$, which is correct classification. With the datum $\boldsymbol{X}_3 = (X_{31},\ X_{32}) = (-25.87,\ 9.29)$ and $y_3 = 0$, the probability of $\boldsymbol{X}_3$ belonging to BMG ($y_3 = 0$) is calculated to be $P_3(y = 0|\boldsymbol{X}_3) = \frac{1}{1+\mathrm{e}^{X_3 W^1}} = \frac{1}{1+\mathrm{e}^{-25.87\times 0.46+9.29\times 2.8}} = 7.4 \times 10^{-7} < \frac{1}{2}$, which is wrong classification. Then, the $\boldsymbol{W}^1$ value shall be updated to $\boldsymbol{W}^2 = \left(w_1^2, w_2^2\right)^{\mathrm{T}} = (26.3,\ 2.8)^{\mathrm{T}}$ via

$$w_1^2 = w_1^1 + r \times X_{31}\left[y_3 - \frac{\mathrm{e}^{X_3 W^1}}{1+\mathrm{e}^{X_3 W^1}}\right]$$
$$= 0.46 + 1 \times (-25.87) \times \left[0 - \frac{\mathrm{e}^{-25.87\times 0.46+9.29\times 2.8}}{1+\mathrm{e}^{-25.87\times 0.46+9.29\times 2.8}}\right] = 26.3,$$
$$w_2^2 = w_2^1 + r \times X_{32}\left[y_3 - \frac{\mathrm{e}^{X_3 W^1}}{1+\mathrm{e}^{X_3 W^1}}\right]$$
$$= 2.8 + 1 \times 9.29 \times \left[0 - \frac{\mathrm{e}^{-25.87\times 26.3+9.29\times 2.8}}{1+\mathrm{e}^{-25.87\times 26.3+9.29\times 2.8}}\right] = 2.8.$$

The decision boundary is moved to $\boldsymbol{X}\boldsymbol{W}^2 = 0$ as shown in Fig. 3.4c. With $\boldsymbol{W}^2$, the datum $\boldsymbol{X}_4 = (-40.10,\ 10.52)$ and $y_4 = 0$ are correctly classified by $P_4(y = 0|\boldsymbol{X}_4) = \frac{1}{1+\mathrm{e}^{X_4 W_2}} = \frac{1}{1+\mathrm{e}^{-40.1\times 26.3+10.52\times 2.8}} \approx 1 > \frac{1}{2}$. At this step, all data are used once, and a cycle of iterations over all data once is called an epoch.

After the first epoch, the iteration starts from datum $\boldsymbol{X}_1$ again in the similar way. During the second epoch, $\boldsymbol{W}^3 = \left(w_1^3,\ w_2^3\right)^{\mathrm{T}} = (21.2,\ 8.2)^{\mathrm{T}}$ and $\boldsymbol{W}^4 = \left(w_1^4,\ w_2^4\right)^{\mathrm{T}} = (17.7,\ 12.5)^{\mathrm{T}}$ are shown up, and after the second epoch, the decision boundary is moved to $\boldsymbol{X}\boldsymbol{W}^4 = 0$ as shown in Fig. 3.4e. The third epoch leads to $\boldsymbol{W}^5 = \left(w_1^5,\ w_2^5\right)^{\mathrm{T}} = (12.6,\ 13.8)^{\mathrm{T}}$ and the decision boundary is $\boldsymbol{X}\boldsymbol{W}^5 = 0$ as shown in Fig. 3.4f. Until now, the local optimal parameter $\boldsymbol{W}^5$ is found by the logistic regression with the coordinate gradient descent method. All data of $\boldsymbol{X}_1$, $\boldsymbol{X}_2$, $\boldsymbol{X}_3$, and $\boldsymbol{X}_4$ are correctly classified by the decision boundary of $\boldsymbol{X}\boldsymbol{W}^5 = \boldsymbol{0}$.

Example 3.3 Table 3.4 lists the data of 34 CRA and BMG alloys. Conduct the logistic regression without any regularization and with L_2 regularization.

With the method demonstrated in Example 3.2, the logistic regression without any regularization is conducted, and Fig. 3.5 and the confusion matrix below (Table 3.5) show the classification result.

Systematically varying C values change the proportion of L_2 regularization term, and subsequently alters the decision boundary, which was shown below (Fig. 3.6).

Figure 3.7 indicates that with the increase of C from 10^{-2} to 10, the classification performance is getting closer and closer to that without regularization. The appropriate value of hyperparameter C should be determined by cross-validation. Figure 3.7 shows the plot of classification accuracy versus hyperparameter C in ten-fold cross-validation. As hyperparameter C increases from 10^{-4} to $10^{-3.1}$, the classification accuracy increases very slowly, and when hyperparameter C increases from $10^{-3.1}$ to $10^{-1.4}$, the classification accuracy increases fast. After the classification accuracy peak with $C = 10^{-1.4}$, the classification accuracy decreases very slowly with further increasing C. Therefore, the value of $C = 10^{-1.4}$ is the appropriate for the balance between classification accuracy and generalization of the logistic regression.

Table 3.4 Phases of 34 alloys

Alloys	H_{mix}	S_{mix}	Phases	Alloys	H_{mix}	S_{mix}	Phases
$Ag_{10.5}Cu_{80.5}P_9$	−5.09	5.22	CRA	$(Zr_{0.55}Al_{0.20}Co_{0.20}Cu_{0.05})_{95}Ag_5$	−40.10	10.52	BMG
$Ag_{10}Ce_3Cu_{87}$	−1.86	3.80	CRA	$(Zr_{0.55}Al_{0.20}Co_{0.20}Cu_{0.05})_{97}Ag_3$	−41.16	10.17	BMG
$Ag_{10}Ce_5Cu_{85}$	−3.49	4.31	CRA	$Fe_{41}Co_7Cr_{15}Mo_{14}C_{15}B_6Y_2$	−33.35	13.66	BMG
$Ag_{10}Mg_{50}Y_{40}$	−11.44	7.84	CRA	$La_{62}Al_{14}(Cu_5Ag_1)_{12}(NiCo)_{12}$	−27.50	10.12	BMG
$Ag_{10}Mg_{85}Y_5$	−5.00	4.31	CRA	$La_{62}Al_{14}(Cu_5Ag_1)_{16}(NiCo)_8$	−27.15	9.93	BMG
$Ag_{12.2}Cu_{81.5}P_{6.3}$	−3.37	4.97	CRA	$La_{62}Al_{14}(Cu_5Ag_1)_{20}(NiCo)_4$	−26.89	9.48	BMG
$Ag_{14}Cu_{78.5}P_{7.5}$	−4.02	5.48	CRA	$Zr_{65}Al_{7.5}Cu_{17.5}Ni_{10}$	−32.22	8.39	BMG
$Ag_{15}Ce_4Cu_{81}$	−2.47	4.86	CRA	$Au_2Zr_{48}Cu_{34}Al_8Ag_8$	−28.10	9.99	BMG
$Ag_{15}Cu_{83}Fe_2$	2.20	4.30	CRA	$Au_4Zr_{48}Cu_{32}Al_8Ag_8$	−30.46	10.39	BMG
$Ag_{20.9}Pd_{62.6}Si_{16.5}$	−29.15	7.63	CRA	$Cu_{38}Zr_{46}Ag_8Al_8$	−25.48	9.39	BMG
$Ag_{20}Ce_3Cu_{77}$	−1.43	5.22	CRA	$Hf_4Zr_{44}Cu_{36}Al_8Ag_8$	−25.22	10.49	BMG
$Ag_{20}Ce_4Cu_{76}$	−2.30	5.48	CRA	$La_{62}Al_{14}(Cu_5Ag_1)_{14}(NiCo)_{10}$	−27.31	10.06	BMG
$Ag_{20}Ce_5Cu_{75}$	−3.15	5.72	CRA	$Pd_6Zr_{48}Cu_{30}Al_8Ag_8$	−35.59	10.70	BMG
$Ag_{20}Mg_{20}Y_{60}$	−18.40	7.90	CRA	$Zr_{46}Cu_{37.64}Ag_{8.36}Al_8$	−25.46	9.43	BMG
$Cu_{34}Zr_{50}Ag_8Al_8$	−25.87	9.29	BMG	$Hf_2Zr_{46}Cu_{36}Al_8Ag_8$	−25.47	10.04	BMG
$Cu_{40}Zr_{44}Ag_8Al_8$	−25.18	9.41	BMG	$Ni_2Zr_{48}Cu_{34}Al_8Ag_8$	−26.66	9.99	BMG
$Hf_6Zr_{42}Cu_{36}Al_8Ag_8$	−24.97	10.85	BMG	$Ni_6Zr_{48}Cu_{30}Al_8Ag_8$	−28.58	10.70	BMG

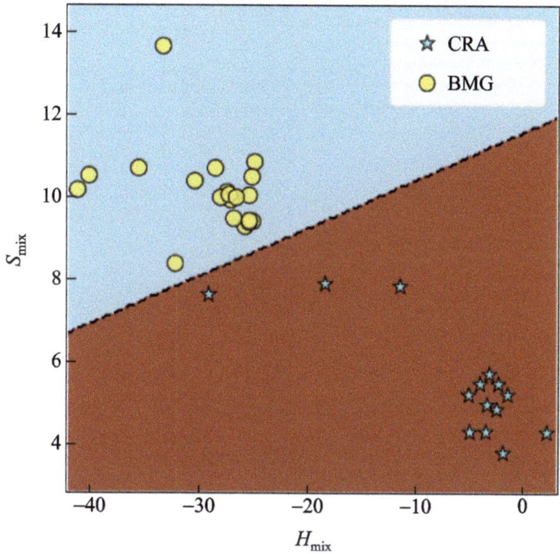

Fig. 3.5 The data in the (H_{mix}, S_{mix}) feature space. *Note* Cyan and yellow dots represent the CRA and BMG data, respectively; the dashed line is the separating hyperplane calculated by logistic regression

Table 3.5 Confusion matrix showing the classification result

Real phase	Predicted phase	
	CRA	BMG
CRA	14	0
BMG	0	20

Example 3.4 Table 3.6 shows 18 high entropy alloys collected from Xiong et al. (2021), with which the ROC curve with logistic regression will be plotted. The 18 alloys belong to single-phase (FCC) or multi-phases (FCC + BCC), and only two features are considered here, i.e., the average valance electron concentration ($\overline{\text{VEC}}$), or just called the valance electron concentration (VEC) for simplicity, and electronegativity difference ($\delta\chi$). The VEC and $\delta\chi$ are calculated by $\text{VEC} = \sum_{i=1} a_i \text{VEC}_i$ and $\delta\chi = \sqrt{\sum_{i=1} a_i \left(1 - \frac{\chi_i}{\sum_{i=1} a_i \chi_i}\right)^2}$, respectively, where a_i, VEC_i, and χ_i are the atomic percentage, valance electron concentration, and Pauling electronegativity of element i, respectively.

The phases FCC and FCC + BCC are labeled as class 1 and class 2, respectively. The logistic regression is conducted with LOOCV. In LOOCV, each datum is used for validation and the other rest data for training, and the average of 18 LOOCVs will be used as the classification result. Table 3.7 lists the probability values of 18 samples. Usually, the threshold is 0.5 in logistic regression. To plot the ROC curve, the value of threshold will be varied, and the probability values, after ranking from

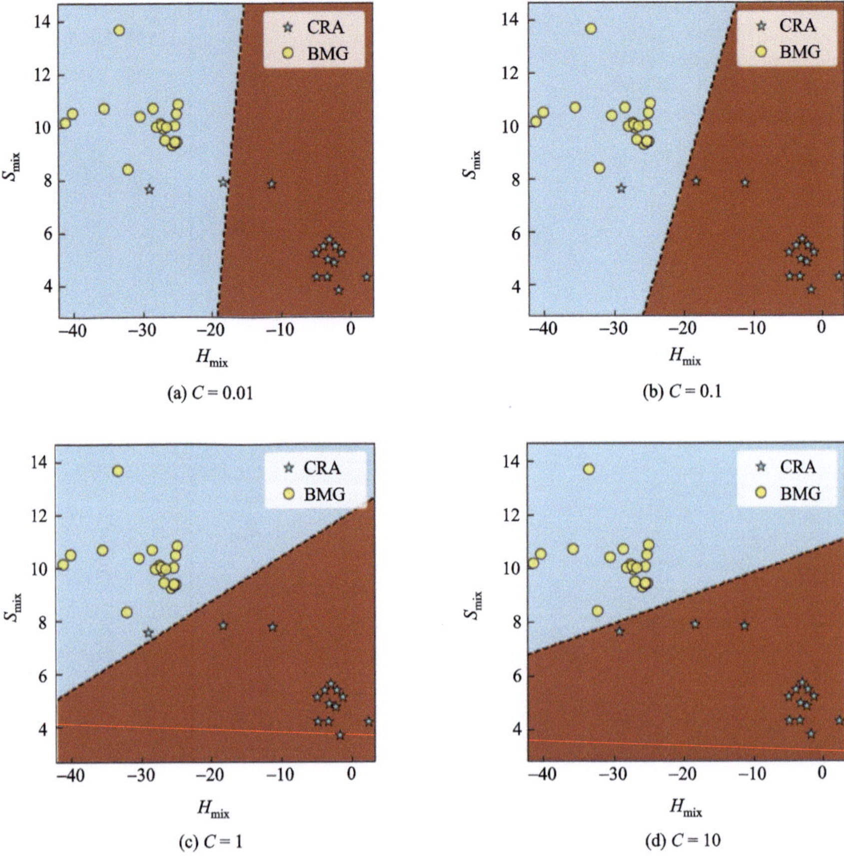

(a) $C = 0.01$ (b) $C = 0.1$ (c) $C = 1$ (d) $C = 10$

Fig. 3.6 The performance of logistic regression with penalty parameters of $C = 0.01$, 0.1, 1, and 10

high to low, either for class 1, $P(y = 1)$, or for class 2, $P\ (y = 2)$, can be taken as threshold values. Here, the ranked $P(y = 1)$ values plus an extreme $P(y = 1) = 1$ as h_0 are taken as the threshold values $\text{TH} = (h_0,\ h_1, \cdots,\ h_{18})$, as shown in Table 3.8. Then, the classification criterion is that if $P(y = 1) \geq \text{TH}$, the data are classified as class 1, otherwise, class 2. Thus, a given value of TH generates a classification result, which can be expressed by a confusion matrix, leading to the values of true positive rate (TPR) and false positive rate (FPR). After that, plot TPR versus FPR yields the ROC curve based on the LOOCV predictions, which is shown in Fig. 3.8. The size of AUC, which evaluates the classifier performance, is AUC = 0.779.

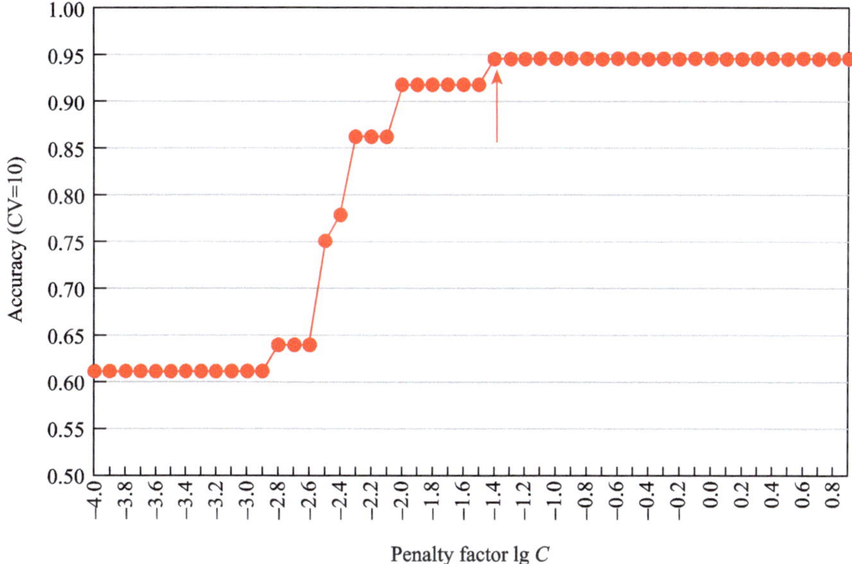

Fig. 3.7 Classification accuracy versus penalty parameter *C*. *Note* The arrow indicates cut-off position of penalty factor

3.3 Linear Discriminant Analysis

Linear Discriminant Analysis (LDA) of various classes is developed by Fisher (1936) and often called Fisher Discriminant Analysis as well. The ideal of LDA is that all data of various classes are projected on a straight line with the following goals. The data in each class should get together as closely as possible on the line, and the distance between different classes along the line should be as far as possible. For binary classification, consider $D = \{(\boldsymbol{X}_1, y_1), (\boldsymbol{X}_2, y_2), \cdots, (\boldsymbol{X}_n, y_n)\}$ of n separable data, where $\boldsymbol{X}_i = (x_{i1}, x_{i2}, \cdots, x_{im})(i = 1, 2, \cdots, n) \in \mathbf{R}^m$ denotes an $(1 \times m)$ input vector and $y \in \{0,\ 1\}$ is the label of the data. In the class 0 or 1, the features have their $(1 \times m)$ row mean vectors $\boldsymbol{\mu}_0$ and $\boldsymbol{\mu}_1$, and $(m \times m)$ covariance matrixes $\boldsymbol{\Sigma}_0$ and $\boldsymbol{\Sigma}_1$. The projections of mean vectors $\boldsymbol{\mu}_0$ and $\boldsymbol{\mu}_1$ on the straight line are located on the points of $\boldsymbol{\mu}_0\boldsymbol{w}$ and $\boldsymbol{\mu}_1\boldsymbol{w}$, where $\boldsymbol{w}$ is an $(m \times 1)$ vector of the line direction. When all n data are projected on the line, the n data will distribute on the one-dimensional line with the means of $\boldsymbol{\mu}_0\boldsymbol{w}$ and $\boldsymbol{\mu}_1\boldsymbol{w}$ at two points on the line, and the covariance matrixes $\boldsymbol{w}^{\mathrm{T}}\boldsymbol{\Sigma}_0\boldsymbol{w}$ and $\boldsymbol{w}^{\mathrm{T}}\boldsymbol{\Sigma}_1\boldsymbol{w}$ on the line, where $\boldsymbol{w}^{\mathrm{T}}\boldsymbol{\Sigma}_0\boldsymbol{w}$ and $\boldsymbol{w}^{\mathrm{T}}\boldsymbol{\Sigma}_1\boldsymbol{w}$ are scalars, representing lengths on the line. The requirement that the data in each class should get together as closely as possible on the line is expressed by each value of $\boldsymbol{w}^{\mathrm{T}}\boldsymbol{\Sigma}_0\boldsymbol{w}$ and $\boldsymbol{w}^{\mathrm{T}}\boldsymbol{\Sigma}_1\boldsymbol{w}$ as small as possible, and the requirement that the distance between different classes along the line should be as far as possible is expressed by the distance square of $\boldsymbol{w}^{\mathrm{T}}(\boldsymbol{\mu}_0 - \boldsymbol{\mu}_1)^{\mathrm{T}}(\boldsymbol{\mu}_0 - \boldsymbol{\mu}_1)\boldsymbol{w}$ as large as possible. Thus, LDA requires the maximization of the generalized Rayleigh quotient

Table 3.6 Eighteen high entropy alloys

Sample No.	Alloys	VEC	$\delta\chi$	Phases
1	$Co_1Cr_1Fe_1Mn_1Ni_1$	8.00	0.0783	FCC (1)
2	$Co_1Cr_1Fe_1Ni_1$	8.25	0.0531	FCC (1)
3	$Co_1Cr_1Mn_1Ni_1$	8.00	0.0860	FCC (1)
4	$Co_1Fe_1Mn_1Ni_1$	8.50	0.0797	FCC (1)
5	$Co_1Cr_1Ni_1$	8.33	0.0614	FCC (1)
6	$Co_1Fe_1Ni_1$	9.00	0.0176	FCC (1)
7	$Co_1Mn_1Ni_1$	8.67	0.0916	FCC (1)
8	$Fe_1Mn_1Ni_1$	8.33	0.0875	FCC (1)
9	$Co_1Cu_1Fe_1Ni_1$	9.50	0.0164	FCC (1)
10	$Co_1Fe_1Ni_1$	9.00	0.0176	FCC (1)
11	$Al_{0.6}Co_1Fe_1Ni_1Ti_{0.4}$	7.60	0.0725	FCC (1)
12	$Co_1Cr_2Fe_1Ni_1$	7.80	0.0602	FCC + BCC (2)
13	$Cr_1Fe_1Ni_1$	8.00	0.0579	FCC + BCC (2)
14	$Al_1Co_1Cr_1Cu_1Fe_1$	7.40	0.0667	FCC + BCC (2)
15	$Co_1Cr_1Cu_1Fe_1Ni_1$	8.80	0.0502	FCC + BCC (2)
16	$Co_1Cr_1Fe_1Ni_1$	8.25	0.0531	FCC + BCC (2)
17	$Al_{0.6}Co_1Cr_1Fe_1Ni_1Ti_{0.4}$	7.28	0.0731	FCC + BCC (2)
18	$Co_{0.5}Cr_1Fe_1Ni_1Ti_{0.5}$	7.63	0.0731	FCC + BCC (2)

$$\Re = \frac{\boldsymbol{w}^{\mathrm{T}}(\boldsymbol{\mu}_0 - \boldsymbol{\mu}_1)^{\mathrm{T}}(\boldsymbol{\mu}_0 - \boldsymbol{\mu}_1)\boldsymbol{w}}{\boldsymbol{w}^{\mathrm{T}}\boldsymbol{\Sigma}_0\boldsymbol{w} + \boldsymbol{w}^{\mathrm{T}}\boldsymbol{\Sigma}_1\boldsymbol{w}} = \frac{\boldsymbol{w}^{\mathrm{T}}(\boldsymbol{\mu}_0 - \boldsymbol{\mu}_1)^{\mathrm{T}}(\boldsymbol{\mu}_0 - \boldsymbol{\mu}_1)\boldsymbol{w}}{\boldsymbol{w}^{\mathrm{T}}(\boldsymbol{\Sigma}_0 + \boldsymbol{\Sigma}_1)\boldsymbol{w}}. \tag{3.10a}$$

Introducing the symmetric between-class scatter matrix $\boldsymbol{S}_{\mu} = \frac{1}{4}(\boldsymbol{\mu}_0 - \boldsymbol{\mu}_1)^{\mathrm{T}}(\boldsymbol{\mu}_0 - \boldsymbol{\mu}_1)$ and symmetric within-class scatter matrix $\boldsymbol{S}_{\Sigma} = \boldsymbol{\Sigma}_0 + \boldsymbol{\Sigma}_1$, we simplify Eq. (3.10a) to

$$\Re = \frac{\boldsymbol{w}^{\mathrm{T}}\boldsymbol{S}_{\mu}\boldsymbol{w}}{\boldsymbol{w}^{\mathrm{T}}\boldsymbol{S}_{\Sigma}\boldsymbol{w}}. \tag{3.10b}$$

The projection line is estimated by the maximization of the generalized Rayleigh quotient, i.e.,

$$\hat{\boldsymbol{w}} = \underset{\boldsymbol{w}}{\operatorname{argmax}} \frac{\boldsymbol{w}^{\mathrm{T}}\boldsymbol{S}_{\mu}\boldsymbol{w}}{\boldsymbol{w}^{\mathrm{T}}\boldsymbol{S}_{\Sigma}\boldsymbol{w}}. \tag{3.11a}$$

Since the constant factor of $\frac{1}{4}$ in the between-class scatter matrix does not affect the maximization, it will be ignored and $\boldsymbol{S}_{\mu} = (\boldsymbol{\mu}_0 - \boldsymbol{\mu}_1)^{\mathrm{T}}(\boldsymbol{\mu}_0 - \boldsymbol{\mu}_1)$ will be used. The maximization of the generalized Rayleigh quotient requires

$$\frac{\partial \Re}{\partial \boldsymbol{w}} = 2\frac{\boldsymbol{S}_{\mu}\boldsymbol{w}}{\boldsymbol{w}^{\mathrm{T}}\boldsymbol{S}_{\Sigma}\boldsymbol{w}} - 2\left[\frac{(\boldsymbol{\mu}_0 - \boldsymbol{\mu}_1)\boldsymbol{w}}{\boldsymbol{w}^{\mathrm{T}}\boldsymbol{S}_{\Sigma}\boldsymbol{w}}\right]^2 \boldsymbol{S}_{\Sigma}\boldsymbol{w} = 0. \tag{3.11b}$$

Table 3.7 The LOOCV prediction of logistic regression

Sample No.	Alloy compositions	$P(y = 1)$	$P(y = 2)$	Measured class	Predicted class
1	$Co_1Cr_1Fe_1Mn_1Ni_1$	0.5279	0.4721	FCC (1)	FCC (1)
2	$Co_1Cr_1Fe_1Ni_1$	0.5689	0.4311	FCC (1)	FCC (1)
3	$Co_1Cr_1Mn_1Ni_1$	0.5279	0.4721	FCC (1)	FCC (1)
4	$Co_1Fe_1Mn_1Ni_1$	0.6634	0.3366	FCC (1)	FCC (1)
5	$Co_1Cr_1Ni_1$	0.6213	0.3787	FCC (1)	FCC (1)
6	$Co_1Fe_1Ni_1$	0.7664	0.2336	FCC (1)	FCC (1)
7	$Co_1Mn_1Ni_1$	0.7016	0.2984	FCC (1)	FCC (1)
8	$Fe_1Mn_1Ni_1$	0.6214	0.3786	FCC (1)	FCC (1)
9	$Co_1Cu_1Fe_1Ni_1$	0.8455	0.1545	FCC (1)	FCC (1)
10	$Co_1Fe_1Ni_1$	0.7664	0.2336	FCC (1)	FCC (1)
11	$Al_{0.6}Co_1Fe_1Ni_1Ti_{0.4}$	0.3908	0.6092	FCC (1)	FCC + BCC (2)
12	$Co_1Cr_2Fe_1Ni_1$	0.5475	0.4525	FCC + BCC (2)	FCC (1)
13	$Cr_1Fe_1Ni_1$	0.5978	0.4022	FCC + BCC (2)	FCC (1)
14	$Al_1Co_1Cr_1Cu_1Fe_1$	0.4552	0.5448	FCC + BCC (2)	FCC + BCC (2)
15	$Co_1Cr_1Cu_1Fe_1Ni_1$	0.8180	0.1820	FCC + BCC (2)	FCC (1)
16	$Co_1Cr_1Fe_1Ni_1$	0.6647	0.3353	FCC + BCC (2)	FCC + BCC (2)
17	$Al_{0.6}Co_1Cr_1Fe_1Ni_1Ti_{0.4}$	0.4294	0.5706	FCC + BCC (2)	FCC (1)
18	$Co_{0.5}Cr_1Fe_1Ni_1Ti_{0.5}$	0.5071	0.4929	FCC + BCC (2)	FCC + BCC (2)

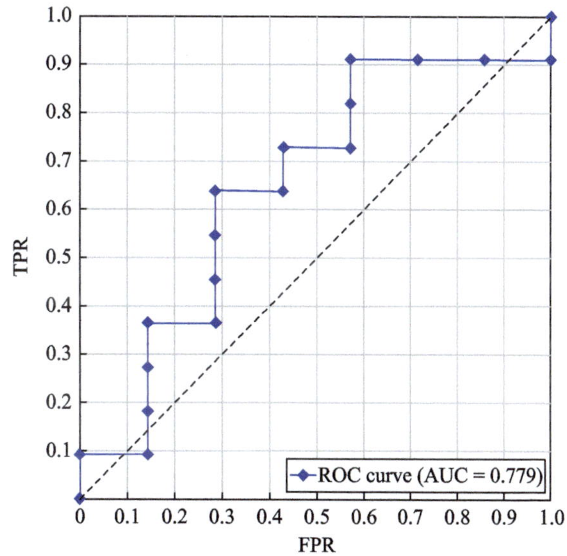

Fig. 3.8 The ROC curve of LOOCV predictions

Table 3.8 Calculation details of the ROC curve

Sample No.	Messured class	P(y = 1)	Predicted class of samples with different thresholds																		
			h_0	h_1	h_2	h_3	h_4	h_5	h_6	h_7	h_8	h_9	h_{10}	h_{11}	h_{12}	h_{13}	h_{14}	h_{15}	h_{16}	h_{17}	h_{18}
9	1	0.84549 (h_1)	2	1	1	1	1	1	1	1	1	1	1	1	1	1	1	1	1	1	1
15	2	0.81796 (h_2)	2	2	1	1	1	1	1	1	1	1	1	1	1	1	1	1	1	1	1
10	1	0.76644 (h_3)	2	2	2	1	1	1	1	1	1	1	1	1	1	1	1	1	1	1	1
6	1	0.76644 (h_4)	2	2	2	2	1	1	1	1	1	1	1	1	1	1	1	1	1	1	1
7	1	0.70165 (h_5)	2	2	2	2	2	1	1	1	1	1	1	1	1	1	1	1	1	1	1
16	2	0.66468 (h_6)	2	2	2	2	2	2	1	1	1	1	1	1	1	1	1	1	1	1	1
4	1	0.66336 (h_7)	2	2	2	2	2	2	2	1	1	1	1	1	1	1	1	1	1	1	1
8	1	0.62143 (h_8)	2	2	2	2	2	2	2	2	1	1	1	1	1	1	1	1	1	1	1
5	1	0.62134 (h_9)	2	2	2	2	2	2	2	2	2	1	1	1	1	1	1	1	1	1	1
13	2	0.59783 (h_{10})	2	2	2	2	2	2	2	2	2	2	1	1	1	1	1	1	1	1	1
2	1	0.56887 (h_{11})	2	2	2	2	2	2	2	2	2	2	2	1	1	1	1	1	1	1	1
12	2	0.54753 (h_{12})	2	2	2	2	2	2	2	2	2	2	2	2	1	1	1	1	1	1	1

(continued)

Table 3.8 (continued)

Sample No.	Messured class	P(y = 1)	Predicted class of samples with different thresholds																		
			h_0	h_1	h_2	h_3	h_4	h_5	h_6	h_7	h_8	h_9	h_{10}	h_{11}	h_{12}	h_{13}	h_{14}	h_{15}	h_{16}	h_{17}	h_{18}
3	1	0.52793 (h_{13})	2	2	2	2	2	2	2	2	2	2	2	2	2	1	1	1	1	1	1
1	1	0.52792 (h_{14})	2	2	2	2	2	2	2	2	2	2	2	2	2	2	1	1	1	1	1
18	2	0.50712 (h_{15})	2	2	2	2	2	2	2	2	2	2	2	2	2	2	2	1	1	1	1
14	2	0.45524 (h_{16})	2	2	2	2	2	2	2	2	2	2	2	2	2	2	2	2	1	1	1
17	2	0.42939 (h_{17})	2	2	2	2	2	2	2	2	2	2	2	2	2	2	2	2	2	1	1
11	1	0.39075 (h_{18})	2	2	2	2	2	2	2	2	2	2	2	2	2	2	2	2	2	2	1
	FPR		0.00	0.00	0.14	0.14	0.14	0.14	0.29	0.29	0.29	0.29	0.43	0.43	0.57	0.57	0.57	0.71	0.86	1.00	1.00
	TPR		0.00	0.09	0.09	0.18	0.27	0.36	0.36	0.45	0.55	0.64	0.64	0.73	0.73	0.82	0.91	0.91	0.91	0.91	1.00

Equation (3.11b) yields

$$w = \frac{w^{\mathrm{T}} S_{\Sigma} w}{(\mu_0 - \mu_1) w} (\Sigma_0 + \Sigma_1)^{-1} (\mu_0 - \mu_1)^{\mathrm{T}}. \tag{3.11c}$$

For a given dataset, $(\mu_0 - \mu_1)w$, $w^{\mathrm{T}} S_{\Sigma} w$, and $\frac{w^{\mathrm{T}} S_{\Sigma} w}{(\mu_0 - \mu_1) w}$ are all scalars, which do not have any influence on the line direction. Therefore, the line direction is determined by

$$w = (\Sigma_0 + \Sigma_1)^{-1} (\mu_0 - \mu_1)^{\mathrm{T}}. \tag{3.11d}$$

Since only the line direction is interested, it usually takes the normalized w in LDA calculation, i.e.,

$$w = \frac{w}{|w|}, \tag{3.11e}$$

where the normalized w is also denoted by w for simplicity.

Example 3.5 Table 3.9 lists 4 data of CRA and BMG alloys. Recall the $n \times n$ symmetric centering matrix $H = \left(I - \frac{1}{n} \mathbf{1}_n \mathbf{1}_n^{\mathrm{T}}\right)$ with I of the $n \times n$ unit matrix and $\mathbf{1}_n = \begin{pmatrix} 1 \\ \vdots \\ 1 \end{pmatrix}_{n \times 1}$ has the idempotent property of $H^n = H$. The $m \times m$ covariance matrix S is given by

$$S = \frac{1}{n-1} (HX)^{\mathrm{T}} HX = \frac{1}{n-1} X^{\mathrm{T}} HX.$$

The original data for feature inputs are normalized into the (0, 1) range, as listed in Table 3.10.

Take the data in Table 3.10 to calculate the decision boundary of LDA.

Table 3.9 Formed phases of 4 alloys

Data	Alloys	H_{mix}	S_{mix}	Phases
X_1	$Ag_{10.5}Cu_{80.5}P_9$	−5.09	5.22	CRA (class 0)
X_2	$Ag_{10}Ce_5Cu_{85}$	−3.49	4.31	CRA (class 0)
X_3	$Cu_{34}Zr_{50}Ag_8Al_8$	−25.87	9.29	BMG (class 1)
X_4	$(Zr_{0.55}Al_{0.20}Co_{0.20}Cu_{0.05})_{95}Ag_5$	−40.10	10.52	BMG (class 1)

Table 3.10 The data with normalized feature inputs

Data	Alloys	H_{mix}	S_{mix}	Phases
X_1	$Ag_{10.5}Cu_{80.5}P_9$	0.96	0.15	CRA (class 0)
X_2	$Ag_{10}Ce_5Cu_{85}$	1	0	CRA (class 0)
X_3	$Cu_{34}Zr_{50}Ag_8Al_8$	0.39	0.80	BMG (class 1)
X_4	$(Zr_{0.55}Al_{0.20}Co_{0.20}Cu_{0.05})_{95}Ag_5$	0	1	BMG (class 1)

The data matrix is $\boldsymbol{X} = \begin{bmatrix} X_{11} & X_{12} \\ X_{21} & X_{22} \\ X_{31} & X_{32} \\ X_{41} & X_{42} \end{bmatrix} = \begin{bmatrix} 0.96 & 0.15 \\ 1 & 0 \\ 0.39 & 0.80 \\ 0 & 1 \end{bmatrix}$, and the label vector is $\boldsymbol{Y} = \begin{pmatrix} y_1 \\ y_2 \\ y_3 \\ y_4 \end{pmatrix} = \begin{pmatrix} 0 \\ 0 \\ 1 \\ 1 \end{pmatrix}$. For each class, there are only two data, and hence the centering matrix for $n = 2$ takes the explicit form.

For CRA with labelled $Y = 0$, the covariance matrix is $\boldsymbol{\Sigma}_0 = \frac{1}{n_{\mathrm{CRA}}-1}\boldsymbol{X}_{\mathrm{CRA}}^{\mathrm{T}}\boldsymbol{H}\boldsymbol{X}_{\mathrm{CRA}} = \frac{1}{2-1}\begin{bmatrix} 0.96 & 0.15 \\ 1 & 0 \end{bmatrix}^{\mathrm{T}}\begin{bmatrix} 0.5 & -0.5 \\ -0.5 & 0.5 \end{bmatrix}\begin{bmatrix} 0.96 & 0.15 \\ 1 & 0 \end{bmatrix} = \begin{bmatrix} 0.0008 & -0.003 \\ -0.003 & 0.01125 \end{bmatrix}$, and the mean vector is $\boldsymbol{\mu}_0 = \frac{1}{2}(X_{11} + X_{21}\ X_{12} + X_{22}) = (0.98\ 0.075)$.

For BMG with labelled $Y = 1$, the covariance matrix is

$$\begin{aligned} \boldsymbol{\Sigma}_1 &= \frac{1}{n_{\mathrm{BMG}} - 1}\boldsymbol{X}_{\mathrm{BMG}}^{\mathrm{T}}\boldsymbol{H}\boldsymbol{X}_{\mathrm{BMG}} \\ &= \frac{1}{2-1}\begin{bmatrix} 0.39 & 0.80 \\ 0 & 1 \end{bmatrix}^{\mathrm{T}}\begin{bmatrix} 0.5 & -0.5 \\ -0.5 & 0.5 \end{bmatrix}\begin{bmatrix} 0.39 & 0.80 \\ 0 & 1 \end{bmatrix} \\ &= \begin{bmatrix} 0.076 & -0.039 \\ -0.039 & 0.02 \end{bmatrix}, \end{aligned}$$

and the mean vector is $\boldsymbol{\mu}_1 = \frac{1}{2}(X_{31} + X_{41}\ X_{32} + X_{42}) = (0.195\ 0.9)$.

With Eq. (3.7d), we have $\boldsymbol{w} = (\boldsymbol{\Sigma}_0 + \boldsymbol{\Sigma}_1)^{-1}(\boldsymbol{\mu}_0 - \boldsymbol{\mu}_1)^{\mathrm{T}} = \left(\begin{bmatrix} 0.0008 & -0.003 \\ -0.003 & 0.01125 \end{bmatrix} + \begin{bmatrix} 0.076 & -0.039 \\ -0.039 & 0.02 \end{bmatrix}\right)^{-1} \cdot ((0.98\ 0.075) - (0.195\ 0.9))^{\mathrm{T}} = \begin{pmatrix} -15.9 \\ -47.7 \end{pmatrix}$. $\boldsymbol{w}$ is then normalized as a unit vector, $\boldsymbol{w} = \begin{pmatrix} -0.32 \\ -0.95 \end{pmatrix}$. The mean points of class CRA and class BMG on the projected line are calculated as

$$\dot{\mu}_0 = \boldsymbol{\mu}_0\boldsymbol{w} = (0.98\ 0.075)\begin{pmatrix} -0.32 \\ -0.95 \end{pmatrix} = -0.38,$$

$$\dot{\mu}_1 = \boldsymbol{\mu}_1 \boldsymbol{w} = (0.195\ 0.9)\begin{pmatrix} -0.32 \\ -0.95 \end{pmatrix} = -0.92.$$

The decision boundary based on LDA usually takes the plane in which the normal direction is given by $\boldsymbol{w}$ and passes the midpoint B between the two means on $\boldsymbol{w}$ (James et al., 2013):

$$B = \frac{1}{2}(\dot{\mu}_0 + \dot{\mu}_1) = -\frac{1}{2}(0.38 + 0.92) = -0.65.$$

Figure 3.9 shows the LDA result.

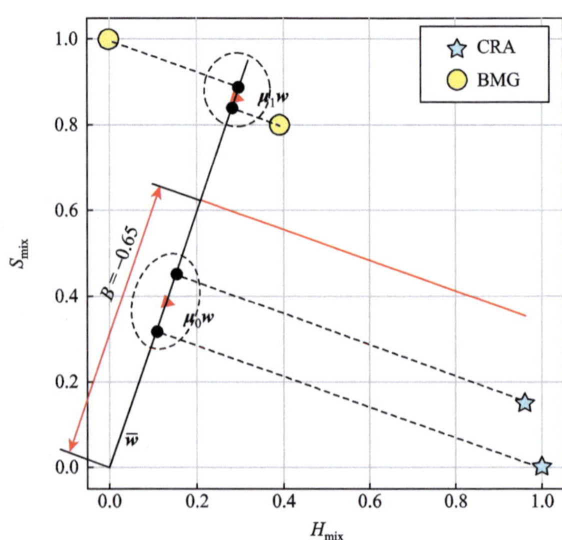

Fig. 3.9 The LDA result in the normalized (H_{mix}, S_{mix}) feature space. *Note* Cyan and yellow dots represent the CRA and BMG data, respectively; the black and red lines are the projection line and the decision boundary, respectively

---

If there are K classes, the mean of class means is defined by

$$\boldsymbol{\mu} = \frac{1}{K}\sum_{k=1}^{K} \boldsymbol{\mu}_k. \tag{3.12a}$$

The whole between-class scatter matrix is defined by

$$\mathbf{S}_{\mu} = \sum_{k=1}^{K} \omega_k (\boldsymbol{\mu}_k - \boldsymbol{\mu})^{\mathrm{T}} (\mu_k - \mu), \tag{3.12b}$$

where ω_k is the class weight for class k, which is the ratio of data number N_k in class k over the total data number N, viz., $\omega_k = \frac{N_k}{N}$. The whole within-class scatter matrix is defined by

$$S_{\Sigma} = \sum_{k=1}^{K} \Sigma_k = \sum_{k=1}^{K} \sum_{i=1}^{N_k} (x_{ki} - \mu_k)^{\mathrm{T}} (x_{ki} - \mu_k), \tag{3.12c}$$

where x_{ki} $(i = 1, 2, \cdots, N_k)$ are the data in class k, and $\boldsymbol{\Sigma}_k$ is the covariance matrix of data in class k $(k = 1, 2, \cdots, K)$.

In analogy with binary classification LDA, the generalized Rayleigh quotient for multiple classes LDA is still given by Eq. (3.10b), except that the whole between-class scatter matrix and the whole within-class scatter matrix replace the between-class scatter matrix and the within-class scatter matrix, respectively. Maximizing the generalized Rayleigh quotient means that the data in each class should get together as closely as possible on the projection line, and the distance of each class along the line should be as far as possible from the mean of class means.

In addition to projecting the data on a straight line, we may consider to project the original feature data $\boldsymbol{X}_i = (x_{i1}, x_{i2}, \cdots, x_{im})(i = 1, 2, \cdots, n) \in \mathbf{R}^m$ in the m-dimensional space to an M-dimensional space $M \leq m$. In this case, the $(m \times 1)$ vector $\boldsymbol{w}$ is expanded to an $(m \times M)$ matrix $\boldsymbol{W}$, which is named as the projection matrix. Then, the generalized Rayleigh quotient is given by

$$\Re = \frac{\mathrm{tr}\left(\boldsymbol{W}^{\mathrm{T}} \boldsymbol{S}_{\mu} \boldsymbol{W}\right)}{\mathrm{tr}\left(\boldsymbol{W}^{\mathrm{T}} \boldsymbol{S}_{\Sigma} \boldsymbol{W}\right)}. \tag{3.13a}$$

The projection matrix is determined by the maximal generalized Rayleigh quotient, i.e.,

$$\hat{\boldsymbol{W}} = \underset{W}{\mathrm{argmax}} \frac{\mathrm{tr}\left(\boldsymbol{W}^{\mathrm{T}} \boldsymbol{S}_{\mu} \boldsymbol{W}\right)}{\mathrm{tr}\left(\boldsymbol{W}^{\mathrm{T}} \boldsymbol{S}_{\Sigma} \boldsymbol{W}\right)}. \tag{3.13b}$$

The maximal generalized Rayleigh quotient requires

$$\frac{\partial \Re}{\partial \boldsymbol{W}} = \frac{\partial\, \mathrm{tr}\left(\left(\boldsymbol{W}^{\mathrm{T}} \boldsymbol{S}_{\Sigma} \boldsymbol{W}\right)^{-1} \boldsymbol{W}^{\mathrm{T}} \boldsymbol{S}_{\mu} \boldsymbol{W}\right)}{\partial \boldsymbol{W}} = 0, \tag{3.13c}$$

which gives

$$-2\boldsymbol{S}_{\Sigma} \hat{\boldsymbol{W}} \left(\hat{\boldsymbol{W}}^{\mathrm{T}} \boldsymbol{S}_{\Sigma} \hat{\boldsymbol{W}}\right)^{-1} \hat{\boldsymbol{W}}^{\mathrm{T}} \boldsymbol{S}_{\mu} \hat{\boldsymbol{W}} \left(\hat{\boldsymbol{W}}^{\mathrm{T}} \boldsymbol{S}_{\Sigma} \hat{\boldsymbol{W}}\right)^{-1} + 2\boldsymbol{S}_{\mu} \hat{\boldsymbol{W}} \left(\hat{\boldsymbol{W}}^{\mathrm{T}} \boldsymbol{S}_{\Sigma} \hat{\boldsymbol{W}}\right)^{-1} = 0. \tag{3.13d}$$

Equation (3.13d) is rewritten as

$$\boldsymbol{S}_{\Sigma} \hat{\boldsymbol{W}} \left(\hat{\boldsymbol{W}}^{\mathrm{T}} \boldsymbol{S}_{\Sigma} \hat{\boldsymbol{W}}\right)^{-1} \hat{\boldsymbol{W}}^{\mathrm{T}} \boldsymbol{S}_{\mu} \hat{\boldsymbol{W}} = \boldsymbol{S}_{\mu} \hat{\boldsymbol{W}}. \tag{3.13e}$$

The solution to Eq. (3.13e) is given by

$$\lambda S_{\Sigma} \hat{W} = S_{\mu} \hat{W}. \tag{3.14}$$

Equation (3.14) is a generalized eigenvalue equation (Horn & Johnson, 2012) of matrix $S_{\Sigma}^{-1} S_{\mu}$, where $\hat{W}$ is the eigenvector and λ is the eigenvalue.

Example 3.6 Table 3.11 lists the normalized feature inputs for two classes, where are three features and six observations normalized in the (0, 1) range. Calculate the decision boundary of LDA in two-dimensional space.

The three features mean that the original data are 3-dimensional. The original data matrix is $X = \begin{bmatrix} X_{11} & X_{12} & X_{13} \\ X_{21} & X_{22} & X_{23} \\ X_{31} & X_{32} & X_{33} \\ X_{41} & X_{42} & X_{43} \\ X_{51} & X_{52} & X_{53} \\ X_{61} & X_{62} & X_{63} \end{bmatrix} = \begin{bmatrix} 0 & 0.17 & 0 \\ 0.49 & 0 & 0.23 \\ 0.86 & 0.69 & 0.23 \\ 0.88 & 0.52 & 0.91 \\ 1 & 1 & 0.58 \\ 0.99 & 0.92 & 1 \end{bmatrix}$, and the label vector is $Y = \begin{pmatrix} y_1 \\ y_2 \\ y_3 \\ y_4 \\ y_5 \\ y_6 \end{pmatrix} = \begin{pmatrix} 0 \\ 0 \\ 0 \\ 1 \\ 1 \\ 1 \end{pmatrix}$.

For class $Y = 0$, the mean vector is calculated as $\boldsymbol{\mu}_0 = \frac{1}{3}(X_{11} + X_{21} + X_{31} \;\; X_{12} + X_{22} + X_{32} \;\; X_{13} + X_{23} + X_{33}) = (0.45\ 0.29\ 0.15)$. In the same way, we have $\boldsymbol{\mu}_1 = (0.96\ 0.81\ 0.83)$. The mean of class means is calculated with Eq. (3.12a) to be $\boldsymbol{\mu} = \frac{1}{2}\sum_{k=0}^{1} \boldsymbol{\mu}_k = \frac{1}{2}(\boldsymbol{\mu}_0 + \boldsymbol{\mu}_1) = (0.71\ 0.55\ 0.49)$. Then, the whole between-class scatter matrix is derived by Eq. (3.12b) as $S_{\mu} = \sum_{k=0}^{1} \omega_i (\boldsymbol{\mu}_k - \boldsymbol{\mu})^{\mathrm{T}} (\boldsymbol{\mu}_k - \boldsymbol{\mu}) = \frac{3}{6}(\boldsymbol{\mu}_0 - \boldsymbol{\mu})^{\mathrm{T}}(\boldsymbol{\mu}_0 - \boldsymbol{\mu}) + \frac{3}{6}(\boldsymbol{\mu}_1 - \boldsymbol{\mu})^{\mathrm{T}}(\boldsymbol{\mu}_1 - \boldsymbol{\mu}) = \begin{bmatrix} 0.065\ 0.066\ 0.087 \\ 0.066\ 0.068\ 0.088 \\ 0.087\ 0.088\ 0.116 \end{bmatrix}$. The whole within-class scatter matrix is calculated by

Table 3.11 Six data with normalized feature inputs belonging to two classes

Datum	Feature 1	Feature 2	Feature 3	Class
X_1	0	0.17	0	0
X_2	0.49	0	0.23	0
X_3	0.86	0.69	0.23	0
X_4	0.88	0.52	0.91	1
X_5	1	1	0.58	1
X_6	0.99	0.92	1	1

$$
\begin{aligned}
\boldsymbol{S}_{\boldsymbol{\Sigma}} &= \sum_{k=1}^{2}\sum_{i=1}^{N_k}(\boldsymbol{x}_{ki}-\boldsymbol{\mu}_k)^{\mathrm{T}}(\boldsymbol{x}_{ki}-\boldsymbol{\mu}_k) \\
&= [(\boldsymbol{X}_1-\boldsymbol{\mu}_0)^{\mathrm{T}}(\boldsymbol{X}_1-\boldsymbol{\mu}_0)+(\boldsymbol{X}_2-\boldsymbol{\mu}_0)^{\mathrm{T}}(\boldsymbol{X}_2-\boldsymbol{\mu}_0)+(\boldsymbol{X}_3-\boldsymbol{\mu}_0)^{\mathrm{T}}(\boldsymbol{X}_3-\boldsymbol{\mu}_0)] \\
&\quad +[(\boldsymbol{X}_4-\boldsymbol{\mu}_1)^{\mathrm{T}}(\boldsymbol{X}_4-\boldsymbol{\mu}_1)+(\boldsymbol{X}_5-\boldsymbol{\mu}_1)^{\mathrm{T}}(\boldsymbol{X}_5-\boldsymbol{\mu}_1)+(\boldsymbol{X}_6-\boldsymbol{\mu}_1)^{\mathrm{T}}(\boldsymbol{X}_6-\boldsymbol{\mu}_1)] \\
&= \begin{bmatrix} 0.381 & 0.241 & 0.092 \\ 0.241 & 0.391 & -0.025 \\ 0.092 & -0.025 & 0.133 \end{bmatrix}.
\end{aligned}
$$

With $\boldsymbol{S}_{\mu}$ and $\boldsymbol{S}_{\boldsymbol{\Sigma}}$, the projection matrix is the eigenvector matrix of $\boldsymbol{S}_{\boldsymbol{\Sigma}}^{-1}\boldsymbol{S}_{\mu}$. The first and second unit vectors are solved as $\hat{\boldsymbol{w}}_1 = (-0.31\ 0.40\ 0.87)$ and $\hat{\boldsymbol{w}}_2 = (0.85\ -0.44\ -0.30)$, corresponding to the eigenvalues of 1.41 and 0.00014, respectively. The vectors $\hat{\boldsymbol{w}}_1$ and $\hat{\boldsymbol{w}}_2$ construct a two-dimensional space, and the six data are projected on the two-dimensional space, expressed by $\boldsymbol{X}$.

$$
\boldsymbol{X} = \boldsymbol{X} \times \hat{\boldsymbol{W}} = \begin{bmatrix} 0 & 0.17 & 0 \\ 0.49 & 0 & 0.23 \\ 0.86 & 0.69 & 0.23 \\ 0.88 & 0.52 & 0.91 \\ 1 & 1 & 0.58 \\ 0.99 & 0.92 & 1 \end{bmatrix} \times \begin{bmatrix} -0.31 & 0.85 \\ 0.40 & -0.44 \\ 0.87 & -0.30 \end{bmatrix} = \begin{bmatrix} 0.068 & -0.0748 \\ 0.0482 & 0.3475 \\ 0.2095 & 0.3584 \\ 0.7269 & 0.2462 \\ 0.5946 & 0.236 \\ 0.9311 & 0.1367 \end{bmatrix}.
$$

Figure 3.10 shows the LDA result.

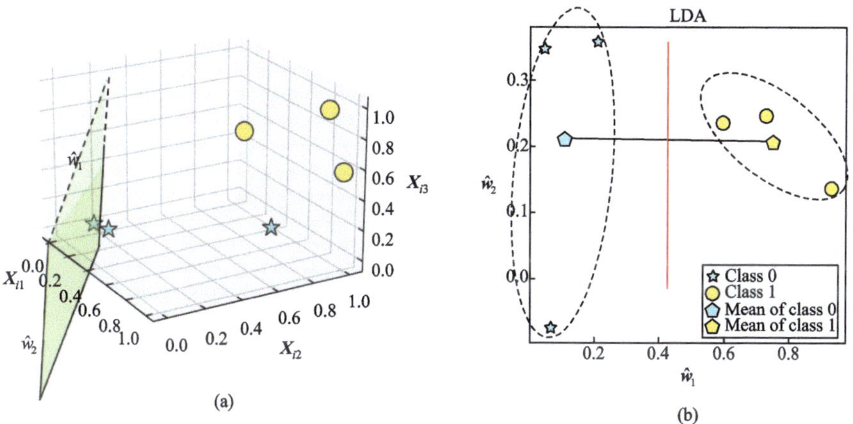

Fig. 3.10 **(a)** The six data in the three-dimensional space. **(b)** the six data projected in the LDA two-dimensional space. *Note* Cyan and yellow dots represent the class 0 and class 1 data, respectively; the light green plane is the projection plane of LDA; the red line is the intersection line of decision boundary plane

Homework

Table 3.12 shows 36 high-entropy alloys (HEA), including 18 body-centered cubic HEAs (BCC) and 18 face-centered cubic HEAs (FCC). The valence electron concentration (VEC) and enthalpy of mixing (H_{mix}) are the two features considered in the example, and VEC is calculated by $\text{VEC} = \sum_{i=1}^{N} a_i \text{VEC}_i$, where a_i and VEC_i are the atomic fraction and VEC of the i-th constituent, respectively.

1. With VEC and H_{mix}, please use perceptron, logistic regression, and LDA to classify the structures of FCC and BCC, respectively, and give the confusion matrixes.
2. Following the above question, try to change the regularization factor C in logistic regression, and figure out how C influences the classification results, e.g., the confusion matrix.
3. Following the above two questions, compare and analyze the similarities and differences of the three decision-making boundaries, and write down your comprehension.
4. The logical "Exclusive OR" of true is defined as that an event is true ($y = 1$) if only one feature is true, viz., data of (1, 0) and (0, 1) are in true class $y = 1$, and data of (0, 0) and (1, 1) are in false class $y = 0$. Try to draw an easy structure of perceptron to realize this logical operator.
5. Please code a small program to print out the complete iteration path of coordinate descent in Example 3.2 following its derivation.
6. Suppose we transform the original predictors $\boldsymbol{X}$ to $\hat{\boldsymbol{Y}}$ via linear regression. In detail, let $\hat{\boldsymbol{Y}} = \boldsymbol{X}(\boldsymbol{X}^{\text{T}}\boldsymbol{X})^{-1}\boldsymbol{X}^{\text{T}}\boldsymbol{Y} = \boldsymbol{X}\hat{\boldsymbol{B}}$, where $\boldsymbol{Y}$ is the response matrix. Show that LDA using $\hat{\boldsymbol{Y}}$ is identical to LDA in the original space, and give you derivation (Trevor et al., 2005).

Table 3.12 Solid solution structure of 36 high-entropy alloys

Alloys	VEC	H_{mix}	Structures	Alloys	VEC	H_{mix}	Structures
$Al_{13.64}Co_{18.18}Cr_{18.18}Cu_{4.55}Fe_{18.18}Ni_{18.18}Ti_{9.09}$	7.27	−14.39	BCC	$Al_{7.69}Co_{30.77}Fe_{30.77}Ni_{30.77}$	8.54	−6.06	FCC
$Al_{12.70}Cr_{15.87}Cu_{15.87}Fe_{15.87}Mn_{23.81}Ni_{15.87}$	7.60	−4.23	BCC	$Al_{3.85}Cr_{19.23}Cu_{19.23}Fe_{19.23}Ni_{38.46}$	8.77	0.12	FCC
$Al_{17.39}Co_{17.39}Cr_{17.39}Cu_{4.35}Fe_{17.39}Ni_{17.39}Ti_{8.70}$	7.09	−15.50	BCC	$Co_5Cr_2Fe_{40}Mn_{27}Ni_{26}$	8.26	−3.58	FCC
$Al_{16.67}Co_{16.67}Cr_{16.67}Cu_{8.33}Fe_{16.67}Ni_{16.67}Ti_{8.33}$	7.25	−13.42	BCC	$Co_{25}Cr_{25}Cu_{25}Fe_{25}$	8.50	6.25	FCC
$Al_{18.52}Co_{18.52}Cr_{18.52}Fe_{18.52}Ni_{18.52}Si_{7.41}$	6.96	−19.84	BCC	$Co_{25}Cr_{25}Fe_{25}Ni_{25}$	8.25	−3.75	FCC
$Al_{18.18}Co_{18.18}Cr_{18.18}Fe_{18.18}Ni_{18.18}Ti_{9.09}$	6.91	−17.92	BCC	$Co_{25}Cr_{25}Mn_{25}Ni_{25}$	8.00	−5.50	FCC
$Al_{16.67}Cr_{16.67}Cu_{16.67}Fe_{16.67}Mn_{16.67}Ni_{16.67}$	7.50	−5.11	BCC	$Co_{33.33}Cr_{33.33}Ni_{33.33}$	8.33	−4.89	FCC
$Al_{16.67}Hf_{16.67}Nb_{16.67}Ta_{16.67}Ti_{16.67}Zr_{16.67}$	4.17	−14.78	BCC	$Co_{25}Cu_{25}Fe_{25}Ni_{25}$	9.50	5.00	FCC
$Cr_{28.57}Mo_{14.29}Nb_{14.29}Ta_{14.29}V_{14.29}W_{14.29}$	5.57	−4 .82	BCC	$Co_{24.88}Cu_{24.88}Fe_{24.88}Ni_{24.88}Sn_{0.50}$	9.47	5.02	FCC
$Al_{5.66}Hf_{18.87}Nb_{18.87}Ta_{18.87}Ti_{18.87}Zr_{18.87}$	4.32	−3.99	BCC	$Co_{25}Fe_{25}Mn_{25}Ni_{25}$	8.50	−4.00	FCC
$Al_{9.09}Hf_{18.18}Nb_{18.18}Ta_{18.18}Ti_{18.18}Zr_{18.18}$	4.27	−7.67	BCC	$Co_{33.33}Fe_{33.33}Ni_{33.33}$	9.00	−1.33	FCC
$Al_{13.04}Hf_{17.39}Nb_{17.39}Ta_{17.39}Ti_{17.39}Zr_{17.39}$	4.22	−11.55	BCC	$Co_{30.77}Fe_{30.77}Ni_{30.77}Si_{7.69}$	8.62	−11.83	FCC

(continued)

Table 3.12 (continued)

Alloys	VEC	H_{mix}	Structures	Alloys	VEC	H_{mix}	Structures
$Al_{20}Co_{10}Cr_{10}Cu_{10}Fe_{10}Mn_{10}Ni_{10}Ti_{10}V_{10}$	6.60	−15.44	BCC	$Co_{25}Fe_{25}Ni_{25}V_{25}$	8.00	−10.50	FCC
$Cr_{9.09}Mo_{18.18}Nb_{18.18}Ta_{18.18}V_{18.18}W_{18.18}$	5.45	−4.83	BCC	$Co_{33.33}Mn_{33.33}Ni_{33.33}$	8.67	−5.78	FCC
$Cr_{16.67}Mo_{16.67}Nb_{16.67}Ta_{16.67}V_{16.67}W_{16.67}$	5.50	−4.89	BCC	$Cr_{18.03}Fe_{27.32}Mn_{27.32}Ni_{27.32}$	7.91	−4.17	FCC
$Hf_{18.18}Mo_{9.09}Nb_{18.18}Ta_{18.18}Ti_{18.18}Zr_{18.18}$	4.55	0.60	BCC	$Cr_{25}Cu_{25}Fe_{25}Ni_{25}$	8.57	3.01	FCC
$Hf_{17.39}Mo_{13.04}Nb_{17.39}Ta_{17.39}Ti_{17.39}Zr_{17.39}$	4.61	−0.21	BCC	$Fe_{33.33}Mn_{33.33}Ni_{33.33}$	8.33	−4.44	FCC
$Hf_{16.67}Nb_{16.67}Ta_{16.67}Ti_{16.67}V_{16.67}Zr_{16.67}$	4.50	0.78	BCC	$Co_9Cr_7Cu_{36}Mn_{25}Ni_{23}$	9.24	2.05	FCC

References

Boyd, S., Boyd, S. P., & Vandenberghe, L. (2004). *Convex optimization*. Cambridge University Press.

Fisher, R. A. (1936). The use of multiple measurements in taxonomic problems. *Annals of Eugenics, 7*(2), 179–188.

Horn, R. A., & Johnson, C. R. (2012). *Matrix analysis*. Cambridge University Press.

James, G., Witten, D., Hastie, T., et al. (2013). *An introduction to statistical learning*. Springer.

Li, H. (2012). *Statistical learning method (统计学习方法)*. Tsinghua University Press.

Nocedal, J., & Wright, S. (2006). *Numerical optimization*. Springer Science & Business Media.

Rosenblatt, F. (1958). The perceptron: A probabilistic model for information storage and organization in the brain. *Psychological Review, 65*(6), 386.

Rosenblatt, F., & Papert, S. (1957). The perceptron. A perceiving and recognizing automation. *Cornell Aeronautical Laboratory Report, 85*, 460.

Spackman, K. A. (1989). Signal detection theory: Valuable tools for evaluating inductive learning. In *Proceedings of the sixth international workshop on machine learning* (pp. 160–163). Morgan Kaufmann.

Takeuchi, A., & Inoue, A. (2005). Classification of bulk metallic glasses by atomic size difference, heat of mixing and period of constituent elements and its application to characterization of the main alloying element. *Materials Transactions, 46*(12), 2817–2829.

Trevor, H., Robert, T., & Jerome, F. (2005). The elements of statistical learning. *Mathematical Intelligencer, 27*(2), 83–85.

Xiong, J., Shi, S. Q., & Zhang, T. Y. (2020). A machine-learning approach to predicting and understanding the properties of amorphous metallic alloys. *Materials & Design, 187*, 108378.

Xiong, J., Shi, S. Q., & Zhang, T. Y. (2021). Machine learning of phases and mechanical properties in complex concentrated alloys. *Journal of Materials Science & Technology, 87*, 133–142.

Chapter 4
Support Vector Machine

Support vector machines (SVMs) are widely used in classification and regression, and thus are called Support Vector Classification/Classifier (SVC) and Support Vector Regression/Regressor (SVR) (Cortes & Vapnik, 1995; Suykens & Vandewalle, 1999). The boundary between two different classes is, in SVC, determined by support vectors, as described in the following section, and therefore SVC might be appropriate for small data. The characteristic of SVR lies in the tolerance of data error, whose degree is controlled by a preset positive parameter ε.

4.1 SVC

Consider a dataset $D = \{(\boldsymbol{X}_1, y_1), (\boldsymbol{X}_2, y_2), \cdots, (\boldsymbol{X}_n, y_n)\}$ of n separable data, where $\boldsymbol{X}_i = (x_{i1}, x_{i2}, \cdots, x_{im})(i = 1, 2, \cdots, n) \in \mathbf{R}^m$ denotes an $(1 \times m)$ input vector, and $y \in \{-1, +1\}$ is the label of the data. Figure 4.1 schematically shows the two-category data with each occupying one side. There are many hyperplanes in the m-feature space that can separate the two categories of data. SVC determines the optimal separating hyperplane that has the largest separation or margin between the two categories, as shown in Fig. 4.1.

In the m-feature space, a separating hyperplane has a normal vector $\boldsymbol{W} = (w_1, w_2, \cdots, w_m)^{\mathrm{T}}$ and an intercept b. Any point on the hyperplane in the m-feature space is mathematically described by

$$\hat{\boldsymbol{X}}\boldsymbol{W} + b = 0, \tag{4.1}$$

where $\hat{\boldsymbol{X}}$ denotes all points on this hyperplane. In the m-feature space, there exist many hyperplanes satisfying Eq. (4.1) separating the two categories of data, and thus called separating hyperplanes. Equation (4.1) indicates that the values of ($\boldsymbol{W}$, b) determine a hyperplane in the m-feature space. If a point $\boldsymbol{X}$ is out of the ($\boldsymbol{W}$, b)

T. Zhang, *An Introduction to Materials Informatics*,
https://doi.org/10.1007/978-981-99-7992-9_4

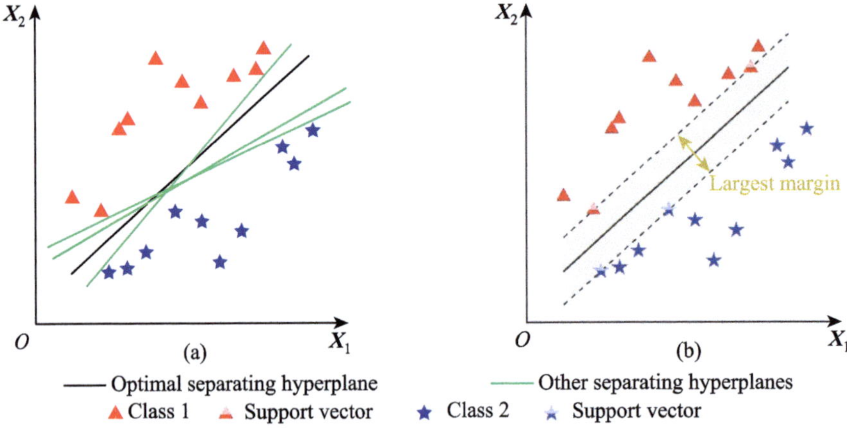

Fig. 4.1 Schematic plot of SVC. *Note* Red triangles denote one category of data and the blue stars denote another category of data. **(a)** without margins, and **(b)** with the largest margin

hyperplane, the distance between the point and the hyperplane is given by

$$r = \frac{|\boldsymbol{X}\boldsymbol{W} + b|}{\|\boldsymbol{W}\|}. \tag{4.2}$$

A separating hyperplane separates the training data, meaning that for one datum $(\boldsymbol{X}_i, y_i)$, $y_i = +1$ corresponds to $\boldsymbol{X}_i\boldsymbol{W} + b > 0$ and $y_i = -1$ corresponds to $\boldsymbol{X}_i\boldsymbol{W} + b < 0$. Then, the classification rule is given by

$$\boldsymbol{f}_{\boldsymbol{W},\,b}(\boldsymbol{X}_i) = \text{sign}(\boldsymbol{X}_i\boldsymbol{W} + b). \tag{4.3}$$

The optimal hyperplane is determined by maximizing margin (M) between the training points (Rosset et al., 2003), i.e.,

$$\max_{\boldsymbol{W},\,b} M \tag{4.4a}$$

$$\text{subject to } y_i \tfrac{\boldsymbol{X}_i\boldsymbol{W}+b}{\|\boldsymbol{W}\|} \geq M,\ i = 1,\, 2, \cdots, n,$$

equivalently,

$$\text{subject to } y_i(\boldsymbol{X}_i\boldsymbol{W} + b) \geqslant M\|\boldsymbol{W}\|,\ i = 1, 2, \cdots, n.$$

With an appropriate scale, we have

$$\boldsymbol{X}_i\boldsymbol{W} + b \geqslant 1 \text{ for } y_i = +1, \tag{4.4b}$$

$$\boldsymbol{X}_i\boldsymbol{W} + b \leqslant -1 \text{ for } y_i = -1. \tag{4.4c}$$

Then, the distance of the closest datum to the hyperplane is $r = 1/\|W\|$, and the margin is given by $2/\|W\|$, as shown in Fig. 4.1. Thus, maximizing margin (M) is converted to minimize $\|W\|^2$.

$$\min_{W,\,b} \frac{1}{2}\|W\|^2 \tag{4.5a}$$

$$\text{subject to } y_i(X_i W + b) \geqslant 1, \quad i = 1, 2, \cdots, n.$$

Equation (4.5a) is a convex quadratic programming problem (Boyd & Vandenberghe, 2004), and can be written in the Lagrange function to be minimized as

$$\min L_\text{P} = \min\left\{\frac{1}{2}\|W\|^2 - \sum_{i=1}^{n} \lambda_i[y_i(X_i W + b) - 1]\right\}, \tag{4.5b}$$

where $\{\lambda_1, \lambda_2, \cdots, \lambda_n\}$ are the Lagrange multipliers, one for one datum and $\lambda_i \geqslant 0$. The following Karush–Kuhn–Tucker (KKT) conditions (Nocedal & Wright, 1999) (complementary conditions) must be satisfied in solving the minimization problem of Eq. (4.5b)

$$\begin{cases} \lambda_i \geq 0, \\ y_i(X_i W + b) \geq 1, \\ \lambda_i[y_i(X_i W + b) - 1] = 0 \ \text{ for any } i. \end{cases}$$

The minimization of L_P requires $\frac{\partial L_\text{P}}{\partial W} = 0$ and $\frac{\partial L_P}{\partial b} = 0$, which yield

$$W^\text{T} = \sum_{i=1}^{n} \lambda_i X_i y_i, \tag{4.6a}$$

$$\sum_{i=1}^{n} \lambda_i y_i = 0. \tag{4.6b}$$

When $\lambda_i \neq 0$, we get $y_i(X_i W + b) = 1$. The vectors fulfilling this requirement are called support vectors since they lie closest to the separating hyperplane, as shown in Fig. 4.1. Substituting Eq. (4.6a, b) into Eq. (4.5b) yields the dual objective function

$$L_\text{D} = \sum_{i=1}^{n} \lambda_i - \frac{1}{2}\sum_{i=1}^{n}\sum_{j=1}^{n} \lambda_i \lambda_j y_i y_j X_j X_i^\text{T} (\lambda_i \geqslant 0). \tag{4.7a}$$

The values of λ_i ($i = 1, 2, \cdots, n$) are determined by maximizing the dual objective function, i.e.,

$$\hat{\lambda}_i = \underset{\lambda}{\text{argmax}}\, L_\text{D},$$

$$\text{subject to} \sum_{i=1}^{n} \lambda_i y_i = 0 \text{ and } \lambda_i \geqslant 0, i = 1, 2, \cdots, n. \tag{4.7b}$$

The KKT conditions described above must also be satisfied in solving the strong dual problems (Nocedal & Wright, 1999).

Consequently, the values of $\boldsymbol{W}$ are determined by Eq. (4.6a) with the data of support vectors. The value of b is also determined by the data of support vectors. It holds on every datum of support vectors, $y_s(\boldsymbol{X}_s\boldsymbol{W} + b) = 1$, i.e.,

$$y_s\left(\sum_i \lambda_i y_i \boldsymbol{X}_s \boldsymbol{X}_i^{\mathrm{T}} + b\right) = 1, \tag{4.8a}$$

where S being the data subset of support vectors. Let $|S|$ denote the number of support vectors, and $|S|$ b values will be determined by Eq. (4.8a). The average of $|S|$ b values is taken as the final b value, i.e.,

$$b = \frac{1}{|S|}\sum_{s\in S}\left(\frac{1}{y_s} - \sum_i \lambda_i y_i \boldsymbol{X}_s \boldsymbol{X}_i^{\mathrm{T}}\right). \tag{4.8b}$$

Example 4.1 Table 4.1 shows 4 high-entropy alloys (HEAs), including 2 face-centered cubic (FCC) and 2 not FCC (N-FCC) HEAs, collected from Xiong et al. (2020).

The average valence electron concentration (VEC) and enthalpy of mixing (H_{mix}) (Takeuchi & Inoue, 2005) are the two features considered in the example, which are calculated by, respectively,

$$\text{VEC} = \sum_{i=1}^{N} a_i \text{VEC}_i, \tag{4.9}$$

$$H_{\text{mix}} = 4\sum_{j=i}^{N}\sum_{i=1}^{N} \Delta H_{ij} a_i a_j. \tag{4.10}$$

Table 4.1 Formed phases of 4 HEAs

Samples	VEC (x_1)	H_{mix} (x_2, kJ·mol^{-1})	Formed phases	Class
$\text{Co}_{0.25}\text{Cr}_{0.25}$FeMn	7.5	−1.0	FCC	FCC (1)
CoCuFeMnNi	9	1.8	FCC	FCC (1)
DyGdLuTbTm	3	0	HCP	N-FCC (−1)
NbTaTiVW	5	−3.7	BCC	N-FCC (−1)

where a_i and VEC_i are the atomic fraction and VEC of the i-th constituent, respectively. ΔH_{ij} is the molar mixing enthalpy for binary liquid alloys. With the two features of VEC and H_{mix}, conduct SVM classification.

The values of Lagrange multipliers $\hat{\lambda}_i$ are determined by the maximum of the dual objective function, as stated by Eq. (4.7b). Equation (4.7a) shows the dual objective function L_{D}. In this case, $(\boldsymbol{X}_j, y_j)$ are $\begin{bmatrix} \boldsymbol{X}_1 & y_1 \\ \boldsymbol{X}_2 & y_2 \\ \boldsymbol{X}_3 & y_3 \\ \boldsymbol{X}_4 & y_4 \end{bmatrix} = \begin{bmatrix} 7.5 & -1.0 & 1 \\ 9 & 1.8 & 1 \\ 3 & 0 & -1 \\ 5 & -3.7 & -1 \end{bmatrix}$ and $\begin{pmatrix} \boldsymbol{X}_i^{\text{T}} \\ y_i \end{pmatrix} = \begin{bmatrix} 7.5 & 9 & 3 & 5 \\ -1.0 & 1.8 & 0 & -3.7 \\ 1 & 1 & -1 & -1 \end{bmatrix}$ and hence L_{D} is $L_{\text{D}} = \lambda_1 + \lambda_2 + \lambda_3 + \lambda_4 - \frac{1}{2}(57.3\lambda_1^2 + 84.2\lambda_2^2 + 9\lambda_3^2 + 38.7\lambda_4^2 + 131.4\lambda_1\lambda_2 - 45\lambda_1\lambda_3 - 82.4\lambda_1\lambda_4 - 54\lambda_2\lambda_3 - 76.7\lambda_2\lambda_4 + 30\lambda_3\lambda_4)$.

Equation (4.6b) shows that the values of λ_i should satisfy the condition of $\sum_{i=1}^{n} \lambda_i y_i = 0$, which gives $\lambda_3 = \lambda_1 + \lambda_2 - \lambda_4$. With this relationship, the dual objective function L_{D} is rewritten as $L_{\text{D}} = 2\lambda_1 + 2\lambda_2 - 10.65\lambda_1^2 - 19.6\lambda_2^2 - 8.85\lambda_4^2 - 25.2\lambda_1\lambda_2 + 12.7\lambda_1\lambda_4 + 5.35\lambda_2\lambda_4$.
The maximization of L_{D} requires

$$\begin{cases} \dfrac{\partial L_{\text{D}}}{\partial \lambda_1} = 2 - 21.3\lambda_1 - 25.2\lambda_2 + 12.7\lambda_4 = 0, \\ \dfrac{\partial L_{\text{D}}}{\partial \lambda_2} = 2 - 25.2\lambda_1 - 39.2\lambda_2 + 5.35\lambda_4 = 0, \\ \dfrac{\partial L_{\text{D}}}{\partial \lambda_4} = 12.7\lambda_1 + 5.35\lambda_2 - 17.7\lambda_4 = 0, \end{cases}$$

which yields

$$\begin{cases} \lambda_1 = 18.65, \\ \lambda_2 = -10.55, \\ \lambda_4 = 10.20. \end{cases}$$

However, this set of solutions do not satisfy the condition of $\lambda_i \geqslant 0$. Therefore, one Lagrange multiplier must be zero, viz., $\lambda_1 = 0, \lambda_2 = 0, \lambda_3 = 0$, or $\lambda_4 = 0$. If $\lambda_1 = 0$, then

$$\begin{cases} \lambda_3 = \lambda_2 - \lambda_4, \\ \dfrac{\partial L_{\text{D}}}{\partial \lambda_2} = -39.2\lambda_2 + 5.35\lambda_4 + 2 = 0, \\ \dfrac{\partial L_{\text{D}}}{\partial \lambda_4} = 5.35\lambda_2 - 17.7\lambda_4 = 0, \end{cases}$$

and the values of Lagrange multipliers are

$$\begin{cases} \lambda_1 = 0, \\ \lambda_2 = 0.05322, \\ \lambda_3 = 0.03714, \\ \lambda_4 = 0.01608, \text{ and the maximum } L_{\mathrm{D}}\text{value } (L_{\mathrm{D}})_{\max|_{\lambda_1=0}} = 0.05322. \end{cases}$$

Similarly, the other three cases give the solutions of

$$\begin{cases} \lambda_1 = 0.1641, \\ \lambda_2 = 0, \\ \lambda_3 = 0.0464, \\ \lambda_4 = 0.1177, \text{ and the maximum } L_{\mathrm{D}}\text{value } (L_{\mathrm{D}})_{\max|_{\lambda_2=0}} = 0.1641. \end{cases}$$

However, $\lambda_3 = 0$ and $\lambda_4 = 0$ lead to negative values of λ_2, violating the condition of $\lambda_i \geqslant 0$. This means that the optimized solution should be one with $\lambda_1 = 0$ or $\lambda_2 = 0$. Since $(L_{\mathrm{D}})_{\max|_{\lambda_2=0}} = 0.1641 > (L_{\mathrm{D}})_{\max|_{\lambda_1=0}} = 0.05322$, the four values of Lagrange multipliers are determined to be $\lambda_2 = 0$, $\lambda_1 = 0.1641$, $\lambda_3 = 0.0464$, and $\lambda_4 = 0.1177$. Since there are three nonzero Lagrange multipliers, the corresponding three data of $\boldsymbol{X}_1$, $\boldsymbol{X}_3$, and $\boldsymbol{X}_4$ are support vectors. Then, the separating hyperplane is determined by

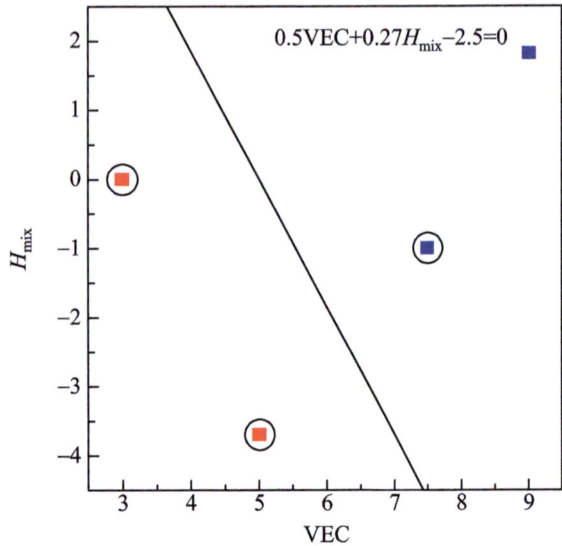

Fig. 4.2 The data in the (VEC and H_{mix}) feature space. *Note* Blue and red dots represent the N-FCC and FCC data, respectively; the solid line is the optimal separating hyperplane; the support vectors are marked by circles

$$
\begin{aligned}
\boldsymbol{W}^{\mathrm{T}} &= \sum_{i=1}^{4} \lambda_i \boldsymbol{X}_i y_i \\
&= 0.1641 \times 1 \times (7.5, -1) + 0.0464 \times (-1) \\
&\times (3, 0) + 0.1177 \times (-1) \times (5, -3.7) \\
&= (0.503, 0.271), \\
b &= \frac{(y_1 - X_1 \boldsymbol{W}) + (y_3 - X_3 \boldsymbol{W}) + (y_4 - X_4 \boldsymbol{W})}{3} \approx -2.5,
\end{aligned}
$$

$$
\text{i.e., } 0.50x_1 + 0.27x_2 - 2.5 = 0.
$$

Figure 4.2 shows the classification results with the separating hyperplane.

4.2 Kernel Functions

If data are not separable in the direct-feature space $\boldsymbol{X}$, but separable in the high-dimensional space $\boldsymbol{F}$ mapped by $\phi(\boldsymbol{X})$, the SVC model will be

$$
f_{\boldsymbol{W},b}(\boldsymbol{X}_i) = \operatorname{sign}(\phi(\boldsymbol{X}_i)\boldsymbol{W} + b). \tag{4.11}
$$

Then, Eqs. (4.5a) and (4.7b) are rewritten as

$$
\begin{gathered}
\min_{\boldsymbol{W},\, b} \frac{1}{2} \|\boldsymbol{W}\|^2 \\
\text{subject to } y_i(\phi(\mathbf{X}_i)\mathbf{W} + b) \geqslant 1, \quad i = 1, 2, \cdots, n.
\end{gathered} \tag{4.12a}
$$

$$
\begin{gathered}
\max \left[\sum_{i=1}^{n} \lambda_i - \frac{1}{2} \sum_{i=1}^{n} \sum_{j=1}^{n} \lambda_i \lambda_j y_i y_j \phi(\boldsymbol{X}_j)(\phi(\boldsymbol{X}_i))^{\mathrm{T}} \right] \\
\text{subject to } \sum_{i=1}^{n} \lambda_i y_i = 0 \text{ and } \lambda_i \geqslant 0, \quad i = 1, 2, \cdots, n,
\end{gathered} \tag{4.12b}
$$

respectively. Introducing the kernel function by

$$
K(\boldsymbol{X}_i, \boldsymbol{X}_j) = \phi(\boldsymbol{X}_i)(\phi(\boldsymbol{X}_j))^{\mathrm{T}}, \tag{4.13a}
$$

we have the final solution

$$
f(\boldsymbol{X}) = \operatorname{sign}\left(\sum_{i=1}^{n} \lambda_i y_i K(\boldsymbol{X}, \boldsymbol{X}_i) + b \right). \tag{4.13b}
$$

Table 4.2 lists some commonly-used kernel functions (Hofmann et al., 2008).

Table 4.2 Commonly-used kernel functions

Names	Functions	Parameters
Linear kernel	$K(\boldsymbol{X}_i, \boldsymbol{X}_j) = \boldsymbol{X}_i \boldsymbol{X}_j^{\mathrm{T}}$	
Polynomial kernel	$K(\boldsymbol{X}_i, \boldsymbol{X}_j) = \left(\boldsymbol{X}_i \boldsymbol{X}_j^{\mathrm{T}}\right)^d$	$d \geqslant 1$, the polynomial number
Gaussian radial basis function (RBF) kernel	$K(\boldsymbol{X}_i, \boldsymbol{X}_j) = \exp\left(-\frac{\Vert\boldsymbol{X}_i - \boldsymbol{X}_j\Vert^2}{2\sigma^2}\right)$	$\sigma > 0$, the width of RBF kernel
Laplace kernel	$K(\boldsymbol{X}_i, \boldsymbol{X}_j) = \exp\left(-\frac{\Vert\boldsymbol{X}_i - \boldsymbol{X}_j\Vert}{\sigma}\right)$	$\sigma > 0$
Sigmoid kernel	$K(\boldsymbol{X}_i, \boldsymbol{X}_j) = \tanh\left(\beta \boldsymbol{X}_i \boldsymbol{X}_j^{\mathrm{T}} + \theta\right)$	tanh is the hyperbolic tangent function, $\beta > 0, \theta < 0$.

Note The $\Vert\cdot\Vert$ represents the Euclidean distance of argument vector

With n data, the kernel function $K(\boldsymbol{X}_i, \boldsymbol{X}_j)(i, j = 1, 2, \cdots, n)$ can expressed by the kernel matrix of

$$K(\boldsymbol{X}_i, \boldsymbol{X}_j) = \begin{bmatrix} k(\boldsymbol{X}_1, \boldsymbol{X}_1) & k(\boldsymbol{X}_1, \boldsymbol{X}_2) \cdots & k(\boldsymbol{X}_1, \boldsymbol{X}_n) \\ k(\boldsymbol{X}_2, \boldsymbol{X}_1) & k(\boldsymbol{X}_2, \boldsymbol{X}_2) \ldots & k(\boldsymbol{X}_2, \boldsymbol{X}_n) \\ \vdots & \vdots & \vdots \\ k(\boldsymbol{X}_n, \boldsymbol{X}_1) & k(\boldsymbol{X}_n, \boldsymbol{X}_2) \cdots & k(\boldsymbol{X}_n, \boldsymbol{X}_n) \end{bmatrix}.$$

New kernel functions can be generated from available kernel functions. For example, a new kernel function can be formed from the linear combination of other kernel functions as, $K_\eta(\boldsymbol{X}_i, \boldsymbol{X}_j) = \gamma_1 K_1(\boldsymbol{X}_i, \boldsymbol{X}_j) + \gamma_2 K_2(\boldsymbol{X}_i, \boldsymbol{X}_j)$, where $K_1(\boldsymbol{X}_i, \boldsymbol{X}_j)$ and $K_2(\boldsymbol{X}_i, \boldsymbol{X}_j)$ are kernel functions, and γ_1 and γ_2 are positive coefficients, viz.,

$$K_\eta(\boldsymbol{X}_i, \boldsymbol{X}_j) = \begin{bmatrix} \begin{matrix}\gamma_1 k_1(\boldsymbol{X}_1, \boldsymbol{X}_1) \\ +\gamma_2 k_2(\boldsymbol{X}_1, \boldsymbol{X}_1)\end{matrix} & \begin{matrix}\gamma_1 k_1(\boldsymbol{X}_1, \boldsymbol{X}_2) \\ +\gamma_2 k_2(\boldsymbol{X}_1, \boldsymbol{X}_2)\end{matrix} & \cdots & \begin{matrix}\gamma_1 k_1(\boldsymbol{X}_1, \boldsymbol{X}_n) \\ +\gamma_2 k_2(\boldsymbol{X}_1, \boldsymbol{X}_n)\end{matrix} \\ \begin{matrix}\gamma_1 k_1(\boldsymbol{X}_2, \boldsymbol{X}_1) \\ +\gamma_2 k_2(\boldsymbol{X}_2, \boldsymbol{X}_1)\end{matrix} & \begin{matrix}\gamma_1 k_1(\boldsymbol{X}_2, \boldsymbol{X}_2) \\ +\gamma_2 k_2(\boldsymbol{X}_2, \boldsymbol{X}_2)\end{matrix} & \cdots & \begin{matrix}\gamma_1 k_1(\boldsymbol{X}_2, \boldsymbol{X}_n) \\ +\gamma_2 k_2(\boldsymbol{X}_2, \boldsymbol{X}_n)\end{matrix} \\ \vdots & \vdots & & \vdots \\ \begin{matrix}\gamma_1 k_1(\boldsymbol{X}_n, \boldsymbol{X}_1) \\ +\gamma_2 k_2(\boldsymbol{X}_n, \boldsymbol{X}_1)\end{matrix} & \begin{matrix}\gamma_1 k_1(\boldsymbol{X}_n, \boldsymbol{X}_2) \\ +\gamma_2 k_2(\boldsymbol{X}_n, \boldsymbol{X}_2)\end{matrix} & \cdots & \begin{matrix}\gamma_1 k_1(\boldsymbol{X}_n, \boldsymbol{X}_n) \\ +\gamma_2 k_2(\boldsymbol{X}_n, \boldsymbol{X}_n)\end{matrix} \end{bmatrix}.$$

The Hadamard product (element-wise product) between two kernel matrices $K_1(\boldsymbol{X}_i, \boldsymbol{X}_j)$ and $K_2(\boldsymbol{X}_i, \boldsymbol{X}_j)$ gives a new kernel function $K_\theta(\boldsymbol{X}_i, \boldsymbol{X}_j) = K_1(\boldsymbol{X}_i, \boldsymbol{X}_j) \odot K_2(\boldsymbol{X}_i, \boldsymbol{X}_j)$, which is expressed by

$$K_\theta(\boldsymbol{X}_i,\boldsymbol{X}_j)=\begin{bmatrix} k_1(\boldsymbol{X}_1,\boldsymbol{X}_1) & k_1(\boldsymbol{X}_1,\boldsymbol{X}_2) & \cdots & k_1(\boldsymbol{X}_1,\boldsymbol{X}_n) \\ \times k_2(\boldsymbol{X}_1,\boldsymbol{X}_1) & \times k_2(\boldsymbol{X}_1,\boldsymbol{X}_2) & & \times k_2(\boldsymbol{X}_1,\boldsymbol{X}_n) \\ k_1(\boldsymbol{X}_2,\boldsymbol{X}_1) & k_1(\boldsymbol{X}_2,\boldsymbol{X}_2) & \cdots & k_1(\boldsymbol{X}_2,\boldsymbol{X}_n) \\ \times k_2(\boldsymbol{X}_2,\boldsymbol{X}_1) & \times k_2(\boldsymbol{X}_2,\boldsymbol{X}_2) & & \times k_2(\boldsymbol{X}_2,\boldsymbol{X}_n) \\ \vdots & \vdots & & \vdots \\ k_1(\boldsymbol{X}_n,\boldsymbol{X}_1) & k_1(\boldsymbol{X}_n,\boldsymbol{X}_2) & \cdots & k_1(\boldsymbol{X}_n,\boldsymbol{X}_n) \\ \times k_2(\boldsymbol{X}_n,\boldsymbol{X}_1) & \times k_2(\boldsymbol{X}_n,\boldsymbol{X}_2) & & \times k_2(\boldsymbol{X}_n,\boldsymbol{X}_n) \end{bmatrix}.$$

With a kernel function, the dual objective function takes the form

$$L_{\mathrm{D}}=\sum_{i=1}^{n}\lambda_i-\frac{1}{2}\sum_{i=1}^{n}\sum_{j=1}^{n}\lambda_i\lambda_j y_i y_j K\left(\mathbf{X}_i,\mathbf{X}_j\right)\quad(\lambda_i\geqslant 0). \tag{4.14a}$$

The values of λ_i $(i=1,2,\cdots,n)$ are determined by maximizing the dual objective function, i.e.,

$$\hat{\lambda}_i=\underset{\lambda}{\operatorname{argmax}}\, L_{\mathrm{D}},$$

$$\text{subject to}\ \sum_{i=1}^{n}\lambda_i y_i=0\ \text{and}\ \lambda_i\geqslant 0\quad i=1,2,\cdots,n. \tag{4.14b}$$

The KKT conditions described above must also be satisfied in solving the strong dual problems (Nocedal & Wright, 1999). With the data, Eq. (4.14a, b) can be solved, and every one of support vectors satisfies

$$y_s\left(\sum_i\lambda_i y_i K(\boldsymbol{X}_s,\boldsymbol{X}_i)+b\right)=1, \tag{4.15a}$$

where S being the data subset of support vectors. Let $|S|$ denote the number of support vectors, and $|S|$ b values will be determined by Eq. (4.8a). The average of $|S|$ b values is taken as the final b value, i.e.,

$$b=\frac{1}{|S|}\sum_{s\in S}\left(\frac{1}{y_s}-\sum_i\lambda_i y_i K(\boldsymbol{X}_s,\boldsymbol{X}_i)\right). \tag{4.15b}$$

To illustrate the common kernel functions, we consider the original n data of $\boldsymbol{X}_i=(x_{i1})(i=1,2,\cdots,n)$ with only one feature. In this case, $K(\boldsymbol{X}_i,\boldsymbol{X}_j)=\left(\boldsymbol{X}_i\boldsymbol{X}_j^{\mathrm{T}}\right)^d=(x_{i1}x_{j1})^d=x_{i1}^d\left(x_{j1}^d\right)^{\mathrm{T}}$, indicating that the data are mapped from the original feature space to a new space $\phi(\boldsymbol{X})$ with $\phi(\boldsymbol{X}_i)=\left(x_{i1}^d\right)(i=1,2,\cdots,n)$. If the original data have two features, $\boldsymbol{X}_i=(x_{i1},x_{i2})(i=1,2,\cdots,n)$. Considering the simple polynomial kernel of $d=2$, we have $K(\boldsymbol{X}_i,\boldsymbol{X}_j)=\left(\boldsymbol{X}_i\boldsymbol{X}_j^{\mathrm{T}}\right)^2=$

$\left((x_{i1}\ x_{i2})\begin{pmatrix}x_{j1}\\x_{j2}\end{pmatrix}\right)^2 = (x_{i1}x_{j1} + x_{i2}x_{j2})^2 = x_{i1}^2x_{j1}^2 + x_{i2}^2x_{j2}^2 + 2x_{i1}x_{j1}x_{i2}x_{j2} = \left(x_{i1}^2\ x_{i2}^2\ \sqrt{2}x_{i1}x_{i2}\right)\left(x_{j1}^2\ x_{j2}^2\ \sqrt{2}x_{j1}x_{j2}\right)^{\mathrm{T}}(i, j = 1, 2, \cdots, n)$. In this case, the new space $\phi(\boldsymbol{X})$ is three-dimensional, and $\phi(\boldsymbol{X}_i) = \left(x_{i1}^2\ x_{i2}^2\ \sqrt{2}x_{i1}x_{i2}\right)$ $(i = 1, 2, \cdots, n)$. If the exponent number is $d = 3$, the polynomial kernel $K(\boldsymbol{X}_i, \boldsymbol{X}_j) = \left(\boldsymbol{X}_i\boldsymbol{X}_j^{\mathrm{T}}\right)^3$ will be $\left(x_{i1}^3\ \sqrt{3}x_{i1}^2x_{i2}\ \sqrt{3}x_{i1}x_{i2}^2\ x_{i2}^3\right)\left(x_{i1}^3\ \sqrt{3}x_{i1}^2x_{i2}\ \sqrt{3}x_{i1}x_{i2}^2\ x_{i2}^3\right)^{\mathrm{T}}$ with $\phi(\boldsymbol{X}_i) = \left(x_{i1}^3\ \sqrt{3}x_{i1}^2x_{i2}\ \sqrt{3}x_{i1}x_{i2}^2\ x_{i2}^3\right)(i = 1, 2, \cdots, n)$ becoming four-dimensional. Similarly, we shall have the high space with polynomial kernel $d \geqslant 4$. In general, the high space associated with the Gaussian radial basis function (RBF) kernel, Laplace kernel, and Sigmoid kernel has infinite and continuous dimensions, like that in Laplace transformation and Fourier transformation. Since we are interested in kernel $K(\boldsymbol{X}_i, \boldsymbol{X}_j)$, rather than the mapping function of $\phi(\boldsymbol{X})$, the explicit form of $\phi(\boldsymbol{X})$ will not be discussed in detail in SVM. The parameter (or parameters) in a kernel is called hyperparameter, and an SVM model is developed, with a preset value of hyperparameter, from a training dataset. The optimal value of hyperparameter is usually determined by cross-validation.

Example 4.2 The polynomial kernel function $\left(K(\boldsymbol{X}_i, \boldsymbol{X}_j) = \left(\boldsymbol{X}_i\boldsymbol{X}_j^{\mathrm{T}}\right)^d, d = 2\right)$ maps the original two-dimensional data listed in Table 4.3 to a three-dimensional space with the mapping function $\phi(\boldsymbol{X}_i) = \left(x_{i1}^2\ x_{i2}^2\ \sqrt{2}x_{i1}x_{i2}\right)(i = 1, 2, \cdots, n)$. Figure 4.3 shows that the inseparable data in the original two-dimensional space become separable in the three-dimensional space.

Table 4.3 Original data with two features

Sample No.	x_1	x_2	Class	Sample No.	x_1	x_2	Class
1	0	0.1	−1	15	−1.2	0.1	1
2	0.4	0.2	−1	16	−1.4	0.15	1
3	0.3	0.3	−1	17	−0.2	1.2	1
4	−0.1	0.2	−1	18	−0.5	0.9	1
5	−0.4	0.1	−1	19	−1.2	−0.1	1
6	−0.4	−0.4	−1	20	−1	−1	1
7	−0.2	−0.1	−1	21	−0.2	−1.5	1
8	0.1	−0.3	−1	22	−0.3	−1.2	1
9	0.3	−0.4	−1	23	−0.5	−0.9	1
10	0.8	1.2	1	24	0.5	−0.9	1
11	1.2	1.2	1	25	1	−1	1
12	1.4	0.2	1	26	0.2	−1.2	1
13	0.2	1.8	1	27	0.3	−1.5	1
14	0.1	1.7	1	28	1.2	−0.1	1

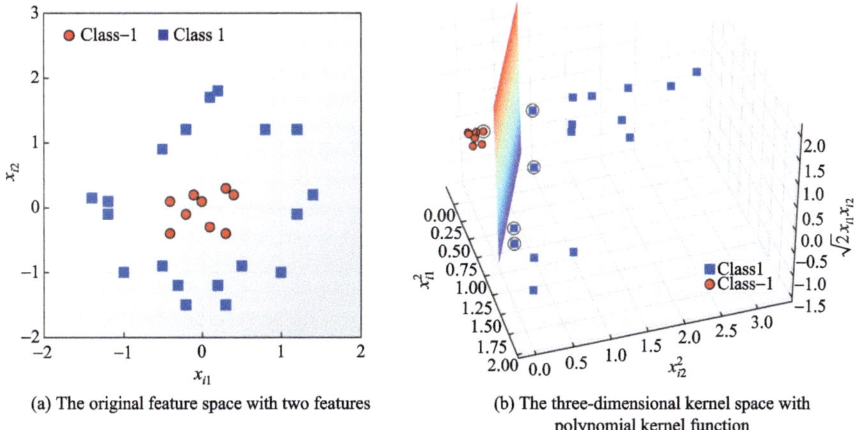

(a) The original feature space with two features

(b) The three-dimensional kernel space with polynomial kernel function

Fig. 4.3 **(a)** The original two-dimensional feature space with two features. **(b)** the three-dimensional space mapped by the polynomial kernel function ($d = 2$) with the highlighted separating hyperplane and support vectors in circles

The following shows the calculation of the SVC hyperplane in the three-dimensional space. With the 28 samples in Table 4.3, the dual objective function is

$$L_D = \sum_{i=1}^{28} \lambda_i - \frac{1}{2} \sum_{i=1}^{28} \sum_{j=1}^{28} \lambda_i \lambda_j y_i y_j \boldsymbol{X}_j \boldsymbol{X}_i^{\mathrm{T}} \quad (\lambda_i \geq 0).$$

With a similar way to Example 4.1, we can obtain

$$\lambda_6 = 5.751, \quad \lambda_{19} = 1.091, \quad \lambda_{23} = 3.184, \quad \lambda_{24} = 1.412, \quad \lambda_{26} = 0.064,$$
$$\text{and } \lambda_i = 0 \ \ (i = 1, 2, \cdots, 28 \text{ and } i \neq 6, 19, 23, 24, 26).$$

Since there are five nonzero Lagrange multipliers, the corresponding five data of $\boldsymbol{X}_6$, $\boldsymbol{X}_{19}$, $\boldsymbol{X}_{23}$, $\boldsymbol{X}_{24}$, and $\boldsymbol{X}_{26}$ are support vectors, which are marked by circles in Figure 4.3b. Then, the separating hyperplane is determined by

$$\begin{aligned}\boldsymbol{W}^{\mathrm{T}} &= \sum_{i=1}^{28} \lambda_i y_i \boldsymbol{X}_i = 5.751 \times (-1) \times (0.16 \ \ 0.16 \ \ 0.226) \\ &+ 1.091 \times 1 \times (1.44 \ \ 0.01 \ \ 0.170) \\ &+ 3.184 \times 1 \times (0.25 \ \ 0.81 \ \ 0.636) + 1.412 \times 1 \\ &\times (0.25 \ \ 0.81 \ \ -0.636) + 0.064 \times 1 \\ &\times (0.04 \ \ 1.44 \ \ -0.340)\end{aligned}$$

$$= (1.892 \quad 2.814 \quad -9.024 \times 10^{-3})$$

and

$$b = \frac{(y_6 - \boldsymbol{X}_6\boldsymbol{W}) + (y_{19} - \boldsymbol{X}_{19}\boldsymbol{W}) + (y_{23} - \boldsymbol{X}_{23}\boldsymbol{W}) + (y_{24} - \boldsymbol{X}_{24}\boldsymbol{W}) + (y_{26} - \boldsymbol{X}_{26}\boldsymbol{W})}{5} = -1.753.$$

Finally, the separating hyperplane is $1.892x_1^2 + 2.814x_2^2 - 9.024 \times 10^{-3}x_1x_2 - 1.753 = 0$, which is shown in Fig. 4.3b.

Example 4.3 This example shows the calculation detail of SVM with an RBF kernel function on the dataset listed in Table 4.1. The normalized data are expressed in the matrix form of

$$(\boldsymbol{X},\ \boldsymbol{y}) = \begin{bmatrix} \boldsymbol{X}_1 & y_1 \\ \boldsymbol{X}_2 & y_2 \\ \boldsymbol{X}_3 & y_3 \\ \boldsymbol{X}_4 & y_4 \end{bmatrix} = \begin{bmatrix} x_{11} & x_{12} & y_1 \\ x_{21} & x_{22} & y_2 \\ x_{31} & x_{32} & y_3 \\ x_{41} & x_{42} & y_4 \end{bmatrix} = \begin{bmatrix} 0.75 & 0.491 & 1 \\ 1 & 1 & 1 \\ 0 & 0.673 & -1 \\ 0.333 & 0 & -1 \end{bmatrix}. \tag{4.16a}$$

With the data, the kernel matrix can be expressed by

$$K(\boldsymbol{X}_i, \boldsymbol{X}_j) = \begin{bmatrix} k(\boldsymbol{X}_1, \boldsymbol{X}_1) & k(\boldsymbol{X}_1, \boldsymbol{X}_2) & \cdots & k(\boldsymbol{X}_1, \boldsymbol{X}_n) \\ k(\boldsymbol{X}_2, \boldsymbol{X}_1) & k(\boldsymbol{X}_2, \boldsymbol{X}_2) & \ldots & k(\boldsymbol{X}_2, \boldsymbol{X}_n) \\ \vdots & \vdots & & \vdots \\ k(\boldsymbol{X}_n, \boldsymbol{X}_1) & k(\boldsymbol{X}_n, \boldsymbol{X}_2) & \cdots & k(\boldsymbol{X}_n, \boldsymbol{X}_n) \end{bmatrix}. \tag{4.16b}$$

The RBF kernel is rewritten as

$$K\left(\boldsymbol{X}_i, \boldsymbol{X}_j\right) = \exp\left(-\gamma ||\boldsymbol{X}_{\mathrm{i}} - \boldsymbol{X}_{\mathrm{j}}||^2\right),\quad \gamma = \frac{1}{2\sigma^2}. \tag{4.16c}$$

The original data have two features, i.e., $\boldsymbol{X}_i = (x_{i1}, x_{i2})$; thus, we have

$$\boldsymbol{X}_i - \boldsymbol{X}_j = \begin{pmatrix} x_{i1} \\ x_{i2} \end{pmatrix}^{\mathrm{T}} - \begin{pmatrix} x_{j1} \\ x_{j2} \end{pmatrix}^{\mathrm{T}} = \begin{pmatrix} x_{i1} - x_{j1} \\ x_{i2} - x_{j2} \end{pmatrix}^{\mathrm{T}} \tag{4.16d}$$

$(i, j = 1, 2, \cdots, n)$.

Substituting Eq. (4.16d) into the RBF kernel Eq. (4.16c) yields

$$\exp\left(-\gamma\|\boldsymbol{X}_i-\boldsymbol{X}_j\|^2\right)=\exp(-\gamma(\boldsymbol{X}_i-\boldsymbol{X}_j)(\boldsymbol{X}_i-\boldsymbol{X}_j)^{\mathrm{T}})\quad(i,j=1,2,\cdots,n).$$

Let $\gamma = 1$, with the data in Eq. (4.16a), the detail calculation for $k(\boldsymbol{X}_1, \boldsymbol{X}_1)$ and $k(\boldsymbol{X}_1, \boldsymbol{X}_2)$ are shown below.

$$\begin{aligned}
k(\boldsymbol{X}_1,\boldsymbol{X}_1)&=\exp(-\gamma(\boldsymbol{X}_1-\boldsymbol{X}_1)(\boldsymbol{X}_1-\boldsymbol{X}_1)^{\mathrm{T}})=1,\\
k(\boldsymbol{X}_1,\boldsymbol{X}_2)&=\exp(-\gamma(\boldsymbol{X}_1-\boldsymbol{X}_2)(\boldsymbol{X}_1-\boldsymbol{X}_2)^{\mathrm{T}})\\
&=\exp\left(-(0.75-1\ 0.491-1)\begin{pmatrix}0.75-1\\0.491-1\end{pmatrix}\right)\\
&=\exp(-0.321581)=0.725.
\end{aligned}$$

Similarly, we can get the value of each element in the kernel matrix $K(\boldsymbol{X}_i, \boldsymbol{X}_j)(i, j = 1, 2, \cdots, n)$. Then, the dual objective function is

$$\begin{aligned}
L_{\mathrm{D}}&=\sum_{i=1}^{n}\lambda_i-\frac{1}{2}\sum_{i=1}^{n}\sum_{j=1}^{n}\lambda_i\lambda_j y_i y_j K(\boldsymbol{X}_i,\boldsymbol{X}_j)\\
&=\lambda_1+\lambda_2+\lambda_3+\lambda_4-(0.7250\lambda_1\lambda_2+0.5512\lambda_1\lambda_3+0.6604\lambda_1\lambda_4\\
&\quad+0.3306\lambda_2\lambda_3+0.2358\lambda_2\lambda_4+0.5690\lambda_3\lambda_4).
\end{aligned}$$

Equation (4.6b) shows that the values of λ_i should satisfy the condition of $\sum_{i=1}^{n}\lambda_i y_i = 0$, which gives $\lambda_3 = \lambda_1 + \lambda_2 - \lambda_4$. With this relationship, the dual objective function L_{D} is rewritten as

$$\begin{aligned}
L_{\mathrm{D}}=2\lambda_1+2\lambda_2&-1.6068\lambda_1\lambda_2-0.5512\lambda_1^2-0.6782\lambda_1\lambda_4\\
&-0.3306\lambda_2^2-0.4742\lambda_2\lambda_4+0.5690\lambda_4^2.
\end{aligned}$$

The maximization of L_{D} requires

$$\begin{cases}
\dfrac{\partial L_{\mathrm{D}}}{\partial\lambda_1}=2-1.1024\lambda_1-1.6068\lambda_2-0.6782\lambda_4=0,\\
\dfrac{\partial L_{\mathrm{D}}}{\partial\lambda_2}=2-1.6068\lambda_1-0.6612\lambda_2-0.4742\lambda_4=0,\\
\dfrac{\partial L_{\mathrm{D}}}{\partial\lambda_4}=-0.6782\lambda_1-0.4742\lambda_2+1.1380\lambda_4=0,
\end{cases}$$

$$\text{which yields} \begin{cases} \lambda_1 = 0.9056, \\ \lambda_2 = 0.3364, \text{ and } \lambda_3 = \lambda_1 + \lambda_2 - \lambda_4 = 0.5621. \\ \lambda_4 = 0.6799, \end{cases}$$

Since there are four nonzero Lagrange multipliers, all data points are support vectors. According to the Eq. (4.15b), the b value is

$$\begin{aligned} b &= \frac{1}{4}\sum_{s\in S}\left(\frac{1}{y_s} - \sum_{i\in S}\lambda_i y_i K(\boldsymbol{X}_s, \boldsymbol{X}_i)\right) \\ &= -0.7250(\lambda_1 + \lambda_2) - 0.5512(\lambda_1 - \lambda_3) \\ &\quad - 0.6604(\lambda_1 - \lambda_4) - 0.3306(\lambda_2 - \lambda_3) \\ &\quad - 0.2358(\lambda_2 - \lambda_4) + 0.5690(\lambda_3 + \lambda_4) = -0.337. \end{aligned}$$

Finally, the classification rule is

$$f(\boldsymbol{X}) = \text{sign}\left(\sum_{i=1}^{4}\lambda_i y_i \exp(-\|\boldsymbol{X} - \boldsymbol{X}_i\|^2) - 0.337.\right)$$

Example 4.4 Table 4.4 shows 34 alloys, including 14 crystalline alloys (CRA) and 20 bulk metallic glasses (BMG) collected from Xiong et al. (2020). The enthalpy of mixing (H_{mix}, in kJ·mol^{-1}) and entropy of mixing (S_{mix}, in J·K^{-1}·mol^{-1}) (Takeuchi & Inoue, 2005) are the only two features considered in the example. S_{mix} is calculated by

$$S_{\text{mix}} = -R\sum_{i=1}^{N} a_i \ln a_i, \tag{4.17}$$

where R is the gas constant, a_i is the atomic fraction of the i-th constituent. With H_{mix} and S_{mix}, the classification of CRA and BMG is done with the SVC algorithms combined with the linear kernel, polynomial kernel, and RBF kernel.

As expected, the optimal separating hyperplane and margins are straight lines with the linear kernel in the feature space, because the linear kernel varies only in the scale and origin. The polynomial and RBF kernels nonlinearly map the feature space to a high-dimensional space, and make the optimal separating hyperplane; margins become curves. Moreover, the number of support vectors is six, including four BMGs and two CRAs, as shown in Fig. 4.4c.

Table 4.4 Formed phases of 34 alloys

Alloys	H_{mix}	S_{mix}	Phases	Alloys	H_{mix}	S_{mix}	Phases
$Ag_{10.5}Cu_{80.5}P_9$	−5.09	5.22	CRA	$(Zr_{0.55}Al_{0.20}Co_{0.20}Cu_{0.05})_{95}Ag_5$	−40.10	10.52	BMG
$Ag_{10}Ce_3Cu_{87}$	−1.86	3.80	CRA	$(Zr_{0.55}Al_{0.20}Co_{0.20}Cu_{0.05})_{97}Ag_3$	−41.16	10.17	BMG
$Ag_{10}Ce_5Cu_{85}$	−3.49	4.31	CRA	$Fe_{41}Co_7Cr_{15}Mo_{14}C_{15}B_6Y_2$	−33.35	13.66	BMG
$Ag_{10}Mg_{50}Y_{40}$	−11.44	7.84	CRA	$La_{62}Al_{14}(Cu_5Ag_1)_{12}(NiCo)_{12}$	−27.50	10.12	BMG
$Ag_{10}Mg_{85}Y_5$	−5.00	4.31	CRA	$La_{62}Al_{14}(Cu_5Ag_1)_{16}(NiCo)_8$	−27.15	9.93	BMG
$Ag_{12.2}Cu_{81.5}P_{6.3}$	−3.37	4.97	CRA	$La_{62}Al_{14}(Cu_5Ag_1)_{20}(NiCo)_4$	−26.89	9.48	BMG
$Ag_{14}Cu_{78.5}P_{7.5}$	−4.02	5.48	CRA	$Zr_{65}Al_{7.5}Cu_{17.5}Ni_{10}$	−32.22	8.39	BMG
$Ag_{15}Ce_4Cu_{81}$	−2.47	4.86	CRA	$Au_2Zr_{48}Cu_{34}Al_8Ag_8$	−28.10	9.99	BMG
$Ag_{15}Cu_{83}Fe_2$	2.20	4.30	CRA	$Au_4Zr_{48}Cu_{32}Al_8Ag_8$	−30.46	10.39	BMG
$Ag_{20.9}Pd_{62.6}Si_{16.5}$	−29.15	7.63	CRA	$Cu_{38}Zr_{46}Ag_8Al_8$	−25.48	9.39	BMG
$Ag_{20}Ce_3Cu_{77}$	−1.43	5.22	CRA	$Hf_4Zr_{44}Cu_{36}Al_8Ag_8$	−25.22	10.49	BMG
$Ag_{20}Ce_4Cu_{76}$	−2.30	5.48	CRA	$La_{62}Al_{14}(Cu_5Ag_1)_{14}(NiCo)_{10}$	−27.31	10.06	BMG
$Ag_{20}Ce_5Cu_{75}$	−3.15	5.72	CRA	$Pd_6Zr_{48}Cu_{30}Al_8Ag_8$	−35.59	10.70	BMG
$Ag_{20}Mg_{20}Y_{60}$	−18.40	7.90	CRA	$Zr_{46}Cu_{37.64}Ag_{8.36}Al_8$	−25.46	9.43	BMG
$Cu_{34}Zr_{50}Ag_8Al_8$	−25.87	9.29	BMG	$Hf_2Zr_{46}Cu_{36}Al_8Ag_8$	−25.47	10.04	BMG
$Cu_{40}Zr_{44}Ag_8Al_8$	−25.18	9.41	BMG	$Ni_2Zr_{48}Cu_{34}Al_8Ag_8$	−26.66	9.99	BMG
$Hf_6Zr_{42}Cu_{36}Al_8Ag_8$	−24.97	10.85	BMG	$Ni_6Zr_{48}Cu_{30}Al_8Ag_8$	−28.58	10.70	BMG

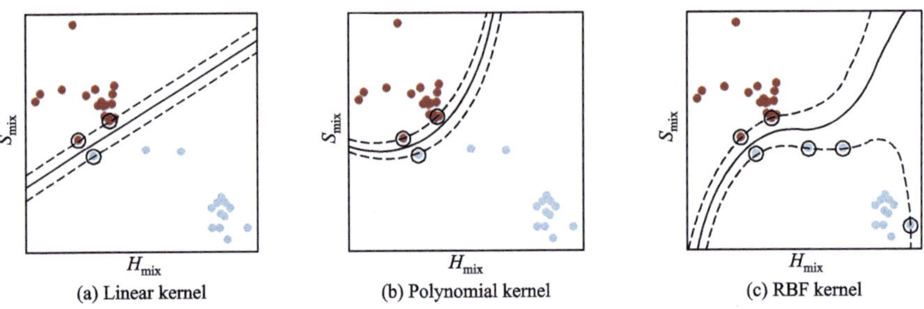

(a) Linear kernel (b) Polynomial kernel (c) RBF kernel

Fig. 4.4 The data in the $(H_{\mathrm{mix}}, S_{\mathrm{mix}})$ feature space, where blue and red dots represent the CRA and BMG data, respectively; the solid and dashed curves (lines) are the optimal separating hyperplanes and margins; and the support vectors are marked by circles. The SVCs are with **(a)** linear kernel, **(b)** polynomial kernel of $d = 3$, and **(c)** RBF kernel of $\sigma = 12$

4.3 Soft Margin

Linearly separable data mean that there exists a separating hyperplane in feature space or in mapped high-dimensional space, which clearly and completely separates the data, and no data will be within the margin region at all. For such kind of separation, the margin is called hard margin (Tsochantaridis et al., 2005). If the classes of data

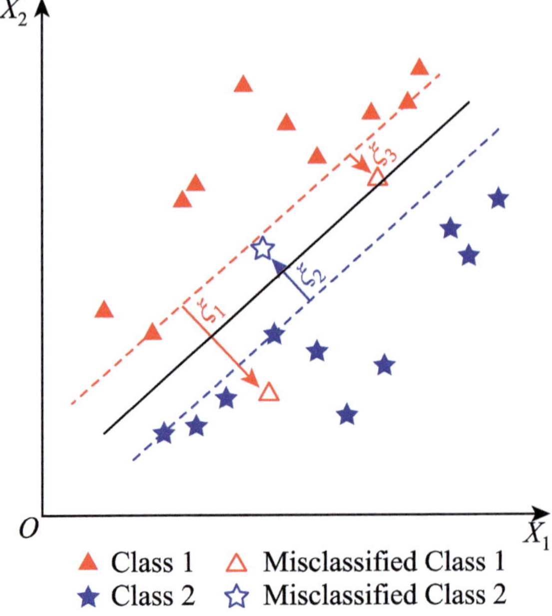

Fig. 4.5 Schematic plot of SVC with soft margin, where red triangles denote one category of data, and the blue stars denote another category of data. A slack variable $\xi_i \geqslant 0\ (i = 1, 2, \cdots, n)$ is introduced to represent the distance from the class margin to a datum located in the wrong side, if a datum is within the right side, $\xi_i = 0$

overlap slightly in feature space or mapped high-dimensional space, the classification of slightly overlapped data will allow few data to stay within the margin region, and this type of margin is termed as soft margin, as shown by Fig. 4.5.

Soft margin tolerates few data, which unsatisfy the constraint of $y_i(\boldsymbol{X}_i\boldsymbol{W} + b) \geqslant 1$. Of course, the number of unsatisfied data should be as small as possible. Therefore, a step loss function of $L_{0/1}(z)$ is introduced to minimize the number of unsatisfied data in classification. In practice, the following surrogate loss functions are widely used in margin maximization.

$$
\begin{aligned}
\text{0/1 loss: } L_{0/1}(z) &= \begin{cases} 1, & \text{if } z < 0 \\ 0, & \text{if } z > 0 \end{cases} \\
\text{Hinge loss: } L_{\text{hinge}}(z) &= \max(0,\ 1 - z), \\
\text{Exponential loss: } L_{\exp}(z) &= \exp(-z), \\
\text{Logistic loss: } L_{\log}(z) &= \log(1 + \exp(-z)).
\end{aligned}
$$

Note that the loss functions for regression might be different from those for classification, e.g., Residual Sum-of-Squares (RSS) is a widely-used loss function for regression (Steel et al., 1997). With a loss function listed above, Eq. (4.5a) can be revised to

$$
\min_{\boldsymbol{W},b}\left[\frac{1}{2}\|\boldsymbol{W}\|^2 + C\sum_{i=1}^{n} L(f(X_i), y_i)\right] \tag{4.18a}
$$

$$
\text{or generally} \quad \min_{f}\left[\Omega(f) + C\frac{1}{n}\sum_{i=1}^{n} L(f(X_i), y_i)\right] \tag{4.18b}
$$

$$
\text{or equivalently,} \quad \min_{f}\left[\lambda\Omega(f) + \frac{1}{n}\sum_{i=1}^{n} L(f(X_i), y_i)\right] \tag{4.18c}
$$

where $\Omega(f)$ denotes an ML model to be minimized, like $\frac{1}{2}\|\boldsymbol{W}\|^2$ in SVM, C is a preset positive constant and called a "cost" parameter, and $\lambda = 1/C$ is usually called a penalty parameter or regularization parameter (Hastie et al., 2008). Equation (4.18c) or Eq. (4.18b) represents a common regularized problem. Without the penalty term, Eq. (4.18c) is reduced to

$$
\min_{f}\left[\frac{1}{n}\sum_{i=1}^{n} L(f(X_i), y_i)\right]. \tag{4.18d}
$$

Equation (4.18d) is called the empirical risk minimization (Vapnik & Chervonenkis, 1991), while Eqs. (4.18b) or (4.18c) is termed as the structural risk minimization, where the value of regularized parameter C determines the tradeoff between the model prediction accuracy and the model complexity.

More generally, slack variables $\xi_i \geqslant 0$ ($i = 1, 2, \cdots, n$) are introduced for soft margin. With slack variables, Eq. (4.5a) will be revised to

$$\min_{\boldsymbol{W},b}\left[\frac{1}{2}\|\boldsymbol{W}\|^2 + C\sum_{i=1}^{n}\xi_i\right],$$

$$\text{subject to } \xi_i \geqslant 0 \text{ and } y_i(\boldsymbol{X}_i\boldsymbol{W} + b) \geqslant 1 - \xi_i,\, i = 1, 2, \cdots, n. \tag{4.19a}$$

A sufficiently large C value makes the values of $\xi_i (i = 1, 2, \cdots, n)$ smaller and even approach zero; consequently, Eq. (4.15a) is reduced to Eq. (4.5a). The Lagrange function of Eq. (4.5b) is revised to

$$L_{\mathrm{P}} = \frac{1}{2}\|\boldsymbol{W}\|^2 + C\sum_{i=1}^{n}\xi_i - \sum_{i=1}^{n}\lambda_i[y_i(\boldsymbol{X}_i\boldsymbol{W} + b) - (1 - \xi_i)] - \sum_{i=1}^{n}\mu_i\xi_i, \tag{4.19b}$$

where $\mu_i (i = 1, 2, \cdots, n)$ are also the Lagrange multipliers, $\mu_i \geqslant 0$. The minimization of L_{P} requires $\frac{\partial L_{\mathrm{P}}}{\partial \boldsymbol{W}} = 0$, $\frac{\partial L_{\mathrm{P}}}{\partial b} = 0$, and $\frac{\partial L_{\mathrm{P}}}{\partial \xi_i} = 0$, which yield

$$\boldsymbol{W}^{\mathrm{T}} = \sum_{i=1}^{n}\lambda_i\boldsymbol{X}_i y_i, \tag{4.20a}$$

$$\sum_{i=1}^{n}\lambda_i y_i = 0, \tag{4.20b}$$

$$\lambda_i = C - \mu_i. \tag{4.20c}$$

Substituting Eq. (4.20a–c) into Eq. (4.19b) yields the dual objective function

$$L_{\mathrm{D}} = \left[\sum_{i=1}^{n}\lambda_i - \frac{1}{2}\sum_{i=1}^{n}\sum_{j=1}^{n}\lambda_i\lambda_j y_i y_j \boldsymbol{X}_j\boldsymbol{X}_i^{\mathrm{T}}\right]. \tag{4.21a}$$

Equation (4.21a) appears the same as Eq. (4.7a). The values of λ_i ($i = 1, 2, \cdots, n$) are determined by maximizing the dual objective function, i.e.,

$$\max_{\lambda_i} L_{\mathrm{D}}$$

$$\text{subject to } \sum_{i=1}^{n} \lambda_i y_i = 0 \text{ and } C \geqslant \lambda_{\mathrm{i}} \geqslant 0,\ i = 1, 2, \cdots, n. \tag{4.21b}$$

The KKT conditions must be satisfied in solving the maximization problem.

$$\begin{cases} \lambda_i \geq 0, \quad \mu_{\mathrm{i}} \geq 0, \\ y_i(\boldsymbol{X}_i \boldsymbol{W} + b) \geq 1 - \xi_{\mathrm{i}}, \\ \lambda_i [y_i(\boldsymbol{X}_i \boldsymbol{W} + b) - 1 + \xi_i] = 0, \\ \xi_i \geq 0, \quad \mu_{\mathrm{i}} \xi_{\mathrm{i}} = 0. \end{cases}$$

The nonzero values of $\lambda_i (i = 1, 2, \cdots, n)$ hold only for those data, called support vectors, meeting $y_i(\boldsymbol{X}_i \boldsymbol{W} + b) - 1 + \xi_i = 0$. Any of these support vectors $(\lambda_i > 0, \xi_i = 0)$ can be used to determine the b value. In the soft margin SVC, the value of cost parameter C should be turned to have the best solution.

Example 4.5 To illustrate the soft margin and the influence of penalty parameter C, SVC algorithms with linear and RBF kernels are conducted on the data in Table 4.5. Table 4.5 shows 36 HEAs, including 18 body-centered cubic (BCC) HEAs and 18 FCC HEAs. The VEC and enthalpy of mixing (H_{mix}, in kJ·mol^{-1}) are the two features considered in the example.

With VEC and H_{mix}, the classification of BCC and FCC is done with the SVC algorithms by using linear kernel and RBF kernel with different penalty parameters of $C = 1$, 10 and 100 for linear kernel, and $C = 100$, 1000 and 10,000 for the RBF kernel. The results are illustrated in Figs. 4.6 and 4.7 with solid and dashed lines (curves) representing optimal separating hyperplanes and margins, respectively, and support vectors circled.

On the one hand, the results indicate that a small value of penalty parameter C leads to a large soft margin, and many data are misclassified within the soft margin zone; thereby the model accuracy is low. On the other hand, a large value of penalty parameter C leads to a small soft margin, and all data are correctly classified, so the model accuracy is high. However, a too accurate model may cause over-fitting, meaning the power of the model prediction on new data will be low. Therefore, the cross-validation method is used to determine the proper value of C to avoid over-fitting in practice. Here, the LOOCV is conducted to determine the appropriate C values for SVCs with linear kernel and RBF kernel ($\sigma = 10$). As excepted, when the value of C increases, the training accuracy increases, while the validation accuracy increases to a maximum and then decreases. The validation maximum accuracy and the training accuracy determine the C to be 10 for linear kernel and 1000 for RBF kernel, respectively, which are shown in Table 4.6.

Table 4.5 Solid solution structure of 36 HEAs

Alloys	VEC	H_{mix}	Structures	Alloys	VEC	H_{mix}	Structures
$Al_{13.64}Co_{18.18}Cr_{18.18}Cu_{4.55}Fe_{18.18}Ni_{18.18}Ti_{9.09}$	7.27	−14.39	BCC	$Al_{7.69}Co_{30.77}Fe_{30.77}Ni_{30.77}$	8.54	−6.06	FCC
$Al_{12.70}Cr_{15.87}Cu_{15.87}Fe_{15.87}Mn_{23.81}Ni_{15.87}$	7.60	−4.23	BCC	$Al_{3.85}Cr_{19.23}Cu_{19.23}Fe_{19.23}Ni_{38.46}$	8.77	0.12	FCC
$Al_{17.39}Co_{17.39}Cr_{17.39}Cu_{4.35}Fe_{17.39}Ni_{17.39}Ti_{8.70}$	7.09	−15.50	BCC	$Co_5Cr_2Fe_{40}Mn_{27}Ni_{26}$	8.26	−3.58	FCC
$Al_{16.67}Co_{16.67}Cr_{16.67}Cu_{8.33}Fe_{16.67}Ni_{16.67}Ti_{8.33}$	7.25	−13.42	BCC	$Co_{25}Cr_{25}Cu_{25}Fe_{25}$	8.50	6.25	FCC
$Al_{18.52}Co_{18.52}Cr_{18.52}Fe_{18.52}Ni_{18.52}Si_{7.41}$	6.96	−19.84	BCC	$Co_{25}Cr_{25}Fe_{25}Ni_{25}$	8.25	−3.75	FCC
$Al_{18.18}Co_{18.18}Cr_{18.18}Fe_{18.18}Ni_{18.18}Ti_{9.09}$	6.91	−17.92	BCC	$Co_{25}Cr_{25}Mn_{25}Ni_{25}$	8.00	−5.50	FCC
$Al_{16.67}Cr_{16.67}Cu_{16.67}Fe_{16.67}Mn_{16.67}Ni_{16.67}$	7.50	−5.11	BCC	$Co_{33.33}Cr_{33.33}Ni_{33.33}$	8.33	−4.89	FCC
$Al_{16.67}Hf_{16.67}Nb_{16.67}Ta_{16.67}Ti_{16.67}Zr_{16.67}$	4.17	−14.78	BCC	$Co_{25}Cu_{25}Fe_{25}Ni_{25}$	9.50	5.00	FCC
$Cr_{28.57}Mo_{14.29}Nb_{14.29}Ta_{14.29}V_{14.29}W_{14.29}$	5.57	−4.82	BCC	$Co_{24.88}Cu_{24.88}Fe_{24.88}Ni_{24.88}Sn_{0.50}$	9.47	5.02	FCC
$Al_{5.66}Hf_{18.87}Nb_{18.87}Ta_{18.87}Ti_{18.87}Zr_{18.87}$	4.32	−3.99	BCC	$Co_{25}Fe_{25}Mn_{25}Ni_{25}$	8.50	−4.00	FCC
$Al_{9.09}Hf_{18.18}Nb_{18.18}Ta_{18.18}Ti_{18.18}Zr_{18.18}$	4.27	−7.67	BCC	$Co_{33.33}Fe_{33.33}Ni_{33.33}$	9.00	−1.33	FCC
$Al_{13.04}Hf_{17.39}Nb_{17.39}Ta_{17.39}Ti_{17.39}Zr_{17.39}$	4.22	−11.55	BCC	$Co_{30.77}Fe_{30.77}Ni_{30.77}Si_{7.69}$	8.62	−11.83	FCC
$Al_{20}Co_{10}Cr_{10}Cu_{10}Fe_{10}Mn_{10}Ni_{10}Ti_{10}V_{10}$	6.60	−15.44	BCC	$Co_{25}Fe_{25}Ni_{25}V_{25}$	8.00	−10.50	FCC

(continued)

Table 4.5 (continued)

Alloys	VEC	H_{mix}	Structures	Alloys	VEC	H_{mix}	Structures
$Cr_{9.09}Mo_{18.18}Nb_{18.18}Ta_{18.18}V_{18.18}W_{18.18}$	5.45	−4.83	BCC	$Co_{33.33}Mn_{33.33}Ni_{33.33}$	8.67	−5.78	FCC
$Cr_{16.67}Mo_{16.67}Nb_{16.67}Ta_{16.67}V_{16.67}W_{16.67}$	5.50	−4.89	BCC	$Cr_{18.03}Fe_{27.32}Mn_{27.32}Ni_{27.32}$	7.91	−4.17	FCC
$Hf_{18.18}Mo_{9.09}Nb_{18.18}Ta_{18.18}Ti_{18.18}Zr_{18.18}$	4.55	0.60	BCC	$Cr_{25}Cu_{25}Fe_{25}Ni_{25}$	8.57	3.01	FCC
$Hf_{17.39}Mo_{13.04}Nb_{17.39}Ta_{17.39}Ti_{17.39}Zr_{17.39}$	4.61	−0.21	BCC	$Fe_{33.33}Mn_{33.33}Ni_{33.33}$	8.33	−4.44	FCC
$Hf_{16.67}Nb_{16.67}Ta_{16.67}Ti_{16.67}V_{16.67}Zr_{16.67}$	4.50	0.78	BCC	$Co_{9}Cr_{7}Cu_{36}Mn_{25}Ni_{23}$	9.24	2.05	FCC

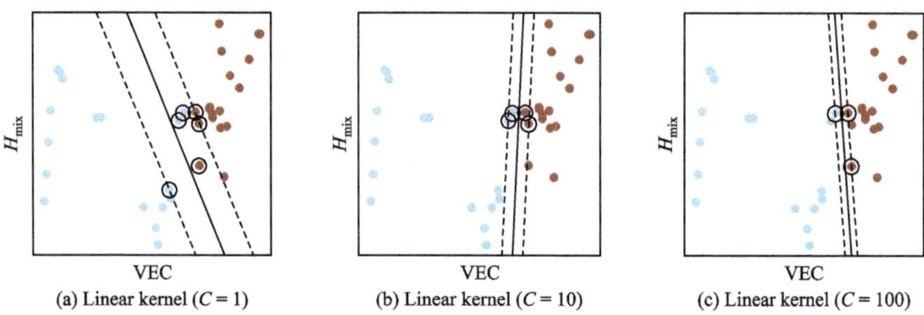

Fig. 4.6 The (VEC, H_{mix}) space with blue and red dots stand for BCC and FCC, respectively, and SVC with the linear kernel and **(a)** $C = 1$, **(b)** $C = 10$, and **(c)** $C = 100$

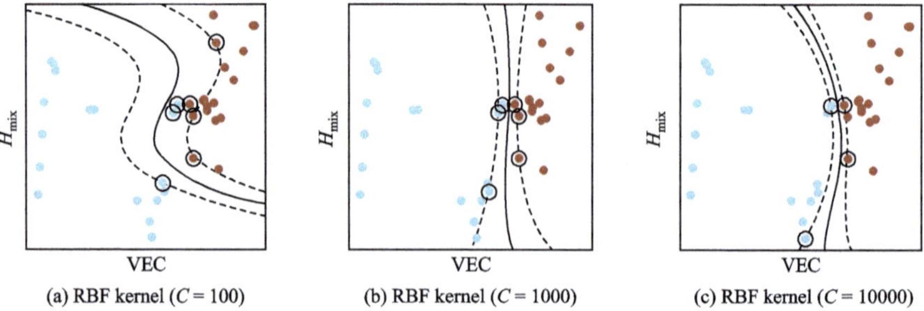

Fig. 4.7 The (VEC, H_{mix}) space with blue and red dots stand for BCC and FCC, respectively, and SVC with the RBF kernel ($\sigma = 10$) and **(a)** $C = 100$, **(b)** $C = 1000$, and **(c)** $C = 10{,}000$

Table 4.6 Training and validation performance of different kernel, and values of C

Kernels	Penalty parameter C	Training accuracy (%)	Validation accuracy (%)
Linear kernel	1	94.4	94.4
	10	100	100
	100	100	97.2
RBF kernel ($\sigma = 10$)	100	94.4	94.4
	1000	100	100
	10,000	100	97.2

4.4 SVR

SVR is based on linear regression. In regression, the input feature vector is $\boldsymbol{X} = (x_1, x_2, \cdots, x_m) \in \mathbf{R}^m$, the response is $y \in \mathbf{R}$, the dataset $D = \{(\boldsymbol{X}_1, y_1), (\boldsymbol{X}_2, y_2), \cdots, (\boldsymbol{X}_n, y_n)\}$ of n data, and the linear ML model is denoted by $f(\boldsymbol{X}_i)$, which is a hyperplane in the m-feature space $\boldsymbol{X}$ or in mapped high-dimensional space $\phi(\boldsymbol{X})$. In the classical least squares linear regression, RSS is taken as the loss function, in which the loss is zero only when $f(\boldsymbol{X}_i) = y_i$, indicating no errors are tolerated. SVR, however, tolerates error ε on each side of the $f(\boldsymbol{X})$ hyperplane (Drucker et al., 1996), as shown in Fig. 4.8, where ε is a preset positive parameter. SVR builds up a linear model:

$$f(\boldsymbol{X}) = \sum_{i=1}^{n} \boldsymbol{W}\phi(\boldsymbol{X}_i) + b. \tag{4.22}$$

The values of $\boldsymbol{W}$ and b are determined by the minimization of the regularized risk function, as described in Eq. (4.18a) and (4.18b)

$$\min_{\boldsymbol{W},\, b}\left(C\sum_{i=1}^{n} L_\varepsilon(f(\boldsymbol{X}_i) - y_i) + \frac{1}{2}\|\boldsymbol{W}\|^2\right), \tag{4.23a}$$

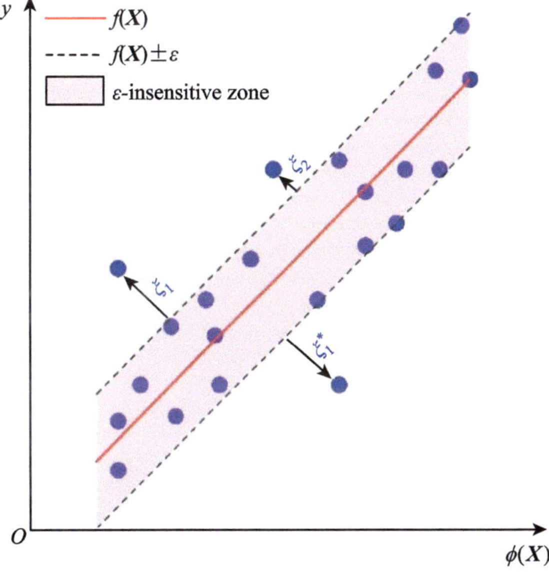

Fig. 4.8 Schematic plot of SVR with ε-insensitive zone, where the data outside the ε-insensitive zone determine the regression hyperplane

where $L_\varepsilon(f(\boldsymbol{X}_i) - y_i)$ is the ε-insensitive loss, and its value, like constant C, is prescribed

$$L_\varepsilon(z) = \begin{cases} |z| - \varepsilon, & \text{for } z = y_i - f(X_i) > \varepsilon \\ 0, & \text{for } |z| \leqslant \varepsilon \\ |z| - \varepsilon, & \text{for } z = f(X_i) - y_i > \varepsilon \end{cases}, \quad z \equiv f(X_i) - y_i, \tag{4.23b}$$

where $z = y_i - f(\boldsymbol{X}_i) > \varepsilon$ means the data $(\boldsymbol{X}_i, y_i)$ lies above the ε-toleration zone; $z \leq \varepsilon$ means the data $(\boldsymbol{X}_i, y_i)$ lies within the ε-toleration zone; and $z = f(\boldsymbol{X}_i) - y_i > \varepsilon$ means the data $(\boldsymbol{X}_i, y_i)$ lies below the ε- toleration zone.

The ε-insensitive loss indicates that it does not penalize errors below ε. Figure 4.8 schematically shows the concept of SVR.

With slack variables ξ_i and $\xi_i^*(i = 1, 2, \cdots, n)$, Eq. (4.23a) is revised to

$$\begin{gathered} \min_{\boldsymbol{W},\, b,\, \xi_i,\, \xi_i^*} \left(C \sum_{i=1}^{n} (\xi_i + \xi_i^*) + \frac{1}{2} \|\boldsymbol{W}\|^2 \right), \\ \text{subject to} \begin{cases} y_i - f(\boldsymbol{X}_i) \leq \varepsilon + \xi_i, \\ f(\boldsymbol{X}_i) - y_i \leq \varepsilon + \xi_i^*, \\ \xi_i \geq 0, \xi_i^* \geq 0. \end{cases} \end{gathered} \tag{4.24a}$$

With a similar approach to soft margin problems and using Lagrange multipliers $\lambda_i \geqslant 0$, $\lambda_i^* \geqslant 0$, $\mu_i \geqslant 0$, and $\mu_i^* \geqslant 0$, we have the Lagrange function

$$\begin{gathered} L_P = \frac{1}{2} \|\boldsymbol{W}\|^2 + C \sum_{i=1}^{n} (\xi_i + \xi_i^*) - \sum_{i=1}^{n} \mu_i \xi_i - \sum_{i=1}^{n} \mu_i^* \xi_i^* \\ + \sum_{i=1}^{n} \lambda_i [-f(\boldsymbol{X}_i + y_i - \varepsilon - \xi_i] + \sum_{i=1}^{n} \lambda_i^* [f(\boldsymbol{X}_i) - y_i - \varepsilon - \xi_i^*]. \end{gathered} \tag{4.24b}$$

The minimization of L_P requires $\frac{\partial L_P}{\partial \boldsymbol{W}} = 0$, $\frac{\partial L_P}{\partial b} = 0$, $\frac{\partial L_P}{\partial \xi_i} = 0$, and $\frac{\partial L_P}{\partial \xi_i^*} = 0$, which yield

$$\boldsymbol{W} = \sum_{i=1}^{n} (\lambda_i^* - \lambda_i) \phi(\boldsymbol{X}_i), \tag{4.25a}$$

$$\sum_{i=1}^{n} (\lambda_i^* - \lambda_i) = 0, \tag{4.25b}$$

$$C = \lambda_i + \mu_i, \tag{4.25c}$$

$$C = \lambda_i^* + \mu_i^*. \tag{4.25d}$$

Substituting Eq. (4.25a–d) into Eq. (4.24b) yields the dual objective function, which will be maximized as

$$\max_{\lambda_i,\lambda_i^*}\left[\sum_{i=1}^{n}\left[\left(\lambda_i-\lambda_i^*\right)y_i-\varepsilon\left(\lambda_i^*+\lambda_i\right)\right]-\frac{1}{2}\sum_{i=1}^{n}\sum_{j=1}^{n}\left(\lambda_i^*-\lambda_i\right)\left(\lambda_j^*-\lambda_j\right)K\left(\boldsymbol{X}_i,\boldsymbol{X}_j\right)\right] \tag{4.26}$$

$$\text{subject to}\begin{cases}\sum_{i=1}^{n}\left(\lambda_i^*-\lambda_i\right)=0,\\ 0\le\lambda_i,\lambda_i^*\le C.\end{cases}$$

The following KKT conditions must be satisfied in solving the maximization problem.

$$\begin{cases}\lambda_i[-f(\boldsymbol{X}_i)+y_i-\varepsilon-\xi_i]=0,\\ \lambda_i^*[f(\boldsymbol{X}_i)-y_i-\varepsilon-\xi_i^*]=0,\\ \lambda_i\lambda_i^*=0,\quad \xi_i\xi_i^*=0,\\ (C-\lambda_i)\xi_i=0,\quad (C-\lambda_i^*)\xi_i^*=0.\end{cases}$$

The KKT conditions indicate that $\lambda_i \neq 0$ only when $-f(\boldsymbol{X}_i)+y_i-\varepsilon-\xi_i=0$; $\lambda_i^* \neq 0$ only when $f(\boldsymbol{X}_i)-y_i-\varepsilon-\xi_i^*=0$; if $\lambda_i \neq 0$, λ_i^* must be zero, and vice versa; if $\xi_i \neq 0$, ξ_i^* must be zero, and vice versa; if $\xi_i=0$, $C=\lambda_i$; and if $\xi_i^*=0$, $C=\lambda_i^*$. The SVR model is finally given by

$$f(\boldsymbol{X})=\sum_{i=1}^{n}\left(\lambda_i^*-\lambda_i\right)K(\boldsymbol{X}_i,\boldsymbol{X})+b. \tag{4.27a}$$

If datum $\boldsymbol{X}_i$ $(i=1,2,\cdots,n)$ is within the ε-insensitive zone, $\lambda_i^*-\lambda_i=0$, only those data outside the ε-insensitive zones determine the regression hyperplane, and might be called support vectors in SVR, which are completely different from those in SVC. Let S denote the data set containing support vectors, and let $|S|$ represent the number of support vectors. The mean value of the intercept b is calculated by the following formula:

$$b=\frac{1}{|S|}\sum_{s\in S}(y_s\pm\varepsilon-\sum_{i}\left(\lambda_i^*-\lambda_i\right)K(\boldsymbol{X}_i,\boldsymbol{X}_s)), \tag{4.27b}$$

where $-\varepsilon$ is used for the support vectors that are above the upper bound (λ_i>0), and $+\varepsilon$ is used for the support vectors that are below the lower bound (λ_i*>0).

In the computation of SVM model, it should be noted that the selection of an appropriate value for the regularization parameter C is very important because it affects both trained and predicted results, since it controls the tradeoff between maximizing the margin and minimizing the training error. Usually, C should be optimized for fear of neither under-fitting nor over-fitting. It is also noticed that the predicted results are largely affected by the kernel functions and associated parameters. The advantage of SVM is workable with a small size of a sample set. In many cases, obtaining sufficiently large experimental samples is still time-consuming and costly in the development of novel materials. Therefore, efficient learning from a limited number of samples becomes increasingly important in the data-driven development of materials science and engineering.

Example 4.6 SVR algorithms with RBF kernels of three hyperparameter sets are conducted on the data of 219 samples given in Xiong et al. (2019). The average electronegativity (AEN), average atomic volume (V_{A}), and mixing entropy (S_{m}) are considered as the input features, and the target property is the shear modulus (G).

$$\mathrm{AEN} = \sum a_i \mathrm{EN}_i, \tag{4.28a}$$

$$V_{\mathrm{A}} = \sum a_i \cdot \frac{4}{3}\pi r_i^3, \tag{4.28b}$$

$$S_{\mathrm{m}} = -R\sum_{i=1}^{n} a_i \ln\phi_i, \tag{4.28c}$$

where a_i is the atomic percentage of the i-th constituent; EN_i and r_i are the Pauling electronegativity and metallic radius of the i-th constituent, respectively; R is the ideal gas constant; ϕ_i is the volume percentages of i-th component. The replacement of a_i by ϕ_i in the logarithmic term of S_{m} is due to the effect of dissimilar sizes of atoms (Jiang et al., 2003).

In the example, N samples are randomly taken out from the 219 samples to form a data subset. Then, the N samples are randomly divided by ten folds, nine folds for training and one fold for validation, viz., ten-fold validation. The performance of an SVR model is evaluated by the coefficient of determination R^2.

Figure 4.9a shows that the training and validation R^2 both are extremely bad with the values of hyperparameters $C = 0.001$, $\varepsilon = 0.1$, $\sigma = 0.5$, indicating the underfitting case, although the performance is improved when the number of samples is equal to and larger than 150. With the hyperparameters $C = 10000, \varepsilon = 0.1$, $\sigma = 0.16$, the SVR model has a good training performance and a poor validation performance, as shown in Fig. 4.9b, indicating the overfitting case. Figure 4.9c shows the model performance with the hyperparameters $C = 10, \varepsilon = 0.1, \sigma = 0.71$. In these values of hyperparameters, the SVR model is a good-fitted model.

In general, the validation performances are worse than or the same as the training performances. The tuning of hyperparameters means selecting the hyperparameter values that maximize the validation performance.

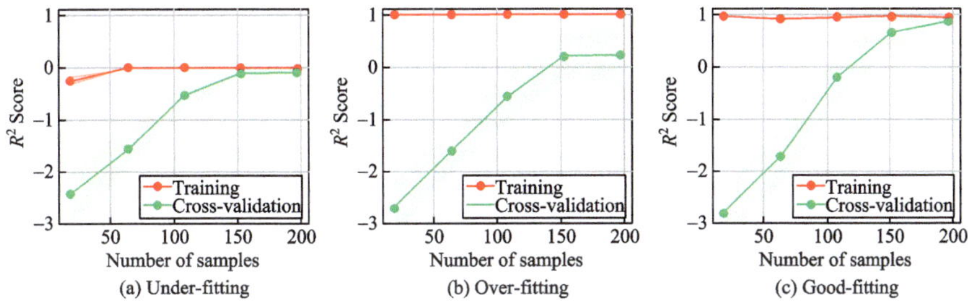

Fig. 4.9 The R^2 score versus the training samples in the SVR with RBF kernel with the values of hyperparameters **(a)** $C = 0.001$, $\varepsilon = 0.1$, $\sigma = 0.5$ (under-fitting), **(b)** $C = 10{,}000$, $\varepsilon = 0.1$, $\sigma = 0.16$ (over-fitting), and **(c)** $C = 10$, $\varepsilon = 0.1$, $\sigma = 0.71$ (good-fitting)

Example 4.7 The GridSearch-CV function in python library scikit-learn (Pedregosa et al., 2011) can be employed to tune hyperparameters. We take the 20 NIMS fatigue data in Table 4.8 as an example to illustrate the search process of hyperparameters. Take fatigue strength as the regression target. Tables 4.7 and 4.8 list the used 16 features and 20 data of steels, respectively.

The radial basis function $K(\boldsymbol{X}_i, \boldsymbol{X}_j) = \exp\left(-\gamma \left\| \boldsymbol{X}_i - \boldsymbol{X}_j \right\|^2\right)$ is used in the SVM regression, and the hyperparameter space in the GridSearch-CV function is listed in Table 4.9.

Table 4.7 Feature description of 20 NIMS fatigue data

Features	Descriptions
NT	Normalizing temperature (°C)
QT	Quenching temperature (°C)
TT	Tempering temperature (°C)
C	wt.% of carbon
Si	wt.% of silicon
Mn	wt.% of manganese
P	wt.% of phosphorus
S	wt.% of sulphur
Ni	wt.% of nickel
Cr	wt.% of chromium
Cu	wt.% of copper
Mo	wt.% of molybdenum
RR	Reduction ratio (%)
dA	Fraction of plastic work-inclusions (%)
dB	Fraction of discontinuous array-inclusions (%)
dC	Fraction of isolated inclusions (%)

Table 4.8 Twenty NIMS fatigue data

Sample No.	NT	QT	TT	C	Si	Mn	P	S	Ni	Cr	Cu	Mo	RR	dA	dB	dC	Fatigue
1	900	845	600	0.32	0.24	0.58	0.009	0.010	2.65	0.71	0.06	0.00	700	0.01	0.01	0.03	510
2	870	855	550	0.41	0.25	0.74	0.013	0.019	0.05	0.99	0.08	0.16	700	0.03	0.00	0.00	566
3	870	855	600	0.33	0.27	0.69	0.015	0.008	0.04	0.99	0.05	0.18	640	0.03	0.00	0.00	535
4	870	845	630	0.40	0.23	0.70	0.011	0.010	1.82	0.78	0.09	0.18	740	0.05	0.00	0.01	543
5	865	865	650	0.35	0.25	0.82	0.016	0.023	0.03	0.01	0.01	0.00	825	0.07	0.04	0.00	341
6	870	845	550	0.38	0.21	1.59	0.017	0.015	0.03	0.20	0.05	0.00	480	0.02	0.00	0.00	432
7	870	855	650	0.40	0.22	0.70	0.010	0.004	0.04	0.94	0.12	0.17	820	0.02	0.00	0.02	517
8	845	845	550	0.45	0.24	0.71	0.010	0.009	0.12	0.06	0.13	0.00	610	0.06	0.00	0.01	465
9	870	855	550	0.40	0.26	0.77	0.018	0.005	0.05	0.98	0.13	0.17	610	0.05	0.00	0.01	562
10	870	845	550	0.40	0.22	1.56	0.011	0.019	0.06	0.09	0.05	0.00	530	0.02	0.00	0.01	451
11	825	825	650	0.52	0.21	0.80	0.021	0.021	0.02	0.10	0.02	0.00	1740	0.09	0.00	0.00	410
12	825	825	650	0.52	0.29	0.83	0.016	0.025	0.04	0.04	0.02	0.00	825	0.12	0.02	0.00	402
13	870	855	600	0.37	0.29	0.76	0.017	0.009	0.12	1.01	0.12	0.16	500	0.05	0.00	0.02	542
14	900	845	550	0.32	0.30	0.52	0.013	0.009	2.67	0.77	0.06	0.00	1740	0.03	0.00	0.00	570
15	870	855	600	0.41	0.25	0.74	0.013	0.019	0.05	0.99	0.08	0.16	700	0.03	0.00	0.00	527
16	870	855	550	0.40	0.25	0.87	0.011	0.015	0.07	1.06	0.11	0.18	700	0.07	0.01	0.00	603
17	870	845	680	0.40	0.23	0.70	0.011	0.010	1.82	0.78	0.09	0.18	740	0.05	0.00	0.01	482
18	870	845	550	0.41	0.24	1.51	0.020	0.008	0.04	0.22	0.08	0.00	610	0.02	0.01	0.02	425
19	870	855	600	0.38	0.27	0.80	0.015	0.012	0.07	1.07	0.12	0.16	1120	0.03	0.00	0.02	592
20	845	845	550	0.45	0.25	0.79	0.018	0.016	0.02	0.13	0.02	0.00	1740	0.07	0.00	0.00	513

Table 4.9 Parameters, search space

Parameters	Min	Max	Step	Num.
C	1	241	12	21
ε	1	55	5.4	11
γ	0.01	0.58	0.03	20

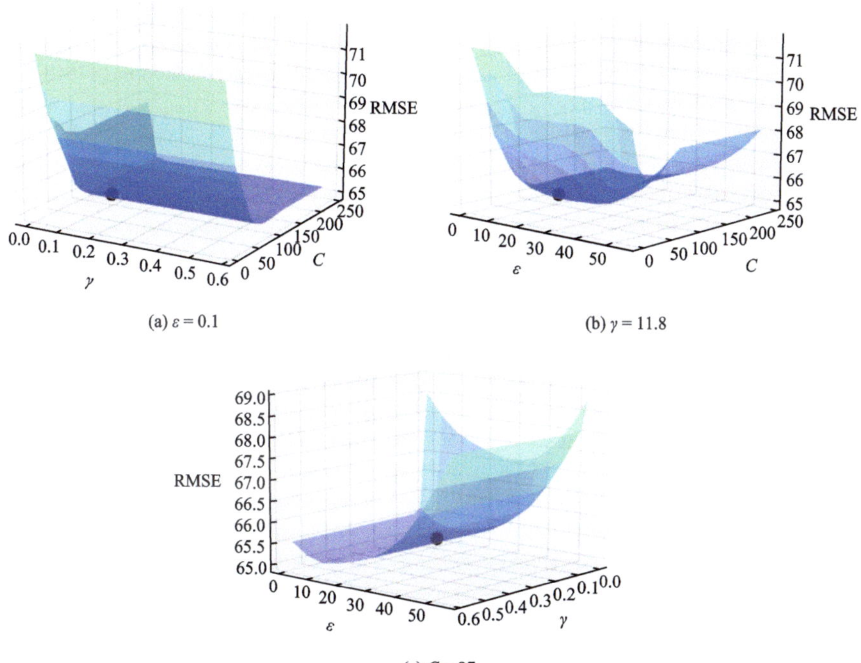

Fig. 4.10 GridSearch-CV (CV = 2) results. *Note* The black dots denote the optimal hyperparameter combination

With twofold CV, the greedy GridSearch yields $C = 97$, $\varepsilon = 0.1$, and $\gamma = 11.8$, and the RMSE $= 65.2$. Figure 4.10 shows the GridSearch, where the solid dots indicate the locations of tuned γ and ε.

Homework

Lead-free solder alloy data collected from literature include four chemical elements of Tin (Sn), Silver (Ag), Copper (Cu), and Bismuth (Bi) in wt.% and one property of yield strength in MPa, as shown in Tables 4.10 and 4.11. Alloys with yield strengths greater than 44 MPa are marked as high-strength alloys of Class 1, while the rest alloys are low-strength alloys of Class 0.

Table 4.10 Training set of 19 data

No.	Sn (wt.%)	Ag (wt.%)	Cu (wt.%)	Bi (wt.%)	Class	Yield Strength(MPa)
1	96.205	3.03	0.76	0.005	1	46.54
2	96	3.5	0.5	0	1	46.3
3	95.5	2.5	0	2	1	45.46
4	95.5	3.8	0.7	0	0	43.37
5	97	3	0	0	0	39.65
6	95	3	0	2	0	37.68
7	95.5	2.5	0	2	1	48.76
8	93.75	2.5	0.75	3	1	82.8
9	95.6	3.7	0.7	0	0	42.39
10	95.48	3.8	0.72	0	0	42.38
11	95.56	3.8	0.64	0	0	37.59
12	95.4	3.9	0.7	0	0	37.58
13	95	3	0	2	1	47.71
14	95	3	2	0	1	46.10
15	93	3	4	0	0	43.28
16	92.8	3.4	0.5	3.3	0	36.3
17	92.4	3.8	0.8	3	1	63
18	91.8	3.4	0	4.8	1	46.29
19	92.5	3.8	0.7	3	0	34.9

Table 4.11 Testing set of 5 data

No.	Sn (wt.%)	Ag (wt.%)	Cu (wt.%)	Bi (wt.%)	Class	Yield Strength(MPa)
1	95.55	3.8	0.65	95.55	0	42.39
2	95.1	2.9	0	95.1	1	47.71
3	95.5	2.5	2	95.5	1	46.10
4	93.1	2.9	4	93.1	0	43.28
5	89.8	3.4	2	89.8	1	46.29

1. Considering only two features of Sn and Ag, with the data of No. 1 to No. 4 in Table 4.10, use SVC to find the optimal separating hyperplane and the support vectors.
2. Develop an SVC model to classify alloys with the accuracy on the testing set higher than 90%. The accuracy is defined by $\mathrm{ACC} = \frac{T}{T+F}$, where T and F denote the numbers of correctly and wrongly classified alloys, respectively.
3. Develop an SVR model to fit the yield strength of the alloy in the training set, and predict the tensile strength of the alloy in the testing set. The R^2 on the testing set is required to reach 90% at least.
4. With the data of Table 4.3, build an SVC model combined with the Sigmoid kernel, and give the number of support vectors when the LOOCV validation R^2 reaches the maximum.
5. Build up the SVC models on the dataset listed in Table 4.10 by using linear kernel with different penalty parameters of $C = 0.001$, 1 and 1000, and show the variations of optimal separating hyperplanes, margins and support vectors with C.
6. With the dataset of Table 4.10, train a good-fitted SVR model by selecting the hyperparameter values that maximize the LOOCV validation R^2.

References

Boyd, S., & Vandenberghe, L. (2004). *Convex optimization*. Cambridge University Press.

Cortes, C., & Vapnik, V. (1995). Support-vector networks. *Machine Learning, 20*(3), 273–297.

Drucker, H., Burges, C. J., Kaufman, L., et al. (1996). Support vector regression machines. In *Proceedings of the 9th International Conference on Neural Information Processing Systems* (*NIPS*'96) (pp. 155–161). The MIT Press.

Hastie, T. J., Tibshirani, R. J., & Friedman, J. H. (2008). *The elements of statistical learning: Data mining, inference and predication* (2nd ed.). Springer.

Hofmann, T., Schölkopf, B., & Smola, A. J. (2008). Kernel methods in machine learning. *The Annals of Statistics, 36*(3), 1171–1220.

Jiang, Q., Chi, B. Q., & Li, J. C. (2003). A valence electron concentration criterion for glass-formation ability of metallic liquids. *Applied Physics Letters, 82*(18), 2984–2986.

Nocedal, J., & Wright, S. (1999). *Numerical optimization*. Springer Science.

Pedregosa, F., Varoquaux, G., Gramfort, A., et al. (2011). Scikit-learn: Machine learning in Python. *The Journal of Machine Learning Research, 12*, 2825–2830.

Rosset, S., Zhu, J. & Hastie, T. (2003). *Margin maximizing loss functions*. Paper presented at the Advances in Neural Information Processing Systems 16 (NIPS 2003). Vancouver and Whistler, British Columbia, Canada.

Steel, R. G. D., Torrie, J. H., & Dickey, D. A. (1997). *Principles and procedures of statistics: A biometrical approach*. McGraw-Hill.

Suykens, J. A., & Vandewalle, J. (1999). Least squares support vector machine classifiers. *Neural Processing Letters, 9*(3), 293–300.

Takeuchi, A., & Inoue, A. (2005). Classification of bulk metallic glasses by atomic size difference, heat of mixing and period of constituent elements and its application to characterization of the main alloying element. *Materials Transactions, 46*(12), 2817–2829.

Tsochantaridis, I., Joachims, T., Hofmann, T., et al. (2005). Large margin methods for structured and interdependent output variables. *Journal of Machine Learning Research, 6*(9), 1453–1484.
Vapnik, V., & Chervonenkis, A. (1991). The necessary and sufficient conditions for consistency in the empirical risk minimization method. *Pattern Recognition and Image Analysis, 1*(3), 283–305.
Xiong, J., Shi, S. Q., & Zhang, T. Y. (2020). A machine-learning approach to predicting and understanding the properties of amorphous metallic alloys. *Materials & Design, 187*, 108378.
Xiong, J., Zhang, T. Y., & Shi, S. Q. (2019). Machine learning prediction of elastic properties and glass-forming ability of bulk metallic glasses. *MRS Communications, 9*(2), 576–585.

Chapter 5
Decision Tree and *K*-Nearest-Neighbors (KNN)

5.1 Introduction

Decision tree is a typical "divide-and-conquer" approach. It uses a tree-like graph or model to partition the feature space, also called the search space. A decision tree is composed of nodes and lines linking a node to its subsequent nodes. A decision tree is formed from a starting node called the root of the tree, which includes all data of interest and is split into subsequent nodes. Then, each of the subsequent nodes is split again into their successive nodes. The decision tree grows up in such a recursive manner until reaching the terminal nodes called leaves. The data splitting is usually based on one feature associated with a critical value of inputs or a combination of a few features with each associated a critical value of inputs and generates some new branches. From the root to leaves, there are many layers of nodes, and at each layer there are also many nodes (branches). The layer number is called the depth of the tree. A tree is huge if the tree depth is large and the number of terminal nodes is big. There are many decision-tree-based algorithms, such as Iterative Dichotomiser (ID3) (Quinlan, 1986), C4.5 (Quinlan, 1993), Classification and Regression Tree (CART) (Breiman et al., 1984), etc. In general, information gain is used as the information-theoretic split criterion.

5.2 Classification Trees

For simplicity, consider a node $\boldsymbol{D}$ containing n data that are classified into two classes of class 1 and class 2, with n_1 data in class 1 and n_2 data in class 2 so that $n_1 + n_2 = n$. The proportions of classes 1 and 2 in node $\boldsymbol{D}$ are $p_1 = n_1/n$ and $p_2 = n_2/n$, respectively. The number of data in node $\boldsymbol{D}$ is often denoted by $|D|$, and those in classes 1 and 2 by $|D_1|$ and $|D_2|$, respectively. Generally, for a node $\boldsymbol{D} = \{(\boldsymbol{X}_1, y_1), (\boldsymbol{X}_2, y_2), \cdots, (\boldsymbol{X}_n, y_n)\}$ that has n data (i.e., $\boldsymbol{D}$ is an $n \times m$ matrix) with

T. Zhang, *An Introduction to Materials Informatics*,
https://doi.org/10.1007/978-981-99-7992-9_5

datum i having m input features $\boldsymbol{X}_i = (x_{i1}, x_{i2}, \cdots, x_{im})(i = 1, 2, \cdots, n) \in \mathbf{R}^m$ (i.e., $\boldsymbol{X}_i$ is an $1 \times m$ vector) and belonging to class $y_i = k \in \{1, 2, \cdots, K\}$, the proportion of class k in node D is defined by

$$p_k = \frac{n_k}{n}, \quad k = 1, 2, \cdots, K, \tag{5.1}$$

where n_k is the number of data in class k in node $\boldsymbol{D}$. Obviously, $\sum_{k=1}^{K} p_k = 1$. When $n_k \gg n_{j \neq k}$ or $p_k \gg p_{j \neq k}$, class k is regarded as the majority class in node $\boldsymbol{D}$, and the majority class k is also often named solvent and other classes solute or impurity.

The following three parameters are often used to measure the class mixing degree in a node: (1) misclassification error $1 - \max_{k=1,\cdots,k} \{p\}$; (2) The Gini index $\text{GI} \equiv \sum_{k=1}^{K} p_k(1 - p_k)$; and (3) information entropy $S \equiv -\sum_{k=1}^{K} p_k \ln(p_k)$. For binary classification, the three parameters reduce to $1 - \max(p, 1 - p)$, $2p(1 - p)$, and $-p\ln(p) - (1 - p)\ln(1 - p)$, respectively. Obviously, the smaller the misclassification error is, the higher the majority class portion will be, and equivalently the less the class mixing degree is. Similarly, the smaller the Gini index and information entropy are, the less the class mixing degree will be. The decrease in the class mixing degree means information gain. The Gini index and information entropy are more sensitive in the description of class mixing degree than the misclassification error, and thus are more often used in decision tree classification.

A parent node is split into V child nodes based on one or several of the m input features, and the split must cause information gain. The change in information entropy ΔS induced by the split is calculated by

$$\Delta S(x_{ij}) = S^p - \sum_{v=1}^{V} \frac{n_v^c}{n^p} S_v^c, \tag{5.2}$$

where x_{ij} denotes feature x_j and its value i used in the split; the superscripts "p" and "c" denote parent and child, respectively; n^p is the number of data in the parent node; and n_v^c and S_v^c are the numbers of data and information entropy in child node v ($v = 1, 2, \cdots, V$), respectively. $\Delta S(x_{ij})$ quantitatively characterizes the change in information entropy for the split at x_{ij} and indicates information gain when $\Delta S(x_{ij}) > 0$. The information gain varies with the feature and its value used in the split. The optimal $x_{ij} \in \boldsymbol{X}$ to use in the split is determined by maximizing the information gain:

$$x_{IJ} = \underset{x_{ij} \in \boldsymbol{X}}{\text{argmax}}\, \Delta S(x_{ij}), \tag{5.3}$$

where x_{IJ} denotes that when feature J and its value I are used in the split, the information gain is maximized. The maximization of information gain is adopted as the split criterion in the ID3 decision tree algorithm for binary classification.

In analogy with the definition of information entropy, the entropy induced by purely splitting a parent node into V child nodes is called the intrinsic value (IV), which is defined by

$$\mathrm{IV}(x_{ij}) = -\sum_{v=1}^{V} \frac{n_v^c}{n^p} \ln\left(\frac{n_v^c}{n^p}\right). \tag{5.4}$$

The intrinsic value will be bigger if the number of child nodes gets larger. As shown in Eq. (5.4), the information gain also increases with the number of child nodes. To minimize or eliminate the influence of the number of child nodes, the gain ratio (GR) is introduced and defined by

$$\mathrm{GR}(x_{ij}) = \frac{\Delta S(x_{ij})}{\mathrm{IV}(x_{ij})}. \tag{5.5}$$

Maximizing the gain ratio determines which feature and its value will be used in the split, i.e.,

$$x_{IJ} = \underset{x_{ij} \in X}{\mathrm{argmax}}\, \mathrm{GR}(x_{ij}). \tag{5.6}$$

Both information gain and gain ratio are employed in the C4.5 (Quinlan, 1993) algorithm.

Similar to information entropy, the change in the Gini index is defined by

$$\Delta\mathrm{GI}(x_{ij}) = \mathrm{GI}^p(x_{ij}) - \mathrm{GI}^c(x_{ij}), \tag{5.7a}$$

where $\mathrm{GI}^c(x_{ij})$ is the Gini index of the child nodes calculated by

$$\mathrm{GI}^c(x_{ij}) = \sum_{v=1}^{V} \frac{n_v^c}{n^p} \mathrm{GI}_v^c(x_{ij}). \tag{5.7b}$$

The smaller the Gini index is, the higher the node (dataset) purity will be. The feature and its value used in the split can also be determined by maximizing the change in the Gini index, i.e.,

$$x_{IJ} = \underset{x_{ij} \in X}{\mathrm{argmax}}\, \Delta\mathrm{GI}(x_{ij}), \tag{5.7c}$$

or equivalently, by

$$x_{IJ} = \underset{x_{ij} \in \mathbf{X}}{\operatorname{argmin}} \, \mathrm{GI}^c(x_{ij}). \tag{5.7d}$$

The Gini index is used in the CART decision tree algorithm (Breiman et al., 1984).

In practice, the search for the best splitting point is usually conducted on the available features one by one, meaning that each time an optimal point is searched along one feature in the m feature space, then the best splitting point is determined by the maximum of information gain in the m optimal points. A full-size classification tree is defined as the data in each leaf of the tree belongs to only one class.

Example 5.1 Apply the CART decision tree algorithm to classify 22 catalysts listed in Table 5.1 into the two classes of high and low nitrogen oxide (NO_x) conversion, based on the two features of gas hour space velocity (GHSV, the ratio of the volume of gas entering the reactor per hour to the volume of catalysts) and reaction temperature (T). The catalysts with high and low NO_x conversion are labeled as class 1 and class 2, respectively.

Table 5.1 Activity of 22 catalysts

Catalysts	GHSV (h^{-1})	T (°C)	NO_x conversion
Mn/Glass-fiber	50,000	160	Low
Mn_8Ce_1/Glass-fiber	50,000	140	Low
Mn_5Ce_1/Glass-fiber	50,000	150	Low
Mn_1Ce_2/Graphene	24,000	80	Low
$Mn_{20}Gd_1/MnO_x$	36,000	120	Low
$Mn_{10}Gd_1/MnO_x$	100,000	175	Low
$V_1Mn_4Fe_1$/Attapulgite	40,000	100	Low
Mn_5Ce_1/TiO_2	120,000	220	Low
Mn_5Eu_1/TiO_2	108,000	150	Low
Mn/ZSM-5	36,000	180	High
Mn_5Ce_2/Titanate Nanotubes	100,000	300	High
Fe_1Mn_8/MnO_x	30,000	135	High
Fe_1Mn_5/MnO_x	30,000	120	High
Mn/TiO_2	108,000	300	High
Mn_2Ce_1/Graphene	24,000	140	High
Mn_4Ce_1/Graphene	24,000	100	High
Mn_8Ti_2	60,000	250	High
$Fe_1Mn_8Ti_1$	60,000	200	High
$Mn_{10}Gd_3/MnO_x$	36,000	240	High
$Mn_{10}Gd_1/MnO_x$	36,000	200	High
MnO_x	100,000	200	High
Ce/MnO_x	40,000	200	High

In binary classification, Eq. (5.1) is reduced to

$$p = p_1 = \frac{|D_1|}{|D|}, \quad p_2 = \frac{|D_2|}{|D|} = 1 - p, \tag{5.8}$$

where $|D_1|$ and $|D_2|$ denote the sample numbers in class 1 and class 2, respectively, and $|D|$ is the total sample number in node D. The Gini index of node D is $\mathrm{GI}(D) = 2p(1-p)$. When a parent node D is split into two child nodes D^1 and D^2, which are further classified into D_1^1 and D_2^1, and D_1^2 and D_2^2, respectively (we follow the convention that superscript indexes child nodes, while subscript indexes class throughout), Eq. (5.7a) is reduced to

$$\begin{aligned}\Delta\mathrm{GI} = {} & 2\frac{|D_1|}{|D|}\left(1 - \frac{|D_1|}{|D|}\right) - \left[2\frac{|D^1|}{|D|}\frac{|D_1^1|}{|D^1|}\left(1 - \frac{|D_1^1|}{|D^1|}\right)\right. \\ & \left. + 2\frac{|D^2|}{|D|}\frac{|D_1^2|}{|D^2|}\left(1 - \frac{|D_1^2|}{|D^2|}\right)\right],\end{aligned} \tag{5.9}$$

where $|D^1|$ and $|D^2|$ are the sample numbers in nodes D^1 and D^2, respectively (note: Not to confuse them with the sample numbers in classes 1 and 2, namely $|D_1|$ and $|D_2|$). Obviously, $|D^1| + |D^2| = |D|$, $|D_1^1| + |D_1^2| = |D_1|$, and $|D_2^1| + |D_2^2| = |D_2|$, and also $|D_1^1| + |D_2^1| = |D^1|$ and $|D_1^2| + |D_2^2| = |D^2|$. The feature and its value used in the split are determined by maximizing the change in the Gini index of Eq. (5.7c) or equivalently by minimizing the Gini index of the child nodes of Eq. (5.7d). For example, minimizing Eq. (5.7d) in this case means

$$x_{IJ} = \underset{x_{ij} \in X}{\operatorname{argmin}} \left[\frac{|D^1|}{|D|}\frac{|D_1^1|}{|D^1|}\left(1 - \frac{|D_1^1|}{|D^1|}\right) + \frac{|D^2|}{|D|}\frac{|D_1^2|}{|D^2|}\left(1 - \frac{|D_1^2|}{|D^2|}\right)\right]. \tag{5.10}$$

The Gini index of the parent node is

$$\mathrm{GI}^p = 2\frac{|D_1|}{|D|}\left(1 - \frac{|D_1|}{|D|}\right) = 2 \times \frac{9}{22}\left(1 - \frac{9}{22}\right) = 0.4835. \tag{5.11}$$

To calculate $\mathrm{GI}^c(x_{ij})$ and hence $\Delta\mathrm{GI}$, let us first take temperature T as the feature for split, and re-arrange the data by T in the order of low to high, as shown in Table 5.2. If we further take the mean of each adjacent pair of two different T values as the value for split, we calculate the Gini index of child nodes by

$$\mathrm{GI}^c = \frac{|D^1|}{|D|}\frac{|D_1^1|}{|D^1|}\left(1 - \frac{|D_1^1|}{|D^1|}\right) + \frac{|D^2|}{|D|}\frac{|D_1^2|}{|D^2|}\left(1 - \frac{|D_1^2|}{|D^2|}\right), \tag{5.12}$$

and the change in the Gini index by

$$\Delta\text{GI} = \text{GI}^p - \text{GI}^c. \tag{5.13}$$

Table 5.2 shows the calculation results of GI^c and ΔGI for all possible splitting points, indicating obviously that when $T = 177.5\,°\text{C}$, $\text{GI}^c = 0.162$ is the smallest, and $\Delta\text{GI} = 0.3215$ is the largest. $T = 177.5\,°\text{C}$ is therefore the optimal splitting point for feature T. Similar calculation is done for feature GHSV to find the optimal splitting point, which yields $\text{GHSV} = 38000\text{h}^{-1}$, at which $\text{GI}^c = 0.218$, and $\Delta\text{GI} = 0.2655$. Since $\Delta\text{GI} = 0.3215$ at $T = 177.5\,°\text{C}$ is higher than $\Delta\text{GI} = 0.2655$ at $\text{GHSV} = 38000\text{h}^{-1}$, $T = 177.5\,°\text{C}$ is selected as the first splitting point and the tree grows one layer. Subsequently, each child node is divided in the same way until the tree grows up to a designed size.

Table 5.2 All possible splitting points x_{IJ} for feature T and the corresponding Gini indexes

T (°C)	NO_x conversion	x_{IJ}(°C)	GI^c	ΔGI
80	Low	80	0.483	0
100	Low	90	0.225	0.2585
100	High			
120	Low	110	0.231	0.2525
120	High			
135	High	127.5	0.231	0.2525
140	Low	137.5	0.239	0.2445
140	High			
150	Low	145	0.237	0.2465
150	Low			
160	Low	155	0.211	0.2725
175	Low	167.5	0.190	0.2935
180	High	177.5	0.162	0.3215
200	High	190	0.180	0.3035
200	High			
200	High			
200	High			
220	Low	210	0.229	0.2545
240	High	230	0.205	0.2785
250	High	245	0.215	0.2685
300	High	275	0.225	0.2585
300	High			

Overfitting is an often-encountered problem for decision tree. To combat overfitting, the so-called pruning is often performed by reducing the depth of the tree or/and requiring a minimum amount of data at a leaf. To demonstrate the effect of pruning, here we explore the influence of the maximum depth (max_depth) of the tree on the prediction accuracy of the above case of NO_x catalyst. Five-fold cross validation was performed to determine the model hyperparameters. Table 5.3 compares the accuracy of the hence obtained models with different maximum depths. Figure 5.1 shows the details of the two classification tree models and the corresponding regions of different classes in the feature space. These data show that the model obtained with max_depth = 3 is overfitted, even though reaching 100% accuracy on the training set. By pruning the tree to max_depth = 2, the accuracy of the validation set is improved from 77% to 82%.

Table 5.3 Training and validation sets, performance of different values of max_depth

max_depth	Training accuracy (%)	Validation accuracy (%)
3	100	77
2	95.5	82

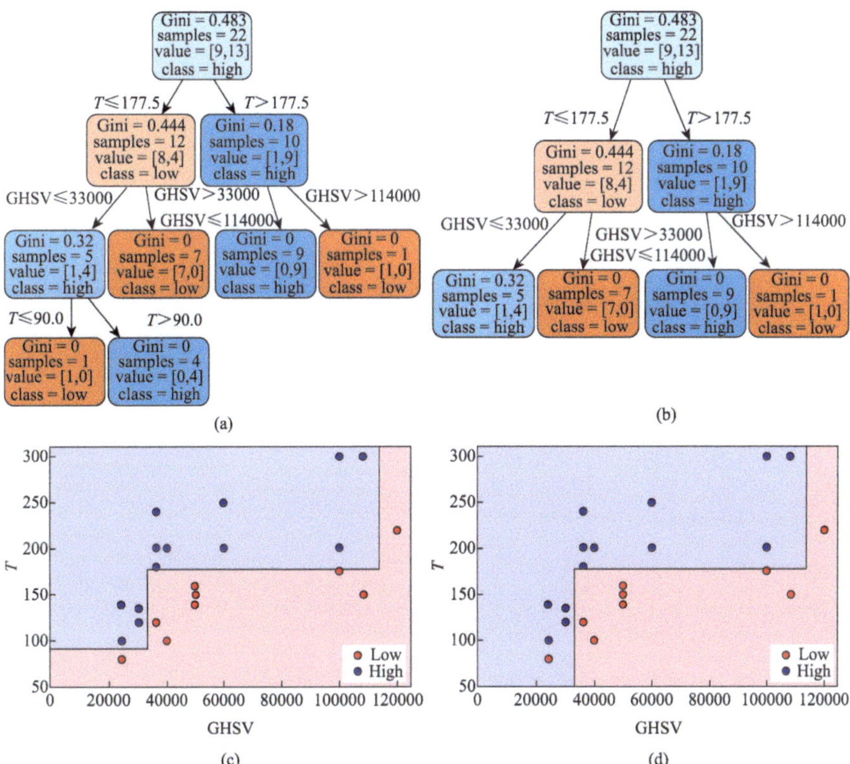

Fig. 5.1 Effect of the max tree depth on the decision tree of the NO_x catalysts. The classification tree with **(a)** max_depth = 3 and **(b)** max_depth = 2; the division of feature space when **(c)** max_depth = 3 and **(d)** max_depth = 2

Example 5.2 CART decision tree algorithm is used to partition the search space of 797,482 alloys of $Ti_{50}Ni_{50\text{-}x-y\text{-}z}Cu_xFe_yPd_z$ with $x \le 20\%$, $y \le 5\%$, $z \le 20\%$, $50\% - x - y - z \ge 30\%$ and the variation step of 0.1% in x, y, and z; the output is the thermal hysteresis (Xue et al., 2016). The goal of the research is to design shape memory alloys with the smallest thermal hysteresis, which is characterized by the difference in the phase transition temperature, ΔT, between those measured by differential scanning calorimetry in heating and cooling cycles. The 53 data used for training in CART are listed in Table 5.4. In binary classification, the 53 data are categorized into class 1 of $\Delta T < 4$K and class 2 of $\Delta T \ge 4$K, respectively, where K is the

Table 5.4 53 alloys of $Ti_{50}Ni_{50\text{-}x-y\text{-}z}Cu_xFe_yPd_z$

Cu (at. %)	Pd (at. %)	ΔT (K)	Class	Cu (at. %)	Pd (at. %)	ΔT (K)	Class
0.8	0.2	1.83898	1	0	0	3.88616	1
0.2	0.2	2.09427	1	2	0	3.88841	1
1.1	0.2	2.31896	1	0	0	3.92136	1
1.9	0.1	2.52766	1	3.6	0.8	4.12627	2
0.9	0.3	2.64172	1	1.5	0	4.13238	2
1.9	0	2.71949	1	0	0	4.20952	2
1.2	0.1	2.74548	1	1.5	1	4.25957	2
0	0.1	2.872	1	2.3	0	4.30849	2
1.6	0.2	3.05015	1	2.1	0	4.54446	2
0	0	3.06755	1	13	3	4.7	2
1.9	0.1	3.0689	1	10	6	5.32	2
0	0.6	3.11654	1	12	4	5.8	2
1.2	0	3.11754	1	5	0	5.83235	2
1.9	0.2	3.1468	1	12	3	5.92619	2
1	0	3.15427	1	0	10	6.04315	2
1.1	0.1	3.23977	1	14	2	6.65	2
2.6	0	3.2483	1	16	0	7.4	2
1.4	0	3.29346	1	1	4	8.36027	2
2.3	0.1	3.30293	1	0	16	8.53	2
2	0	3.4002	1	12	0	8.6171	2
2	0	3.45827	1	4.6	4	9.10042	2
2.1	0	3.49218	1	6	0	10.1638	2
4	0	3.71258	1	10	0	10.33648	2
1.7	0.1	3.72292	1	14	0	10.79427	2
2	0.1	3.73474	1	0.6	0.3	11.04894	2
3	0	3.77995	1	8	0	12.6594	2
2	0.1	3.82909	1				

unit of absolute temperature. Figure 5.2a shows a decision tree, where the split feature and the discriminant value [either X (Cu) or Z (Pd) in the unit of at. %] are given near the branches, and the information about data is given atop the nodes. Figure 5.2b illustrates how the tree in Fig. 5.2a partitions the search space. The Gini index decreases from the original value of 0.4913 to the value of 0.2449 in the left terminal node, which indicates information gain. There are 35 data (alloys) in the left terminal node, among which 30 data and 5 data belong to class 1 and class 2, respectively. The 35 data contain an alloy $Ti_{50}Ni_{48.2}Cu_{0.6}Fe_{0.9}Pd_{0.3}$, of which $\Delta T = 11.05$ K, which is much higher than others. This datum with $\Delta T = 11.05$ K is deemed an outlier and removed in further processing. Special attention should be paid to outliers. Repeated tests or/and calculations should be conducted to make sure the data of outliers are reliable. After that, systematic and comprehensive investigations should be conducted on reliable outliers to explore the hidden science and technology picture in order to gain new knowledge. After the removal of the outlier, the data number is 34 in the left terminal node. The decision tree shown in Fig. 5.2 shrinks the original search space of 797,482 alloys of $Ti_{50}Ni_{50\text{-}x\text{-}y\text{-}z}Cu_xFe_yPd_z$ with $x \leq 20\%$, $y \leq 5\%$, $z \leq 20\%$, $50\% - x - y - z \geq 30\%$ to a much small search space of 17,952 alloys of $Ti_{50}Ni_{50-x-y-z}Cu_xFe_yPd_z$ with $x \leq 4.3\%$, $y \leq 5\%$, $z \leq 0.7\%$, $50\% - x - y - z \geq 30\%$. The shrinkage in the search space and the information gain bring in great advantages in the following machine learning.

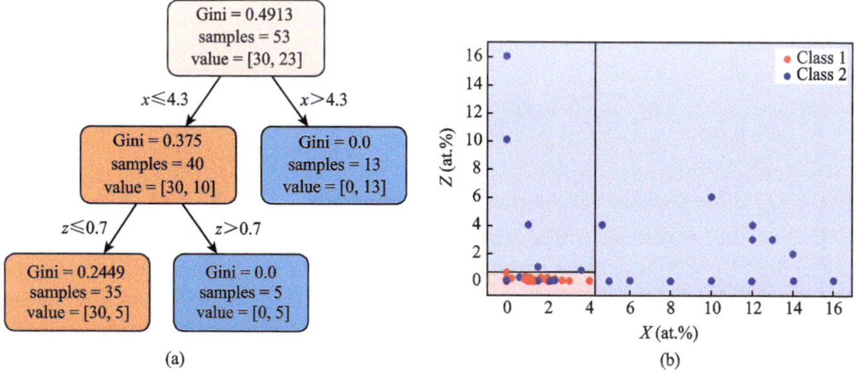

Fig. 5.2 **(a)** A decision tree partitions the 53 shape memory alloys into two classes of $\Delta T < 4$K and $\Delta T \geq 4$K. n_1 and n_2 in "value = $[n_1, n_2]$" indicate the sample numbers with $\Delta T < 4$K and $\Delta T \geq 4$K, respectively; **(b)** partition of the two-dimensional search space by the decision tree leads to a great reduction in the remaining search space for further machine learning

5.3 Regression Tree

Besides classification, decision tree can also be used for regression tasks. Regression tries to find the best correlation function, implicitly or explicitly, between responses and inputs (features). For n data of m-dimensional inputs $\{(\boldsymbol{X}_1, y_1), (\boldsymbol{X}_2, y_2), \cdots, (\boldsymbol{X}_n, y_n)\}$, where each datum input $\boldsymbol{X}_i = (x_{i1}, x_{i2}, \cdots, x_{im})(i = 1, 2, \cdots, n) \in \mathbf{R}^m$ is an $1 \times m$ vector and the one-dimensional response is a scalar $y_i \in \mathbf{R}$, the regression tree algorithm decides the variable(s) and value(s) of the splitting point(s) in order to obtain a function of $\hat{y} = f(\boldsymbol{X})$, which estimates y as accurately as possible without overfitting. Training data are partitioned into K regions of $R_1, R_2, \cdots, R_K$, viz., K terminal nodes. When each of the terminal regions is small enough, the output can be approximately represented by a constant. Otherwise, some functions might be used, or more generally, machine learning with other algorithms can be conducted in each of such large terminal regions. The regression function is expressed by

$$f(X) = \sum_{k=1}^{K} c_k I(\boldsymbol{X}, \boldsymbol{y} \in R_k), \tag{5.14}$$

where $I(\boldsymbol{X}, \boldsymbol{y} \in R_k)$ is the identity function that returns to 1 if the datum $(\boldsymbol{X}_i, y_i)$ is in R_k and 0, otherwise. If the accuracy is measured by the minimum of RSS, we have

$$\hat{c}_k = \underset{\boldsymbol{X}, \boldsymbol{y} \in R_k}{\operatorname{argmin}} \sum_{i=1}^{n_{R_k}} (c_k - y_i)^2 = \overline{y}_{n_{R_k}}, \tag{5.15}$$

where n_{R_k} is the number of data in region R_k, and $\overline{y}_{n_{R_k}} = \frac{1}{n_{R_k}} \sum_{i=1}^{n_{R_k}} y_i$ is their mean.

During the growth of a tree, the feature and value at each splitting point are determined by the global minimalization of RSS in the set of minimal RSSs for all features. The minimal RSS of a feature is calculated by different values of that feature while keeping the values of other features unchanged. More specifically, the splitting point for each feature should yield the minimal RSS but which feature is eventually used is determined by the minimum of the set of minimal RSSs. If value s of feature j is selected as the splitting point, then the two regions $R_1(j, s)$ and $R_2(j, s)$ are separated, as given by

$$R_1(j, s) = \{x | x_{ij} \le s\},$$

$$R_2(j, s) = \{x | x_{ij} > s\}.$$

The optimal splitting point minimizes the mean square error,

$$\min_{j,s}\left[\min_{c_1}\frac{1}{n_{R_1}}\sum_{x_{ij}\in R_1(j,s)}(y_i-c_1)^2+\min_{c_2}\frac{1}{n_{R_2}}\sum_{x_{ij}\in R_2(j,s)}(y_i-c_2)^2\right]. \tag{5.16a}$$

When the optimal output values of c_1 and c_2 are the means of y in the respective regions, Eq. (5.16a) is rewritten as

$$\min_{j,s}\left[\frac{1}{n_{R_1}}\sum_{x_{ij}\in R_1(j,s)}(y_i-\hat{c}_1)^2+\frac{1}{n_{R_2}}\sum_{x_{ij}\in R_2(j,s)}(y_i-\hat{c}_2)^2\right], \tag{5.16b}$$

where $\hat{c}_k=\overline{y}_{n_{R_k}}$. A full-size regression tree is defined as that the RSS in each leaf of the tree nulls.

In prediction, if a new datum with input X^* is classified into one of the trained terminal nodes $X^*\in R_k, k=1,2,\cdots,K$, the predicted y^* will be $\overline{y}_{n_{R_k}}$. Obviously, if a terminal node contains less data, the averaged value of $\overline{y}_{n_{R_k}}$ will fit the data better, and the function $f(X)$ will have a higher value of goodness of fit. However, if each terminal node contains too less data, the model will be too complicated and causes overfit. Many methods have been developed to avoid potential overfit (Breiman et al., 1984). Widely-used methods can be categorized into two general approaches. The first approach is to control the tree size in advance by presetting the values of the maximum depth of the tree, the maximum terminal nodes (leaves), the minimum data in each terminal node (leaf), and/or the minimum impurity decrease, etc. The second approach is to prune the tree after a tree is fully formed without any restriction, which includes reduced error pruning and cost-complexity pruning. Denote a completely-full-size tree by T_0 and a subset tree by $T\subset T_0$, which is obtained by pruning the full-size tree. Reduced error pruning simply compares the errors before and after pruning, viz., comparing the error of T with that of T_0. Cost-complexity pruning works similarly as LASSO, of which the criterion is given by

$$T_\alpha=\underset{T\subset T_0}{\operatorname{argmin}}\,C_\alpha(T), \tag{5.17a}$$

$$C_\alpha(T)=\sum_{k=1}^{|T|}\sum_{i=1}^{n_{R_k}}(c_k-y_{ki})^2+\alpha|T|, \tag{5.17b}$$

where $|T|$ is the number of terminal nodes of tree T, and $\alpha\geq 0$ is a preset constant. Obviously, the number of terminal nodes of tree T is smaller than the original number K of tree T_0, i.e., $|T|\leq K$. For a given value of α, there is a unique smallest subtree T_α that is determined by minimizing $C_\alpha(T)$. $\alpha|T|$ is the penalty term. When the α value is large, the subtree will be small, and with $\alpha=0$, the tree will be the full size.

Example 5.3 Decision tree is easily overfitted if growing without any restriction. Overfitting means the model performs well on the training dataset, but poorly on the testing dataset. Therefore, the tree must be pruned to improve its generalization performance. Tree pruning can be done by setting restrictions on tree depth, viz.,

Table 5.5 The efficiency of nitrogen oxide conversion of 29 catalysts

Catalysts	Mn (at.%)	Ce (at.%)	Fe (at.%)	GHSV (h^{-1})	H_2O (vol%)	SO_2 (ppm)	T(°C)	NO_x conversion (%)
Mn_8Ce_1/MnO_x	0.1407	0.075	0	24,000	0	0	80	0.99
$Mn_{16}Ce_1/MnO_x$	0.1723	0.0693	0	24,000	0	0	100	0.99
Mn_2Ce_1/MnO_x	0.0789	0.1155	0	24,000	10	0	140	0.93
Mn_2Ce_1/MnO_x	0.0789	0.1155	0	24,000	0	200	140	0.73
Mn_2Ce_1/MnO_x	0.0789	0.1155	0	24,000	10	200	140	0.65
Mn_8Ce_1/MnO_x	0.1407	0.075	0	24,000	10	0	140	0.98
Mn_8Ce_1/MnO_x	0.1407	0.075	0	24,000	0	200	140	0.77
Mn/Ce-acetic acid	0.0391	0	0	60,000	5	100	200	0.67
Mn/Ce-ethanol	0.0383	0	0	60,000	5	100	200	0.65
$Ce_{0.2}/MnO_x$	0.116	0.023	0	40,000	0	100	200	0.70
Mn/Ce-acetic acid	0.0391	0	0	60,000	0	0	200	0.90
$Fe_{0.2}Mn_{0.8}/TiO_2$	0.139	0	0.023	60,000	0	0	200	0.96
$Fe_{0.3}Mn_{0.8}/TiO_2$	0.132	0	0.033	60,000	0	0	200	0.97
$Fe_{0.1}Mn_{0.8}/TiO_2$	0.153	0	0	60,000	0	0	250	0.90
$Fe_{0.2}Mn_{0.8}/TiO_2$	0.148	0	0.012	60,000	0	0	200	0.92
Mn/Ce-oxalic acid	0.0496	0	0	60,000	0	0	175	0.93
$Ce_{0.2}/TiO_2$	0	0.046	0	50,000	0	0	300	0.96
$Ce_{0.6}/TiO_2$	0	0.053	0	50,000	0	0	300	0.99
Pure CeO_2	0	0.328	0	50,000	0	0	400	0.85
$Ce_{0.6}/TiO_2$	0	0.053	0	100,000	0	0	350	0.98
$Ce_{0.6}/TiO_2$	0	0.053	0	30,000	0	0	250	0.96
$Ce_{0.6}/TiO_2$	0	0.053	0	12,000	0	0	200	0.99

(continued)

Table 5.5 (continued)

Catalysts	Mn (at.%)	Ce (at.%)	Fe (at.%)	GHSV (h^{-1})	H_2O (vol%)	SO_2 (ppm)	T(°C)	NO_x conversion (%)
$Ce_{0.01}$/ TiO_2	0	0.0001	0	50,000	0	0	400	0.81
$Ce_{0.5}/TiO_2$	0	0.082	0	30,000	0	0	275	1
$Fe_{0.02}Ce$/ TiO_2	0	0.079	0.006	30,000	0	0	250	1
$Fe_{0.1}Ce$/ TiO_2	0	0.075	0.011	30,000	0	0	250	1
$Fe_{0.2}Ce$/ TiO_2	0	0.074	0.015	30,000	0	0	225	1
$Fe_{0.7}Ce$/ TiO_2	0	0.068	0.037	30,000	0	0	225	1
$Ce_{0.6}/TiO_2$	0	0.053	0	50,000	0	0	350	0.98

setting the maximum of tree depth, max_depth. Nitrogen oxide (NO_x) conversion efficiencies of 29 catalysts are listed in Table 5.5. The 7 input features include the at.% of elements Mn, Ce and Fe in the catalysts, the test conditions of GHSV, water content (H_2O), sulfur dioxide content (SO_2), and reaction temperature (T), while the target property is NO_x conversion efficiency in percentage.

The data for different features differ significantly in their spans, so they are first normalized to the range of (0, 1) with

$$\hat{x}_{ij} = \frac{x_{ij} - x_{\min,j}}{x_{\max,j} - x_{\min,j}}, \quad j = 1, 2, \cdots, m. \tag{5.18}$$

Take feature SO_2 as an example. Its normalized data have three values of 0, 0.5, and 1 only. Thus, there are only two potential splitting points, 0.25 and 0.75. Note that when one feature and its value are used as the splitting point, the average of other features should be used in the calculation of mean square error (MSE). Table 5.6 shows the splitting detail for feature SO_2. The MSE in each child node is calculated by $\frac{1}{n_{R_r}} \sum\limits_{x_{ij} \in R_r(j,s)} (y_i - \hat{c}_r)^2, r = 1, 2.$

Summing up the MSEs in the two child nodes gives the MSE of Eq. (5.17b), which is listed in Table 5.7.

Table 5.6 Predictions and errors when the splitting point is normalized $SO_2 = 0.25$ ppm

SO_2 (ppm)	Normalized SO_2	NO_x conversion (%)	$\hat{c}_1, \hat{c}_2$ values	Predicted value	MSE in each child node
200	1	0.73	$\hat{c}_1 = 0.695$	0.695	0.002
200	1	0.65		0.695	
200	1	0.77		0.695	
100	0.5	0.67		0.695	
100	0.5	0.65		0.695	
100	0.5	0.70		0.695	
0	0	0.90	$\hat{c}_2 = 0.956$	0.956	0.003
0	0	0.96		0.956	
0	0	0.97		0.956	
0	0	0.90		0.956	
0	0	0.92		0.956	
0	0	0.93		0.956	
0	0	0.96		0.956	
0	0	0.99		0.956	
0	0	0.85		0.956	
0	0	0.98		0.956	
0	0	0.96		0.956	
0	0	0.99		0.956	
0	0	0.81		0.956	
0	0	1		0.956	
0	0	1		0.956	
0	0	1		0.956	
0	0	1		0.956	
0	0	1		0.956	
0	0	0.98		0.956	
0	0	0.99		0.956	
0	0	0.99		0.956	
0	0	0.93		0.956	
0	0	0.98		0.956	

Table 5.7 MSEs of the two splitting points of feature SO_2

Splitting points	0.25	0.75
MSE	0.005	0.013

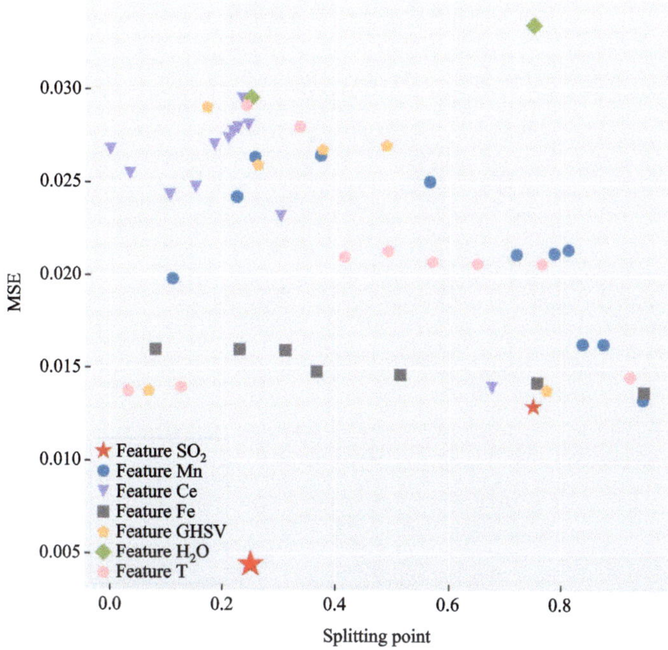

Fig. 5.3 MSE values for all potential splitting points

Using the above method, all potential splitting points in the feature space are searched in a greedy manner, and the MSEs are shown in Fig. 5.3, indicating that the first optimal splitting point is $(j, s) = (\mathrm{SO_2}, 0.25)$. Subsequently, the tree grows from the root to the first layer of two nodes. Repeating the splitting process without any pruning grows a full-size tree, as shown in Fig. 5.4a.

Ten-fold cross validation with the coefficient of determination R^2 is used to examine the effect of tree size on the accuracy of the regression tree model. For instance, Fig. 5.5 shows the variations of R^2 with max_depth on the training set and the validation set. As expected, the R^2 value approaches monotonically to one as max_depth increases on the training set. On the validation set, however, the R^2 increases with max_depth until max_depth = 5 and decreases afterwards. Therefore, the model is underfitting when max_depth < 5, overfitting when max_depth > 5, and good-fitting when max_depth = 5. Similarly, a decision tree can also be pruned by controlling the number of samples in each leaf node (min_samples_leaf) or/and the maximum number of leaf nodes (max_leaf_nodes). Figure 5.4 shows the tree shapes (the model complexity) before and after pruning.

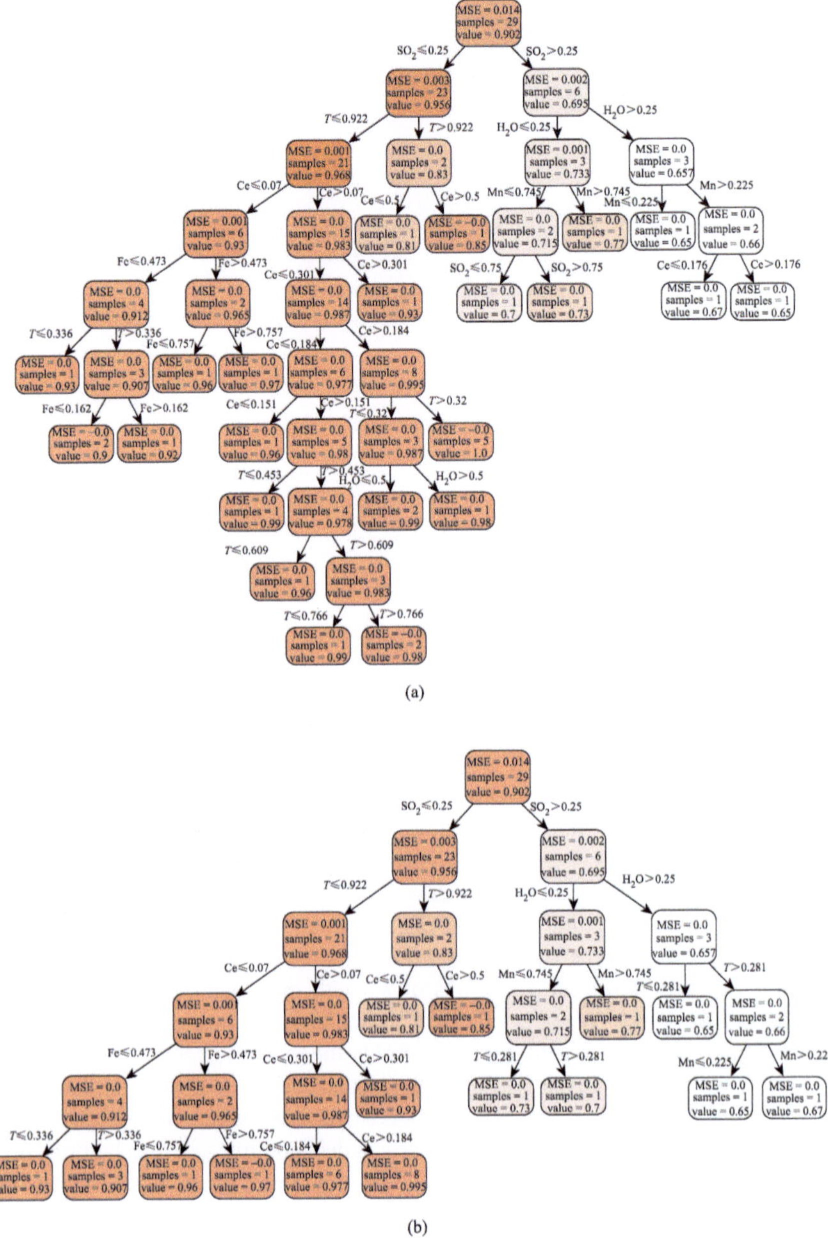

Fig. 5.4 Tree shapes before and after pruning. **(a)** Full size, and tenfold CV optimization subject to the constraints of **(b)** max_depth = 5, **(c)** max_depth = 5 and min_samples_leaf = 2, **(d)** max_depth = 5 and max_leaf_nodes = 6

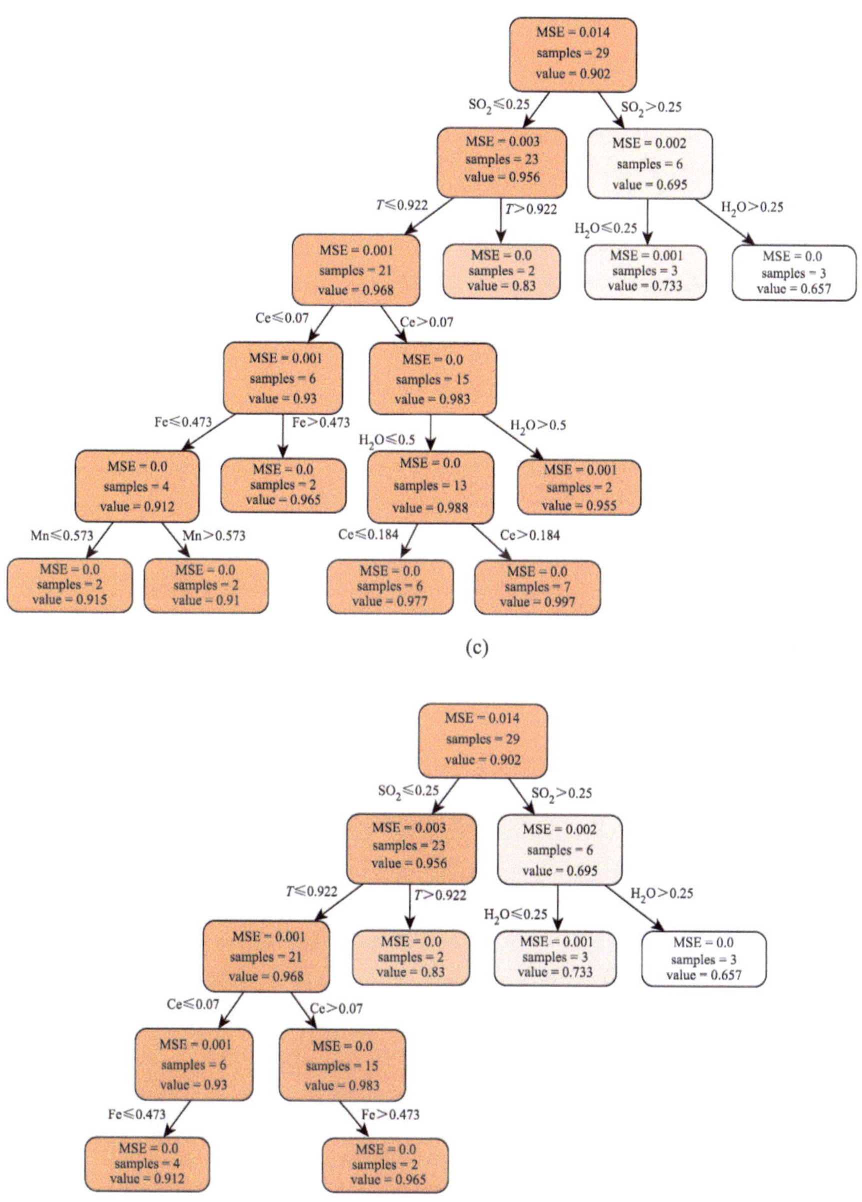

Fig. 5.4 (continued)

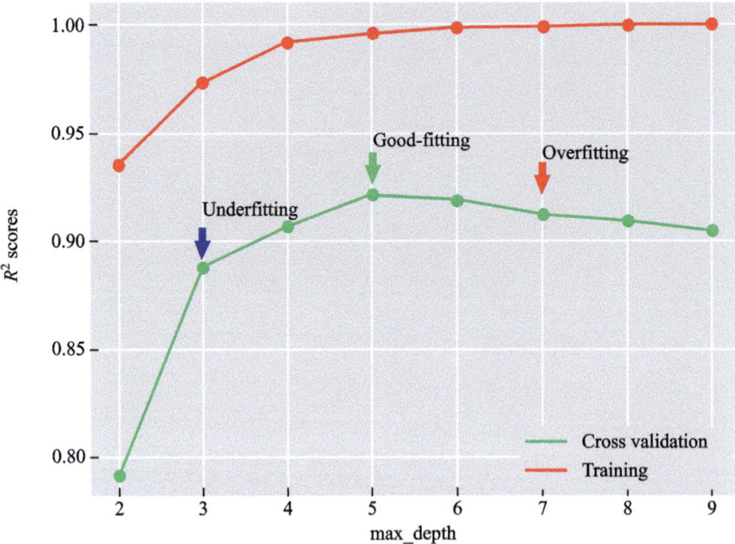

Fig. 5.5 R^2 of the training set and the validation set under different values of max_depth

---

5.4 K-Nearest-Neighbors (KNN) Methods

Principally, the KNN method predicts the response of a sample (datum) based on the major or averaged response of its KNN data in a dataset. If there are n data, $(x_{i1}, x_{i2}, \cdots, x_{im}, y_i)(i = 1, 2, \cdots, n)$, $\boldsymbol{X} \in \mathbf{R}^m$ and $\boldsymbol{y} \in \mathbf{R}$, the predicted response $\hat{y}_u$ of sample $\boldsymbol{X}_u = (x_{u1}, x_{u2}, \cdots, x_{um}, y_u)$ is given by

$$\hat{y}_u = \frac{1}{k} \sum_{i=1}^{k} y_i(\boldsymbol{X}_i \in R_k(\boldsymbol{X})) \tag{5.19a}$$

where $R_k(\boldsymbol{X})$ denotes the KNN data of sample $\boldsymbol{X}_u$. When a tested datum is taken as the center, the top K data closest to the tested datum in a training dataset are the nearest neighbors of the tested datum. In KNN regression, the accuracy is measured by the minimum of RSS. Then, the predicted value of a tested datum is determined by

$$\hat{y}(x) = \underset{x,y \in R_k}{\operatorname{argmin}} \sum_{i=1}^{k} (\hat{y}(x) - y_i)^2 = \overline{y}_k, \tag{5.19b}$$

where R_k denotes the region of the nearest neighbors, k is the number of the nearest neighbors. In classification, the prediction may take the major response by voting or weighted voting. Taking binary classification for example, we have $y \in \{-1, +1\}$ and

$$\hat{y}(x) = \underset{x,y \in R_k}{\text{sign}} \left\{ \sum_{i=1}^{k} y_i \right\}, \tag{5.19c}$$

where sign(·) is a signal function defined $\text{sign}(M) = \begin{cases} 1, & M > 0, \\ 0, & M = 0, \\ -1, & M < 0. \end{cases}$

The number of the nearest neighbors plays an important role in the KNN method. When $K = 1$, the KNN method is called **the 1-nearest-neighbor method**.

The nearest neighbors of a point are measured by a critical distance to the point, or the critical radius of a sphere centered in high-dimensional space, within which samples in a training dataset are the nearest neighbors of the point. The critical distance, in turn, is determined by a preset critical number of the nearest neighbors. Thus, the critical distance determined by the critical number may vary from point to point. The general distance measure is the Minkowski distance, defined by

$$\text{dist}_{\text{mk}}(\boldsymbol{x}_j, \boldsymbol{x}_u) = \left(\sum_{i=1}^{m} \left| \boldsymbol{x}_{ji} - \boldsymbol{x}_{ui} \right|^p \right)^{1/p} \quad \text{with} \quad p \geq 1. \tag{5.20a}$$

When $p = 2$, the distance is the widely-used Euclidean distance,

$$\text{dist}_{\text{ed}}(\boldsymbol{x}_j, \boldsymbol{x}_u) = \left\| \boldsymbol{x}_j - \boldsymbol{x}_u \right\|_2 = \sqrt{\sum_{i=1}^{m} (\boldsymbol{x}_{ji} - \boldsymbol{x}_{ui})^2}. \tag{5.20b}$$

When $p = 1$, the distance is called the Manhattan distance,

$$\text{dist}_{\text{man}}(\boldsymbol{x}_j, \boldsymbol{x}_u) = \left\| \boldsymbol{x}_j - \boldsymbol{x}_u \right\|_1 = \sum_{i=1}^{m} \left| \boldsymbol{x}_{ji} - \boldsymbol{x}_{ui} \right|. \tag{5.20c}$$

Example 5.4 Table 5.8 shows 4 alloys, including 2 crystalline alloys (CRA) and 2 bulk metallic glasses (BMG), collected from reference Xiong et al. (2020). The enthalpy of mixing (H_{mix}) and entropy of mixing (S_{mix}) are the only two features considered in the example, which are calculated respectively by

$$S_{\text{mix}} = -R \sum_{i=1}^{N} a_i \ln a_i, \tag{5.21}$$

$$H_{\text{mix}} = 4\sum_{j=i}^{N}\sum_{i=1}^{N}\Delta H_{ij}a_i a_j, \tag{5.22}$$

where R is the gas constant, a_i is the atomic fraction of the i-th constituent, and ΔH_{ij} is the molar mixing enthalpy for binary liquid alloys.

Table 5.8 Phases of 4 alloys in the training dataset

Data	Alloys	H_{mix}	S_{mix}	Phases	Classes
X_1	$Ag_{10.5}Cu_{80.5}P_9$	−5.09	5.22	CRA	+1
X_2	$Ag_{10}Ce_5Cu_{85}$	−3.49	4.31	CRA	+1
X_3	$Cu_{34}Zr_{50}Ag_8Al_8$	−25.87	9.29	BMG	−1
X_4	$(Zr_{0.55}Al_{0.20}Co_{0.20}Cu_{0.05})_{95}Ag_5$	−40.10	10.52	BMG	−1

The distances between the tested datum and the four training data are (Table 5.9).
$D_1(X^*, X_1) = \sqrt{((-5.09) - (-3.15))^2 + (5.22 - 5.72)^2} = 2,$
$D_2(X^*, X_2) = \sqrt{((-3.49) - (-3.15))^2 + (4.31 - 5.72)^2} = 1.45,$
$D_3(X^*, X_3) = \sqrt{((-25.87) - (-3.15))^2 + (9.29 - 5.72)^2} = 23.0,$
$D_4(X^*, X_4) = \sqrt{((-40.10) - (-3.15))^2 + (10.52 - 5.72)^2} = 37.3.$

Table 5.9 Phases of one alloy in the test dataset

Datum	Alloy	H_{mix}	S_{mix}	Phase
X^*	$Ag_{20}Ce_5Cu_{75}$	−3.15	5.72	CRA

If $k = 1$, the shortest distance between the tested datum and one datum in the training dataset determines the center that the tested datum belongs to. Clearly, X_2 ($Ag_{10}Ce_5Cu_{85}$) is the nearest datum to the tested datum X^* with $D_2 = 1.45$. According to Eq. (5.19c), we have

$$\hat{y}(x) = \underset{x,y\in R_1}{\text{sign}}\left\{\sum_{i=1}^{1} y_i\right\} = \text{sign}\{\ (+1)\} = +1,$$

where $R_1 = \{X_2\}$. Therefore, X^* belongs to the same class as X_2, a CRA alloy.

If $k = 2$, the top two nearest neighbors of X^* are X_1 ($Ag_{10.5}Cu_{80.5}P_9$, with distance D_1) and X_2 ($Ag_{10}Ce_5Cu_{85}$, with distance D_2). Then, applying Eq. (5.19c) gives

$$\hat{y}(x) = \underset{x,y\in R_2}{\text{sign}}\left\{\sum_{i=1}^{2} y_i\right\} = \text{sign}\{\ (+1) + (+1)\} = +1,$$

where $R_2 = \{X_1, X_2\}$. Thus X^* is a CRA alloy labelled " +1".

If $k = 3$, the top three nearest neighbors of X^* are X_1 ($Ag_{10.5}Cu_{80.5}P_9$, with distance D_1), X_2 ($Ag_{10}Ce_5Cu_{85}$, with distance D_2) and X_3 ($Cu_{34}Zr_{50}Ag_8Al_8$, with distance D_3). Then, applying Eq. (5.19c) gives

$$\hat{y}(x) = \underset{x,y \in R_3}{\text{sign}} \left\{ \sum_{i=1}^{3} y_i \right\} = \text{sign}\{ (+1) + (+1) + (-1)\} = +1,$$

where $R_3 = \{X_1, X_2, X_3\}$. Thus X^* is a CRA alloy labelled " +1".

Example 5.5 Lead-free solder alloy data include four chemical elements of Tin (Sn), Silver (Ag), Copper (Cu), and Bismuth (Bi) in wt.% and one property of yield strength in MPa (Tables 5.10 and 5.11).

Table 5.10 Lead-free solder alloys in the training dataset

Sn (wt.%)	Ag (wt.%)	Cu (wt.%)	Bi (wt.%)	Yield (MPa)	Data
95	3	0	2	47.7	X_1
95	3	2	0	46.1	X_2
93	3	4	0	43.3	X_3
91.8	3.4	0	4.8	46.3	X_4

Table 5.11 Lead-free solder alloy in the test datum

Sn (wt.%)	Ag (wt.%)	Cu (wt.%)	Bi (wt.%)	Yield (MPa)	Data
95.5	3.8	0.7	0	46.4	X^*

The distances between the tested datum and the four training data are

$$D_1(X^*, X_1) = \sqrt{(95 - 95.5)^2 + (3 - 3.8)^2 + (0 - 0.7)^2 + (2 - 0)^2} = 2.32,$$
$$D_2(X^*, X_2) = \sqrt{(95 - 95.5)^2 + (3 - 3.8)^2 + (2 - 0.7)^2 + (0 - 0)^2} = 1.61,$$
$$D_3(X^*, X_3) = \sqrt{(93 - 95.5)^2 + (3 - 3.8)^2 + (4 - 0.7)^2 + (0 - 0)^2} = 4.54,$$
$$D_4 = \sqrt{(91.8 - 95.5)^2 + (3.4 - 3.8)^2 + (0 - 0.7)^2 + (4.8 - 0)^2} = 6.11.$$

The next next nearest neighbor, the next nearest neighbor, and the nearest neighbor of the tested datum are X_3, X_1, and X_2, respectively. Thus, if $k = 1, 2$, and 3, the predicted yield strengths of the tested datum X^* are $\hat{y}(X^*) = y(X_2) = 46.1\text{MPa}$, $\hat{y}(X^*) = \frac{1}{2}\sum_{i=1}^{2} y(X_i) = \frac{1}{2} \times (46.1 + 47.7) = 46.9\text{MPa}$, and $\hat{y}(X^*) = \frac{1}{3}\sum_{i=1}^{3} y(X_i) = \frac{1}{3} \times (46.1 + 47.7 + 43.3) = 45.7\text{MPa}$, respectively.

Homework

Part 1: In Table 5.12, 20 data of Mn-Ce-Fe catalysts are given with the same input features and target property as in Example 5.3.

1. Use the above data to perform the following tasks.
(1) Use the decision tree regressor to fit the NO_x conversion of the catalysts in the dataset, and R^2 on the validation set in tenfold cross validation should be no less than 0.75.
(2) Determine the feature and its value used in the first splitting point?
2. Use the alloy data in Table 5.4 to calculate and answer the following questions.
(1) How many splitting points are there for feature Pd (at.%)?
(2) What is the information gain when divided by each splitting point for feature Pd (at.%)?
(3) Is it reasonable to take feature Pd (at.%) as the basis for the first division? If not, which feature should be used?

Table 5.12 The data of Mn-Ce-Fe catalysts

Catalysts	Mn (at.%)	Ce (at.%)	Fe (at.%)	GHSV (h^{-1})	H_2O (vol%)	SO_2 (ppm)	T (°C)	NO_x conversion(%)
MnCe/ TiO_2	0.0969	0.052	0	100,000	0	0	300	0.80
Ce_1Mn_4/ Glass-fiber	0.0554	0.0163	0	50,000	0	0	150	0.84
Ce_1Mn_5/ Glass-fiber	0.0858	0.018	0	50,000	0	0	150	0.86
Mn/ ZSM-5 (200 °C)	0.165	0	0	36,000	0	0	180	0.99
Mn/ ZSM-5 (300 °C)	0.1676	0	0	36,000	0	0	180	0.99

(continued)

Table 5.12 (continued)

Catalysts	Mn (at.%)	Ce (at.%)	Fe (at.%)	GHSV (h^{-1})	H_2O (vol%)	SO_2 (ppm)	T (°C)	NO_x conversion(%)
Mn/ZSM-5 (400 °C)	0.1621	0	0	36,000	0	0	180	0.99
Mn/ZSM-5 (500 °C)	0.1579	0	0	36,000	0	0	210	0.99
Mn/ZSM-5 (600 °C)	0.1545	0	0	36,000	0	0	240	0.99
$FeMnO_x$	0.2227	0	0.1083	30,000	10	100	120	0.75
$Ce_5/FeMnO_x$	0.1809	0.0038	0.0751	30,000	0	0	135	0.98
$Ce_5/FeMnO_x$	0.1809	0.0038	0.0751	30,000	0	100	120	0.92
$Ce_{10}/FeMnO_x$	0.1701	0.0056	0.0704	30,000	0	0	135	0.98
$Ce_{10}/FeMnO_x$	0.1701	0.0056	0.0704	30,000	0	100	120	0.914
$Ce_{12.5}/FeMnO_x$	0.1937	0.0068	0.0823	30,000	0	0	135	0.99
$Ce_{12.5}/FeMnO_x$	0.1937	0.0068	0.0823	30,000	0	100	120	0.956
$Ce_{12.5}/FeMnO_x$	0.1937	0.0068	0.0823	30,000	5	0	120	0.925
$Ce_{12.5}/FeMnO_x$	0.1937	0.0068	0.0823	30,000	10	0	120	0.81
$MnCe_2/MnO_x$	0.0345	0.147	0	24,000	0	0	140	0.99
Mn_2Ce/MnO_x	0.0789	0.1155	0	24,000	0	0	140	0.99
Mn_4Ce/MnO_x	0.1305	0.0782	0	24,000	0	0	100	0.99

Part 2:

1. Please discuss the similarities and differences between the decision tree method and the KNN method.
2. Use the KNN method to perform the regression task using the data in Table 5.5.

References

Breiman, L., Friedman, J. H., Stone, C. J., et al. (1984). *Classification and regression trees.* Chapman and Hall/CRC.

Quinlan, J. R. (1993). C4.5: Programs for machine learning. Morgan Kaufmann.

Quinlan, J. R. (1986). Introduction of decision trees. *Machine Learning, 1*(1), 81–106.

Xue, D., Balachandran, P. V., Hogden, J., et al. (2016). Accelerated search for materials with targeted properties by adaptive design. *Nature Communications, 7*(1), 1–9.

Chapter 6
Ensemble Learning

Ensemble learning is committee-based learning, in which each committee member is an individual learner, called a base learner with a base learning algorithm. The prediction of ensemble learning is given by voting or weighted voting of base learners. If all base learning algorithms in an ensemble learner are the same, the ensemble learning is homogeneous, otherwise, heterogeneous, and there a base learner is then often called a component learner. Homogeneous ensemble learning is taken in this chapter to illustrate the working principle of ensemble learning.

Consider an ensemble learner $G_M(x)$, which includes M base learners $g_j(x)$, $j = 1, 2, \cdots, M$. With equal-weight base learners and weighted base learners, we have

$$G_M(x) = \sum_{j=1}^{M} g_j(x) \tag{6.1a}$$

and

$$G_M(x) = \sum_{j=1}^{M} \alpha_j g_j(x), \tag{6.1b}$$

respectively, where α_j is the weight of the j-th base learner $g_j(x)$, $j = 1, 2, \cdots, M$. Base learners $g_j(x)$ are often expressed by $T_j(x)$ as well.

There are two major types of ensemble learning: boosting and bagging. The boosting method of ensemble learning behaves like many base learners in series, and all data in a training dataset are used by each base learner. Thus, the boosting ensemble learning is conducted in an iterative manner, and each iteration is termed as a base learner, as shown in Fig. 6.1a. As the iteration goes, the weights of base learners are optimized sequentially, and the weighted values are updated accordingly. The ensemble learner assembles the outputs from all base learners with optimal weights. The bagging method is such an ensemble learning that all base learners act

T. Zhang, *An Introduction to Materials Informatics*,
https://doi.org/10.1007/978-981-99-7992-9_6

in parallel, as shown in Fig. 6.1b, where the data for a base learner are randomly sampled with replacement (using the bootstrap sampling method) from a training dataset. The data already used for a base learner are replaced in the training dataset, and have the same probability for being sampled by another base learner. In addition to data, features are also randomly chosen in the bootstrap sampling manner. The output of a bagging-based homogeneous ensemble learner usually averages the outputs of its base learners.

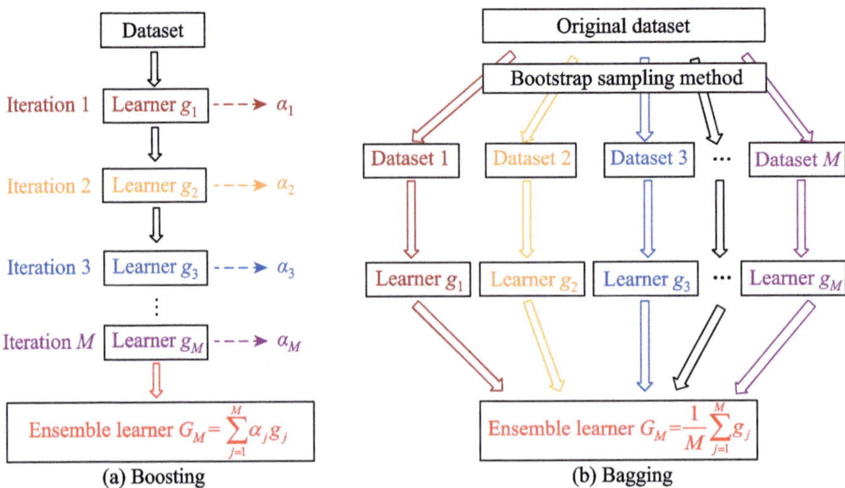

Fig. 6.1 Schematics of **(a)** boosting and **(b)** bagging

6.1 Boosting

6.1.1 AdaBoost Classification

AdaBoost, or "AdaBoost M1", named by its inventors Freund and Schapire (1997) is one of the most famous boosting algorithms. The training dataset contains n data of (x_i, y_i), $i = 1, 2, \cdots, n$ and $y_i \in (-1, +1)$. The AdaBoost algorithm is developed based on the exponential loss function of

$$\mathcal{L}\big(y, G_j(x)\big) = \sum_{i=1}^{n} \exp\big(-y_i G_j(x_i)\big), \tag{6.2a}$$

where $G_j(x_i)$ denotes an ensemble learner with j base learners, which is obtained after j iterations. The exponential loss function is widely used in classification. The ensemble learner has the probabilities of $P\big(G_j(x_i) = 1|x_i\big)$ and $P\big(G_j(x_i) = -1|x_i\big)$ to classify a given datum x_i to class $+1$ and class -1, respectively. To find the general property of $G_j(x_i)$, we differentiate the exponential loss function with respect to $G_j(x_i)$ and let it be zero, which is the requirement of the minimum of loss function:

$$\begin{aligned}\frac{\partial \mathcal{L}\big(y_i, G_j(x_i)\big)}{\partial G_j(x_i)} &= -y_i \exp\big(-y_i G_j(x_i)\big)\\ &= -\exp\big(-G_j(x_i)\big)P\big(G_j(x_i) = 1|x_i\big)\\ &\quad + \exp\big(G_j(x_i)\big)P\big(G_j(x_i) = -1|x_i\big) = 0,\end{aligned} \tag{6.2b}$$

Equation (6.2b) gives

$$G_j(x_i) = \frac{1}{2}\ln\frac{P\big(G_j(x_i) = 1|x_i\big)}{P\big(G_j(x_i) = -1|x_i\big)}. \tag{6.2c}$$

Therefore, we have

$$\begin{aligned}\operatorname{sign}\big(G_j(x_i)\big) &= \begin{cases}1, & P\big(G_j(x_i) = 1|x_i\big) > P\big(G_j(x_i) = -1|x_i\big)\\ -1, & P\big(G_j(x_i) = 1|x_i\big) < P\big(G_j(x_i) = -1|x_i\big)\end{cases}\\ &= \operatorname{argmax}_{y_i\in\{-1,+1\}} P\big(G_j(x_i) = y_i|x_i\big).\end{aligned} \tag{6.2d}$$

By definition, we have

$$\sum_{i=1}^{n} \exp\big(-y_i G_j(x_i)\big) = \sum_{i=1}^{n} \exp\big(-y_i\big(G_{j-1}(x_i) + \alpha_j g_j(x_i)\big)\big), \tag{6.3a}$$

where base learner $g_j(x_i)$ has weight α_j. Equation (6.3a) can be rewritten as

$$\sum_{i=1}^{n} \exp\big(-y_i G_{j-1}(x_i)\big)\exp\big(-y_i\big(\alpha_j g_j(x_i)\big)\big) = \sum_{i=1}^{n} w_{i,j}\exp\big(-y_i\big(\alpha_j g_j(x_i)\big)\big) \tag{6.3b}$$

in which the weight $w_{i,j}$ of base learner $g_j(x_i)$ for datum x_i is defined by

$$w_{i,j} \equiv \exp\big(-y_i G_{j-1}(x_i)\big). \tag{6.3c}$$

The $w_{i,j}$ definition of Eq. (6.3c) indicates

$$w_{i,j} = \begin{cases}\mathrm{e}, & \text{if } y_i \neq G_{j-1}(x_i),\\ \frac{1}{\mathrm{e}}, & \text{if } y_i = G_{j-1}(x_i).\end{cases} \tag{6.3d}$$

Therefore, misclassified data by the ensemble classifier $G_{j-1}(x_i)$ are assigned a higher value of weight so that misclassification in $G_{j-1}(x_i)$ can be corrected in the following base classifier $G_j(x_i)$. Initially, the equal-weight $w_{i,1} = 1/n$ is assigned to every one of the data.

The optimal of both the base learner g_j and its weight α_j are determined by the minimum of the loss function Eq. (6.3b):

$$\left(\alpha_j, g_j\right) = \text{argmin} \sum_{i=1}^{n} w_{i,j} \exp\left(-y_i \alpha_j g_j(x_i)\right). \tag{6.4a}$$

The weight $w_{i,j}$ of base learner $g_j(x_i)$ for datum x_i is determined by the performance of ensemble learner G_{j-1} of the previous iteration according to Eq. (6.3d), and is constant in the current iteration. Therefore, the value of weight α_j is determined by the minimalization of $\exp(-y_i\alpha_j g_j(x_i))$, which can be rewritten as

$$\begin{aligned} \exp\left(-y_i\left(\alpha_j g_j(x_i)\right)\right) &= e^{-\alpha_j} P\left(g_j(x_i) = y_i\right) + e^{\alpha_j} P\left(g_j(x_i) \neq y_i\right) \\ &= e^{-\alpha_j} \cdot \left(1 - \varepsilon_j\right) + e^{\alpha_j} \cdot \varepsilon_j, \end{aligned} \tag{6.4b}$$

where $\varepsilon_j = P\left(g_j(x_i) \neq y_i\right)$ is the error rate of the base learner $g_j(x_i)$, and is calculated by

$$\varepsilon_j = \frac{\sum_{i=1}^{n} w_{i,j} I\left(y_i \neq g_j(x_i)\right)}{\sum_{i=1}^{n} w_{i,j}}. \tag{6.4c}$$

In Eq. (6.4c), the term $I(y_i \neq g_j(x_i))$ takes one if the prediction fails, otherwise zero. Obviously, minimizing each term $\exp(-y_i\alpha_j g_j(x_i))$ in Eq. (6.4a) minimizes the loss function. The minimization of term $\exp(-y_i\alpha_j g_j(x_i))$ requires its differentiation with respect to α_j to be zero, viz.,

$$\frac{\partial \exp\left(-y_i\alpha_j g_j(x_i)\right)}{\partial \alpha_j} = -e^{-\alpha_j} \cdot \left(1 - \varepsilon_j\right) + e^{\alpha_j} \cdot \varepsilon_j = 0, \tag{6.4d}$$

which yields

$$\alpha_j = \frac{1}{2} \ln\left(\frac{1 - \varepsilon_j}{\varepsilon_j}\right). \tag{6.4e}$$

Equation (6.4e) generally holds for a base learner, and indicates that for $0 < \varepsilon_j < 0.5$, the lower the error rate is, the higher the weight of the base learner will be.

Once α_j is optimized, Eq. (6.4a) can be reduced to

$$g_j = \text{argmin} \sum_{i=1}^{n} w_{i,j} I\left(y_i \neq g_j(x_i)\right). \tag{6.5}$$

Although the value of $w_{i,j}$ in the misclassification data is higher, the value of g_j is determined by the lowest error rate, because the loss function in Eq. (6.5) is just the divisor of the error rate defined in Eq. (6.4c). Alternately speaking, the error rate of Eq. (6.4c) is the lowest error rate because the value of g_j shown in Eq. (6.4c) is determined by the minimization of the error rate, as shown in Eq. (6.5).

Once g_j and α_j are determined, the weight $w_{i,j+1}$ of base learner g_{j+1} for datum x_i can be calculated according to Eq. (6.3c) as

$$\begin{aligned} w_{i,j+1} &= \exp\left(-y_i G_j(x_i)\right) = \exp\left(-y_i\left(G_{j-1}(x_i) + \alpha_j g_j(x_i)\right)\right) \\ &= \exp\left(-y_i G_{j-1}(x_i)\right)\exp\left(-y_i \alpha_j g_j(x_i)\right) \\ &= w_{i,j}\exp\left(-y_i \alpha_j g_j(x_i)\right). \end{aligned} \tag{6.6a}$$

Often $w_{i,j+1}$ is normalized by dividing the right-hand side of Eq. (6.6a) with a constant $\sum_{i=1}^{n} w_{i,j}\exp\left(-y_i \alpha_j g_j(x_i)\right)$:

$$w_{i,j+1} = \frac{w_{i,j}\exp\left(-y_i \alpha_j g_j(x_i)\right)}{\sum_{i=1}^{n} w_{i,j}\exp\left(-y_i \alpha_j g_j(x_i)\right)}, \quad i = 1, 2, \cdots, n. \tag{6.6b}$$

Since $-y_i g_j(x_i) = 2 \cdot I\left(y_i \neq g_j(x_i)\right) - 1$, Eq. (6.6a) can be rewritten as

$$w_{i,j+1} = w_{i,j}\exp(-\alpha_j)\exp\left(2\alpha_j \cdot I\left(y_i \neq g_j(x_i)\right)\right), \quad i = 1, 2, \cdots, n. \tag{6.6c}$$

Ignoring the common factor of $\exp(-\alpha_j)$ reduces Eq. (6.6c) to

$$w_{i,j+1} = w_{i,j}\exp\left(2\alpha_j \cdot I\left(y_i \neq g_j(x_i)\right)\right), \quad i = 1, 2, \cdots, n. \tag{6.6d}$$

Equations (6.2)–(6.6) show the workflow of the AdaBoost algorithm, and Eq. (6.1b) is the prediction of the AdaBoost ensemble classifier when $j = M$.

Example 6.1 Table 6.1 lists the data of ten high-entropy alloys, where the feature is the valence electron concentration (VEC) of the alloys, and +1 and −1 represent their crystal structure being body-centered cubic (BCC) and face-centered cubic (FCC), respectively. Employ the AdaBoost algorithm to classify the data.

Table 6.1 The training dataset

Data	1	2	3	4	5	6	7	8	9	10
VEC	6.5	6.6	6.8	7	7.2	7.25	7.3	7.5	7.6	7.8
Crystal structure	+1	+1	+1	−1	−1	−1	+1	+1	+1	−1

Notice

Based on Eq. (6.6b), we know that

$$w_{i,j+1} = \frac{w_{i,j}\exp\big(-y_i\alpha_j g_j(x_i)\big)}{Z_j},$$

where Z_j is the normalizing constant,

$$Z_j = \sum_{i=1}^{n} w_{i,j}\exp\big(-y_i\alpha_j g_j(x_i)\big).$$

This can be rewritten as

$$w_{i,j+1} = \begin{cases} \frac{w_{i,j}\exp(-\alpha_j)}{Z_j}, & g_j(x_i) = y_i, \\ \frac{w_{i,j}\exp(\alpha_j)}{Z_j}, & g_j(x_i) \neq y_i. \end{cases}$$

Iteration 1

The initial weight of each datum in Table 6.1 is calculated by the following equation

$$w_{i,1} = \frac{1}{10} = 0.1, \quad i = 1, 2, 3, \cdots, 10, \text{ and } \mathrm{Z}_1 = 1.$$

Then, the Eq. (6.5) gives $g_1(x) = \text{argmin}\left[\frac{1}{10}\sum_{i=1}^{10} I\big(y_i \neq g_j(x_i)\big)\right]$, with the data in Table 6.1, we have the base learner with the lowest error rate as

$$g_1(x) = \begin{cases} +1, & x \leq 6.9, \\ -1, & x > 6.9. \end{cases}$$

The base learner $g_1(x)$ misclassified Data 7, Data 8, and Data 9, with the corresponding error rate being

$$\varepsilon_1 = \sum_{i=1}^{10} w_{i,1} I\big(y_i \neq g_j(x_i)\big) = 0.1 + 0.1 + 0.1 = 0.3.$$

The weight of $g_1(x)$ is

$$\alpha_1 = \frac{1}{2}\ln\left(\frac{1-0.3}{0.3}\right) = 0.4236.$$

The boosted learner is

$$G_1(x) = 0.4236 g_1(x).$$

The predicted results (misclassified samples are highlighted with bold) are as follows.

Data	1	2	3	4	5	6	7	8	9	10
VEC	6.5	6.6	6.8	7	7.2	7.25	7.3	7.5	7.6	7.8
Crystal structure	+1	+1	+1	−1	−1	−1	+1	+1	+1	−1
$G_1(x)$	0.4236	0.4236	0.4236	−0.4236	−0.4236	−0.4236	0.4236	0.4236	0.4236	−0.4236
sign $(G_1(x))$	+1	+1	+1	−1	−1	−1	**−1**	**−1**	**−1**	−1

Iteration 2

Applying Eq. (6.6b) leads to

$$w_{i,2} = \begin{cases} \frac{w_{i,1}\exp(-\alpha_1)}{Z_1} = \frac{0.1\cdot\exp(-0.4236)}{Z_1} = 0.0715, & g_1(x_i) = y_i, \\ \frac{w_{i,1}\exp(\alpha_1)}{Z_1} = \frac{0.1\cdot\exp(0.4236)}{Z_1} = 0.1667, & g_1(x_i) \neq y_i. \end{cases}$$

The weight of misclassified data increases from 0.1 to 0.167, while the weight of correctly classified data decreases from 0.1 to 0.071.

The Eq. (6.5) gives $g_2(x) = \operatorname{argmin}\left[\sum_{i=1}^{10} w_{i,2}\ I(y_i \neq g_2(x_i))\right]$, and the base learner with the lowest error rate is

$$g_2(x) = \begin{cases} +1, & x \leq 7.7, \\ -1, & x > 7.7. \end{cases}$$

The second base learner $g_2(x)$ misclassified Data 4, Data 5, Data 6, with the corresponding error rate being

$$\varepsilon_2 = \sum_{i=1}^{10} w_{i,1}\ I(y_i \neq g_2(x_i)) = 3 \times 0.071429 = 0.2143.$$

The weight of $g_2(x)$ is

$$\alpha_2 = \frac{1}{2}\ln\left(\frac{1 - 0.2143}{0.2143}\right) = 0.6496.$$

The boosted learner is

$$G_2(x) = 0.4236 g_1(x) + 0.6496 g_2(x).$$

The predicted results (misclassified samples are highlighted with bold) are as follows.

Data	1	2	3	4	5	6	7	8	9	10
VEC	6.5	6.6	6.8	7	7.2	7.25	7.3	7.5	7.6	7.8
Crystal structure	+1	+1	+1	−1	−1	−1	+1	+1	+1	−1
$G_2(x)$	1.073	1.073	1.073	0.226	0.226	0.226	0.226	0.226	0.226	−1.073
sign($G_2(x)$)	+1	+1	+1	**+1**	**+1**	**+1**	+1	+1	+1	−1

Iteration 3

The third iteration is performed similarly, applying Equation 6.1 leads to

$$w_{i,3} = \begin{cases} \frac{w_{i,2}\ \exp\left(-\alpha_2\right)}{Z_2} = \frac{0.0714\ \exp\left(-0.6496\right)}{Z_2} = 0.045455, & g_2(x_i) = y_i, \\ \frac{w_{i,2}\ \exp\left(\alpha_2\right)}{Z_2} = \frac{0.1667\ \exp\left(0.6496\right)}{Z_2} = 0.22725, & g_2(x_i) \neq y_i. \end{cases}$$

The base learner with the lowest error rate is

$$g_3(x) = \begin{cases} +1, & x > 7.28, \\ -1, & x \leq 7.28. \end{cases}$$

The third base learner $g_3(x)$ misclassified Data 10, Data 1, Data 2 and Data 3, with the corresponding error rate being

$$\varepsilon_3 = 0.045455 \times 4 = 0.1818.$$

The weight of $g_3(x)$ is

$$\alpha_3 = \frac{1}{2}\ln\left(\frac{1-0.1818}{0.1818}\right) = 0.7520.$$

The boosted learner is

$$G_3(x) = 0.4236g_1(x) + 0.6496g_2(x) + 0.7520g_3(x).$$

All samples are correctly classified by the boosted learner $G_3(x)$ as seen in the following table, so the iteration ends.

Data	1	2	3	4	5	6	7	8	9	10
VEC	6.5	6.6	6.8	7	7.2	7.25	7.3	7.5	7.6	7.8
Crystal structure	+1	+1	+1	−1	−1	−1	+1	+1	+1	−1
$G_3(x)$	0.32	0.32	0.32	−0.53	−0.53	−0.53	0.98	0.98	0.98	−0.32
sign($G_3(x)$)	+1	+1	+1	−1	−1	−1	+1	+1	+1	−1

The final boosted learner is

$$G_3(x) = 0.4236g_1(x) + 0.6496g_2(x) + 0.7520g_3(x).$$

Notice

Similarly, if we ignore the normalizing constant, we can get the following iteration.

Data	1	2	3	4	5	6	7	8	9	10
VEC	6.5	6.6	6.8	7	7.2	7.25	7.3	7.5	7.6	7.8
Crystal structure	+1	+1	+1	−1	−1	−1	+1	+1	+1	−1
$w'_{i,1}$	0.1	0.1	0.1	0.1	0.1	0.1	0.1	0.1	0.1	0.1
$G'_1(x)$	0.42	0.42	0.42	−0.42	0.42	−0.42	−0.42	−0.42	−0.42	−0.42
$\text{sign}(G'_1(x))$	+1	+1	+1	−1	−1	−1	−1	−1	−1	−1
$w'_{i,2}$	0.065	0.065	0.065	0.065	0.065	0.065	0.153	0.153	0.153	0.065
$G'_2(x)$	1.13	1.13	1.13	0.28	0.28	0.28	0.28	0.28	0.28	−1.13
$\text{sign}(G'_2(x))$	+1	+1	+1	+1	+1	+1	+1	+1	+1	−1
$w'_{i,3}$	0.032	0.032	0.032	0.132	0.132	0.132	0.076	0.076	0.076	0.032
$G'_3(x)$	0.18	0.18	0.18	−0.67	−0.67	−0.67	1.23	1.23	1.23	−0.18
$\text{sign}(G'_3(x))$	+1	+1	+1	−1	−1	−1	+1	+1	+1	−1

The boosted learner thus is:

$$G'_3(x) = 0.4236g_1(x) + 0.7045g_2(x) + 0.9529g_3(x).$$

Example 6.2 Boosting learning with a decision tree classifier as the base learner is used to classify the BCC and FCC high-entropy alloys listed in the reference of Xiong et al. (2021) with the two features of H_{mix} and S_{mix} only. The decision tree and AdaBoost models are developed by using the scikit-learn library. A constant learning rate usually multiplies all weights of base learners to avoid potential overfitting and the learning rate 0.2 is used here. The classification accuracy is used to evaluate the model performance, which is defined by $\text{Acc} = \frac{\text{TP+TN}}{\text{TP+TN+FP+FN}}$, where T, F, P, and N denote true, false, positive, and negative, respectively; TP, TN, FP, and FN stand for the results of a classifier with P (positive) and N (negative) being correctly and incorrectly classified, respectively. The result from the first iteration using only one decision tree classifier is taken here as reference. Figure 6.2 shows that the training accuracy of the boosting learning maintains unchanged when increasing the iteration number from 1 to 2, and then increases quickly from 0.640 to 0.677 as the iteration number increases from 2 to 3; after that, further increasing iteration number from 3 to 10 does not change the accuracy significantly. The validation accuracy (obtained with ten-fold cross-validation) initially increases quickly from 0.610 to 0.640 as the iteration goes from 1 to 4, and then almost maintains unchanged at Acc = 0.640, forming a plateau during the iteration number range (4, 7), and finally decreases to 0.634 in iteration 8. The maximal validation accuracy gives the optimal boosting learning model with 4 base learners.

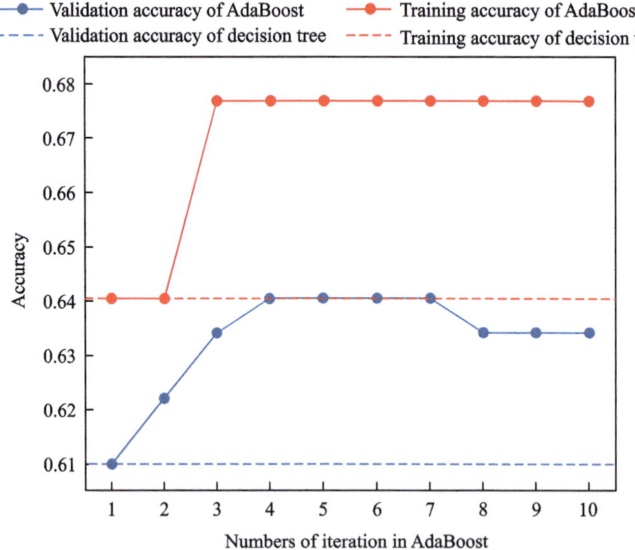

Fig. 6.2 The training and validation accuracy of AdaBoost versus iteration number

---

6.1.2 AdaBoost Regression and Gradient Boosting Machine (GBM)

The AdaBoost algorithm is classical and typical in binary classification. In addition to classification, the strategy of boosting ensemble learning can be applied to regression when other loss functions, e.g., squared error, likelihood, etc., are adopted. Usually, there are parameters $\beta_j = \left(\beta_{j1}, \beta_{j2}, \cdots, \beta_{jk}\right)$ involved in base learners $g_j\left(x, \beta_j\right), \quad j = 1, 2, \cdots, M$. In general, the weight α_j could be $\alpha_j(x)$ or independent of x, and the weights of α_j and the parameters β_j should be obtained from the minimization of a loss function, i.e.,

$$\left(\alpha_{i,j}, \beta_j\right) = \operatorname{argmin} \sum_{i=1}^{n} \mathcal{L}\left(y_i, \sum_{j=1}^{M} \alpha_j(x_i) g_j\left(x_i, \beta_j\right)\right), \quad j = 1, 2, \cdots, M. \tag{6.7a}$$

The forward stagewise additive approach is an efficient and practical method to solve Eq. (6.7a) numerically. As illustrated in the AdaBoost algorithm, Eq. (6.7a) is solved step by step via iteration. Taking the squared error loss function as an example, we have

$$\mathcal{L}\big(y, G_j(x)\big) = \sum_{i=1}^{n} \big(y_i - G_{j-1}(x_i) - \alpha_j(x_i) g_j\big(x_i, \beta_j\big)\big)^2$$

$$= \sum_{i=1}^{n} \big(R_{j-1}(x_i) - \alpha_j(x_i) g_j\big(x_i, \beta_j\big)\big)^2, \tag{6.7b}$$

$$\big(\alpha_{i,j}, \beta_j\big) = \text{argmin} \sum_{i=1}^{n} \big(R_{j-1}(x_i) - \alpha_j(x_i) g_j\big(x_i, \beta_j\big)\big)^2 \text{ only for a certain } j, \tag{6.7c}$$

where $R_{j-1}(x_i) = y_i - G_{j-1}(x_i)$ is termed the residual after $j-1$ rounds of iterations. Equations (6.7b) and (6.7c) indicate that the values of $(\alpha_{i,j}, \beta_j)$ are determined step by step (iteration by iteration), and at each step (iteration), the residual of its previous ensemble learner is used to determine the values of $(\alpha_{i,j}, \beta_j)$ at this iteration, thereby implying that the residual will gradually be reduced towards zero as the iteration goes on. In this sense, the boosting ensemble learning is able to reduce bias, rather than variance.

Friedman (2001) delivered 1999 REITZ Lecture by providing an overview on Gradient Boosting Machine (GBM). The steepest descent method is a widely-used method to solve minimization (maximization) problems. With the steepest descent method, the loss function is approximately expressed by:

$$\mathcal{L}\big(y, G_j(x)\big) = \sum_{i=1}^{n} \mathcal{L}(y_i, (G_{j-1}(x_i) + \alpha_{i,j} g_j(x_i, \beta_j))), \tag{6.8a}$$

where $\alpha_{i,j} = \frac{\partial \mathcal{L}(y_i, G_j(x_i))}{\partial G_{j-1}((x_i))}$ is the first order gradient of the loss function and the increment of $g_j(x_j, \beta_j)$ is termed the step length. Many steps of steepest descents are needed to reach the minimum of loss function, and each step is regarded as a base learner. For squared error loss function, $\mathcal{L}\big(y_i, G_{j-1}(x_i)\big) = 1/2 \sum_{i=1}^{n} \big(y_i - G_{j-1}(x_i)\big)^2$, the negative first order gradient of the loss function is given by

$$\tilde{y}_i = -\frac{\partial \mathcal{L}(y_i, G_{j-1}(x_i))}{\partial G_{j-1}((x_i))} = y_i - G_{j-1}(x_i) = R_{j-1}(x_i). \tag{6.8b}$$

The involved parameters $\beta_j = \big(\beta_{j1}, \beta_{j2}, \cdots, \beta_{jk}\big)$ in the base learner $g_j(x, \beta_j)$ are determined by

$$\beta_j = \text{argmin} \sum_{i=1}^{n} \big(R_{j-1}(x_i) - \alpha_{i,j} g_j\big(x_i, \beta_j\big)\big)^2. \tag{6.8c}$$

De Jong et al. (2016) developed the GBM multivariate local polynomial regression, denoted as GBM-Locfit. Locfit (Loader, 1999) represents local regression algorithms. GBM-Locfit implements the Locfit in base-learner within the GBM framework, meanwhile GBM uses a gradient descent algorithm to iteratively assemble a predictor by minimizing an appropriate, typically squared error, loss function (Natekin and Knoll, 2013). As described above, the main idea behind GBM is to construct the next base learner to be maximally correlated with the negative gradient of the loss function with the current step (status) of the ensemble learner. The complete form of the GBM algorithm was originally proposed by Friedman (2001).

Example 6.3 Use the decision stump as the base learner of GBM to predict the critical casting diameter ($D_{\max}$) of BMG with the super-cooled liquid region (ΔT). The dataset contains 10 BMGs listed in Table 6.2.

Table 6.2 Training data for BMG

Alloys	ΔT (K)	D_{max} (cm)
$Ca_{65}Mg_{15}Cu_{20}$	26	4
$Ti_{34}Zr_{11}Cu_{47}Ni_8$	28.8	4.5
$Fe_{68}Mo_4Y_6B_{22}$	29	6.5
$Mg_{65}Cu_{7.5}Ni_{7.5}Ag_5Zn_5Gd_{2.5}Y_{7.5}$	40	9.5
$Mg_{65}Cu_{15}Ag_{10}Gd_{10}$	43	7.5
$Ti_{41.5}Zr_{2.5}Hf_5Cu_{37.5}Ni_{7.5}Si_1Sn_5$	64.2	6
$Zr_{61.5}Al_{10.7}Cu_{13.65}Ni_{14.15}$	67.6	5.5
$La_{62.5}Al_{12.5}Ag_5Cu_{17.5}Fe_{2.5}$	73	7
$Ce_{69.8}Al_{10}Cu_{20}Co_{0.2}$	75	8
$Ce_{69.5}Al_{10}Cu_{20}Co_{0.5}$	82	10

In general, the loss function of the regression problem in GBM is the squared error:

$$\mathcal{L}(y, G(x)) = \frac{1}{2}\sum_{i=1}^{n}(y_i - G(x_i))^2.$$

For this loss function, the negative descent is

$$\tilde{y}_i = -\frac{\partial \mathcal{L}\big(y_i, G_{j-1}(x_i)\big)}{\partial G_{j-1}((x_i))} = y_i - G_{j-1}(x_i) = R_{j-1}(x_i),$$

i.e., the negative descent is the residual for squared error loss function. A constant weight $\alpha_j = 1$ is used here for all base learners for simplicity (Li, 2012) and the base learners g_j are often expressed by T_j as well.

Iteration 1

The base learner with the lowest least squared error should be

$$G_1(x) = T_1(x) = \begin{cases} 4.25\,\text{cm}, & x \le 28.9, \\ 7.5\,\text{cm}, & x > 28.9. \end{cases}$$

The squared error loss function of the initial GBM, i.e., the base learner, is

$$\mathcal{L}(y, G_1(x)) = \frac{1}{2}\sum_{i=1}^{10}(y_i - G_1(x_i))^2 = 9.065\ \left(\text{cm}^2\right).$$

Iteration 2

The corresponding residual errors given by $G_1(x)$ are listed in Table 6.3.

Table 6.3 The residual error of each datum with $G_1(x)$

ΔT (K)	26	28.8	29	40	43	64.2	67.6	73	75	82
$R_1(x_i)$(cm)	−0.25	0.25	−1	2	0	−1.5	−2	−0.5	0.5	2.5

Then, we determine the base learner $T_2(x)$ to best fit the data in Table 6.3. The residual error $R_1(x_i)$ instead of $D_{\max}$ is the target of this base learner.

$$T_2(x) = \begin{cases} -0.278\,\text{cm}, & x \le 80, \\ 2.5\,\text{cm}, & x > 80. \end{cases}$$

Thus, we get a GBM like

$$G_2(x) = G_1(x) + T_2(x) = \begin{cases} 3.972\,\text{cm}, & x \le 28.9, \\ 7.222\,\text{cm}, & 28.9 < x \le 80, \\ 10\,\text{cm}, & x > 80. \end{cases}$$

The squared-error loss function of the GBM is

$$\mathcal{L}(y, G_2(x)) = \frac{1}{2}\sum_{i=1}^{10}(y_i - G_2(x_i))^2 = 5.59\ \left(\text{cm}^2\right).$$

Iteration 3

The corresponding residual errors given by $G_2(x)$ are listed in Table 6.4.

Table 6.4 The residual error of each datum with $G_2(x)$

ΔT (K)	26	28.8	29	40	43	64.2	67.6	73	75	82
$R_2(x_i)$(cm)	0.03	0.53	−0.72	2.28	0.28	−1.22	−1.72	−0.22	0.78	0

Again, we determine the base learner $T_3(x)$ to best fit the data in Table 6.4. The residual error $R_2(x_i)$ instead of $D_{\max}$ is the target.

$$T_3(x) = \begin{cases} 0.478\,\text{cm}, & x \leq 50, \\ -0.478\,\text{cm}, & x > 50. \end{cases}$$

Thus, we update the GBM to

$$G_3(x) = G_2(x) + T_3(x) = \begin{cases} 4.45\,\text{cm}, & x \leq 28.9, \\ 7.7\,\text{cm}, & 28.9 < x \leq 50, \\ 6.74\,\text{cm}, & 50 < x \leq 80, \\ 9.52\,\text{cm}, & x > 80. \end{cases}$$

The squared error loss function of the GBM is

$$\mathcal{L}(y, G_3(x)) = \frac{1}{2}\sum_{i=1}^{10}(y_i - G_3(x_i))^2 = 4.45\left(\text{cm}^2\right).$$

Iteration 4
Similarly, we can get a GBM with the fourth iteration:

$$T_4(x) = \begin{cases} -0.284\,\text{cm}, & x \leq 70, \\ 0.663\,\text{cm}, & x > 70. \end{cases}$$

$$G_4(x) = G_3(x) + T_4(x) = \begin{cases} 4.166\,\text{cm}, & x \leq 28.9, \\ 7.416\,\text{cm}, & 28.9 < x \leq 50, \\ 6.460\,\text{cm}, & 50 < x \leq 70, \\ 7.401\,\text{cm}, & 70 < x \leq 80, \\ 10.185\,\text{cm}, & x > 80. \end{cases}$$

The loss function is

$$\mathcal{L}(y, G_4(x)) = \frac{1}{2}\sum_{i=1}^{10}(y_i - G_4(x_i))^2 = 3.51\left(\text{cm}^2\right).$$

Assuming that the stop criterion is the loss function smaller than 4 cm^2, we can obtain the final GBM $G_4(x)$.

6.1.3 The Second-Order Expansion of Loss Function

Generally, the second-order expansion of the loss function is a more efficient way to quickly optimize the loss function in the general form of

$$\begin{aligned}\mathcal{L}(y, G_j(x)) &= \sum_{i=1}^{n} \mathcal{L}(y_i, G_j(x_i)) \\ &= \sum_{i=1}^{n} \left\{\mathcal{L}(y_i, G_{j-1}(x_i)) + \alpha_{i,j} g_j(x_i, \beta_j) + \frac{1}{2} h_{i,j} \left[g_j(x_i, \beta_j)\right]^2\right\},\end{aligned} \tag{6.9a}$$

where $\alpha_{i,j} = \frac{\partial \mathcal{L}(y_i, G_{j-1}(x))}{\partial [G_{j-1}(x)]}$ and $h_{i,j} = \frac{\partial^2 \mathcal{L}(y_i, G_{j-1}(x))}{\partial [G_{j-1}(x)]^2}$ are the first and second order derivatives of the loss function, respectively, and $g_j(x_i, \ \beta_i)$ is the step length. Parameters β_j are determined by

$$\beta_j = \operatorname{argmin} \sum_{i=1}^{n} \left\{\mathcal{L}(y_i, G_{j-1}(x_i)) + \alpha_{i,j} g_j(x_i, \beta_j) + \frac{1}{2} h_{i,j} \left[g_j(x_i, \beta_j)\right]^2\right\}. \tag{6.9b}$$

After removing the constant, Eq. (6.9b) is equivalent to

$$\beta_j = \operatorname{argmin} \sum_{i=1}^{n} \left\{\frac{1}{2} h_{i,j} \left[g_j(x_i, \beta_j) + \frac{\alpha_{i,j}}{h_{i,j}}\right]^2\right\}. \tag{6.9c}$$

In general, the squared error loss function is employed for regression problem, the negative of its first derivative is the residual, viz., $-\frac{\partial \mathcal{L}(y, G(x))}{\partial (G(x))} = y - G(x) = R$ and its second derivative is constant one. In this circumstance, Eq. (6.9c) will be reduced to Eq. (6.8c), viz., the second-order expansion boosting learning (SOB) is the same as the GBM with the weight one of base learner. SOB can also use other loss function, which second derivative is not a constant. For example, the fourth power loss function, $\mathcal{L}(y, G(x)) = \frac{1}{4} \sum_{i=1}^{n} (y_i - G_4(x_i))^4$, has its first derivative $\frac{\partial \mathcal{L}(y, G(x))}{\partial (G(x))} = -(y - G(x))^3$ and second derivative $\frac{1}{2} \cdot \frac{\partial^2 \mathcal{L}(y, G(x))}{\partial (G(x))^2} = 1.5(y - G(x))^2$. The Log-Cosh loss function $\mathcal{L}(y, G(x)) = \sum_{i=1}^{n} \ln(\cosh(G(x_i) - y_i))$ is also used in SOB. Its first and second derivative are $\frac{\partial \mathcal{L}(y, G(x))}{\partial (G(x))} = \tanh(G(x_i) - y_i)$ and $\frac{1}{2} \cdot \frac{\partial^2 \mathcal{L}(y, G(x))}{\partial (G(x))^2} = 0.5 \operatorname{sech}^2(G(x_i) - y_i)$, respectively.

6.1.4 EXtreme Gradient Boosting (XGBoost)

As described in Chap. 5, a decision tree partitions the data in a training dataset in a recursive manner into K terminal nodes, which are denoted by $R_k, k = 1, 2, \cdots, K$. The represented values of all responses in node R_k are denoted by $c_k, k = 1, 2, \cdots, K$. The combination of $\Theta = \{R_k, c_k\}$ is termed the parameters of a tree with K being a meta-parameter. A tree is called a stump if $K = 2$, and generally, a K node tree has K terminal nodes. The information-entropy-based loss functions and squared error loss function are used in classification and regression trees, respectively. A boosting tree model $f_M(x)$ is a sum of trees in a forward stagewise manner:

$$f_M(x) = \sum_{j=1}^{M} T_j\left(x, \Theta_j\right), \tag{6.10a}$$

$$f_M(x) = \sum_{j=1}^{M} \alpha_j T_j\left(x, \Theta_j\right), \tag{6.10b}$$

and parameters Θ_j are determined by

$$\Theta_j = \text{argmin} \sum_{i=1}^{n} \mathcal{L}\left(y_i, f_{j-1}(x_i) + T_j\left(x_i, \Theta_j\right)\right). \tag{6.10c}$$

The AdaBoost algorithm can be directly applied to boosting trees with base binary classifiers. In such case, $y_i \in \{-1, 1\}$ and $c_{k,j} \in \{-1, 1\}$. Equations (6.1)–(6.6) describe the working flow and

$$\Theta_j = \text{argmin} \sum_{i=1}^{n} w_{i,j} \exp\left(-y_i \alpha_j T_j\left(x_i, \Theta_j\right)\right). \tag{6.10d}$$

Equations (6.7a)-(6.7c) can be applied to regression boosting trees, where parameters β_j and base learners g_j should be regarded as parameters Θ_j and base trees $T_j(x_i, \Theta_j)$ respectively, and all weights α_j are all equal and thus moved out.

XGBoost (Chen & Guestrin, 2016) means eXtreme Gradient Boosting, which is a sparsity-aware algorithm for sparse-space and weighted quantile sketch for approximate tree learning. XGBoost utilizes M additive trees of Eq. (6.10a) with a regularization (penalty) term $\Omega(T_j)$. Thus, the objective (loss) function is expressed by the quadratic expansion as

$$\begin{aligned} \mathcal{L}_j &= \sum_{i=1}^{n} \mathcal{L}\left(y_i, T_j(x_i)\right) + \Omega(T_j) \\ &= \sum_{i=1}^{n} \left\{ \mathcal{L}\left(y_i, G_{j-1}(x_i)\right) + \alpha_{i,j} g_j\left(x_i, \beta_j\right) + \frac{1}{2} h_{i,j} \left[g_j\left(x_i, \beta_j\right)\right]^2 \right\} \end{aligned}$$

$$+\Omega(T_j), \tag{6.11a}$$

with

$$\Omega(T_j) = \gamma K_j + \frac{1}{2}\lambda \sum_{k=1}^{K_j} c_{k,j}^2, \tag{6.11b}$$

where γ and λ are preset constants, and K_j and $c_{k,j}$ are the number of leaves and the value of leaf R_k in the tree T_j, respectively. The term $\mathcal{L}(y_i, G_{j-1}(x_i))$ is a constant and thus can be ignored in the minimization of the objective function. Equation (6.11a) indicates that the summation is calculated on all data of x_i in the training dataset, which is actually carried out on all leaves of tree T_j. If we define $I_k = \{i|R_k(x_i)\}$ as the instant set of samples in leaf R_k, we can convert the sample sum of Eq. (6.11a) to the leaf sum of

$$\mathcal{L}_j = \sum_{k=1}^{K_j} \left\{ G_j c_{k,j} + \frac{1}{2}(H_j + \lambda) c_{k,j}^2 \right\} + \gamma K_j \tag{6.11c}$$

with $G_j = \sum\limits_{i \in I_k} \alpha_{i,j}$ and $H_j = \sum\limits_{i \in I_k} h_{i,j}$. If the structure of tree T_j is fixed, the minimization of $\mathcal{L}_j$ requires $\frac{\partial \mathcal{L}_j}{\partial c_{k,j}} = 0$, which gives

$$c_{k,j} = -\frac{G_j}{H_j + \lambda}. \tag{6.11d}$$

Substituting Eq. (6.11d) into Eq. (6.11c) yields the minimum loss function

$$\min(\mathcal{L}_j) = -\frac{1}{2} \sum_{k=1}^{K_j} \frac{G_j^2}{H_j + \lambda} + \gamma K_j. \tag{6.11e}$$

Example 6.4 In practice, we often set a learning rate ε between [0, 1] in boosting learning to smooth the learning curve.

$$G_{i+1}(x) = G_i(x) + \varepsilon T_{i+1}(x).$$

Here we utilize the squared error as the loss function to predict shear modulus with the total electronegativity (TEN). The dataset is given in Example 4.6 of Chap. 4 on SVM. The learning rate of XGBoost is set as 0.1, and the iteration number is 100.

We compare the performance of XGBoost with and without the regularization term. The regularization term, also called penalty term, in XGBoost can control the structure of base learners. As seen in Figure 6.3, the model with the regularization term is much smoother than that without the regularization term.

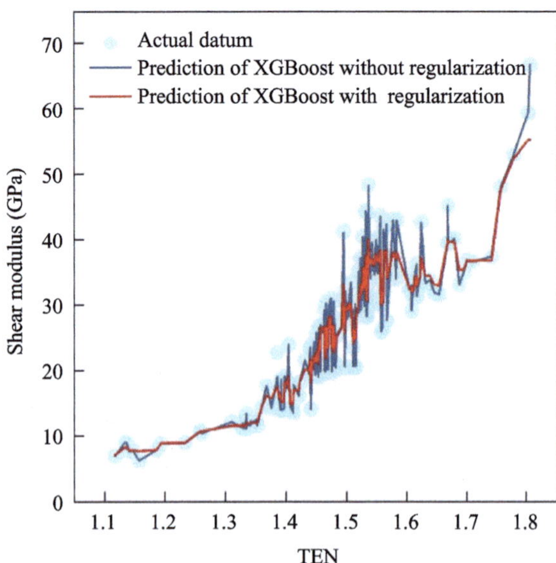

Fig. 6.3 The performance of XGBoost learner with and without regularization term

---

6.2 Bagging

Bootstrap is a statistical method to estimate accuracy, especially the expected prediction error. The basic approach of Bootstrap is to randomly sample data with replacement from an original dataset $(x_{1,i}, \cdots, x_{m,i}, y_i)$ $(i = 1, 2, \cdots, N)$ to generate a randomly drawn dataset with the same size as that of the original dataset. The probability of a datum $(x_{1,i}, \cdots, x_{m,i}, y_i)$ appearing in the Bootstrap dataset is

$$\Pr(\text{datum}(x_i, y_i) \in \text{ Bootstrap dataset }) = 1 - \left(1 - \frac{1}{N}\right)^N \approx 1 - \mathrm{e}^{-1} = 0.632. \tag{6.12}$$

This means that about 36.8% of the data in the original dataset will not be included in the Bootstrap dataset. When a Bootstrap dataset is used as the training set, and the original dataset is used as the testing set, the testing set will have about 36.8% of the data that do not appear in the training set. If only those 36.8% data are used as the testing set, the testing result is called out-of-bag estimate. The Bootstrap method is more suitable for small data, for which the number of data is so small that it is statistically unsound to employ cross-validation.

Bagging (Breiman, 1996) uses the Bootstrap method to generate M Bootstrap datasets with each having the same size as that of the original dataset. One base learner is developed with each of the M Bootstrap datasets, and thus there are M base learners, which form the ensemble learner. The prediction from the ensemble learner is usually taken as the simple voting result from the M base learners for classification and the average of the M predictions for regression. Bagging reduces the variance of ML model prediction, and is more suitable for high-variance and low-bias ML algorithms, such as trees.

- **Random Forest (RF)**

Random Forest (RF) (Breiman, 1996; Ho, 1998) extends the Bagging approach. As described in the decision tree, a split is carried out on such a feature among all d features in a current node that the information gain is maximized for classification, or the accuracy is the highest for regression. This means that each of the d features must be calculated. In RF, however, $k < d$ features are randomly selected from the d features, and the split is conducted based on the best performance feature among the k features. If $k = d$, the approach is the same as in the normal decision tree. If $k = 1$, it selects randomly one feature to carry out the split. Usually, $k = \log_2 d$ is used in RF classification and $k = d/3$ in RF regression (Hastie et al., 2009).

Example 6.5 Use RF ($k = d/3$) and bagging decision tree ($k = d$) to predict the hardness of high, entropy alloys. The features include enthalpy of mixing (H_{mix}), average valence electron concentration ($\overline{\text{VEC}}$), average atomic number ($\overline{\text{AN}}$), average thermal conductivity ($\overline{K}$), the mismatch in electronegativity (δ_{XP}) and in molar heat capacity ($\delta_{\text{C}_\text{m}}$). All features are normalized to [0, 1], and the dataset is given in Xiong et al. (2020), also available from https://github.com/George-JieXIONG/Materials-Dataset/blob/main/Example%206/EX6.csv.

The convergence trend of RF is similar to bagging decision tree. As shown in Fig. 6.4, RF generally converges to a lower generalization error as the number of base learners increases. In addition, the training efficiency of RF is better than bagging decision tree due to the use of "random selected features".

Fig. 6.4 The performance of bagging decision tree and RF

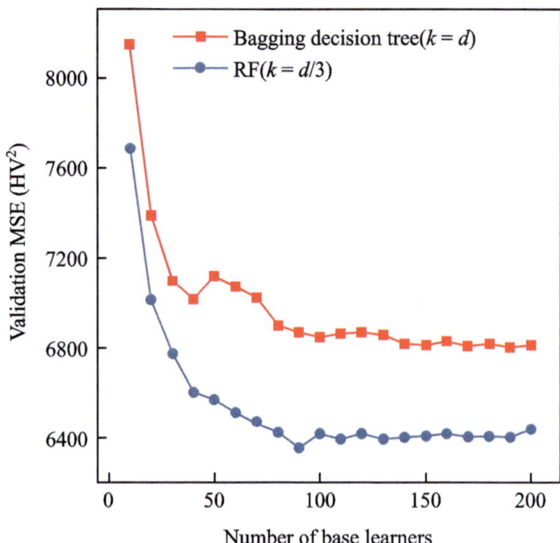

Example 6.6 To illustrate how RF works, dissolution temperatures of 24 Co-base superalloys are listed in Table 6.5. The input features are the at. % of elements Co, Al, W and Ni in the superalloys.

Since the number of features is small (only 4), we use K=d/2=2 and LOOCV to determine the hyperparameters. When there are two trees in RF (n_estimators = 2), the maximum depth of each tree is 3 (max_depth = 3), and the RMSE value of the model on the validation set is the smallest (RMSE = 25.49). Figure 6.6 shows the growth of the two decision trees. Each node of the decision tree is split according to $\sqrt{d}$ features to find the best splitting point.

For the new sample point in Table 6.6, RF works as follows.

The prediction given by the first decision tree is

$$\text{value}(1) = 1101.0\,°\text{C};$$

The prediction given by the second decision tree is

$$\text{value}(2) = 995.5\,°\text{C};$$

The RF prediction is the average of the two predictions, that is

$$\text{value} = \frac{\text{value}(1) + \text{value}(2)}{2} = 1048.25\,°\text{C}.$$

Table 6.5 Composition and dissolution temperatures of 24 Co-base superalloys

Co(at.%)	Al(at.%)	W(at.%)	Ni(at.%)	Dissolution temperature(°C)
65.5	7.3	7	20.2	880
65.5	7.3	7	20.2	881
82	9	7	2	966
73	7.9	8.9	10.2	982
72.5	10	7.5	10	990
42.5	10	7.5	40	1070
36	9	10	45	1079
46.4	8.8	9.8	35	1091
50	10	10	30	1095
32.5	10	7.5	50	1105
56.4	8.8	9.8	25	1113
40	10	10	40	1120
22.5	10	7.5	60	1120
12.5	10	7.5	70	1140
30	10	10	50	1150
20	10	10	60	1170
10	10	10	70	1200
75	10	10	5	1052
70	10	10	10	1062
65	10	10	15	1074
60	10	10	20	1097
50	10	10	30	1125
40	10	10	40	1153
79.4	8.8	9.8	2	1005

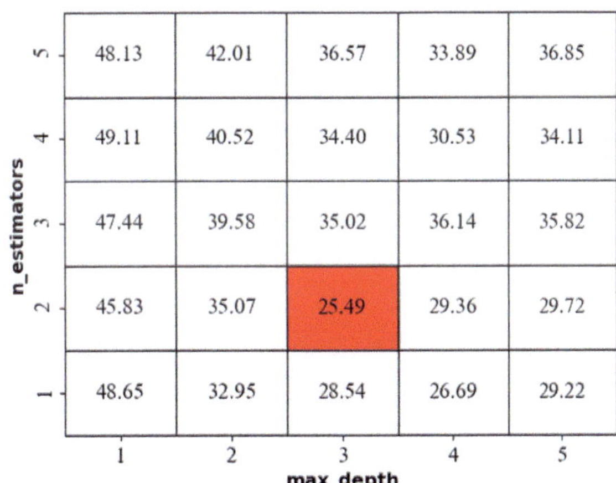

Fig. 6.5 RMSE value of validation set under different parameters

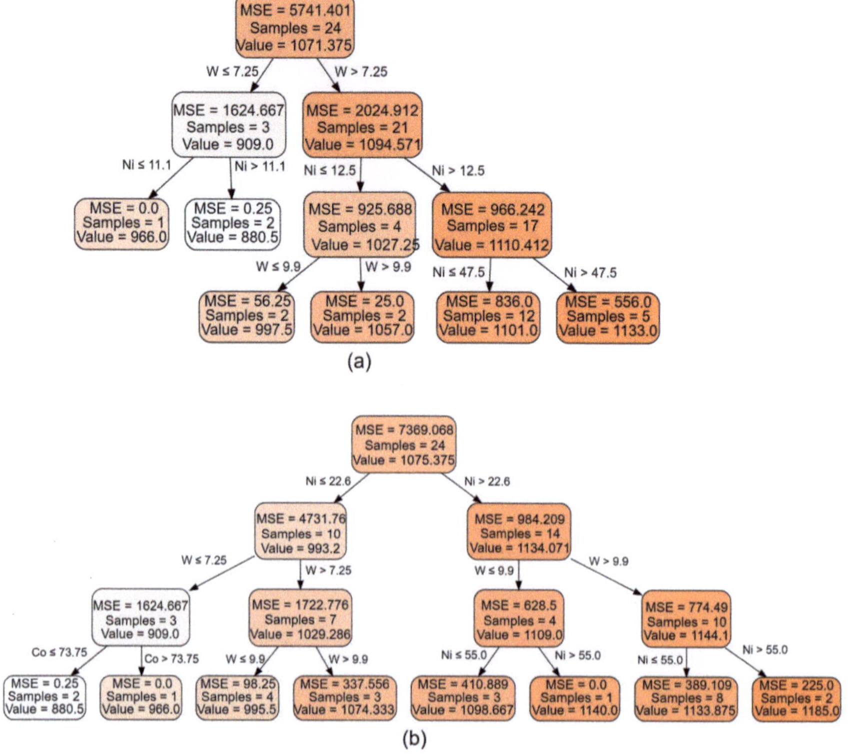

Fig. 6.6 Decision trees in the two-tree random forest. **(a)** Tree one; and **(b)** tree two

Table 6.6 New sample point of Co base superalloys

Element	Co	Al	W	Ni
at.%	62.5	10	7.5	20

Homework

Table 6.7 lists the experimental results of plastic strain rates from stress relaxation tests on Cu at the temperatures -22°C, 30 °C, 40 °C, 50 °C and 75 °C. The plastic strain rate $\dot{\varepsilon}_{\mathrm{p}}(\sigma, T)$ depends on temperature T and applied stress σ. Based

Table 6.7 Plastic strain rates from stress relaxation tests on Cg-Cu at various stress levels and temperatures

$1/kT$ (J^{-1})	Strain rates at different stress levels for Cg-Cu@ 1st cycle (s^{-1})							
	113 MPa	112 MPa	111 MPa	110 MPa	109 MPa	108 MPa	107 MPa	106 MPa
2.46×10^{20}	3.95×10^{-5}	2.51×10^{-5}	1.42×10^{-5}	7.79×10^{-6}	4.59×10^{-6}	2.65×10^{-6}	1.67×10^{-6}	1.06×10^{-6}
2.39×10^{20}	4.35×10^{-5}	2.83×10^{-5}	1.74×10^{-5}	9.66×10^{-6}	5.06×10^{-6}	3.11×10^{-6}	2.07×10^{-6}	1.50×10^{-6}
2.31×10^{20}	4.95×10^{-5}	3.65×10^{-5}	2.51×10^{-5}	1.54×10^{-5}	9.99×10^{-6}	5.97×10^{-6}	4.35×10^{-6}	3.11×10^{-6}
2.24×10^{20}	5.65×10^{-5}	4.05×10^{-5}	2.89×10^{-5}	1.76×10^{-5}	1.19×10^{-5}	8.23×10^{-6}	5.47×10^{-6}	4.12×10^{-6}
2.08×10^{20}	6.67×10^{-5}	5.17×10^{-5}	3.90×10^{-5}	2.79×10^{-5}	1.91×10^{-5}	1.35×10^{-5}	9.65×10^{-6}	8.09×10^{-6}

on domain knowledge, the plastic strain rate follows the equation of $\dot{\varepsilon}_p(\sigma, T) = \dot{\varepsilon}_{00}\left(\frac{\sigma}{\sigma_0}\right)^n \exp\left(-\frac{\Delta G_0}{kT} + \frac{\sigma V}{kT\sqrt{3}}\right)$, where $\dot{\varepsilon}_{00}$ is a constant, σ_0 is a reference stress, n is the stress exponent, k is the Boltzmann constant, ΔG_0 is a stress-independent activation energy, and V is the nominal activation volume conjugate to shear stress. Letting $y = \ln\dot{\varepsilon}_p$, $x_1 = \ln\left(\frac{\sigma}{\sigma_0}\right)$, $x_2 = \frac{1}{kT}$, and $x_3 = \frac{\sigma}{kT}$.

1. With the data listed in Table 6.7, use the three ensemble algorithms AdaBoost (for tree model), XGBoost, Fandom Forest to fit this physical law.
2. Use the three-layer CART tree as the weak estimator, change the number of estimators, and compare the differences between the three ensemble models.
3. Which of the bootstrap and boosting methods has smaller generalization errors when the estimators are the same?
4. Select any sample data and digest the working principle of GBM following Example 6.3.
5. Reproduce Example 6.1 to understand the working principle of AdaBoost.
6. Describe your understanding of the principles based on what you have learned.

References

Breiman, L. (1996). Bagging predictors. *Machine Learning, 24*(2), 123–140.

Chen, T., & Guestrin, C. (2016). XGBoost: A SCALABLE tree boosting system. In *Proceedings of the 22nd ACM SIGKDD international conference on knowledge discovery and data mining* (*KDD*'16)(pp. 785–794). Association for Computing Machinery.

De Jong, M., Chen, W., Notestine, R., Persson, K., Ceder, G., Jain, A., ... & Gamst, A. (2016). A statistical learning framework for materials science: application to elastic moduli of k-nary inorganic polycrystalline compounds. *Scientific Reports, 6*(1), 34256.

Franklin, J. (2005). The elements of statistical learning: Data mining, inference and prediction. *The Mathematical Intelligencer, 27*(2), 83–85.

Freund, Y., & Schapire, R. E. (1997). A decision-theoretic generalization of on-line learning and an application to boosting. *Journal of Computer and System Sciences, 55*(1), 119–139.

Friedman, J. H. (2001). Greedy function approximation: A gradient boosting machine. *Annals of Statistics, 29*(5), 1189–1232.

Friedman, J., Hastie, T., & Tibshirani, R. (2000). Additive logistic regression: A statistical view of boosting. *Annals of Statistics, 28*(2), 337–374.

Hastie, T., Tibshirani, R., & Friedman, J. (2009). *The elements of statistical learning: data mining, inference, and prediction* (2nd ed.). Springer.

Ho, T. K. (1998). The random subspace method for constructing decision forests. *IEEE Transactions on Pattern Analysis and Machine Intelligence, 20*(8), 832–844.

Li, H. (2012). *Statistical learning method*. Tsinghua University Press.

Loader, C. (1999). *Local regression and likelihood*. Springer Science & Business Media.

Natekin, A., & Knoll, A. (2013). Gradient boosting machines: A tutorial. *Frontiers in Neurorobotics, 7*, 21.

Webb, G. I., & Zheng, Z. J. (2004). Multistrategy ensemble learning: Reducing error by combining ensemble learning techniques. *IEEE Transactions on Knowledge and Data Engineering, 16*(8), 980–991.

Witten, I. H., Frank, E., & Hall, M. A. (2011). *Data mining: Practical machine learning tools and techniques* (3rd ed.). Morgan Kaufmann Publisher.

Xiong, J., Zhang, T. Y., & Shi, S. Q. (2020). Machine learning of mechanical properties of steels. *Science China Technological Sciences, 63*(7), 1247–1255.

Xiong, J., Shi, S. Q., & Zhang, T. Y. (2021). Machine learning of phases and mechanical properties in complex concentrated alloys. *Journal of Materials Science and Technology, 87*, 133–142.

Chapter 7
Bayesian Theorem and Expectation–Maximization (EM) Algorithm

7.1 Bayesian Theorem

The probability that both events H and E occur is called the joint probability of H and E, and is denoted by $P(H \cap E)$, where $H \cap E$ denotes both events H and E occur simultaneously (Berger, 2013), or by $P(H, E)$. Events H and E are considered to be independent if the two events are physically unrelated. In this case, the joint probability is the product of independent probabilities, i.e.,

$$P(H \cap E) = P(H)P(E). \tag{7.1}$$

Equation (7.1) is the mathematic definition of independent events or independent distributions.

The probability for event H to occur under the condition that event E has already occurred is called the conditional probability, and is denoted by $P(H|E)$, which is given by

$$P(H|E) = \frac{P(H \cap E)}{P(E)}, \tag{7.2a}$$

if the probability for event E to occur is not zero. Similarly, if the probability for event H to occur is not zero, the conditional probability for event E to occur under the condition that event H has already occurred is given by

$$P(E|H) = \frac{P(H \cap E)}{P(H)}. \tag{7.2b}$$

From Eqs. (7.2a) and (7.2b), we have

$$P(H \cap E) = P(H|E)P(E) = P(E|H)P(H). \tag{7.2c}$$

T. Zhang, *An Introduction to Materials Informatics*,
https://doi.org/10.1007/978-981-99-7992-9_7

Equation (7.2c) leads to Bayesian theorem

$$P(H|E) = \frac{P(H)P(E|H)}{P(E)}. \tag{7.3a}$$

The conditional probability $P(H|E)$ is also called the posterior probability of event H. The probability $P(H)$ is called the prior probability, which is the probability for event H to occur without any conditions. The conditional probability $P(E|H)$ is also called the conditional likelihood. The probability $P(E)$ is the marginal probability, and is also called the total likelihood, i.e., the probability for event E to occur without any conditions. In the calculation of posterior probability under a given value of event $E = e$, i.e.,

$$P(H|E = e) = \frac{P(H)P(E = e|H)}{P(E = e)}. \tag{7.3b}$$

If there are K events denoted by $H_k (k = 1, 2, \cdots, K)$, Bayes' theorem states

$$P(H_k|E) = \frac{P(H_k)P(E|H_k)}{\sum\limits_k P(H_k)P(E|H_k)} = \frac{P(H_k)P(E|H_k)}{P(E)}. \tag{7.3c}$$

Bayes' theorem holds for continuous distributed functions, in which the probabilities are replaced by corresponding probability densities, viz.,

$$p(H_k|E) = \frac{p(H_k)p(E|H_k)}{\int_{-\infty}^{+\infty} p(H_k)p(E|H_k)\mathrm{d}H}, \tag{7.3d}$$

where the term of $\int_{-\infty}^{+\infty} p(H_k)p(E|H_k)\mathrm{d}H$ is the total likelihood. The total likelihood can be regarded as a normalizing constant to normalize the posterior probability (or density), and can be ignored if without considering normalization.

7.2 Naive Bayes Classifier

Naive Bayes classifier is a classification algorithm (McCallum & Nigam, 1998) based on the Bayesian theorem with the "naive" assumption of conditional independence among features. Consider a dataset $D = \{x_{ij}, y_i\}$ $(i = 1, 2, \cdots, n;\ j = 1, 2, \cdots, m)$ of n separable data, where $\boldsymbol{x} = (x_1, x_2, \cdots, x_m) \in \mathbf{R}^m$ denotes an $1 \times m$ feature vector, and $C = \{c_1, c_2, \cdots, c_K\}$ $(K \leqslant n)(k = 1, 2, \cdots, K) \in \mathbf{R}$ is the $1 \times K$ class label vector of the data. Bayes' theorem gives the posterior probability of

$$P(y = c_k|\boldsymbol{x}) = \frac{P(y = c_k)P(\boldsymbol{x}|y = c_k)}{P(\boldsymbol{x})}, \tag{7.4a}$$

where $P(y = c_k)$ is the prior probability, $P(\boldsymbol{x}|y = c_k)$ is the conditional likelihood, and $P(\boldsymbol{x}) = \sum_{k=1}^{K} P(y = c_k)P(\boldsymbol{x}|y = c_k)$ is the total likelihood. Under the conditional independence assumption of features, Eq. (7.4a) can be rewritten as

$$P(y = c_k|\boldsymbol{x}) = \frac{P(y = c_k)}{P(\boldsymbol{x})} \prod_{j=1}^{m} P(x_j|y = c_k). \tag{7.4b}$$

As mentioned above, the total likelihood $P(\boldsymbol{x})$ is a constant and can be ignored. Thus, Eq. (7.4b) leads to

$$P(y = c_k|\boldsymbol{x}) \propto P(y = c_k) \prod_{j=1}^{m} P(x_j|y = c_k). \tag{7.4c}$$

Which class the optimized prediction of an output belongs to is determined by the maximum posterior probability (MAP), which is mathematically expressed as

$$\hat{y} = \underset{y\in\{c_1,c_2,\cdots,c_K\}}{\text{argmax}} \quad P(y = c_k) \prod_{j=1}^{m} P(x_j|y = c_k). \tag{7.5}$$

The prior probability $P(y_k)$ is calculated by

$$P(y = c_k) = \frac{n_k}{n} \ (k = 1, 2, \cdots, K), \tag{7.6a}$$

where n_k is the number of samples in class c_k. The conditional probability $P(x_j|y = c_k)$ is calculated by

$$P(x_j|y = c_k) = \frac{n_{kj}}{n_k}, \tag{7.6b}$$

where n_{kj} is the number of samples with feature x_j in class c_k.

Gaussian Naive Bayes assumes that each continuous feature variable follows a Gaussian distribution. Hence, the conditional probability density $p(x_j|y = c_k)$ is given by

$$p\left(x_j|y = c_k\right) = \frac{1}{\sqrt{2\pi}\sigma_j^{(k)}}\exp\left(-\frac{\left(\boldsymbol{x}_j^{(k)} - \mu_j^{(k)}\right)^2}{2\left(\sigma_j^{(k)}\right)^2}\right), \tag{7.7}$$

where $\boldsymbol{x}_j^{(k)}$ or $(x_j|y = c_k)$ denotes the variable of feature x_j in class $y = c_k$, and $\mu_j^{(k)}$ and $\left(\sigma_j^{(k)}\right)^2$ are the mean value and variance of $\boldsymbol{x}_j^{(k)}$, respectively. For continuous probability distributions of features, the MAP optimized prediction is given by

$$\hat{y} = \underset{y \in \{c_1, c_2, \cdots, c_K\}}{\operatorname{argmax}} P(y = c_k) \prod_{j=1}^{m} p(x_j | y = c_k). \tag{7.8a}$$

If some features are continuous (John, 1995) and some are discrete, a mixture of MAP and MAP density should be used, and for simplicity it is still called MAP. Thus, the MAP optimized prediction is estimated by

$$\hat{y} = \underset{y \in \{c_1, c_2, \cdots, c_K\}}{\operatorname{argmax}} P(y = c_k) \prod_{j=1}^{m_1} p(x_j | y = c_k) \prod_{j=m_1+1}^{m} P(x_j | y = c_k), \tag{7.8b}$$

where m_1 and $m - m_1$ denote the numbers of continuous and discrete features, respectively.

Example 7.1 Table 7.1 shows 18 high-entropy alloys collected from Xiong et al. (2021). The 18 alloys belong to single-phase (FCC) or multi-phases (FCC + BCC). Three features are considered here, which include average valance electron concentration (VEC), electronegativity difference ($\delta\chi$), and processing condition (PC) of

Table 7.1 Eighteen high-entropy alloys

Sample No.	Alloys	PC	VEC	$\delta\chi$	Phases
1	$Co_1Cr_1Fe_1Mn_1Ni_1$	AC	8.00	0.0783	FCC (1)
2	$Co_1Cr_1Fe_1Ni_1$	AC	8.25	0.0531	FCC (1)
3	$Co_1Cr_1Mn_1Ni_1$	AC	8.00	0.0860	FCC (1)
4	$Co_1Fe_1Mn_1Ni_1$	AC	8.50	0.0797	FCC (1)
5	$Co_1Cr_1Ni_1$	AC	8.33	0.0614	FCC (1)
6	$Co_1Fe_1Ni_1$	AC	9.00	0.0176	FCC (1)
7	$Co_1Mn_1Ni_1$	AC	8.67	0.0916	FCC (1)
8	$Fe_1Mn_1Ni_1$	AC	8.33	0.0875	FCC (1)
9	$Co_1Cu_1Fe_1Ni_1$	MA	9.50	0.0164	FCC (1)
10	$Co_1Fe_1Ni_1$	MA	9.00	0.0176	FCC (1)
11	$Al_{0.6}Co_1Fe_1Ni_1Ti_{0.4}$	MA	7.60	0.0725	FCC (1)
12	$Co_1Cr_2Fe_1Ni_1$	AC	7.80	0.0602	FCC + BCC (2)
13	$Cr_1Fe_1Ni_1$	AC	8.00	0.0579	FCC + BCC (2)
14	$Al_1Co_1Cr_1Cu_1Fe_1$	MA	7.40	0.0667	FCC + BCC (2)
15	$Co_1Cr_1Cu_1Fe_1Ni_1$	MA	8.80	0.0502	FCC + BCC (2)
16	$Co_1Cr_1Fe_1Ni_1$	MA	8.25	0.0531	FCC + BCC (2)
17	$Al_{0.6}Co_1Cr_1Fe_1Ni_1Ti_{0.4}$	MA	7.28	0.0731	FCC + BCC (2)
18	$Co_{0.5}Cr_1Fe_1Ni_1Ti_{0.5}$	MA	7.63	0.0731	FCC + BCC (2)

as-cast (AC) and mechanical alloying (MA). The VEC and $\delta\chi$ are calculated by $\mathrm{VEC} = \sum_{i=1} a_i \mathrm{VEC}_i$, $\delta\chi = \sqrt{\sum_{i=1} a_i \left(1 - \frac{\chi_i}{\sum_{i=1} a_i \chi_i}\right)^2}$, where a_i, VEC_i, and χ_i are the atomic percentage, valance electron concentration, and Pauling electronegativity of element i, respectively.

The phases FCC and FCC + BCC are denoted by class 1 and class 2, respectively, and the three features of PC, VEC, and $\delta\chi$ are denoted by x_1, x_2, and x_3, respectively. The Naive Bayes classifier with LOOCV is conducted. The first cycle of LOOCV is described in detail in the following, where datum 1 is used as the validation datum, and data 2–18 form the training set, which gives $n_1 = 10$ and $n_2 = 7$. The prior probabilities $P(y = 1)$ and $P(y = 2)$ are respectively estimated by

$$P(y = 1) = \frac{10}{17},$$
$$P(y = 2) = \frac{7}{17}.$$

Since there are only two kinds of processing conditions, and $n_{11} = 7$ and $n_{21} = 2$, the conditional probabilities $P(x_1|y)$ are estimated below:

$$P(x_1 = \mathrm{AC}|y = 1) = \frac{7}{10},$$
$$P(x_1 = \mathrm{AC}|y = 2) = \frac{2}{7}.$$

The features of $x_j^{(k)} (j = 2,\ 3;\ k = 1,\ 2)$ are continuous variables, whose means and variances are calculated to be

$$\bar{x}_2^{(1)} = \frac{1}{10} \sum_{w=1}^{n_1=10} x^{(1)} w_2 = 8.518, \quad \bar{x}_2^{(2)} = \frac{1}{7} \sum_{w=1}^{n_2=7} x^{(2)} w_2 = 7.880,$$
$$\bar{x}_3^{(1)} = \frac{1}{10} \sum_{w=1}^{n_1=10} x^{(1)} w_3 = 0.058, \quad \bar{x}_3^{(2)} = \frac{1}{7} \sum_{w=1}^{n_2=7} x^{(2)} w_3 = 0.062,$$
$$\sigma_2^{(1)} = \sqrt{\frac{1}{10} \sum_{w=1}^{n_1=10} \left(x^{(1)} w_2 - \bar{x}_2^{(1)}\right)^2} = 0.548,$$
$$\sigma_2^{(2)} = \sqrt{\frac{1}{7} \sum_{w=1}^{n_2=7} \left(x^{(2)} w_2 - \bar{x}_2^{(2)}\right)^2} = 0.526,$$
$$\sigma_3^{(1)} = \sqrt{\frac{1}{10} \sum_{w=1}^{n_1=10} \left(x^{(1)} w_3 - \bar{x}_3^{(1)}\right)^2} = 0.031,$$

$$\sigma_3^{(2)} = \sqrt{\frac{1}{7}\sum_{w=1}^{n_2=7}\left(x^{(2)}w_3 - \bar{x}_3^{(2)}\right)^2} = 0.009.$$

Then, the conditional probability densities $p(x_j|y = c_k)(j = 2,\ 3;\ k = 1,\ 2)$ are calculated with Eq. (7.7) to be

$$p(x_2 = 8.00|y = 1) = \frac{1}{\sqrt{2\pi}\cdot 0.548}\exp\left(-\frac{(8.00 - 8.518)^2}{2\cdot(0.548)^2}\right) = 0.466,$$
$$p(x_2 = 8.00|y = 2) = \frac{1}{\sqrt{2\pi}\cdot 0.526}\exp\left(-\frac{(8.00 - 7.880)^2}{2\cdot(0.526)^2}\right) = 0.740,$$
$$p(x_3 = 0.0783|y = 1) = \frac{1}{\sqrt{2\pi}\cdot 0.031}\exp\left(-\frac{(0.0783 - 0.058)^2}{2\cdot(0.031)^2}\right) = 10.505,$$
$$p(x_3 = 0.0783|y = 2) = \frac{1}{\sqrt{2\pi}\cdot 0.009}\exp\left(-\frac{(0.0783 - 0.062)^2}{2\cdot(0.009)^2}\right) = 9.009.$$

With Eq. (7.8b), we have the MAPs for datum 1 classified into class 1 and class 2, respectively.

$$\begin{aligned}
\text{MAP}(y = 1) &= P(y = 1) \times P(x_1 = \text{AC}|y = 1) \\
&\quad \times p(x_2 = 8.00|y = 1) \times p(x_3 = 0.0783|y = 1) \\
&= \frac{10}{17} \times \frac{7}{10} \times 0.466 \times 10.505 = 2.016, \\
\text{MAP}(y = 2) &= P(y = 2) \times P(x_1 = \text{AC}|y = 2) \\
&\quad \times p(x_2 = 8.00|y = 2) \times p(x_3 = 0.0783|y = 2) \\
&= \frac{7}{17} \times \frac{2}{7} \times 0.740 \times 9.009 = 0.784.
\end{aligned}$$

If the classification depends purely on MAP, the validation datum belongs to class 1. Cycling LOOCV 18 times and letting each datum to be the validation one, the Naive Bayes classifier classifies all the 18 data, and the results are listed in Table 7.2. The confusion matrix in Fig. 7.1 illustrates the performance of the Naive Bayes classifier.

Table 7.2 The LOOCV prediction of Naive Bayes classifier

Sample No.	Alloy compositions	MAP		PP		Measured classes	Predicted classes
		$y = 1$	$y = 2$	$y = 1$	$y = 2$		
1	$Co_1Cr_1Fe_1Mn_1Ni_1$	2.0160	0.7839	0.7199	0.2801	FCC (1)	FCC (1)
2	$Co_1Cr_1Fe_1Ni_1$	3.2688	1.8897	0.6337	0.3663	FCC (1)	FCC (1)
3	$Co_1Cr_1Mn_1Ni_1$	1.6277	0.1267	0.9278	0.0722	FCC (1)	FCC (1)
4	$Co_1Fe_1Mn_1Ni_1$	2.9171	0.3046	0.9055	0.0945	FCC (1)	FCC (1)
5	$Co_1Cr_1Ni_1$	3.5273	2.6787	0.5684	0.4316	FCC (1)	FCC (1)
6	$Co_1Fe_1Ni_1$	0.58505	0.000003	0.99999	0.00001	FCC (1)	FCC (1)
7	$Co_1Mn_1Ni_1$	1.82088	0.00707	0.99613	0.00387	FCC (1)	FCC (1)
8	$Fe_1Mn_1Ni_1$	2.23088	0.05759	0.97483	0.02517	FCC (1)	FCC (1)
9	$Co_1Cu_1Fe_1Ni_1$	0.01294	0.0000004	0.99997	0.00003	FCC (1)	FCC (1)
10	$Co_1Fe_1Ni_1$	0.16716	0.00001	0.99995	0.00005	FCC (1)	FCC (1)
11	$Al_{0.6}Co_1Fe_1Ni_1Ti_{0.4}$	0.1605	4.4218	0.0350	0.9650	FCC (1)	FCC + BCC (2)
12	$Co_1Cr_2Fe_1Ni_1$	2.1607	1.5661	0.5798	0.4202	FCC + BCC (2)	FCC (1)
13	$Cr_1Fe_1Ni_1$	3.1730	1.4266	0.6898	0.3102	FCC + BCC (2)	FCC (1)
14	$Al_1Co_1Cr_1Cu_1Fe_1$	0.2429	3.5373	0.0642	0.9358	FCC + BCC (2)	FCC + BCC (2)
15	$Co_1Cr_1Cu_1Fe_1Ni_1$	1.3679	0.0419	0.9703	0.0297	FCC + BCC (2)	FCC (1)
16	$Co_1Cr_1Fe_1Ni_1$	1.5557	2.8656	0.3519	0.6481	FCC + BCC (2)	FCC + BCC (2)
17	$Al_{0.6}Co_1Cr_1Fe_1Ni_1Ti_{0.4}$	0.1429	1.0506	0.1197	0.8803	FCC + BCC (2)	FCC + BCC (2)
18	$Co_{0.5}Cr_1Fe_1Ni_1Ti_{0.5}$	0.4767	2.1671	0.1803	0.8197	FCC + BCC (2)	FCC + BCC (2)

As mentioned in Chap. 3, a threshold can be put to bias a class. For this purpose, we introduce the pseudo probability (PP), which is the normalized MAP. For binary classification, we

$$\mathrm{PP}(y = i) = \frac{\mathrm{MAP}(y = i)}{\mathrm{MAP}(y = 1) + \mathrm{MAP}(y = 2)}, \qquad i = 1,\ 2.$$

For example, the PPs of datum 1 are

$$\mathrm{PP}(y = 1) = \frac{\mathrm{MAP}(y = 1)}{\mathrm{MAP}(y = 1) + \mathrm{MAP}(y = 2)} = \frac{2.016}{2.016 + 0.784} = 0.7199,$$
$$\mathrm{PP}(y = 2) = \frac{\mathrm{MAP}(y = 2)}{\mathrm{MAP}(y = 1) + \mathrm{MAP}(y = 2)} = \frac{0.784}{2.016 + 0.784} = 0.2801.$$

Fig. 7.1 The confusion matrix of the LOOCV predictions

Confusion Matrix		Predicted class	
		FCC	FCC + BCC
Measured class	FCC	10	1
	FCC + BCC	3	4

Table 7.2 lists all PP values of 18 samples. Either PP ($y = 1$) values or PP ($y = 2$) values can be taken as threshold values. Here the PP ($y = 1$) values are taken as the threshold values TH $= (h_0, \ h_1, \ \cdots, \ h_{18})$ based on the PP ($y = 1$) values from larger to small, and the data are reranked accordingly, as shown in Table 7.3. In addition, an extreme PP ($y = 1$) $= 1$ is also taken as h_0. Then, the classification criterion is that if PP ($y = 1$) $\geqslant$ TH, the data are classified as class 1, otherwise class 2. Thus, a given value of TH generates a classification result, which can be expressed by a confusion matrix, leading to the values of the true positive rate $\left(\text{TPR} = \dfrac{\text{TP}}{\text{TP} + \text{FN}}\right)$ and the false positive rate $\left(\text{FPR} = \dfrac{\text{FP}}{\text{FP} + \text{TN}}\right)$, as described in Chap. 3. After that, plot TPR versus FPR yields the receiver operating characteristic (ROC) curve (Spackman, 1989) based on the LOOCV predictions, which is shown in Fig. 7.2. The size of AUC, which evaluates the classifier performance, is AUC $=$ 0.818.

Table 7.3 The detailed results for the ROC curve

Sample No.	Measured classes	PP ($y = 1$)	Predicted classes of samples with different thresholds																		
			h_0	h_1	h_2	h_3	h_4	h_5	h_6	h_7	h_8	h_9	h_{10}	h_{11}	h_{12}	h_{13}	h_{14}	h_{15}	h_{16}	h_{17}	h_{18}
6	1	0.99999 (h_1)	2	1	1	1	1	1	1	1	1	1	1	1	1	1	1	1	1	1	1
9	1	0.99997 (h_2)	2	2	1	1	1	1	1	1	1	1	1	1	1	1	1	1	1	1	1
10	1	0.99995 (h_3)	2	2	2	1	1	1	1	1	1	1	1	1	1	1	1	1	1	1	1
7	1	0.99613 (h_4)	2	2	2	2	1	1	1	1	1	1	1	1	1	1	1	1	1	1	1
8	1	0.97483 (h_5)	2	2	2	2	2	1	1	1	1	1	1	1	1	1	1	1	1	1	1
15	2	0.97028 (h_6)	2	2	2	2	2	2	1	1	1	1	1	1	1	1	1	1	1	1	1
3	1	0.92779 (h_7)	2	2	2	2	2	2	2	1	1	1	1	1	1	1	1	1	1	1	1
4	1	0.90546 (h_8)	2	2	2	2	2	2	2	2	1	1	1	1	1	1	1	1	1	1	1
1	1	0.71988 (h_9)	2	2	2	2	2	2	2	2	2	1	1	1	1	1	1	1	1	1	1
13	2	0.68984 (h_{10})	2	2	2	2	2	2	2	2	2	2	1	1	1	1	1	1	1	1	1
2	1	0.63368 (h_{11})	2	2	2	2	2	2	2	2	2	2	2	1	1	1	1	1	1	1	1
12	2	0.57977 (h_{12})	2	2	2	2	2	2	2	2	2	2	2	2	1	1	1	1	1	1	1

(continued)

Table 7.3 (continued)

Sample No.	Measured classes	PP ($y = 1$)	Predicted classes of samples with different thresholds																		
			h_0	h_1	h_2	h_3	h_4	h_5	h_6	h_7	h_8	h_9	h_{10}	h_{11}	h_{12}	h_{13}	h_{14}	h_{15}	h_{16}	h_{17}	h_{18}
5	1	0.56837 (h_{13})	2	2	2	2	2	2	2	2	2	2	2	2	2	1	1	1	1	1	1
16	2	0.35186 (h_{14})	2	2	2	2	2	2	2	2	2	2	2	2	2	2	1	1	1	1	1
18	2	0.18032 (h_{15})	2	2	2	2	2	2	2	2	2	2	2	2	2	2	2	1	1	1	1
17	2	0.11973 (h_{16})	2	2	2	2	2	2	2	2	2	2	2	2	2	2	2	2	1	1	1
14	2	0.06424 (h_{17})	2	2	2	2	2	2	2	2	2	2	2	2	2	2	2	2	2	1	1
11	1	0.03502 (h_{18})	2	2	2	2	2	2	2	2	2	2	2	2	2	2	2	2	2	2	1
FPR with different thresholds			0.0	0.00	0.00	0.00	0.00	0.00	0.14	0.14	0.14	0.14	0.29	0.29	0.43	0.43	0.57	0.71	0.86	1.00	1.00
TPR with different thresholds			0.0	0.09	0.18	0.27	0.36	0.45	0.45	0.55	0.64	0.73	0.73	0.82	0.82	0.91	0.91	0.91	0.91	0.91	1.00

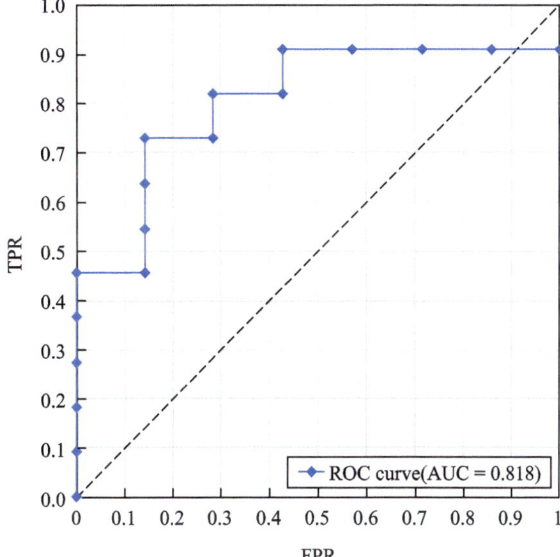

Fig. 7.2 The ROC curve of LOOCV predictions

7.3 Maximum Likelihood Estimation

7.3.1 Gaussian Distribution

In many cases, it is very hard to obtain big experimental data from time-consuming or/and high-cost tests. Unlike big data, small data cannot provide sufficient information about the probability distribution of repeated experimental results. In this circumstance, if a type of probability distribution of data is known, the maximum likelihood is able to estimate, from data, the values of parameters involved in the distribution. Let $p(X|\theta)$ denote the probability distribution density of variable X with θ being involved parameters. For example, in most repeated tests, unknown factors cause scattering in the repeated tests, the scattering is often termed noise, and random noise usually follows Gaussian distribution (Patel & Read, 1996). In this case, $p(X|\theta) \sim N(\mu, \sigma^2)$, where the involved parameters are expectation μ and variance σ. During repeated tests, all tests are independent. Thus, the likelihood of $p(X|\theta)$ is given by

$$L(\theta) = \prod_{i=1}^{n} p(X_i|\theta). \tag{7.9a}$$

Taking natural logarithm of the likelihood gives the log-likelihood

$$\mathrm{LL}(\theta) = \ln[L(\theta)] = \sum_{i=1}^{n} \ln(p(X_i|\theta)). \tag{7.9b}$$

The values of involved parameters are estimated from the maximization of log-likelihood, i.e.,

$$\hat{\theta} = \arg\max_{\theta} LL(\theta) = \arg\max_{\theta} \sum_{i=1}^{n} \ln(p(X_i|\theta)). \tag{7.10}$$

For repeated tests with random noise, Gaussian distribution is taken to estimate expectation μ and variance σ. In this case, the likelihood is

$$L\{\mu, \sigma^2\} = \prod_{i=1}^{n} \frac{1}{\sqrt{2\pi}\sigma} \exp\left(-\frac{(x_i - \mu)^2}{2\sigma^2}\right) = \left(\frac{1}{\sqrt{2\pi}\sigma}\right)^n \exp\left(-\sum_{i=1}^{n} \frac{(x_i - \mu)^2}{2\sigma^2}\right). \tag{7.11a}$$

And the log-likelihood is

$$\mathrm{LL}\{\mu, \sigma^2\} = n \ln \frac{1}{\sqrt{2\pi}} - n \ln \sigma - \sum_{i=1}^{n} \frac{(x_i - \mu)^2}{2\sigma^2}. \tag{7.11b}$$

The values of expectation μ and variance σ are estimated from the maximization of log-likelihood, i.e.,

$$\{\hat{\mu}, \hat{\sigma}^2\} = \underset{\theta}{\mathrm{argmax}}\, \mathrm{LL}\{\mu, \sigma^2\} = \underset{\mu, \sigma^2}{\mathrm{argmax}} \left\{ n \ln \frac{1}{\sqrt{2\pi}} - n \ln \sigma - \sum_{i=1}^{n} \frac{(x_i - \mu)^2}{2\sigma^2} \right\}. \tag{7.11c}$$

The maximum log-likelihood requires

$$\frac{\partial \mathrm{LL}\{\mu, \sigma^2\}}{\partial \mu} = 0, \tag{7.12a}$$

$$\frac{\partial \mathrm{LL}\{\mu, \sigma^2\}}{\partial \sigma^2} = 0, \tag{7.12b}$$

which result in

$$\hat{\mu} = \overline{x} = \frac{1}{n} \sum_{i=1}^{n} x_i, \tag{7.13a}$$

$$\hat{\sigma}^2 = \frac{1}{n}\sum_{i=1}^{n}(x_i - \overline{x})^2. \tag{7.13b}$$

The maximum likelihood estimated expectation is just the mean of the observed data. That is the reason why the mean of repeated testing results is widely used as the approximated expectation. The mean will approach the expectation when the number of repeated tests is sufficiently large. The denominator in the maximum likelihood estimated variance square is n, rather than n–1. When n is sufficiently large, the difference between the two will be negligible.

7.3.2 Weibull Distribution

Weibull distribution (Weibull, 1939) is developed based on the weakest link, and is widely used to statistically describe the failure of materials under uniform loading. The failure probability P_f of a brittle material measured from uniaxial tensile tests on smooth samples with each having the same volume follows Weibull distribution:

$$P_f(\sigma_N) = 1 - \exp\left[-\left(\frac{\sigma_N}{\sigma_0}\right)^m\right], \tag{7.14a}$$

where m is termed the Weibull modulus, and σ_N and σ_0 are the measured strength and the reference strength, respectively. The survival probability $P_s(\sigma_N)$ is calculated from $1–P_f(\sigma_N)$ to be

$$P_s(\sigma_N) = \exp\left[-\left(\frac{\sigma_N}{\sigma_0}\right)^m\right]. \tag{7.14b}$$

The failure probability density p_f of Weibull distribution is then given by

$$p_f(\sigma_N) = \frac{m}{\sigma_0}\left(\frac{\sigma_N}{\sigma_0}\right)^{m-1}\exp\left[-\left(\frac{\sigma_N}{\sigma_0}\right)^m\right]. \tag{7.15}$$

The likelihood of Weibull distribution is

$$\begin{aligned} L\{\sigma_0, m\} &= \prod_{i=1}^{n}\frac{m}{\sigma_0}\left(\frac{\sigma_{N,i}}{\sigma_0}\right)^{m-1}\exp\left[-\left(\frac{\sigma_{N,i}}{\sigma_0}\right)^m\right] \\ &= \left(\frac{m}{\sigma_0}\right)^n\prod_{i=1}^{n}\left(\frac{\sigma_{N,i}}{\sigma_0}\right)^{m-1}\exp\left[-\left(\frac{\sigma_{N,i}}{\sigma_0}\right)^m\right]. \end{aligned} \tag{7.16}$$

The log-likelihood with Weibull distribution is then given by

$$\mathrm{LL}\{\sigma_0, m\} = n \ln \frac{m}{\sigma_0} + \sum_{i=1}^{n} (m-1) \ln \frac{\sigma_{\mathrm{N},i}}{\sigma_0} - \sum_{i=1}^{n} \left(\frac{\sigma_{N,i}}{\sigma_0} \right)^m. \tag{7.17a}$$

The maximum ln-likelihood requires

$$\frac{\partial \mathrm{LL}\{\sigma_0, m\}}{\partial \sigma_0} = -\frac{nm}{\sigma_0} + \frac{m}{\sigma_0} \sum_{i=1}^{n} \left(\frac{\sigma_{\mathrm{N},i}}{\sigma_0} \right)^m = 0, \tag{7.17b}$$

$$\frac{\partial \mathrm{LL}\{\sigma_0, m\}}{\partial m} = \frac{n}{m} + \sum_{i=1}^{n} \ln \frac{\sigma_{\mathrm{N},i}}{\sigma_0} - \sum_{i=1}^{n} \left(\frac{\sigma_{\mathrm{N},i}}{\sigma_0} \right)^m \ln \frac{\sigma_{\mathrm{N},i}}{\sigma_0} = 0. \tag{7.17c}$$

The estimations of m and σ_0 are obtained by solving Eqs. (7.17b) and (7.17c).

Example 7.2 The repeated three-point-bending tests on smooth concrete samples of the same volume of $V = 153{,}600\ \mathrm{mm}^3$ (Hoover et al., 2013) yield the bending strengths of 7.62 MPa, 7.56 MPa, 7.32 MPa, 7.37 MPa, 8.68 MPa, 8.57 MPa, and 7.17 MPa. The failure probability is assumed to follow Weibull distribution of Eq. (7.14a). Estimate the Weibull modulus m and the value of reference strength σ_0.

The log-likelihood of Eq. (7.17a) for this example takes the explicit form of

$$\begin{aligned}
\ln\ L\{\sigma_0, m\} &= n \ln \frac{m}{\sigma_0} + \sum_{i=1}^{n} (m-1) \ln \frac{\sigma_{\mathrm{N},i}}{\sigma_0} - \sum_{i=1}^{n} \left(\frac{\sigma_{\mathrm{N},i}}{\sigma_0} \right)^m \\
&= 7 \ln \frac{m}{\sigma_0} + (m-1) \ln \frac{7.62}{\sigma_0} \\
&\quad - \left(\frac{7.62}{\sigma_0} \right)^m + (m-1) \ln \frac{7.56}{\sigma_0} - \left(\frac{7.56}{\sigma_0} \right)^m \\
&\quad + (m-1) \ln \frac{7.32}{\sigma_0} - \left(\frac{7.32}{\sigma_0} \right)^m \\
&\quad + (m-1) \ln \frac{7.37}{\sigma_0} - \left(\frac{7.37}{\sigma_0} \right)^m + (m-1) \ln \frac{8.68}{\sigma_0} - \left(\frac{8.68}{\sigma_0} \right)^m \\
&\quad + (m-1) \ln \frac{8.57}{\sigma_0} - \left(\frac{8.57}{\sigma_0} \right)^m + (m-1) \ln \frac{7.17}{\sigma_0} - \left(\frac{7.17}{\sigma_0} \right)^m.
\end{aligned}$$

The open software of Scipy.optimize.minimize function with BFGS (Broyden-Fletcher-Goldfarb-Shanno) method (https://docs.scipy.org/doc/scipy/reference/generated/scipy.optimize.minimize.html) is used to maximize the log-likelihood, resulting in $m = 13.64$ and $\sigma_0 = 8.03$ MPa. Taking the natural logarithm twice on the survival probability yields

$$\ln \ln \left(\frac{1}{P_{\mathrm{s}}(\sigma_{\mathrm{N}})} \right) = m \cdot \ln \left(\frac{\sigma_{\mathrm{N}}}{\sigma_0} \right) = m \cdot \ln \sigma_{\mathrm{N}} - m \cdot \ln \sigma_0. \tag{7.18}$$

Once the values of m and σ_0 are known, $\ln\ln\left(\frac{1}{P_s(\sigma_N)}\right)$ versus $\ln\sigma_N$ can be plotted, as shown in Fig. 7.3.

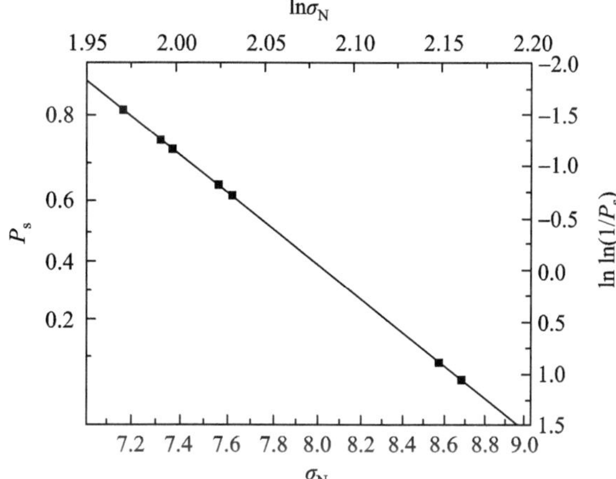

Fig. 7.3 The plot of lnln($P_s(\sigma_N)$) versus lnσ_N for the three-point bending strengths of concrete

---

Maximum likelihood estimation is an effective method to process small data, but the data probability distribution must be known from or suggested by prior knowledge. With the probability distribution and measured data, one can determine the parameters involved in the distribution. Another typical example is the volume-dependent Weibull distribution, which describes the failure probability for various tested sample sizes. The volume-dependent Weibull distribution takes the form of

$$P_f(\sigma_N(V)|m, \sigma_0, V_0) = 1 - \exp\left[-\frac{V}{V_0}\left(\frac{\sigma_N}{\sigma_0}\right)^m\right], \tag{7.19a}$$

where V and V_0 are the sample volume and the reference volume, respectively. This is because for the same materials, the probability of a big defect existing in a large sample is higher than that in a small sample, and hence the measured strength in a large sample is statistically lower than that in a small sample. When all tested samples have the same volume, the reference volume is usually selected as the sample volume, and then Eq. (7.19a) is reduced to Eq. (7.14a). The Weibull modulus for a certain material with the same chemical composition and sample preparation conditions is independent of the sample volume, i.e., the Weibull modulus is constant. Taking advantage of Weibull modulus invariance for a given material, one is able to fully utilize all experimental data from various sample volumes, even if the number of repeated tests for each volume is small. Equation (7.19a) can be rewritten as

$$P_f(\sigma_N(V)|m, \sigma_0(V_0)) = 1 - \exp\left[-\frac{V}{V_0 \times \sigma_0^m}\sigma_N^m\right], \tag{7.19b}$$

which indicates that when m is a constant, $V_0 \times \sigma_0^m$ appears as a constant as well. In practice, the value of V_0 is first preset, and then the values of m and σ_0 are determined by the maximum likelihood. Equation (7.19a) leads to the failure probability density p_f of volume-dependent Weibull distribution

$$p_\mathrm{f}(\sigma_\mathrm{N}(V)|m, \sigma_0(V_0)) = \frac{V}{V_0} \cdot \frac{m}{\sigma_0}\left(\frac{\sigma_\mathrm{N}}{\sigma_0}\right)^{m-1} \exp\left[-\frac{V}{V_0}\left(\frac{\sigma_\mathrm{N}}{\sigma_0}\right)^m\right]. \tag{7.19c}$$

The likelihood and the log-likelihood of volume-dependent Weibull distribution are respectively

$$L\{m, \sigma_0(V_0)\} = \prod_{j=1}^{J}\prod_{i=1}^{n} \frac{V_j}{V_0} \cdot \frac{m}{\sigma_0}\left(\frac{\sigma_{\mathrm{N},ji}}{\sigma_0}\right)^{m-1} \exp\left[-\frac{V_j}{V_0}\left(\frac{\sigma_{\mathrm{N},ji}}{\sigma_0}\right)^m\right], \tag{7.20a}$$

$$\mathrm{LL}\{m, \sigma_0(V_0)\} = \sum_{j=1}^{J}\sum_{i=1}^{n}\left\{\ln\left[\frac{V_j}{V_0} \cdot \frac{m}{\sigma_0}\left(\frac{\sigma_{\mathrm{N},ji}}{\sigma_0}\right)^{m-1}\right] - \frac{V_j}{V_0}\left(\frac{\sigma_{\mathrm{N},ji}}{\sigma_0}\right)^m\right\}, \tag{7.20b}$$

where J is the number of various volumes. The maximum ln-likelihood requires

$$\frac{\partial \mathrm{LL}\{m, \sigma_0(V_0)\}}{\partial \sigma_0} = 0, \tag{7.20c}$$

$$\frac{\partial \mathrm{LL}\{m, \sigma_0(V_0)\}}{\partial m} = 0, \tag{7.20d}$$

The estimations of m and σ_0 are obtained by solving Eqs. (7.20c) and (7.20d).

Example 7.3 The repeated three-point-bending tests on smooth concrete samples of four different sizes (Hoover et al., 2013) generate the volume-dependent strengths, which are listed in the Table 7.4. The failure probability is assumed to follow the volume-dependent Weibull distribution of Eq. (7.19a). Estimate the values of Weibull modulus, reference volume and reference strength.

Table 7.4 Smooth concrete samples

Volume (mm^3)	Strength (MPa)
153,600	7.62, 7.56, 7.32, 7.37, 8.68, 8.57, 7.17
830,304	7.78, 7.86, 7.37, 6.95, 7.4, 7.02, 7.69, 6.74
4,437,600	6.21, 6.23, 6.75, 5.99
24,000,000	5.83, 5.68, 6.36

The log-likelihood of volume-dependent Weibull distribution is given by Eq. (7.20b), and the values of Weibull modulus, reference volume, and reference strength are estimated by the maximization of log-likelihood. Since it is hard to estimate the three values simultaneously, a value of reference volumes is preset, and then the value of reference strength is calculated, i.e.,

$$\{\hat{m}, \hat{\sigma}_0(V_0)\} = \underset{\sigma_0, m}{\text{argmax}} \sum_{j=1}^{J} \sum_{i=1}^{n} \left\{ \ln\left[\frac{V_j}{V_0} \cdot \frac{m}{\sigma_0} \left(\frac{\sigma_{N,ji}}{\sigma_0} \right)^{m-1} \right] - \frac{V_j}{V_0} \left(\frac{\sigma_{N,ji}}{\sigma_0} \right)^m \right\}.$$

Again, the open software of Scipy.optimize.minimize function with BFGS method is used to maximize the log-likelihood. Each of the four sample volumes is used as the reference volume, and the results are tabulated in Table 7.5.

Table 7.5 The results

V_0 (mm^3)	m	σ_0 (MPa)
153,600	16.76	8.16
830,304	16.76	7.38
4,437,600	16.76	6.67
24,000,000	16.76	6.03

Figure 7.4 shows $\ln\ln\left(\frac{1}{P_s}\right) \sim \ln(\sigma_N)$ for the four sample volumes, where $V_0 = 153600$ mm^3, $\sigma_0 = 8.16$ MPa, and $m = 16.76$.

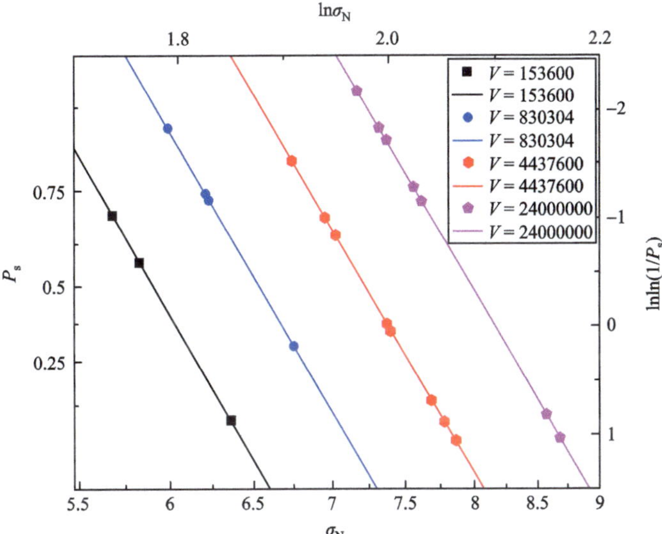

Fig. 7.4 The plot of $\ln\ln(P_s(\sigma_N))$ versus $\ln\sigma_N$ of volume-dependent Weibull distribution for the three-point bending strengths of concrete

7.4 Bayesian Linear Regression

There are generally two kinds of ML models, one giving deterministic predictions and the other providing probabilistic predictions. The Gaussian processes (GPs) are typically probabilistic ML methods based on Bayesian formalism and Gaussian probability distribution (Williams & Rasmussen, 1996). Unlike in other deterministic ML regressions, GP regression utilizes Gaussian probability distribution to regress data and express regression results in terms of mean and covariance of the maximal posterior distribution. Bayesian linear regression is defined in the framework of Bayesian theorem, and lays the foundation of GPs. As described in Chap. 2, linear regression is to find the linear relationship between features and responses, where a set of coefficients determine a linear relationship. Multiple features with the dummy feature of one form a high-dimensional coefficient space, and a point in the coefficient space corresponds to a set of coefficients. Based on the Bayesian theorem, prior distribution of infinite coefficient sets should be proposed without using any data, and these infinite coefficient sets following the prior distribution are the pool, from which the best coefficient set will be selected with data. Usually, normal distribution $N(0, \boldsymbol{\Sigma}_p)$ is proposed to be the prior distribution in linear regression, where $\boldsymbol{\Sigma}_p$ is the covariance matrix of coefficient sets. Since one point in the coefficient space represents a set of coefficients, there are infinite points in the coefficient space, which follow the normal distribution $N(0, \boldsymbol{\Sigma}_p)$. Alternately speaking, the prior distribution suggests infinite linear functions, whose coefficient sets follow the normal distribution $N(0, \boldsymbol{\Sigma}_p)$, although each linear function has a unique set of coefficients. Input data provide the conditional likelihood, and the posterior distribution without normalization can be calculated by ignoring the total likelihood. After that, the maximal posterior distribution selects the best sets of coefficients from the prior pool, or a set of linear functions is selected from the infinite linear functions provided by the prior. The best sets of coefficients also follow a normal distribution, and its mean yields the linear prediction as "close" as possible to the data points. Thus, the uncertainty is reduced close to the observations and still remains at the covariance $\boldsymbol{\Sigma}_p$ in regions far from the data points. Clearly, the prior distribution plays a certain role in the predication. As described below, an optimal prior distribution, balancing the model prediction accuracy and generalization, will be determined by cross-validation, which is actually the maximization of posterior distribution.

Consider a training dataset of $(x_{ij}, y_i)(i = 1, 2, \cdots, n;\ j = 0, 1, 2, \cdots, m)$, where $\{\boldsymbol{x}_j\} = (1, x_1, \cdots, x_m)$ is an $1 \times (m + 1)$ row feature vector, y is a scalar response, and n is the number of data. A linear model is expressed by

$$f(\boldsymbol{x}) = \boldsymbol{x}\boldsymbol{W}, \tag{7.21a}$$

where $\boldsymbol{W} = (w_0, w_1, \cdots, w_m)^{\mathrm{T}}$ is an $(m + 1) \times 1$ coefficient column vector, called a set of coefficients. Then, the scalar response y is given by

$$y = \boldsymbol{x}\boldsymbol{W} + \varepsilon, \tag{7.21b}$$

where ε denotes noise, which is assumed to follow an independent and identical normal distribution with zero mean and variance σ_n^2,

$$\varepsilon \sim N\left(0, \sigma_n^2\right). \tag{7.21c}$$

When each datum (observation or sample) is independent, the likelihood, i.e., the probability density of the observations, given a set of coefficients $\boldsymbol{W}$ (often called weights in GP), is the factored over data in the training dataset, viz.,

$$\begin{aligned} L = p(\boldsymbol{y}|\boldsymbol{X}, \boldsymbol{W}) &= \prod_{i=1}^{n} p(y_i|\boldsymbol{X}_i, \boldsymbol{W}) = \prod_{i=1}^{n} \frac{1}{\sqrt{2\pi}\sigma_n} \exp\left(-\frac{(y_i - \boldsymbol{X}_i \boldsymbol{W})^2}{2\sigma_n^2}\right) \\ &= \frac{1}{\left(2\pi\sigma_n^2\right)^{n/2}} \exp\left(-\frac{|\boldsymbol{y} - \boldsymbol{X}\boldsymbol{W}|^2}{2\sigma_n^2}\right) \sim N(\boldsymbol{X}\boldsymbol{W}, \sigma_n^2 \boldsymbol{I}), \end{aligned} \tag{7.22}$$

where $\boldsymbol{X}_i = (1, x_{i1}, \cdots, x_{im})$ and $\boldsymbol{X} = (\boldsymbol{X}_1, \ \boldsymbol{X}_2, \cdots, \ \boldsymbol{X}_n)^{\mathrm{T}}$ denote original sample i and entire n samples of feature inputs, respectively. Equation (7.22) shows that the likelihood is an n-dimensional normal distribution of response $\boldsymbol{y}$ with the mean of $\boldsymbol{X}\boldsymbol{W}$ and covariance of $\sigma_n^2 \boldsymbol{I}$. The Bayesian formalism requires a prior probability $p(\boldsymbol{W})$ over weights $\boldsymbol{W}$. The prior is preset to be a Gaussian distribution with a zero mean and $(m + 1) \times (m + 1)$ covariance matrix $\boldsymbol{\Sigma}_p$,

$$p(\boldsymbol{W}) \sim N\left(0, \boldsymbol{\Sigma}_p\right). \tag{7.23}$$

On the one hand, if covariance matrix $\boldsymbol{\Sigma}_p$ is small, the prior distribution will form a peak at the zero mean. On the other hand, a large covariance matrix $\boldsymbol{\Sigma}_p$ means a flat prior distribution. For example, consider a two-dimensional case, $\boldsymbol{W} = (w_0, w_1)^{\mathrm{T}}$, the prior distribution $p(\boldsymbol{W})$~N $(0, \boldsymbol{\Sigma}_p)$ indicates that in the (w_0, w_1) coefficient space, one point (w_0, w_1) corresponds to a set of coefficients or a linear function. The infinite points in the (w_0, w_1) coefficient space, representing infinite linear functions, follow the normal distribution $N(0, \boldsymbol{\Sigma}_p)$ and exhibit circle-shaped probabilistic contours, with each contour having the same prior distribution probability in the (w_0, w_1) coefficient space.

Without considering normalization, the Bayes theory gives

$$p(\boldsymbol{W}|\boldsymbol{y}, \boldsymbol{X}) \propto p(\boldsymbol{y}|\boldsymbol{X}, \boldsymbol{W}) p(\boldsymbol{W}). \tag{7.24a}$$

Substituting Eqs. (7.22) and (7.23) into Eq. (7.24a) yields the posterior distribution

$$\begin{aligned} p(\boldsymbol{W}|\boldsymbol{y}, \boldsymbol{X}) &\propto \exp\left(-\frac{(\boldsymbol{y} - \boldsymbol{X}\boldsymbol{W})^{\mathrm{T}}(\boldsymbol{y} - \boldsymbol{X}\boldsymbol{W})}{2\sigma_n^2}\right) \exp\left(-\frac{1}{2}\boldsymbol{W}^{\mathrm{T}} \boldsymbol{\Sigma}_p^{-1} \boldsymbol{W}\right) \\ &\propto \exp\left(-\frac{1}{2}(\boldsymbol{W} - \overline{\boldsymbol{W}})^{\mathrm{T}} \left(\frac{\boldsymbol{X}^{\mathrm{T}}\boldsymbol{X}}{\sigma_n^2} + \boldsymbol{\Sigma}_p^{-1}\right)(\boldsymbol{W} - \overline{\boldsymbol{W}})\right) \sim N(\overline{\boldsymbol{W}}, A^{-1}). \end{aligned} \tag{7.24b}$$

The posterior distribution is a normal distribution in the coefficient space with mean $\overline{\boldsymbol{W}}$ and covariance matrix $\boldsymbol{A}^{-1}$, where

$$\overline{\boldsymbol{W}} = (\sigma_n^{-2}\boldsymbol{X}^{\mathrm{T}}\boldsymbol{X} + \boldsymbol{\Sigma}_p^{-1})^{-1}(\sigma_n^{-2}\boldsymbol{X}^{\mathrm{T}}\boldsymbol{y}), \tag{7.24c}$$

$$\boldsymbol{A} = (\sigma_n^{-2}\boldsymbol{X}^{\mathrm{T}}\boldsymbol{X} + \boldsymbol{\Sigma}_p^{-1}). \tag{7.24d}$$

Equations (7.24c) and (7.24d) indicate that if σ_n^2 is set to be 1, the noise variance will not have any influence on mean $\overline{\boldsymbol{W}}$ and covariance matrix $\boldsymbol{A}^{-1}$. Recalling Eq. (2.19b) of $\hat{w}^{\text{ridge}} = \arg\min_{w}\left[\sum_{i=1}^{n}\left(y_i - \sum_{j=1}^{m} w_j x_{ij}\right)^2 + \lambda\sum_{j=1}^{m} w_j^2\right]$ in Chap. 2, the penalty term of $\lambda\sum_{j=1}^{m} w_j^2$ makes coefficients as close as possible to zero. Thus, Eq. (2.19b) leads to Eq. (2.20b) of $\hat{\boldsymbol{W}}^{\text{ridge}} = (\boldsymbol{X}^{\mathrm{T}}\boldsymbol{X} + \lambda\boldsymbol{I})^{-1}\boldsymbol{X}^{\mathrm{T}}\boldsymbol{y}$ in Chap. 2. Comparing the revised expression of $\hat{\boldsymbol{W}}^{\text{ridge}}$ with Eq. (7.24c) under $\sigma_n^2 = 1$ indicates that the role of $\boldsymbol{\Sigma}_p^{-1}$ is in some way like the penalty term in the L_2 regularization.

As described above, the posterior distribution with prior $p(\boldsymbol{W})\sim N\ (0,\ \boldsymbol{\Sigma}_p)$ is expressed by

$$p(\boldsymbol{W}|\boldsymbol{y}, \boldsymbol{X}, \boldsymbol{\Sigma}_p^{-1}, \sigma_n^2) \propto p(\boldsymbol{y}|\boldsymbol{X}, \boldsymbol{W}, \sigma_n^2)p(\boldsymbol{W}|\boldsymbol{\Sigma}_p^{-1}), \tag{7.25a}$$

which gives the result distribution of

$$\hat{\boldsymbol{W}} \sim N(\overline{\boldsymbol{W}}, \boldsymbol{A}^{-1}), \tag{7.25b}$$

$$\overline{\boldsymbol{W}} = (\sigma_n^{-2}\boldsymbol{X}^{\mathrm{T}}\boldsymbol{X} + \boldsymbol{\Sigma}_p^{-1})^{-1}\sigma_n^{-2}\boldsymbol{X}^{\mathrm{T}}y, \tag{7.25c}$$

$$\boldsymbol{A} = (\sigma_n^{-2}\boldsymbol{X}^{\mathrm{T}}\boldsymbol{X} + \boldsymbol{\Sigma}_p^{-1}). \tag{7.25d}$$

The logarithm of posterior distribution is

$$\lg(p(\boldsymbol{W}|\boldsymbol{y}, \boldsymbol{X}, \boldsymbol{\Sigma}_p^{-1}, \sigma_n^2)) = \lg(p(\boldsymbol{y}|\boldsymbol{X}, \boldsymbol{W}, \sigma_n^2)) + \lg(p(\boldsymbol{W}|\boldsymbol{\Sigma}_p^{-1})), \tag{7.26a}$$

or

$$\lg(p(\boldsymbol{W}|\boldsymbol{y}, \boldsymbol{X}, \boldsymbol{\Sigma}_p^{-1}, \sigma_n^2)) = -\frac{(\boldsymbol{y} - \boldsymbol{X}\boldsymbol{W})^{\mathrm{T}}(\boldsymbol{y} - \boldsymbol{X}\boldsymbol{W})}{2\sigma_n^2} - \frac{1}{2}\boldsymbol{W}^{\mathrm{T}}\boldsymbol{\Sigma}_p^{-1}\boldsymbol{W}. \tag{7.26b}$$

The estimate of $\boldsymbol{W}$ is given by maximizing the posterior distribution or equivalently by maximizing the logarithm of the posterior distribution. With Eq. (7.26b), maximizing the logarithm of posterior distribution is equivalent to minimizing $\frac{(\boldsymbol{y}-\boldsymbol{X}\boldsymbol{W})^{\mathrm{T}}(\boldsymbol{y}-\boldsymbol{X}\boldsymbol{W})}{2\sigma_n^2}$ under the constraint of $\frac{1}{2}\boldsymbol{W}^{\mathrm{T}}\boldsymbol{\Sigma}_p^{-1}\boldsymbol{W}$, i.e.,

$$\overline{\boldsymbol{W}} = \underset{\boldsymbol{W}}{\operatorname{argmin}}\left[\frac{(\boldsymbol{y}-\boldsymbol{X}\boldsymbol{W})^{\mathrm{T}}(\boldsymbol{y}-\boldsymbol{X}\boldsymbol{W})}{2\sigma_n^2} + \frac{1}{2}\boldsymbol{W}^{\mathrm{T}}\boldsymbol{\Sigma}_p^{-1}\boldsymbol{W}\right]. \tag{7.26c}$$

Equation (7.26c) can be rewritten as

$$\overline{\boldsymbol{W}} = \underset{\boldsymbol{W}}{\operatorname{argmin}}\left[(\boldsymbol{y}-\boldsymbol{X}\boldsymbol{W})^{\mathrm{T}}(\boldsymbol{y}-\boldsymbol{X}\boldsymbol{W}) + \frac{1}{2}\boldsymbol{W}^{\mathrm{T}}\lambda\boldsymbol{I}\boldsymbol{W}\right], \tag{7.26d}$$

with $\lambda = \sigma_n^2\boldsymbol{\Sigma}_p^{-1}$. As described in the L_2 regularization, optimal λ is determined by cross-validation (Bishop, 2006), and the Bayesian linear regression hyperparameter λ will be also optimized by cross-validation. With optimal λ and optimal $\overline{\boldsymbol{W}}$, the optimal σ_n^2 is estimated by $\sigma_n^2 = \dfrac{1}{n-1}(\boldsymbol{y} - \boldsymbol{X}\overline{\boldsymbol{W}})^{\mathrm{T}}(\boldsymbol{y} - \boldsymbol{X}\overline{\boldsymbol{W}})$, and finally the optimal $\boldsymbol{\Sigma}_p$ in the prior is determined by the optimal λ and the optimal σ_n^2.

The predictive distribution of feature data $\boldsymbol{X}_*$ is actually the marginal likelihood over coefficients $\boldsymbol{W}$ when all possible linear models with respect to the posterior are considered, i.e.,

$$p(\hat{y}|\boldsymbol{X}_*, \sigma_n^2, \boldsymbol{\Sigma}_p^{-1}) = \int p(\hat{y}, \boldsymbol{W}|\boldsymbol{X}_*, \sigma_n^2, \boldsymbol{\Sigma}_p^{-1})\mathrm{d}\boldsymbol{W}. \tag{7.27a}$$

Once the normal distribution of coefficients is known, the prediction of the linear model will also be normal distribution

$$\hat{y} = \boldsymbol{X}_*\hat{\boldsymbol{W}} \sim N(\boldsymbol{X}_*\overline{\boldsymbol{W}}, \boldsymbol{X}_*\boldsymbol{A}^{-1}\boldsymbol{X}_*^{\mathrm{T}}). \tag{7.27b}$$

Example 7.4 Table 7.6 lists the experimental results of hydrogen diffusion coefficient D in commercially pure iron of group I samples at temperatures of 25 °C, 50 °C, 70 °C, and 80 °C (Zhang & Zheng, 1998). Consider $D = D_0\exp\left(-\dfrac{Q}{RT}\right)$ and $\ln D = \ln D_0 - \dfrac{Q}{RT}$, where D_0 is called the prefactor of diffusion coefficient, Q is the activation energy for diffusion, $R = 8.314\ \mathrm{J}\cdot\mathrm{K}^{-1}\cdot\mathrm{mol}^{-1}$ is the gas constant, and T is the absolute temperature. Letting $y \equiv \ln D$, $x \equiv \dfrac{1}{RT}$, D and D_0 in units of $(\mathrm{cm}^2\cdot\mathrm{s}^{-1})$, and Q and RT in units of $(\mathrm{J}\cdot\mathrm{mol}^{-1})$, we carry out the Bayesian linear regression.

Table 7.6 Hydrogen diffusion coefficient data

$1/RT$ (mol·J^{-1})	0.0004	0.00037	0.00035	0.00034
ln$D$	−10.1	−9.9196	−9.7046	−9.6168

In this example, Bayesian linear regression model is $f(x) = w_0 + w_1 x$ with intercept $w_0 = \ln D_0$ and slope $w_1 = -Q$. First, consider a preset prior with zero mean, i.e., $p(\boldsymbol{W}) \sim N(0, \boldsymbol{\Sigma}_p)$. The prior distribution indicates that in the (w_0, w_1) coefficient space, one point (w_0, w_1) corresponds to a potential sample function, and the infinite points in the (w_0, w_1) coefficient space follow the normal distribution $N(0, \boldsymbol{\Sigma}_p)$ and exhibit circle-shaped probabilistic contours in the (w_0, w_1) coefficient space with each contour having the same prior distribution probability. With the dummy feature of one and feature x_1, the feature data are

$$\mathrm{X} = \begin{bmatrix} 1 & x_1 \\ 1 & x_2 \\ 1 & x_3 \\ 1 & x_4 \end{bmatrix} = \begin{bmatrix} 1 & 0.00040 \\ 1 & 0.00037 \\ 1 & 0.00035 \\ 1 & 0.00034 \end{bmatrix},$$

$$\mathrm{X}^{\mathrm{T}} = \begin{bmatrix} 1 & 1 & 1 & 1 \\ 0.00040 & 0.00037 & 0.00035 & 0.00034 \end{bmatrix},$$

and $\boldsymbol{X}^{\mathrm{T}}\boldsymbol{X} = \begin{bmatrix} 4 & 1.46 \times 10^{-3} \\ 1.46 \times 10^{-3} & 5.35 \times 10^{-7} \end{bmatrix}$.

For the time being, we do not maximize the posterior distribution, take $\sigma_n^{-2} = 1$, and propose prior $\boldsymbol{\Sigma}_p = \boldsymbol{I} = \begin{bmatrix} 1 & 0 \\ 0 & 1 \end{bmatrix}$. Subsequently, according to Eqs. (7.24c) and (7.24d), we have mean $\overline{\boldsymbol{W}}$, i.e.,

$$\begin{aligned} \bar{\boldsymbol{W}} &= \begin{pmatrix} \overline{w_0} \\ \overline{w_1} \end{pmatrix} \\ &= \left[\begin{bmatrix} 4 & 1.46 \times 10^{-3} \\ 1.46 \times 10^{-3} & 5.35 \times 10^{-7} \end{bmatrix} + \begin{bmatrix} 1 & 0 \\ 0 & 1 \end{bmatrix}\right]^{-1} \\ &\quad \begin{bmatrix} 1 & 1 & 1 & 1 \\ 0.00040 & 0.00037 & 0.00035 & 0.00034 \end{bmatrix} \\ &\quad \begin{pmatrix} -10.1 \\ -9.9196 \\ -9.7046 \\ -9.6168 \end{pmatrix} = \begin{pmatrix} -7.86820 \\ -0.00289 \end{pmatrix} \end{aligned}$$

and $\boldsymbol{A} = \begin{bmatrix} 4 & 1.46 \times 10^{-3} \\ 1.46 \times 10^{-3} & 5.35 \times 10^{-7} \end{bmatrix} + \begin{bmatrix} 1 & 0 \\ 0 & 1 \end{bmatrix}$, which yields $\boldsymbol{A}^{-1} = \begin{bmatrix} 0.2 & -0.000292 \\ -0.000292 & 1.0 \end{bmatrix}$.

The results are $Q = -w_1 = 0.00289$ J·mol^{-1}, $\ln D_0 = w_0 = -7.86820$ or $D_0 = \mathrm{e}^{-7.86820}\mathrm{cm}^2 \cdot \mathrm{s}^{-1}$. Once the posterior distribution $p(\boldsymbol{W}|\boldsymbol{y}, \boldsymbol{X})$ is determined, the predictions of $\hat{y}_i(x_i)$ and associated variance $\sigma_{\hat{y}_i}^2$ are calculated with Eq. (7.26b). For the used four feature data, we have

$$\sigma_{y_1}^2 = \boldsymbol{x}_1 \boldsymbol{A}^{-1} \boldsymbol{x}_1^{\mathrm{T}} = (1\ 0.00040) \begin{bmatrix} 0.2 & -0.000292 \\ -0.000292 & 1.0 \end{bmatrix} \begin{pmatrix} 1 \\ 0.00040 \end{pmatrix} = 0.2,$$

$$\sigma_{y_2}^2 = \boldsymbol{x}_2 \boldsymbol{A}^{-1} \boldsymbol{x}_2^{\mathrm{T}} = (1\ 0.00037) \begin{bmatrix} 0.2 & -0.000292 \\ -0.000292 & 1.0 \end{bmatrix} \begin{pmatrix} 1 \\ 0.00037 \end{pmatrix} = 0.2,$$

$$\sigma_{y_3}^2 = \boldsymbol{x}_3 \boldsymbol{A}^{-1} \boldsymbol{x}_3^{\mathrm{T}} = (1\ 0.00035) \begin{bmatrix} 0.2 & -0.000292 \\ -0.000292 & 1.0 \end{bmatrix} \begin{pmatrix} 1 \\ 0.00035 \end{pmatrix} = 0.2,$$

$$\sigma_{y_4}^2 = \boldsymbol{x}_4 \boldsymbol{A}^{-1} \boldsymbol{x}_4^{\mathrm{T}} = (1\ 0.00034) \begin{bmatrix} 0.2 & -0.000292 \\ -0.000292 & 1.0 \end{bmatrix} \begin{pmatrix} 1 \\ 0.00034 \end{pmatrix} = 0.2, \text{ and}$$

$$\hat{y}(\boldsymbol{X}|\overline{\boldsymbol{W}}) = \boldsymbol{X}|\overline{\boldsymbol{W}} = \begin{bmatrix} 1 & x_1 \\ 1 & x_2 \\ 1 & x_3 \\ 1 & x_4 \end{bmatrix} \begin{pmatrix} \overline{w_0} \\ \overline{w_1} \end{pmatrix} = \begin{pmatrix} \overline{w_0} + \overline{w_1} x_1 \\ \overline{w_0} + \overline{w_1} x_2 \\ \overline{w_0} + \overline{w_1} x_3 \\ \overline{w_0} + \overline{w_1} x_4 \end{pmatrix}$$

$$= \begin{bmatrix} -7.86820 - 0.00289 \times 0.00040 \\ -7.86820 - 0.00289 \times 0.00037 \\ -7.86820 - 0.00289 \times 0.00035 \\ -7.86820 - 0.00289 \times 0.00034 \end{bmatrix} = \begin{pmatrix} -7.868 \\ -7.868 \\ -7.868 \\ -7.868 \end{pmatrix}.$$

With $\sigma_n^2 = 1$ and $\boldsymbol{\Sigma}_p = \boldsymbol{I} = \begin{pmatrix} 1 & 0 \\ 0 & 1 \end{pmatrix}$, Fig. 7.5a shows the posterior distribution probability contours in the (w_0, w_1) coefficient space, and Fig. 7.5b plots the prediction and associated variance from the GP regression. Due to the preset prior distribution $p(\boldsymbol{W}) \sim N\ (0, \boldsymbol{I})$ and the data characteristic, the posterior distribution probability has a mean of $(\overline{w}_0 = -7.86820,\ \overline{w}_1 = -0.00289)$, and covariance $\boldsymbol{A}^{-1} = \begin{bmatrix} 0.2 & -0.000292 \\ -0.000292 & 1.0 \end{bmatrix}$ in the (w_0, w_1) coefficient space. The correlation $\rho = \dfrac{-0.000292}{\sqrt{0.2} \times \sqrt{1.0}} \approx -0.00065$ between w_0 and w_1 distributions is so small that its influence on the posterior distribution probability contours can be approximately ignored. Thus, the posterior distribution probability has a variance of $\sigma_{w_0}^2 = 0.2$ and a variance of $\sigma_{w_1}^2 = 1.0$, as shown in Fig. 7.5a. Furthermore, the data and the preset prior distribution $p(\boldsymbol{W}) \sim N\ (0, \boldsymbol{I})$ lead to extremely poor prediction $\hat{y}$, far away from the input data, and a constant variance of $\sigma_y^2 = 0.2$, as shown in Fig. 7.5b. The results indicate clearly that the value of prior covariance $\boldsymbol{\Sigma}_p = 1$ is too small, which gives a too large penalty, so that the fitting goodness is extremely poor.

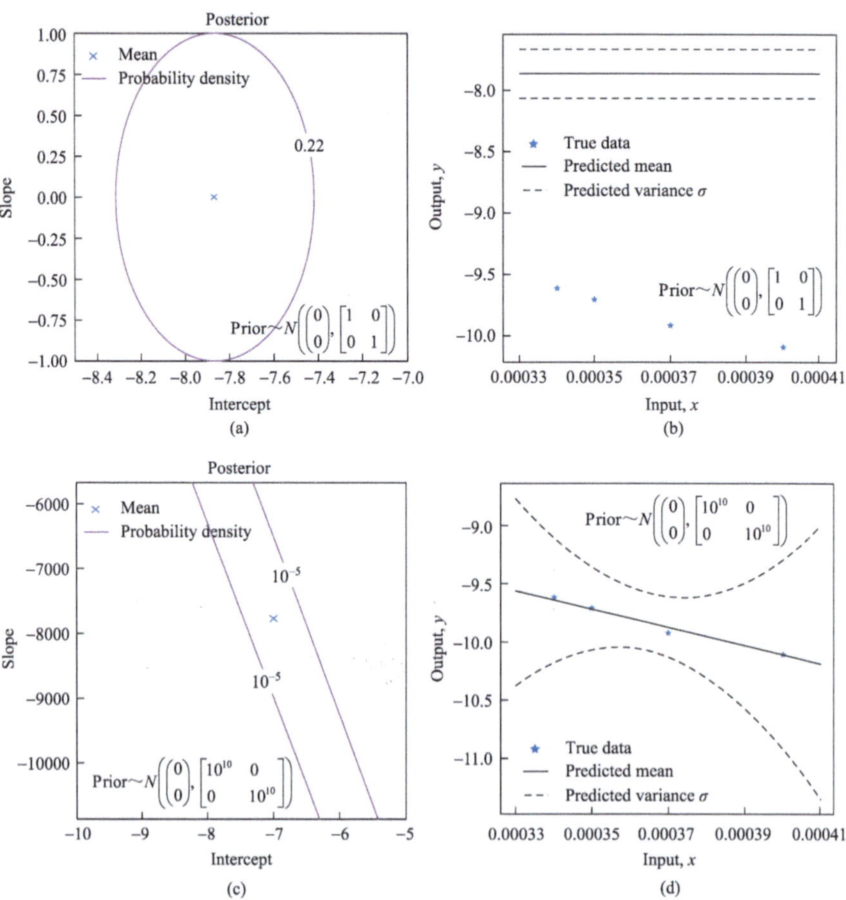

Fig. 7.5 The posterior distribution probability contours in the (w_0,w_1) coefficient space. **(a)** The contour of posterior p.d.f. = 0.22 corresponding to one σ and **(c)** the contour of posterior p.d.f. = 10^{-5}; **(b)** and **(d)** the predictions of $\hat{y}(\boldsymbol{X}|\overline{\boldsymbol{W}})$ and associated variations

To illustrate the effect of prior distribution on the prediction, we do not maximize the posterior distribution again, take $\sigma_n^{-2} = 1$, and propose prior $p(\boldsymbol{W}) \sim N(0, 10^{10}\boldsymbol{I})$, i.e., $\boldsymbol{\Sigma}_p = 10^{10}\boldsymbol{I}$. Consequently, we have

$$\overline{\boldsymbol{W}} = \begin{pmatrix} \overline{w_0} \\ \overline{w_1} \end{pmatrix} = \left[\begin{bmatrix} 4 & 1.46 \times 10^{-3} \\ 1.46 \times 10^{-3} & 5.35 \times 10^{-7} \end{bmatrix} + \begin{bmatrix} 1 \times 10^{-10} & 0 \\ 0 & 1 \times 10^{-10} \end{bmatrix}\right]^{-1}$$

$$\cdot \begin{bmatrix} 1 & 1 & 1 & 1 \\ 0.00040 & 0.00037 & 0.00035 & 0.00034 \end{bmatrix} \begin{pmatrix} -10.1 \\ -9.9196 \\ -9.7046 \\ -9.6168 \end{pmatrix}$$

$$= \begin{pmatrix} -6.99671 \\ -7776.82 \end{pmatrix},$$

$$A = \begin{bmatrix} 4 & 1.46 \times 10^{-3} \\ 1.46 \times 10^{-3} & 5.35 \times 10^{-7} \end{bmatrix} + \begin{bmatrix} 1 \times 10^{-10} & 0 \\ 0 & 1 \times 10^{-10} \end{bmatrix}, \text{ and}$$

$$A^{-1} = \begin{bmatrix} 60.8068 & -1.66 \times 10^{5} \\ -1.66 \times 10^{5} & 4.545 \times 10^{8} \end{bmatrix}.$$

The results are $Q = -w_1 = 7776.82$ J·mol^{-1}, $\ln D_0 = w_0 = -6.99671$ or $D_0 = e^{-6.99671}\text{cm}^2 \cdot \text{s}^{-1}$. When $\boldsymbol{\Sigma}_p = 1 \times 10^{10}\boldsymbol{I}$, the predictions of $\hat{y}(\boldsymbol{X}|\overline{\boldsymbol{W}})$ and associated variance σ_y^2 are calculated for the four data to be

$$\sigma_{y_1}^2 = \boldsymbol{x}_1 \boldsymbol{A}^{-1} \boldsymbol{x}_1^{\mathrm{T}} = \begin{pmatrix} 1 & 0.00040 \end{pmatrix} \begin{bmatrix} 60.8068 & -1.66 \times 10^{5} \\ -1.66 \times 10^{5} & 4.545 \times 10^{8} \end{bmatrix} \begin{pmatrix} 1 \\ 0.00040 \end{pmatrix}$$

$$= 0.8068,$$

$$\sigma_{y_2}^2 = \boldsymbol{x}_2 \boldsymbol{A}^{-1} \boldsymbol{x}_2^{\mathrm{T}} = \begin{pmatrix} 1 & 0.00037 \end{pmatrix} \begin{bmatrix} 60.8068 & -1.66 \times 10^{5} \\ -1.66 \times 10^{5} & 4.545 \times 10^{8} \end{bmatrix} \begin{pmatrix} 1 \\ 0.00037 \end{pmatrix}$$

$$= 0.2614,$$

$$\sigma_{y_3}^2 = \boldsymbol{x}_3 \boldsymbol{A}^{-1} \boldsymbol{x}_3^{\mathrm{T}} = \begin{pmatrix} 1 & 0.00035 \end{pmatrix} \begin{bmatrix} 60.8068 & -1.66 \times 10^{5} \\ -1.66 \times 10^{5} & 4.545 \times 10^{8} \end{bmatrix} \begin{pmatrix} 1 \\ 0.00035 \end{pmatrix}$$

$$= 0.3523,$$

$$\sigma_{y_4}^2 = \boldsymbol{x}_4 \boldsymbol{A}^{-1} \boldsymbol{x}_4^{\mathrm{T}} = \begin{pmatrix} 1 & 0.00034 \end{pmatrix} \begin{bmatrix} 60.8068 & -1.66 \times 10^{5} \\ -1.66 \times 10^{5} & 4.545 \times 10^{8} \end{bmatrix} \begin{pmatrix} 1 \\ 0.00034 \end{pmatrix}$$

$$= 0.5341, \text{ and}$$

$$\hat{y}(\boldsymbol{X}|\overline{\boldsymbol{W}}) = \boldsymbol{X}\overline{\boldsymbol{W}} = \begin{bmatrix} 1 & x_1 \\ 1 & x_2 \\ 1 & x_3 \\ 1 & x_4 \end{bmatrix} \begin{pmatrix} \overline{w_0} \\ \overline{w_1} \end{pmatrix} = \begin{pmatrix} \overline{w_0} + \overline{w_1}x_1 \\ \overline{w_0} + \overline{w_1}x_2 \\ \overline{w_0} + \overline{w_1}x_3 \\ \overline{w_0} + \overline{w_1}x_4 \end{pmatrix} = \begin{pmatrix} -10.11 \\ -9.87 \\ -9.72 \\ -9.64 \end{pmatrix}.$$

With $\sigma_n^2 = 1$ and $\boldsymbol{\Sigma}_p = 1 \times 10^{10}\boldsymbol{I}$, Fig. 7.5c shows the posterior distribution probability contours in the (w_0, w_1) coefficient space, and Fig. 7.5(d) plots the prediction and associated variance from the GP regression. The extremely large covariance in the preset prior distribution $p(\boldsymbol{W}) \sim N(0,\ 1 \times 10^{10}\boldsymbol{I})$ releases the constraint to the posterior distribution probability so that the posterior distribution probability has a mean of ($\overline{w}_0 = -6.99671$, $\overline{w}_1 = -7776.82$) and covariance $\boldsymbol{A}^{-1} = \begin{bmatrix} 60.8068 & -1.66 \times 10^{5} \\ -1.66 \times 10^{5} & 4.545 \times 10^{8} \end{bmatrix}$ in the (w_0, w_1) coefficient space. In this case, it is very hard to find a posterior distribution probability contour with a high value, as shown in Fig. 7.5c. However, the data and the preset prior distribution $p(\boldsymbol{W}) \sim$

$N(0, 1 \times 10^{10}\boldsymbol{I})$ lead to reasonable prediction $\hat{y}$, comparable to that reported in the previous work (Zheng and Zhang, 1998) as indicated in Fig. 7.5d. This is because the value of $\boldsymbol{\Sigma}_p^{-1} = 10^{-10}\boldsymbol{I}$ is so small that nearly no penalty is applied to the regression. The dashed curves in Fig. 7.5d show the variance associated with prediction $\hat{y}$, indicating that the variance is relatively small near the data, the smallest at almost the middle of data, and becomes larger when being away from the data. The result is important to interpolation and extension.

In summary, the predictions of $\hat{\boldsymbol{y}}(\boldsymbol{X}|\overline{\boldsymbol{W}})$ and associated variations highly depend on the characteristics of data and the prior $p(\boldsymbol{W})$. In summary, the posterior distribution must be maximized to find the optimal solution. As described above, the inverse covariance $\boldsymbol{\Sigma}_p^{-1}$ of the prior behaves like the penalty hyperparameter λ in the L_2 regularization. As described in Chap. 2, the value of hyperparameter λ is optimized by cross-validation. Similarly, the prior distribution should be optimized.

LOOCV is conducted to maximize the posterior distribution and optimize the value of $\boldsymbol{\Sigma}_p$ in the prior. Figure 7.6 shows the coefficient of determination (R^2) versus λI during LOOCV, which yields the optimal $\lambda = 1.3 \times 10^{-10}$ and associated $\overline{\boldsymbol{W}} = (-7.0349, \ -7672.197)$. Then, σ_n^2 and $\boldsymbol{\Sigma}_p$ are calculated by $\sigma_n^2 = \frac{1}{n-1}(\boldsymbol{y} - \boldsymbol{X}\overline{\boldsymbol{W}})^{\mathrm{T}}(\boldsymbol{y} - \boldsymbol{X}\overline{\boldsymbol{W}}) = 0.0010272$ and $\boldsymbol{\Sigma}_p = \left(\frac{\sigma_n^2}{\lambda}\right)\boldsymbol{I} = 79015385\boldsymbol{I}$, which yield the results of $Q = 7672.197\ \mathrm{J \cdot mol^{-1}}$ and $D_0 = \mathrm{e}^{-7.0349}\mathrm{cm^2 \cdot s^{-1}}$. Figure 7.7 shows the Gaussian posterior distribution according to Eq. (7.26b).

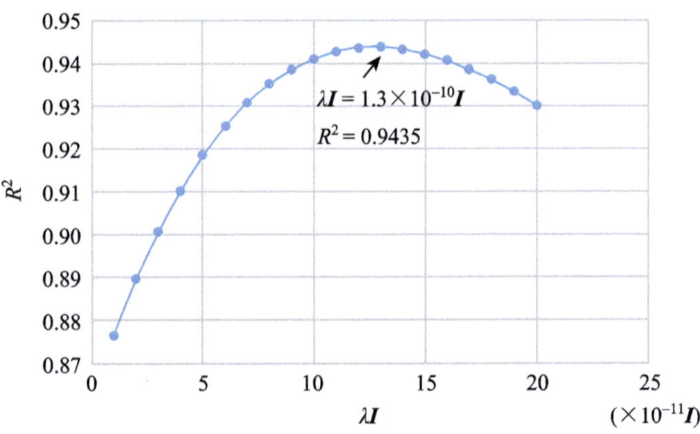

Fig. 7.6 The R^2 value versus $\lambda\boldsymbol{I}$ on leave one out cross-validation

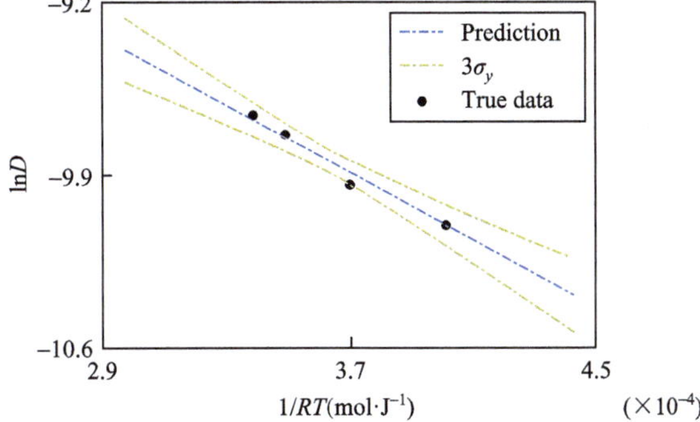

Fig. 7.7 The predicted values and their corresponding standard deviations

7.5 Expectation–Maximization (EM) Algorithm

In powder X-ray diffraction (XRD) pattern, a few peaks may overlap. To accurately determine the peak height, peak position, and half-peak width of each peak, one should clarify the contribution of each peak to the intensity in the overlap region, i.e., the weight of each peak. The overlap region is a mixture of a few XRD peaks, where the total intensity and 2θ are the measured data, and the peak number (e.g., two peaks or three peaks) and the peak weights are missing. Usually, the peak number and peak positions are guessed as initial inputs to fit the observed pattern with a theoretical profile of each peak. Maximizing fitting accuracy determines the peak number, weight, peak height, peak position, and half-peak width of each peak. If there is no overlap at all, the peak height, peak position, and half-peak width will be clearly determined by the observed XRD pattern. Alternately speaking, the peak height, peak position, and half-peak width determine the XRD pattern. In this example, the number of overlapped peaks and the contribution of each peak to the observed intensity in the overlap region are the missing variables, called the latent variables. EM algorithm (Dempster et al., 1977) is a numerical method to step by step fit a measured pattern with overlapped distributions (peaks), provided that the types of distributions are known.

7.5.1 Gaussian Mixture Model (GMM)

The Gaussian distribution is given by

$$p(x) = \frac{1}{\sqrt{2\pi}\sigma}\exp\left(-\frac{(x-\mu)^2}{2\sigma^2}\right). \tag{7.28a}$$

For simplicity, consider the GMM (Newcomb, 1886) with only one measurable variable, where a few Gaussian distributions are mixed together. Independent N observations $\boldsymbol{x} = (x_1, x_2, \cdots, x_N)$ contain only the measurable variable information of total probability density (e.g., XRD intensity) and miss the information of latent variables $\mathbf{Z} = (z_1, z_2, \cdots, z_k)$, where k denotes the number of Gaussian distributions, and z_i $(i = 1, 2, \cdots, k)$ represents the expectation μ_i, variance σ_i and weight w_i (mixture coefficient) of each Gaussian distribution. In practice, the number of Gaussian distributions is 2, 3, or 4, and it is rare to have more than 4 distributions overlapped.

Then, the mixture distribution probability density is given by

$$p_M(\boldsymbol{x}|\mathbf{Z}) = \sum_{i=1}^{k} w_i \cdot p(x|\mu_i, \sigma_i^2) = \sum_{i=1}^{k} w_i \cdot \frac{1}{\sqrt{2\pi}\sigma_i} \exp\left(-\frac{(x-\mu_i)^2}{2\sigma_i^2}\right), \quad (7.28b)$$

with

$$\sum_{i=1}^{k} w_i = 1. \quad (7.28c)$$

The latent variables $\mathbf{Z}$ of w_i, μ_i and σ_i^2 should be determined from the fitting of data. In theory, the mixture distribution probability density also needs to be normalized, i.e.,$\int p_M(x|Z)dx = 1$.

Since all observations are independent, the likelihood and log-likelihood are given, respectively, by

$$L(w_i, \mu_i, \sigma_i^2) = \prod_{j=1}^{N} \sum_{i=1}^{k} w_i \cdot p(x_j|\mu_i, \sigma_i^2)$$

$$= \prod_{j=1}^{N} \sum_{i=1}^{k} w_i \cdot \frac{1}{\sqrt{2\pi}\sigma_i} \exp\left(-\frac{(x_j-\mu_i)^2}{2\sigma_i^2}\right), \quad (7.29a)$$

$$\mathrm{LL}(w_i, \mu_i, \sigma_i^2) = \sum_{j=1}^{N} \ln\left\{\sum_{i=1}^{k} w_i \cdot \frac{1}{\sqrt{2\pi}\sigma_i} \exp\left(-\frac{(x_j-\mu_i)^2}{2\sigma_i^2}\right)\right\}. \quad (7.29b)$$

In practice, the variable x is discretized and the mixture distribution probability density $p_M(x_j|Z)$ is approximately expressed by the frequency I_j at x_j within an infinitesimal range dx. The range dx is dependent on experimental measurement accuracy and/or methods of data pretreatment. In the EM calculation, the frequency I_j is usually an integer so that I_j times of x_j will be calculated with Eq. (7.29b). Equivalently, Eq. (7.29b) can be reduced to

$$LL\left(w_i, \mu_i, \sigma_i^2\right) = \sum_{j=1}^{n} I_j \sum_{i=1}^{k}\left(ln(w_i) - ln\left(\sqrt{2\pi}\sigma_i\right) - \frac{\left(x_j - \mu_i\right)^2}{2\sigma_i^2}\right), \quad (7.29c)$$

where the sum of frequencies is defined as $\sum_{j=1}^{n} I_j = N$. In Eq. (7.29c), $x_j$and $I_j$are known, and the latent variables of $\left\{w_i, \mu_i, \sigma_i^2\right\}$ are estimated from the maximum likelihood estimation as described above, i.e.,

$$\left\{\hat{w}_i, \hat{\mu}_i, \hat{\sigma}_i^2\right\} = \underset{w_i,\mu_i,\sigma_i^2}{\arg\max}\, LL(w_i, \mu_i, \sigma_i^2), \quad (7.29d)$$

$$\text{subject to} \quad \sum_{i=1}^{k} w_i = 1.$$

Numerical iteration is conducted to solve Eq. (7.29d). As described above, each weight w_i represents the contribution of distribution i to the entire observed profile (total data), thereby meaning that the contribution weight of distribution i is the same for every datum of the data. In EM algorithm with frequency I_j, the contribution weight $\alpha_{ji}(j = 1, 2, \cdots, n;\; i = 1, 2, \cdots, k)$ of distribution i to $p_M(x_j|Z)$ is introduced based on Bayes probability theory, which is calculated in the expectation step. With the weight $\alpha_{ji}(j = 1, 2, \cdots, n;\; i = 1, 2, \cdots, k)$, the maximum likelihood estimation of $\left\{w_i, \mu_i, \sigma_i^2\right\}$ is much easier to be calculated with Eq. (7.29d). After that, all weights α_{ji} $(j = 1, 2, \cdots, n;\; i = 1, 2, \cdots, k)$ are up-dated with the estimated $\left\{w_i, \mu_i, \sigma_i^2\right\}(i = 1, 2, \cdots, k)$. The iteration goes on until the convergence of $(\hat{w}_i, \hat{\mu}_i, \hat{\sigma}_i^2)$.

1. Expectation-step (E-step)

Initial input values of latent variables are denoted by $\left\{w_i^{(0)}, \mu_i^{(0)}, \sigma_i^{2,(0)}\right\}$ $(i = 1, 2, \cdots, k)$, and the values of latent variables after t-th iteration are denoted by $\left\{w_i^{(t)}, \mu_i^{(t)}, \sigma_i^{2,(t)}\right\}(t = 1, 2, \cdots)$. The weight expectation in $(t + 1)$-th iteration is given by

$$\alpha_{ji}^{(t+1)} = p_M(w_i^{(t+1)}|x_j) = \frac{P(w_i^{(t)}) \cdot p_M(x_j\Big|w_i^{(t)}, \mu_i^{(t)}, \sigma_i^{2,(t)})}{p_M(x_j)}$$

$$= \frac{w_i^{(t)} \cdot p_M(x_j\Big|\mu_i^{(t)}, \sigma_i^{2,(t)})}{\sum_{i=1}^{k} w_i^{(t)} \cdot p_M(x_j\Big|w_i^{(t)}, \mu_i^{(t)}, \sigma_i^{2,(t)})}$$

$$\text{for } (j = 1, 2, \cdots, n; i = 1, 2, \cdots, k). \quad (7.30a)$$

Explicitly, the expectation takes the following form for a mixture of normal distributions

$$\alpha_{ji}^{(t+1)} = \frac{w_i^{(t)} \cdot \frac{1}{\sqrt{2\pi\sigma_i^{2,(t)}}}\exp\left(-\frac{(x_j-\mu_i^{(t)})^2}{2\sigma_i^{2,(t)}}\right)}{\sum\limits_{i=1}^{k} w_i^{(t)} \cdot \frac{1}{\sqrt{2\pi\sigma_i^{2,(t)}}}\exp\left(-\frac{(x_j-\mu_i^{(t)})^2}{2\sigma_i^{2,(t)}}\right)} \text{ (for } j=1,2,\cdots,n;\ i=1,2,\cdots,k). \tag{7.30b}$$

The α_{ji} parameter represents the probability of the observation x_j belonging to distribution $i (i = 1, 2, \cdots, k)$, or equivalently the contribution weight of distribution i (z_i) to datum $p_M(x_j|Z)$ $(j = 1, 2, \cdots, n)$.

2. Maximization step (M-step)

Following the E-step is the M-step. The Lagrange function of the log-likelihood is

$$\mathrm{LLL}(\lambda, w_i, \mu_i, \sigma_i^2) = \sum_{j=1}^{n} I_j \ln\left\{\sum_{i=1}^{k} w_i \cdot \frac{1}{\sqrt{2\pi}\sigma_i}\exp\left(-\frac{(x_j-\mu_i)^2}{2\sigma_i^2}\right)\right\} + \lambda\left(\sum_{i=1}^{k} w_i - 1\right). \tag{7.31a}$$

The maximization of log-likelihood requires

$$\frac{\partial \mathrm{LLL}}{\partial \mu_i} = \sum_{j=1}^{n} I_j \frac{w_i \cdot p_M(x_j|\mu_i, \sigma_i^2)}{\sum\limits_{i=1}^{k} w_i \cdot p_M(x_j|\mu_i, \sigma_i^2)}(x_j - \mu_i) = 0, \tag{7.31b}$$

$$\frac{\partial \mathrm{LLL}}{\partial \sigma_i} = \sum_{j=1}^{n} I_j \frac{w_i \cdot p_M(x_j|\mu_i, \sigma_i^2)}{\sum\limits_{i=1}^{k} w_i \cdot p_M(x_j|\mu_i, \sigma_i^2)}\left[\frac{(x_j-\mu_i)^2}{\sigma_i^3} - \frac{1}{\sigma_i}\right] = 0, \tag{7.31c}$$

$$\frac{\partial \mathrm{LLL}}{\partial w_i} = \sum_{j=1}^{n} I_j \frac{p_M(x_j|\mu_i, \sigma_i^2)}{\sum\limits_{i=1}^{k} w_i \cdot p_M(x_j|\mu_i, \sigma_i^2)} + \lambda = 0. \tag{7.31d}$$

Equations (7.31b) and (7.31c) yield

$$\mu_i^{(t+1)} = \frac{\sum\limits_{j=1}^{n} I_j \alpha_{ji}^{(t+1)} x_j}{\sum\limits_{j=1}^{n} I_j \alpha_{ji}^{(t+1)}}, \tag{7.32a}$$

$$\sigma_i^{2,(t+1)} = \frac{\sum\limits_{j=1}^{n} I_j \alpha_{ji}^{(t+1)} (x_j - \mu_i^{(t+1)})(x_j - \mu_i^{(t+1)})^{\mathrm{T}}}{\sum\limits_{j=1}^{n} I_j \alpha_{ji}^{(t+1)}}. \tag{7.32b}$$

Multiplying $\sum_{i=1}^{k} w_i$ to Eq. (7.31d) leads to $\sum_{j=1}^{n} I_j \frac{\sum_{i=1}^{k} w_i \cdot p_M(x_j|\mu_i,\sigma_i^2)}{\sum_{i=1}^{k} w_i \cdot p_M(x_j|\mu_i,\sigma_i^2)} + \lambda \sum_{i=1}^{k} w_i = 0$, which gives

$$\lambda = -N, \tag{7.32c}$$

$$w_i^{(t+1)} = \frac{1}{N} \sum_{j=1}^{n} I_j \alpha_{ji}^{(t+1)}. \tag{7.32d}$$

As expected, Eq. (7.32d) indicates that the weight $w_i^{(t+1)}(i = 1, 2, \cdots, k)$ of the distribution i is the average of the weight α_{ji} over all observations in the $(t+1)$-th iteration. Iterating E-step and M-step until convergence gives the final values of parameters $\mathbf{Z}$ of $\hat{w}_i$, $\hat{\mu}_i$ and $\hat{\sigma}_i^2$ $(i = 1, 2, \cdots, k)$.

Example 7.5 This example illustrates how the EM algorithm works in the fitting of a mixed Gaussian distribution. The black curve in Fig. 7.8 shows such a mixed probability distribution density curve of two Gaussian distributions, $p_M(x|Z) \sim 0.30N(8.00, 4.00) + 0.70N(12.00, 8.00)$. Setting the step-size $\Delta x = 1$ and starting with $x = 2.5$ separate the region $x \in [2.5,\ 20.5]$ into 18 sub-regions of $(2.5,\ 3.5)$, $(3.5, 4.5)$, $\cdots$, $(19.5, 20.5)$, and the middle points in the sub-regions are 3, 4, $\cdots$, 20. In this case, $p_M(x_j|Z)\Delta x$ approximately represents $p_M(x_j|Z)$ at $x_j = 3, 4, \cdots, 20$. Then, letting $p_M(x_j|Z)\Delta x$ multiply an appropriate integer gives the frequency at each x_j. Figure 7.8 shows the Gaussian mixture and corresponding histogram, which is listed in Table 7.7 as well.

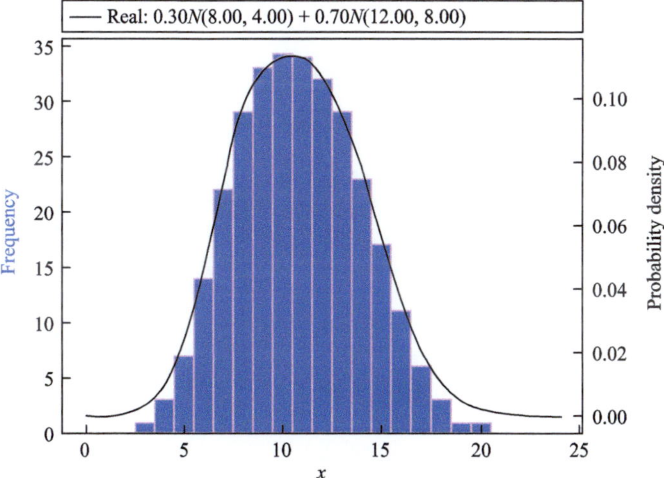

Fig. 7.8 The probability density curve of the mixed Gaussian distribution and the frequency histogram of the dataset

Table 7.7 The 18 data of (x_j, I_j) with $n = 18$ and $N = 300$

j	1	2	3	4	5	6	7	8	9
x_j	3	4	5	6	7	8	9	10	11
I_j	1	3	7	14	22	29	33	34	34
j	10	11	12	13	14	15	16	17	18
x_j	12	13	14	15	16	17	18	19	20
I_j	32	29	23	17	11	6	3	1	1

Step 1. **Initial parameters of Gaussian distributions**

A mixture of two Gaussian distributions is used to fit the histogram shown in Fig. 7.8. The initial values of parameters $\left\{w_i^{(0)}, \mu_i^{(0)}, \sigma_i^{2,(0)}\right\}(i = 1,\ 2)$ are given by

$$\begin{bmatrix} w_1^{(0)} & w_2^{(0)} \\ \mu_1^{(0)} & \mu_2^{(0)} \\ \sigma_1^{2,(0)} & \sigma_2^{2,(0)} \end{bmatrix} = \begin{bmatrix} 0.05 & 0.95 \\ 30 & 60 \\ 100 & 200 \end{bmatrix}.$$

Step 2. **E-step**

In the first iteration, with the data in Table 7.7 and the initial parameter values as well as according to Eq. (7.30b), the contribution weight of Gaussian distribution 1 to datum 1 is calculated by

$$\begin{aligned}\alpha_{11}^{(1)} &= \frac{w_1^{(0)} \cdot \frac{1}{\sqrt{2\pi\sigma_1^{2,(0)}}} \exp\left(-\frac{(x_j-\mu_1^{(0)})^2}{2\sigma_1^{2,(0)}}\right)}{\sum_{i=1}^{2} w_i^{(0)} \cdot \frac{1}{\sqrt{2\pi\sigma_i^{2,(0)}}} \exp\left(-\frac{(x_j-\mu_i^{(0)})^2}{2\sigma_i^{2,(0)}}\right)} \\ &= \frac{w_1^{(0)} \cdot \frac{1}{\sqrt{2\pi\sigma_1^{2,(0)}}} \exp\left(-\frac{(x_j-\mu_1^{(0)})^2}{2\sigma_1^{2,(0)}}\right)}{w_1^{(0)} \cdot \frac{1}{\sqrt{2\pi\sigma_1^{2,(0)}}} \exp\left(-\frac{(x_j-\mu_1^{(0)})^2}{2\sigma_1^{2,(0)}}\right) + w_2^{(0)} \cdot \frac{1}{\sqrt{2\pi\sigma_2^{2,(0)}}} \exp\left(-\frac{(x_j-\mu_2^{(0)})^2}{2\sigma_2^{2,(0)}}\right)} \\ &= \frac{0.05 \times \frac{1}{\sqrt{2\pi\times 100}} \exp\left(-\frac{(3-30)^2}{2\times 100}\right)}{0.05 \times \frac{1}{\sqrt{2\pi\times 100}} \exp\left(-\frac{(3-30)^2}{2\times 100}\right) + 0.95 \times \frac{1}{\sqrt{2\pi\times 200}} \exp\left(-\frac{(3-60)^2}{2\times 200}\right)} = 0.8675.\end{aligned}$$

Similarly, the contribution weight of Gaussian distribution 2 to datum 1 is

$$\alpha_{12}^{(1)} = \frac{w_2^{(0)} \cdot \frac{1}{\sqrt{2\pi\sigma_2^{2,(0)}}} \exp\left(-\frac{(x_j-\mu_2^{(0)})^2}{2\sigma_2^{2,(0)}}\right)}{\sum_{i=1}^{2} w_i^{(0)} \cdot \frac{1}{\sqrt{2\pi\sigma_i^{2,(0)}}} \exp\left(-\frac{(x_j-\mu_i^{(0)})^2}{2\sigma_i^{2,(0)}}\right)}$$

$$= \frac{w_2^{(0)} \cdot \frac{1}{\sqrt{2\pi\sigma_2^{2,(0)}}} \exp\left(-\frac{(x_j-\mu_2^{(0)})^2}{2\sigma_2^{2,(0)}}\right)}{w_1^{(0)} \cdot \frac{1}{\sqrt{2\pi\sigma_1^{2,(0)}}} \exp\left(-\frac{(x_j-\mu_1^{(0)})^2}{2\sigma_1^{2,(0)}}\right) + w_2^{(0)} \cdot \frac{1}{\sqrt{2\pi\sigma_2^{2,(0)}}} \exp\left(-\frac{(x_j-\mu_2^{(0)})^2}{2\sigma_2^{2,(0)}}\right)}$$

$$= \frac{0.95 \times \frac{1}{\sqrt{2\pi\times 200}} \exp\left(-\frac{(3-60)^2}{2\times 200}\right)}{0.05 \times \frac{1}{\sqrt{2\pi\times 100}} \exp\left(-\frac{(3-30)^2}{2\times 100}\right) + 0.95 \times \frac{1}{\sqrt{2\pi\times 200}} \exp\left(-\frac{(3-60)^2}{2\times 200}\right)} = 0.1325.$$

The values of $\alpha_{ji}^{(1)}$ $(j = 2,\ 3, \cdots, 18;\ i = 1,\ 2)$ are calculated in the same way and listed in Table 7.8.

Table 7.8 The values of $\alpha_{ji}^{(1)}$ $(j = 1, 2, \cdots, 18;\ i = 1, 2)$ after the first iteration

x_j	x_1	x_2	x_3	x_4	x_5	x_6	x_7	x_8	x_9
$\alpha_{j1}^{(1)}$	0.8675	0.8655	0.8628	0.8595	0.8556	0.8509	0.8454	0.8391	0.8319
$\alpha_{j2}^{(1)}$	0.1325	0.1345	0.1372	0.1405	0.1444	0.1491	0.1546	0.1609	0.1681
x_j	x_{10}	x_{11}	x_{12}	x_{13}	x_{14}	x_{15}	x_{16}	x_{17}	x_{18}
$\alpha_{j1}^{(1)}$	0.82367	0.8144	0.8040	0.7923	0.7793	0.7647	0.7486	0.7308	0.7112
$\alpha_{j2}^{(1)}$	0.1763	0.1856	0.1960	0.2077	0.2207	0.2353	0.2514	0.2692	0.2888

Then, weights are updated to be $w_1^{(1)} = \frac{1}{300}\sum_{j=1}^{18} I_j\alpha_{j1} = 0.83$, and $w_2^{(1)} = \frac{1}{300}\sum_{j=1}^{18} I_j\alpha_{j2} = 0.17$.

Step 3. **M-step**

The E-step provides the updated values of $\alpha_{ji}(j = 1,\ 2, \cdots, 18;\ i = 1,\ 2)$, which are used in the M-step to update the values of μ_i and σ_i^2 according to Eq. (7.32a) and (7.32b). For Gaussian distribution 1, we have

$$\mu_1^{(1)} = \frac{\sum_{j=1}^{18} I_j\alpha_{j1}^{(1)}x_j}{\sum_{j=1}^{18} I_j\alpha_{j1}^{(1)}} = \frac{I_1\alpha_{11}^{(1)}x_1 + I_2\alpha_{21}^{(1)}x_2 + \cdots + I_{18}\alpha_{18,1}^{(1)}x_{18}}{I_1\alpha_{11}^{(1)} + I_2\alpha_{21}^{(1)} + \cdots + I_{18}\alpha_{18,1}^{(1)}}$$

$$= \frac{1 \times 0.8675 \times 3 + 3 \times 0.8655 \times 4 + \cdots + 1 \times 0.7112 \times 20}{1 \times 0.8675 + 3 \times 0.8655 + \cdots + 1 \times 0.7112} = 10.70 \text{ and}$$

$$\sigma_1^{2,(1)} = \frac{\sum_{j=1}^{18} I_j\alpha_{j1}(x_j - \mu_1^{(1)})(x_j - \mu_1^{(1)})^{\mathrm{T}}}{\sum_{j=1}^{18} I_j\alpha_{j1}^{(1)}}$$

$$= \frac{I_1\alpha_{11}^{(1)}(x_1 - \mu_1^{(1)})(x_1 - \mu_1^{(1)})^{\mathrm{T}} + I_2\alpha_{21}^{(1)}(x_2 - \mu_1^{(1)})(x_2 - \mu_1^{(1)})^{\mathrm{T}} + \cdots + I_{18}\alpha_{18,1}^{(1)}(x_{18} - \mu_1^{(1)})(x_{18} - \mu_1^{(1)})^{\mathrm{T}}}{I_1\alpha_{11}^{(1)} + I_2\alpha_{21}^{(1)} + \cdots + I_{18}\alpha_{18,1}^{(1)}}$$

$$= (1 \times 0.8675 \times (3 - 10.70) \times (3 - 10.70)^{\mathrm{T}} + 3 \times 0.8655 \times (4 - 10.70) \times (4 - 10.70)^{\mathrm{T}} + \cdots$$
$$+ 1 \times 0.7112 \times (20 - 10.70) \times (20 - 10.70)^{\mathrm{T}})/(1 \times 0.8675 + 3 \times 0.8655 + \cdots + 1 \times 0.7112) = 9.81.$$

For Gaussian distribution 2, we have

$$\mu_2^{(1)} = \frac{\sum_{j=1}^{18} I_j\alpha_{j2}^{(1)}x_j}{\sum_{j=1}^{18} I_j\alpha_{j2}^{(1)}} = 11.27, \text{ and}$$

$$\sigma_2^{2,(1)} = \frac{\sum_{j=1}^{18} I_j\alpha_{j2}^{(1)}\left(x_j - \mu_2^{(1)}\right)\left(x_j - \mu_2^{(1)}\right)^{\mathrm{T}}}{\sum_{j=1}^{18} I_j\alpha_{j2}^{(1)}} = 10.44.$$

The first iteration ends here and the updated $\begin{bmatrix} w_1^{(1)} & w_2^{(1)} \\ \mu_1^{(1)} & \mu_2^{(1)} \\ \sigma_1^{2,(1)} & \sigma_2^{2,(1)} \end{bmatrix} = \begin{bmatrix} 0.83 & 0.17 \\ 10.70 & 11.27 \\ 9.81 & 10.44 \end{bmatrix}$.

Step 4. **Iteration until convergence**

The convergence criterion of iteration in this example is defined by having two significant figures in decimal numbers of $\hat{w}_i$, $\hat{\mu}_i$ and $\hat{\sigma}_i^2$ $(i = 1, 2)$. Accordingly, the EM reaches the present convergence criterion after 1902 iterations, and gives the converged values of parameters

$$\begin{bmatrix} \hat{w}_1 & \hat{w}_2 \\ \hat{\mu}_1 & \hat{\mu}_2 \\ \hat{\sigma}_1^2 & \hat{\sigma}_2^2 \end{bmatrix} = \begin{bmatrix} 0.31 & 0.69 \\ 8.01 & 12.07 \\ 3.81 & 7.61 \end{bmatrix}.$$

Figure 7.9 shows the fitting result.

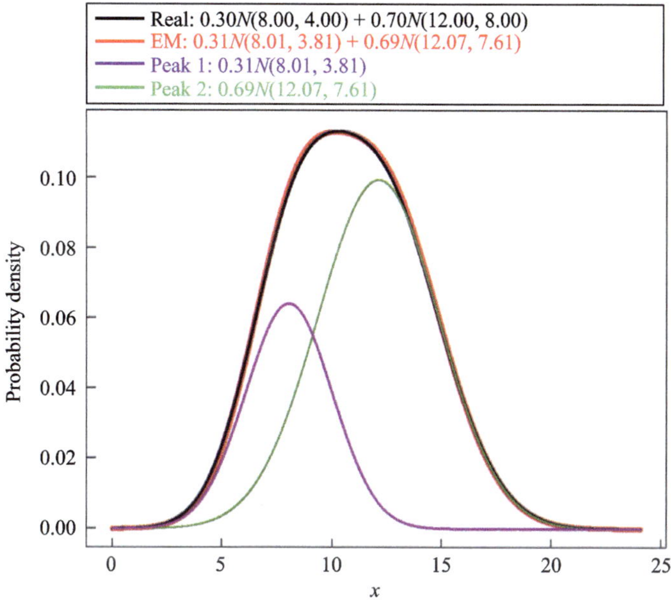

Fig. 7.9 The EM results of the GMM

Example 7.6 Fu and Zhang (1998) conducted three-point bending tests on poled PZT-841 ceramics under 10 kV/cm dc electric field. Table 7.9 lists 55 test results of 15 values of rupture moduli and their frequencies. The data are manually extracted from Fig. 2b in the paper of Fu and Zhang (1998), and the distribution histogram is shown in Fig. 7.10. Fu and Zhang used the mixture of two Weibull distributions to fit their experimental observations and gave the probability density function of rupture modulus (σ_{mr}).

$$p(\sigma_{\mathrm{mr}}) = w_1 \frac{m_1}{\bar{\sigma}_1}\left(\frac{\sigma_{\mathrm{mr}}}{\bar{\sigma}_1}\right)^{m_1-1} \exp\left[-\left(\frac{\sigma_{\mathrm{mr}}}{\bar{\sigma}_1}\right)^{m_1}\right] + w_2 \frac{m_2}{\bar{\sigma}_2}\left(\frac{\sigma_{\mathrm{mr}}}{\bar{\sigma}_2}\right)^{m_2-1} \exp\left[-\left(\frac{\sigma_{\mathrm{mr}}}{\bar{\sigma}_2}\right)^{m_2}\right],$$

with $w_1 = 0.55$, $m_1 = 7.7, \overline{\sigma}_1 = 94.8\text{(MPa)}$, $w_2 = 0.45, m_2 = 4.9$, and $\overline{\sigma}_2 = 52.8\text{(MPa)}$, which is plotted by a black curve in Fig. 7.11. In the following, the EM algorithm is conducted with a mixture of two Gaussian distributions to fit the distribution histogram obtained from the experiment. The initial values of the parameters $\left\{w_i^{(0)}, \mu_i^{(0)}, \sigma_i^{2,(0)}\right\}(i = 1, \ 2)$ for the EM calculation are

$0.1N(10,\ 1000)+0.9N(100, 2000)$. With the same convergence criterion used in Example 7.5, the EM algorithm gives the converged values of parameters, $0.44N(49.46,\ 102.38)+0.56N(92.05,\ 179.28)$ after 87 iterations, as shown in Fig. 7.11. Clearly, both the two-peak Weibull distribution and the two-peak Gaussian distribution can fit the 55 bending strengths well. More accurate and reliable experimental data are required to distinguish the Weibull distribution from the Gaussian distribution.

Table 7.9 The rupture moduli of 55 experimental poled PZT-841 samples under 10 kV·cm^{-1} dc electric field

J	1	2	3	4	5	6	7	8
x_j	31.25	37.5	43.75	50	56.25	62.5	68.75	75
I_j	2	3	5	6	4	4	2	4
j	9	10	11	12	13	14	15	
x_j	81.25	87.5	93.75	100	106.25	112.5	118.75	
I_j	4	4	7	5	2	2	1	

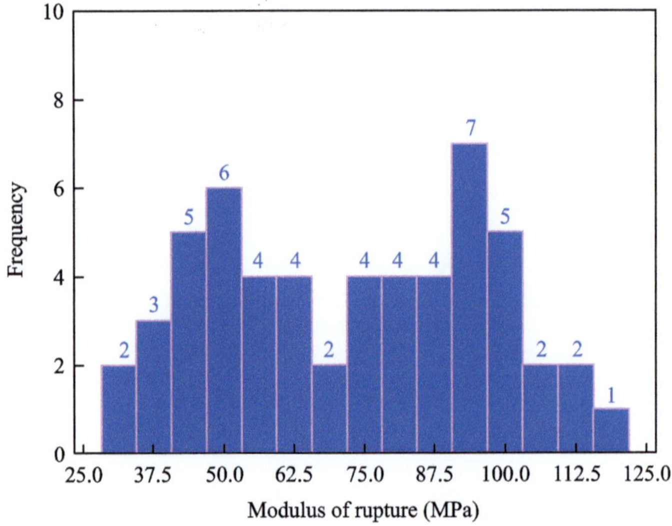

Fig. 7.10 The histogram of experimentally measured rupture moduli (σ_{mr}) of 55 samples

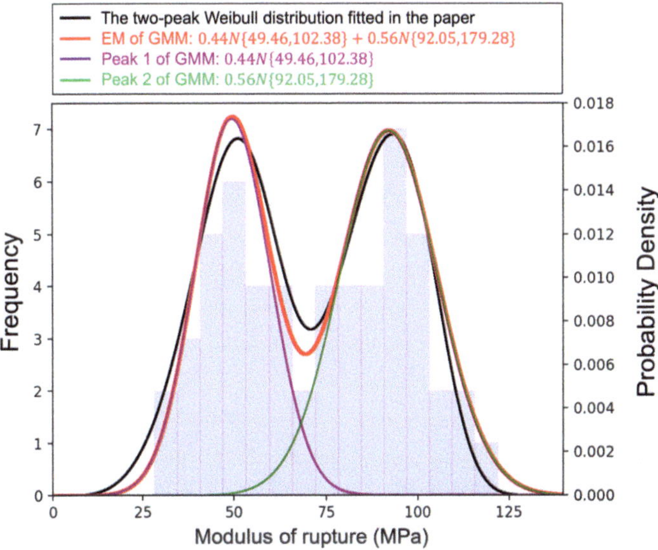

Fig. 7.11 The two-peak Weibull distribution (Fu & Zhang, 1998) and the two-peak Gaussian distribution

---

Example 7.7 The black curve in Fig. 7.12 shows the mixture of two Gaussian distributions, $0.4N(4,\ 1^2) + 0.6N(8,\ 1.5^2)$, which is built up with 3000 x_j, viz., $N = 3000$, by following the mixed distribution in a region of (1, 12). Sampling is conducted with a step size of $\Delta x = 0.2$ ranging from x_j to $x_j + \Delta x$. Since two peaks are visible, the EM fitting is carried out based on a two-Gaussian distribution mixture, and the initial input data for the EM calculation are estimated as $0.8N(2.5,\ 0.8^2)+\ 0.2N(5.5,\ 1.3^2)$. The convergence is reached if the relative error of $\Delta_i = \dfrac{|\theta_{i,t} - \theta_{i,t-1}|}{\theta_{i,t-1}}$ is less than 10^{-3} in the iteration step t, where $\theta_{i,t}$ represents the latent variables of $w_{i,t}$, $\mu_{i,t}$ and $\sigma_{i,t}^2$ $(i = 1, 2)$. The relative error in iteration is preset and called the convergence criterion. The fitting result is shown by the red curve in Fig. 7.12. The fitting quality is evaluated by the weighted profile factor $R_{\text{wp}} = 100 \cdot \sqrt{\dfrac{\sum\limits_{j=1}^{n} \hat{w}_j(\hat{y}_j - y_j)^2}{\sum\limits_{j=1}^{n} \hat{w}_j y_j^2}}$ and profile factor

$R_p = \frac{\sum_j |\hat{y}_j - y_j|}{\sum_j y_j}$, where $\hat{y}_j$ and y_j are the fitting and real probabilities at $x = x_j$, respectively, and $\hat{w}_j = \frac{1}{y_j}$ is the weight of the j-th probability. The EM curve is $0.402N(3.965,\ 1.015^2) + 0.598N(7.967,\ 1.549^2)$ with the weighted profile factor $R_{\text{wp}} = 2.77\%$ and profile factor $R_{\text{p}} = 2.137\%$.

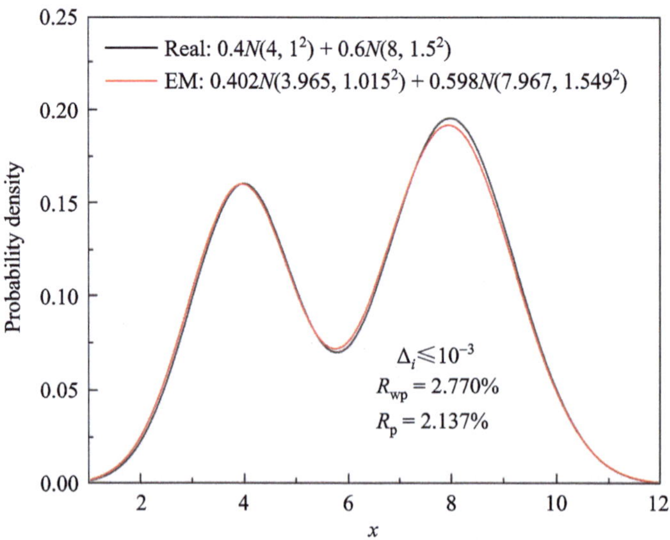

Fig. 7.12 The mixture of two Gaussian distributions, where the black and red curves represent the real and fitted distribution patterns, respectively

Example 7.7 indicates that if individual peaks are obviously separated in a mixture of Gaussian distributions, the values of the latent variables can be roughly estimated as the initial inputs to start the EM fitting. The fitting quality depends on the preset relative error in the iteration, because it sets a criterion for the convergence. It will be a challenge to select the number of Gaussian distributions and the initial values of latent variables, when individual peaks are invisible in a mixture of Gaussian distributions. In this circumstance, purely data fitting with a low convergence criterion might not obtain a unique solution.

Example 7.8 The black curve in Fig. 7.13 shows a mixture of two Gaussian distributions, $0.5N(4,\ 1^2) + 0.5N(6,\ 1.5^2)$, which is built up with 3000 x_j, viz., $N = 3000$, by following the mixed distribution in the region of (1,10). Sampling is conducted with a step size of $\Delta x = 0.2$, ranging from x_j to $x_j + \Delta x$. The two peaks are very close to each other, and it is difficult to guess the peak positions without domain knowledge. More seriously, it is possible to have multiple solutions from purely curve fitting without domain knowledge if the convergence criterion is not high enough. Figure 7.13a, b illustrate two fitting results with the relative error of 10^{-3}, where the initially guessed values are $0.6N(4.4,\ 0.8^2) + 0.4N(5.5,\ 1.3^2)$ and $0.4N(4.4,\ 0.8^2){+}0.6N(5.5,\ 1.3^2)$, respectively, and the fitting results are $0.46N(3.91,\ 0.95^2){+}0.54N(5.88, 1.63^2)$ and $0.37N(3.79,\ 0.86^2){+}0.63N(5.67,\ 1.81^2)$, respectively. The $R_{\text{wp}}(R_{\text{p}})$ values are 2.224% (1.828%) and 2.898% (2.207%) for fitting (a) and fitting (b), respectively.

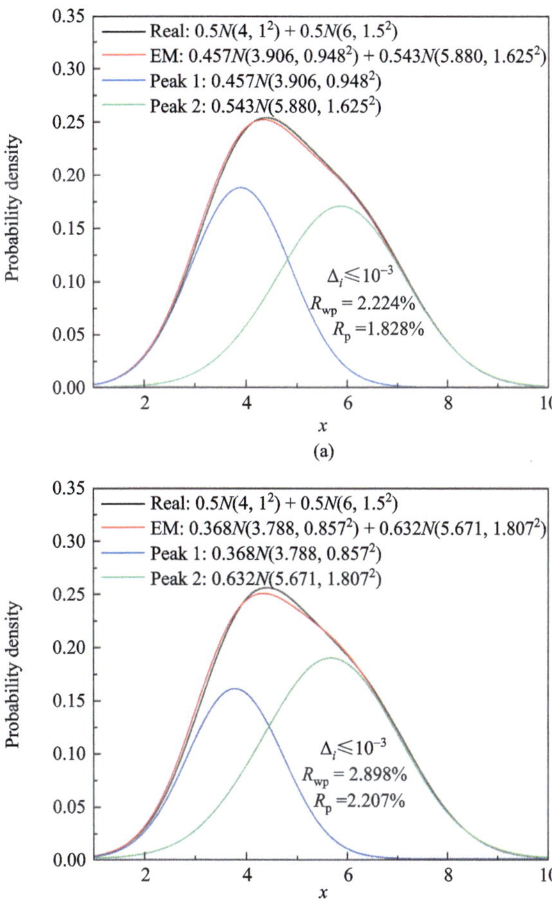

Fig. 7.13 Two fitting results from two sets of initial values. *Note* The black and red curves are the real and fitted curves, respectively, and blue and green curves are for the first and second Gaussian distributions. Obviously, peak 1 is higher than peak 2 in **(a)** and the opposite appears in **(b)**

Table 7.10 The EM fitting results with different convergence criteria for two different initial value sets

Δ_i	Initial values of $0.4N(4.4,\ 0.8^2) + 0.6N(5.8,\ 1.3^2)$				Initial values of $0.3N(4.4,\ 0.8^2) + 0.7N(5.5,\ 1.3^2)$			
	R_{wp} (%)	R_p (%)	Iter #	Results	R_{wp} (%)	R_p (%)	Iter #	Results
10^{-2}	2.754	1.794	8	$0.457N(3.953,\ 1.008^2) + 0.543N(5.839,\ 1.727^2)$	3.345	1.989	11	$0.346N(3.844,\ 0.906^2) + 0.654N(5.578,\ 1.970^2)$
10^{-3}	2.224	1.828	24	$0.457N(3.906,\ 0.948^2) + 0.543N(5.880,\ 1.625^2)$	2.898	2.207	53	$0.368N(3.788,\ 0.857^2) + 0.632N(5.671,\ 1.807^2)$
10^{-4}	2.169	1.817	315	$0.480N(3.934,\ 0.966^2) + 0.520N(5.940,\ 1.567^2)$	2.169	1.817	714	$0.480N(3.935,\ 0.966^2) + 0.520N(5.940,\ 1.567^2)$
10^{-5}	2.133	1.787	1415	$0.496N(3.957,\ 0.981^2) + 0.504N(5.980,\ 1.532^2)$	2.133	1.787	1518	$0.496N(3.957,\ 0.981^2) + 0.504N(5.980,\ 1.532^2)$
10^{-6}	2.131	1.784	2225	$0.497N(3.960,\ 0.983^2) + 0.503N(5.984,\ 1.529^2)$	2.131	1.784	2382	$0.497N(3.960,\ 0.983^2) + 0.503N(5.984,\ 1.529^2)$

When the convergence criterion is high enough, the EM results will be independent of the initial input values of latent variables. Table 7.10 shows the EM results with various values of convergence criterion for the two sets of starting values, respectively, where "Iter #" is the number of iterations. The results indicate that when the convergence criterion $\Delta_i \leqslant 10^{-4}$, the EM result is independent of initial values.

It is obvious that the number of distributions plays an extremely important role in the EM fitting. In the above example, two distributions are selected. If more than two distributions are selected, the curve profile can also be perfectly fit. Therefore, domain knowledge is considerably significant in the curve fitting. In case that no domain knowledge is available, the number of distributions shall increase one by one from two, and, with acceptable fitting accuracy, the fitting result with the smallest number of distributions is the right solution.

A multiple variable Gaussian distribution with variables $\boldsymbol{X} = (x_1, x_2, \cdots, x_m)$ is noted by $N(\boldsymbol{\mu}, \boldsymbol{\Sigma})$ and

$$p(\boldsymbol{X}) = \frac{1}{(2\pi)^{m/2}\sqrt{|\boldsymbol{\Sigma}|}}\exp\left(-\frac{1}{2}(\boldsymbol{X} - \boldsymbol{\mu})^{\mathrm{T}}\boldsymbol{\Sigma}^{-1}(\boldsymbol{X} - \boldsymbol{\mu})\right),$$

where $\boldsymbol{\mu}$ is the expectation vector, $\boldsymbol{\Sigma}$ is the covariance matrix, and $|\cdot|$ denotes the matrix determinant. Clearly, if only one variable is involved, the variable vector and the expectation vector are both reduced to scalars, and the covariance matrix is reduced to variance. The EM algorithm described above can be directly applied to the mixture of multiple variable Gaussian distributions with the replacement of scalars by vectors and the replacement of variance by covariance.

7.5.2 The Mixture of Lorentz and Gaussian Distributions

The normalized Lorentz distribution is given by

$$f(x) = \frac{1}{\pi}\left[\frac{\gamma}{(x - x_0)^2 + \gamma^2}\right], \tag{7.33}$$

where x_0 denotes the peak position, and γ is the half width at the strength of half peak. A single measured XRD peak profile is a mixture of the normalized Lorentz distribution and the normalized Gaussian distribution of

$$F(x) = \frac{A}{\pi}\left[\frac{\gamma}{(x - x_0)^2 + \gamma^2}\right] + \frac{1 - A}{\sqrt{2\pi}\sigma}\exp\left(-\frac{(x - \mu)^2}{2\sigma^2}\right), \tag{7.34}$$

where $A \in (0, 1)$ and 1–A are the weights of Lorentz and Gaussian distributions, respectively; here the integrated intensity of the single peak is also normalized to be one. Since there is only one peak, the mean values of Lorentz and Gaussian distributions are the same, that is, $x_0 = \mu$. In this case, the parameters to be determined are $\{A, \mu, \gamma, \sigma^2\}$. In the XRD pattern, XRD intensity I_j at x_j takes the place of frequency I_j in GMM, and with XRD intensity I_j, each x_j will be calculated only once, viz., n will be used, rather than N, where $N = \sum_{j=1}^{n} I_j$. The E-step calculates

$$\alpha_j^L = \frac{\frac{A}{\pi}\left[\frac{\gamma}{(x_j-\mu)^2+\gamma^2}\right]}{\frac{A}{\pi}\left[\frac{\gamma}{(x_j-\mu)^2+\gamma^2}\right]+\frac{1-A}{\sqrt{2\pi}\sigma}\exp\left(-\frac{(x_j-\mu)^2}{2\sigma^2}\right)} \text{ for } j = 1,\ 2,\cdots,\ n, \tag{7.35a}$$

$$\alpha_j^G = \frac{\frac{1-A}{\sqrt{2\pi}\sigma}\exp\left(-\frac{(x_j-\mu)^2}{2\sigma^2}\right)}{\frac{A}{\pi}\left[\frac{\gamma}{(x_j-\mu)^2+\gamma^2}\right]+\frac{1-A}{\sqrt{2\pi}\sigma}\exp\left(-\frac{(x_j-\mu)^2}{2\sigma^2}\right)} \text{ for } j = 1,\ 2,\cdots,\ n. \tag{7.35b}$$

The XRD intensity at every point j is attributed to the joint contributions from both the Lorentz distribution and the Gaussian distribution. Since α_j^L represents the contribution from the Lorentz distribution and α_j^G represents from the Gaussian distribution,

$$\alpha_j^L + \alpha_j^G = 1 \text{ for } j = 1,\ 2,\cdots,\ n, \tag{7.36}$$

for every point j of the single measured XRD peak profile, as indicated by Eqs. (7.35a) and (7.35b). In the M-step, the log-likelihood is

$$\mathrm{LL}\{A,\mu,\gamma,\sigma^2\} = \sum_{j=1}^{n} I_j \ln\left\{\frac{A}{\pi}\left[\frac{\gamma}{(x_j-\mu)^2+\gamma^2}\right]+\frac{1-A}{\sqrt{2\pi}\sigma}\exp\left(-\frac{(x_j-\mu)^2}{2\sigma^2}\right)\right\}. \tag{7.37a}$$

The maximization of log-likelihood requires

$$\begin{aligned}\frac{\partial \mathrm{LL}}{\partial \mu} &= \sum_{j=1}^{n} I_j \frac{\frac{A}{\pi}\left[\frac{2\gamma}{\left[(x_j-\mu)^2+\gamma^2\right]^2}\right]+\frac{1-A}{\sqrt{2\pi}\sigma^3}\exp\left(-\frac{(x_j-\mu)^2}{2\sigma^2}\right)}{\frac{A}{\pi}\left[\frac{\gamma}{(x_j-\mu)^2+\gamma^2}\right]+\frac{1-A}{\sqrt{2\pi}\sigma}\exp\left(-\frac{(x_j-\mu)^2}{2\sigma^2}\right)}(x_j-\mu)\\ &= \sum_{j=1}^{n} I_j\left(\frac{2\alpha_j^L}{(x_j-\mu)^2+\gamma^2}+\frac{\alpha_j^G}{\sigma^2}\right)(x_j-\mu) = 0,\end{aligned} \tag{7.37b}$$

$$\frac{\partial \mathrm{LL}}{\partial \gamma}=\sum_{j=1}^{n} I_j \frac{\dfrac{A}{\pi}\left[\dfrac{(x_j-\mu)^2-\gamma^2}{\left[(x_j-\mu)^2+\gamma^2\right]^2}\right]}{\dfrac{A}{\pi}\left[\dfrac{\gamma}{(x_j-\mu)^2+\gamma^2}\right]+\dfrac{1-A}{\sqrt{2\pi}\sigma}\exp\left(-\dfrac{(x_j-\mu)^2}{2\sigma^2}\right)}$$

$$=\sum_{j=1}^{n} I_j\alpha_j^L \frac{(x_j-\mu)^2-\gamma^2}{\gamma[(x_j-\mu)^2+\gamma^2]}=0, \tag{7.37c}$$

$$\frac{\partial \mathrm{LL}}{\partial \sigma}=\sum_{j=1}^{n} I_j \frac{\dfrac{1-A}{\sqrt{2\pi}}\exp\left(-\dfrac{(x_j-\mu)^2}{2\sigma^2}\right)\left[\dfrac{(x_j-\mu)^2}{\sigma^4}-\dfrac{1}{\sigma^2}\right]}{\dfrac{A}{\pi}\left[\dfrac{\gamma}{(x_j-\mu)^2+\gamma^2}\right]+\dfrac{1-A}{\sqrt{2\pi}\sigma}\exp\left(-\dfrac{(x_j-\mu)^2}{2\sigma^2}\right)}$$

$$=\sum_{j=1}^{n} I_j\alpha_j^G\left(\frac{(x_j-\mu)^2}{\sigma^3}-\frac{1}{\sigma}\right)=0, \tag{7.37d}$$

$$\frac{\partial \mathrm{LL}}{\partial A}=\sum_{j=1}^{n} I_j \frac{\dfrac{1}{\pi}\left[\dfrac{\gamma}{(x_j-\mu)^2+\gamma^2}\right]-\dfrac{1}{\sqrt{2\pi}\sigma}\exp\left(-\dfrac{(x_j-\mu)^2}{2\sigma^2}\right)}{\dfrac{A}{\pi}\left[\dfrac{\gamma}{(x_j-\mu)^2+\gamma^2}\right]+\dfrac{1-A}{\sqrt{2\pi}\sigma}\exp\left(-\dfrac{(x_j-\mu)^2}{2\sigma^2}\right)}$$

$$=\sum_{j=1}^{n} I_j\left(\frac{\alpha_j^L}{A}-\frac{\alpha_j^G}{1-A}\right)=0. \tag{7.37e}$$

Equation (7.37e) yields

$$A=\frac{1}{N}\sum_{j=1}^{n} I_j\alpha_j^L. \tag{7.38}$$

Equations (7.37b–7.37d) cannot be analytically solved. Numerical iteration is carried to estimate the values. Introducing $\beta_j^L=\alpha_j^L\cdot\frac{\gamma}{\pi[(x_j-\mu)^2+\gamma^2]}$, we rewrite Eqs. (7.37b–7.37d) as

$$\sum_{j=1}^{n} I_j\left(\frac{2\pi}{\gamma}\beta_j^L+\frac{\alpha_j^G}{\sigma^2}\right)(x_j-\mu)=0, \tag{7.39a}$$

$$\frac{\pi}{\gamma^2}\sum_{j=1}^{n} I_j\beta_j^L(x_j-\mu)^2-\pi\sum_{j=1}^{n} I_j\beta_j^L=0, \tag{7.39b}$$

$$\sum_{j=1}^{n} I_j\alpha_j^G\left((x_j-\mu)^2-\sigma^2\right)=0. \tag{7.39c}$$

As indicated in the E-step, the starting values of latent parameters $\{A,\mu,\gamma,\sigma^2\}$ should be input to calculate the parameters α_j^L and α_j^G. In the following iteration, the values of parameters in the t-th iteration are used to calculate the values of parameters $\alpha_j^{L,(t+1)}$ and $\alpha_j^{G,(t+1)}$ in the $(t+1)$-th iteration. In the same manner, the value of parameter $\beta_j^{L,(t)}$ in the t-th iteration is known and will be used to calculate the values of parameters in the $(t+1)$-th iteration. In such an iterative manner, the solutions of Eqs. (7.39a–7.39c) give the values in the $(t+1)$-th iteration

$$\mu^{(t+1)} = \frac{\dfrac{2\pi}{\gamma}\sum_{j=1}^{n} I_j\beta_j^{L,(t)}x_j + \dfrac{1}{\sigma^{2,(t)}}\sum_{j=1}^{n} I_j\alpha_j^{G,(t)}x_j}{\dfrac{2\pi}{\gamma}\sum_{j=1}^{n} I_j\beta_j^{L,(t)} + \dfrac{1}{\sigma^{2,(t)}}\sum_{j=1}^{n} I_j\alpha_j^{G,(t)}}, \tag{7.40a}$$

$$\gamma^{(t+1)} = \left[\frac{\sum_{j=1}^{n} I_j\beta_j^{L,(t)}(x_j-\mu^{(t)})^2}{\sum_{j=1}^{n} I_j\beta_j^{L,(t)}}\right]^{\frac{1}{2}}, \tag{7.40b}$$

$$\sigma^{2,(t+1)} = \frac{\sum_{j=1}^{n} I_j\alpha_j^{G,(t)}(x_j-\mu^{(t)})^2}{\sum_{j=1}^{n} I_j\alpha_j^{G,(t)}}. \tag{7.40c}$$

Note that the values of latent parameters in the right side of each equation in Eqs. (7.40a) to (7.40c) are the values in the t-th iteration. If the full width at half maxima (FWHM, Γ) of Lorentz is considered being the same as that of Gaussian distribution, i.e., $2\gamma = 2\sqrt{2\ln 2}\sigma = \Gamma$, then $\beta_j^L = \alpha_j^L \cdot \dfrac{\gamma}{\pi[(x_j-\mu)^2+\gamma^2]} = \alpha_j^L \cdot \dfrac{(\Gamma/2)}{\pi[(x_j-\mu)^2+(\Gamma/2)^2]}$ and Eqs. (7.37c) and (7.37d) are combined into

$$\sum_{j=1}^{n} I_j\alpha_j^L \frac{(x_j-\mu)^2-(\Gamma/2)^2}{(\Gamma/2)[(x_j-\mu)^2+(\Gamma/2)^2]} + \sum_{j=1}^{n} I_j\alpha_j^G\left(\frac{(x_j-\mu)^2}{\left[\Gamma/\left(2\sqrt{2\ln 2}\right)\right]^3} - \frac{1}{\left[\Gamma/\left(2\sqrt{2\ln 2}\right)\right]}\right) = 0, \tag{7.41a}$$

which gives

$$(\Gamma/2)^2 = \frac{\dfrac{\pi}{(\Gamma/2)^2}\sum_{j=1}^{n} I_j\beta_j^L(x_j-\mu)^2 + \dfrac{1}{\left[\Gamma/\left(2\sqrt{2\ln 2}\right)\right]^3}\sum_{j=1}^{n} I_j\alpha_j^G(x_j-\mu)^2}{\dfrac{\pi}{(\Gamma/2)^2}\sum_{j=1}^{n} I_j\beta_j^L + \dfrac{1}{2\ln 2}\cdot\dfrac{1}{\left[\Gamma/\left(2\sqrt{2\ln 2}\right)\right]^3}\sum_{j=1}^{n} I_j\alpha_j^G}. \tag{7.41b}$$

When a measured XRD pattern is a mixture of multiple XRD peaks, where the number of peaks is k, and each peak profile is a mixture of Lorentz distribution and Gaussian distribution, the following EM algorithm will be adopted, with the variables of $\{w_i, A_i, \mu_i, \gamma_i, \sigma_i^2\}$ for $(i = 1, 2, \cdots, k)$. The XRD pattern is a mixture of each peak profile

$$F_{\mathrm{M}}(x) = \sum_{i=1}^{k} w_i \cdot \left(\frac{A_i}{\pi} \left[\frac{\gamma_i}{(x-\mu_i)^2 + \gamma_i^2} \right] + \frac{1-A_i}{\sqrt{2\pi}\sigma_i} \exp\left(-\frac{(x-\mu_i)^2}{2\sigma_i^2} \right) \right). \quad (7.42)$$

$\frac{A_i}{\pi} \left[\frac{\gamma_i}{(x-\mu_i)^2 + \gamma_i^2} \right] + \frac{1-A_i}{\sqrt{2\pi}\sigma_i} \exp\left(-\frac{(x-\mu_i)^2}{2\sigma_i^2} \right)$ represents a normalized standard peak profile, and the area covered by the single peak is represented by the weight w_i. A whole measured XRD pattern has its integrated intensity calculated by $\int_{-\infty}^{+\infty} F_M(x)\mathrm{d}x = I_{\mathrm{area}}$, viz., the area covered by the pattern. Thus, $\sum_{i=1}^{k} w_i = I_{\mathrm{area}}$ and $w_i \in (0, I_{\mathrm{area}})$. In a measured XRD pattern, the weight w_i represents the weight of diffraction intensity of peak i in the entire XRD spectrum, μ_i denotes the peak position of peak i, and γ_i and σ_i are related to its FWHM. It should be noted that if there are many peaks mixed at the j-th angle (x_j), the observed intensity is attributed to all the mixed peaks.

1. E-step

Equation (7.43) holds for the expectation, i.e., the posterior probability of latent variables, and the explicit expectation takes the following form:

$$\alpha_{ji}^{(t+1)} = \frac{w_i^{(t)} \cdot \left(\frac{A_i^{(t)}}{\pi} \left[\frac{\gamma_i^{(t)}}{(x_j-\mu_i^{(t)})^2+\gamma_i^{2,(t)}} \right] + \frac{1-A_i^{(t)}}{\sqrt{2\pi}\sigma_i^{(t)}} \exp\left(-\frac{(x_j-\mu_i^{(t)})^2}{2\sigma_i^{2,(t)}} \right) \right)}{\sum_{i=1}^{k} w_i^{(t)} \cdot \left(\frac{A_i^{(t)}}{\pi} \left[\frac{\gamma_i^{(t)}}{(x_j-\mu_i^{(t)})^2+\gamma_i^{2,(t)}} \right] + \frac{1-A_i^{(t)}}{\sqrt{2\pi}\sigma_i^{(t)}} \exp\left(-\frac{(x_j-\mu_i^{(t)})^2}{2\sigma_i^{2,(t)}} \right) \right)}$$
$$\text{for } (j = 1, 2, \cdots, n; i = 1, 2, \cdots, k). \quad (7.43)$$

Letting $\alpha_{ji}^{L,(t+1)} = \frac{w_i^{(t)} \cdot \frac{A_i^{(t)}}{\pi} \left[\frac{\gamma_i^{(t)}}{(x_j-\mu_i^{(t)})^2+\gamma_i^{2,(t)}} \right]}{\sum_{i=1}^{k} w_i^{(t)} \cdot \left(\frac{A_i^{(t)}}{\pi} \left[\frac{\gamma_i^{(t)}}{(x_j-\mu_i^{(t)})^2+\gamma_i^{2,(t)}} \right] + \frac{1-A_i^{(t)}}{\sqrt{2\pi}\sigma_i^{(t)}} \exp\left(-\frac{(x_j-\mu_i^{(t)})^2}{2\sigma_i^{2,(t)}} \right) \right)}$, and

$$\alpha_{ji}^{G,(t+1)} = \frac{w_i^{(t)} \cdot \frac{1-A_i^{(t)}}{\sqrt{2\pi}\sigma_i^{(t)}} \exp\left(-\frac{(x_j-\mu_i^{(t)})^2}{2\sigma_i^{2,(t)}} \right)}{\sum_{i=1}^{k} w_i^{(t)} \cdot \left(\frac{A_i^{(t)}}{\pi} \left[\frac{\gamma_i^{(t)}}{(x_j-\mu_i^{(t)})^2+\gamma_i^{2,(t)}} \right] + \frac{1-A_i^{(t)}}{\sqrt{2\pi}\sigma_i^{(t)}} \exp\left(-\frac{(x_j-\mu_i^{(t)})^2}{2\sigma_i^{2,(t)}} \right) \right)}, \text{ we}$$

rewrite Eq. (7.43) as

$$\alpha_{ji}^{(t+1)} = \alpha_{ji}^{L,(t+1)} + \alpha_{ji}^{G,(t+1)} \text{ for } (j = 1, 2, \cdots, n; i = 1, 2, \cdots, k). \quad (7.44)$$

Similarly,

$$\beta_{ji}^{L,(t)} = \alpha_{ji}^{L,(t)} \cdot \frac{\gamma_i^{(t)}}{\pi\left[(x_j - \mu_i^{(t)})^2 + \gamma_i^{2,(t)}\right]} \text{ for } (j = 1, 2, \cdots, n; i = 1, 2, \cdots, k) \tag{7.45}$$

is introduced to simplify the expression in the M-step.

2. M-step

The Lagrange function of the log-likelihood is

$$\begin{aligned} \mathrm{LLL}(\lambda, w_i, A_i, \mu_i, \gamma_i, \sigma_i^2) = \sum_{j=1}^{n} I_j \ln\Bigg\{ \sum_{i=1}^{k} w_i \cdot \Bigg(\frac{A_i}{\pi}\left[\frac{\gamma_i}{(x_j-\mu_i)^2+\gamma_i^2}\right] \\ + \frac{1-A_i}{\sqrt{2\pi}\sigma_i} \exp\left(-\frac{(x_j-\mu_i)^2}{2\sigma_i^2}\right)\Bigg)\Bigg\} \\ + \lambda\left(\sum_{i=1}^{k} w_i - I_{\text{area}}\right), \end{aligned} \tag{7.46a}$$

where the maximization of log-likelihood requires

$$\begin{aligned} \frac{\partial \mathrm{LLL}}{\partial \mu_i} &= \sum_{j=1}^{n} I_j \frac{w_i \cdot \left\{ \frac{A_i}{\pi}\left[\frac{2\gamma_i}{\left[(x_j-\mu_i)^2+\gamma_i^2\right]^2}\right] + \frac{1-A_i}{\sqrt{2\pi}\sigma_i^3} \exp\left(-\frac{(x_j-\mu_i)^2}{2\sigma_i^2}\right)\right\}}{\sum_{i=1}^{k} w_i \cdot \left(\frac{A_i}{\pi}\left[\frac{\gamma_i}{(x_j-\mu_i)^2+\gamma_i^2}\right] + \frac{1-A_i}{\sqrt{2\pi}\sigma_i} \exp\left(-\frac{(x_j-\mu_i)^2}{2\sigma_i^2}\right)\right)} (x_j - \mu_i) \\ &= \sum_{j=1}^{n} I_j \left[\frac{2\pi}{\gamma_i}\beta_{ji}^L + \frac{1}{\sigma_i^2}\alpha_{ji}^G\right](x_j - \mu_i) = 0, \end{aligned} \tag{7.46b}$$

$$\begin{aligned} \frac{\partial \mathrm{LLL}}{\partial \gamma_i} &= \sum_{j=1}^{n} I_j \frac{w_i \cdot \frac{A_i}{\pi}\left[\frac{(x_j-\mu_i)^2-\gamma_i^2}{\left[(x_j-\mu_i)^2+\gamma_i^2\right]^2}\right]}{\sum_{i=1}^{k} w_i \cdot \left(\frac{A_i}{\pi}\left[\frac{\gamma_i}{(x_j-\mu_i)^2+\gamma_i^2}\right] + \frac{1-A_i}{\sqrt{2\pi}\sigma_i} \exp\left(-\frac{(x_j-\mu_i)^2}{2\sigma_i^2}\right)\right)} \\ &= \frac{\pi}{\gamma_i^2} \sum_{j=1}^{n} I_j \beta_{ji}^L (x_j - \mu_i)^2 - \pi \sum_{j=1}^{n} I_j \beta_{ji}^L = 0, \end{aligned} \tag{7.46c}$$

$$\begin{aligned} \frac{\partial \mathrm{LLL}}{\partial \sigma_i} &= \sum_{j=1}^{n} I_j \frac{w_i \cdot \frac{1-A_i}{\sqrt{2\pi}\sigma_i} \exp\left(-\frac{(x_j-\mu_i)^2}{2\sigma_i^2}\right)}{\sum_{i=1}^{k} w_i \cdot \left(\frac{A_i}{\pi}\left[\frac{\gamma_i}{(x_j-\mu_i)^2+\gamma_i^2}\right] + \frac{1-A_i}{\sqrt{2\pi}\sigma_i} \exp\left(-\frac{(x_j-\mu_i)^2}{2\sigma_i^2}\right)\right)} \\ &\quad \left[\frac{(x_j - \mu_i)^2}{\sigma_i^3} - \frac{1}{\sigma_i}\right] = \sum_{j=1}^{n} I_j \alpha_{ji}^G \left(\frac{(x_j - \mu_i)^2}{\sigma_i^3} - \frac{1}{\sigma_i}\right) = 0, \end{aligned} \tag{7.46d}$$

$$\frac{\partial \mathrm{LLL}}{\partial A_i} = \sum_{j=1}^{n} I_j \frac{w_i \cdot \left\{ \frac{1}{\pi}\left[\frac{\gamma_i}{(x_j-\mu_i)^2+\gamma_i^2}\right] - \frac{1}{\sqrt{2\pi}\sigma_i}\exp\left(-\frac{(x_j-\mu_i)^2}{2\sigma_i^2}\right)\right\}}{\sum_{i=1}^{k} w_i \cdot \left(\frac{A_i}{\pi}\left[\frac{\gamma_i}{(x_j-\mu_i)^2+\gamma_i^2}\right] + \frac{1-A_i}{\sqrt{2\pi}\sigma_i}\exp\left(-\frac{(x_j-\mu_i)^2}{2\sigma_i^2}\right)\right)}$$

$$= \sum_{j=1}^{n} I_j \left[\frac{1}{A_i}\alpha_{ji}^{L} - \frac{1}{(1-A_i)}\alpha_{ji}^{G}\right] = 0 \tag{7.46e}$$

and

$$\frac{\partial \mathrm{LLL}}{\partial w_i} = \sum_{j=1}^{n} I_j \frac{\frac{A_i}{\pi}\left[\frac{\gamma_i}{(x_j-\mu_i)^2+\gamma_i^2}\right] + \frac{1-A_i}{\sqrt{2\pi}\sigma_i}\exp\left(-\frac{(x_j-\mu_i)^2}{2\sigma_i^2}\right)}{\sum_{i=1}^{k} w_i \cdot \left(\frac{A_i}{\pi}\left[\frac{\gamma_i}{(x_j-\mu_i)^2+\gamma_i^2}\right] + \frac{1-A_i}{\sqrt{2\pi}\sigma_i}\exp\left(-\frac{(x_j-\mu_i)^2}{2\sigma_i^2}\right)\right)} + \lambda = 0. \tag{7.46f}$$

Equation (7.46f) yields

$$\lambda = -\frac{\sum_{j=1}^{n} I_j}{I_{\text{area}}} \text{ and } w_i = \frac{I_{\text{area}}}{\sum_{j=1}^{n} I_j} \sum_{j=1}^{n} I_j \alpha_{ji} \text{ for } (i = 1, 2, \cdots, k). \tag{7.46g}$$

Equations (7.46b) to (7.46e) lead to the similar equations Eqs. (7.47) to (7.50) of numerical iteration, as Eqs. (7.40a) to (7.40c) for the single XRD peak, where the left side is the values in the $(t+1)$-th iteration, while the right side is the values in the t-th iteration.

$$\mu_i^{(t+1)} = \frac{\frac{2\pi}{\gamma_i^{(t)}} \sum_{j=1}^{n} I_j \beta_{ji}^{L,(t)} x_j + \frac{1}{\sigma_i^{2,(t)}} \sum_{j=1}^{n} I_j \alpha_{ji}^{G,(t)} x_j}{\frac{2\pi}{\gamma_i^{(t)}} \sum_{j=1}^{n} I_j \beta_{ji}^{L,(t)} + \frac{1}{\sigma_i^{2,(t)}} \sum_{j=1}^{n} I_j \alpha_{ji}^{G,(t)}}, \tag{7.47}$$

$$\gamma_i^{(t+1)} = \left[\frac{\sum_{j=1}^{n} I_j \beta_{ji}^{L,(t)} (x_j - \mu_i^{(t)})^2}{\sum_{j=1}^{n} I_j \beta_{ji}^{L,(t)}}\right]^{\frac{1}{2}}, \tag{7.48}$$

$$\sigma_i^{2,(t+1)} = \frac{\sum_{j=1}^{n} I_j \alpha_{ji}^{G,(t)} \left(x_j - \mu_i^{(t)}\right)^2}{\sum_{j=1}^{n} I_j \alpha_{ji}^{G,(t)}}, \tag{7.49}$$

$$A_i^{(t+1)} = \frac{\sum_{j=1}^{n} I_j \alpha_{ji}^{L,(t)}}{\sum_{j=1}^{n} I_j\left(\alpha_{ji}^{L,(t)} + \alpha_{ji}^{G,(t)}\right)} = \frac{\sum_{j=1}^{n} I_j \alpha_{ji}^{L,(t)}}{\sum_{j=1}^{n} I_j \alpha_{ji}^{(t)}} \text{ for } (i = 1, 2, \cdots, k). \tag{7.50}$$

Iterating the E-step and the M-step until convergence gives the final values of parameters $\left\{w_i, A_i, \mu_i, \gamma_i, \sigma_i^2\right\}$ for $(i = 1, 2, \cdots, k)$.

Example 7.9 XRD measures the intensity (I) as a function of the rotated angle (2θ). Obviously, 2θ is the variable x, and the intensity (I) is expressed by $I = F_M(x)$ as shown in Eq. (7.42) for the multiple XRD peak profile. The accurate fitting of an XRD pattern profile is to determine precisely the values of the latent variables Z of $\{w_i, A_i, \mu_i, \gamma_i, \sigma_i^2\}$ for $i = 1, 2, \cdots, k$.

Figure 7.14 a shows a segment of the XRD pattern from $2\theta = 81.3°$ to $2\theta = 85.8°$ obtained by the X-ray wavelength Cu $K_{\alpha_1} = 1.54056$ Å and Cu $K_{\alpha_2} = 1.54439$ Å on crystal Tb_2BaCoO_5, whose space group is Immm. Domain knowledge of crystalline structures shall be used to provide the initial input values of peak positions and the peak number k. In the example, they are estimated by software FullProf Suite (https://www.ill.eu/sites/fullprof/) based on crystal Tb_2BaCoO_5 to be lattice constants of $a = 3.7560$ Å, $b = 5.8238$ Å, $c = 11.5518$ Å, and lattice angles of $\alpha = \beta = \gamma = 90°$. According to the initial value of μ_i and the relationship between γ_i, σ_i^2, and FWHM (Γ), the initial values of γ_i and σ_i^2 are estimated from the original pattern. The initial value of A_i is 0.8, and the initial values of w_i are expressed as a function of I_i^{obs}:

$$w_i^0 = \frac{I_i^{\text{obs}}}{\sum_i I_i^{\text{obs}}} \cdot I_{\text{area}} \text{ for } i = 1, 2, \cdots, k, \tag{7.51}$$

where I_i^{obs} and I_{area} are the observed and integrated diffraction intensities, respectively. In the numerical iteration process, the values of γ_i and σ_i^2 should be constrained to larger than 0.0005 to avoid potential failure, if any, in the EM fitting. Of course, the level of constraint might vary from case to case. The convergence criterion is set, in the EM fitting of the XRD spectrum, by the difference between two consequent iteration values of log-likelihood, and in the example, the difference is set to be lower than 0.1.

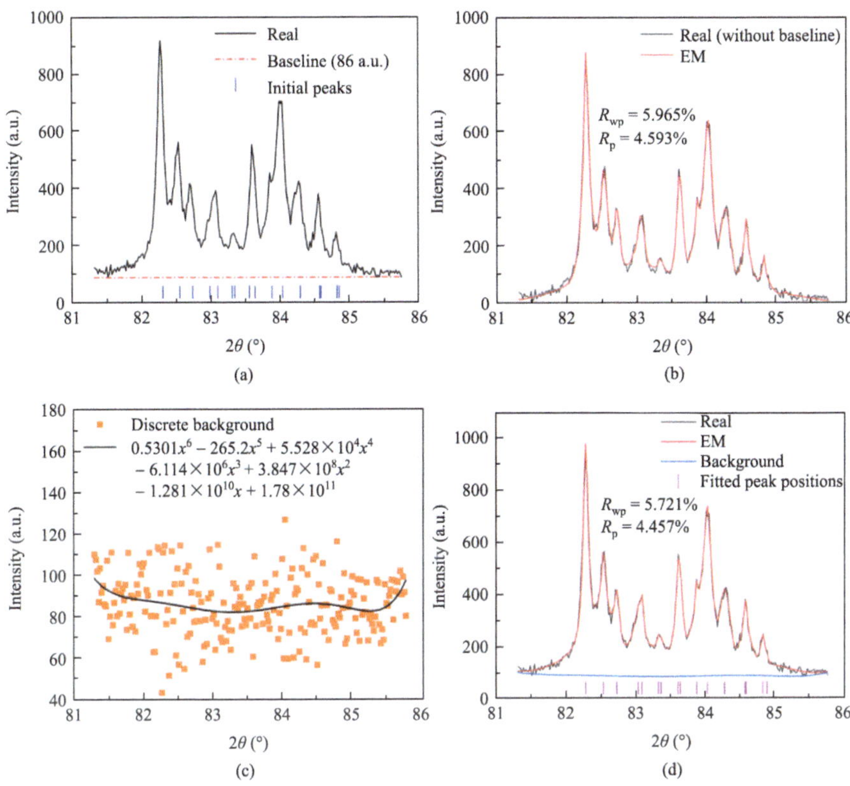

Fig. 7.14 **(a)** The black curve and red dotted line are the real XRD pattern and baseline with a constant value of 86 (a.u.), respectively, and the blue vertical bars are the initial peak positions, which are used as the initial values of μ_i. **(b)** The black and red curves are the real and EM fitted XRD patterns without a baseline, respectively. **(c)** The orange dots are discrete background, and the black curve is the 6-order polynomial fitting of the background. **(d)** The black and red curves are the real and fitted XRD patterns, respectively, the blue curve is the background, and the magenta vertical bars are the fitted peak positions

To eliminate the influence of background on the fitting result, an initial background is estimated and excluded in the following fitting. In this example, a constant intensity of 86 (a.u.) is regarded as a baseline and removed, as shown in Fig. 7.14a, thanks to the high-angle and short segment of the XRD pattern. The iteration of EM fitting can be cycled many times. After each cycle, the difference between the fitting intensity and measured intensity is regarded as discrete background points (noise), and the background is usually assumed to be a continuous function of 2θ and fitted by 6-order polynomial (Rodríguez-Carvajal, 2001). Then, in the next cycle of fitting iteration, the measured intensity minus the background of 6-order polynomial is taken as the XRD pattern to be fitted. The iteration cycle goes on until convergence. In this sample, the iteration cycle is conducted twice to illustrate the fitting procedure. For the first time, the EM fitting is carried out on the original XRD pattern minus a constant baseline, and the constant is the minimum value of intensity ($I_{\min}$), as shown in Fig. 7.14a. Figure 7.14b shows the first EM-fitted pattern with fit accuracy of the weighted profile factor $R_{wp} = 100 \cdot \sqrt{\frac{\sum_{j=1}^{n} \hat{w}_j[(\hat{I}_j + I_{\min}) - I_j]^2}{\sum_{j=1}^{n} \hat{w}_j I_j^2}} = 5.97\%$ and the profile factor $R_p = \frac{\sum_j \left|(\hat{I}_j + I_{\min}) - I_j\right|}{\sum_j I_j} = 4.59\%$, where $\hat{I}_j$ are the fitting values and I_j are the observed values of intensity from the original XRD pattern. The values of $\left[(\hat{I}_j + I_{\min}) - I_j\right]$ are the background, appearing as individual points, as shown in Fig. 7.14c. The discrete background points are then fitted by 6-order polynomial to be

$$\begin{aligned} I_{\text{bac}} &= 0.5301x^6 - 265.2x^5 + 5.528 \times 10^4 x^4 - 6.144 \times 10^6 x^3 + 3.847 \times 10^8 x^2 \\ &\quad - 1.281 \times 10^{10} x + 1.78 \times 10^{11}. \end{aligned} \tag{7.52}$$

The polynomial background is a smooth function covering the entire XRD spectrum. After that, the polynomial background is removed from the original XRD pattern, and the XRD pattern without background is fitted again with the EM algorithm, which yields the final EM results. Table 7.11 lists the fit values of

Table 7.11 The fit values of $(w_i, A_i, \mu_i, \gamma_i, \sigma_i^2)$ for 16 XRD peaks for $K_{\alpha 1}(\lambda_1 = 1.540560 \text{ Å})$ and $K_{\alpha 2}(\lambda_2 = 1.544330 \text{Å})$ X-ray wavelengths

	HKL	w_i	A_i	μ_i	γ_i	σ_i^2
$K_{\alpha 1}$	1 4 5	166.595	0.658	82.280	0.0439	0.1485
$K_{\alpha 2}$	1 4 5	57.386	0.293	82.529	0.0275	0.1548
$K_{\alpha 1}$	2 3 5	18.342	0.158	82.719	0.0224	0.0091
$K_{\alpha 2}$	2 3 5	21.297	0.107	83.028	0.0237	0.0017
$K_{\alpha 1}$	0 4 6	72.224	0.109	83.080	0.0224	0.0043
$K_{\alpha 1}$	0 5 1	48.244	0.068	83.320	0.0224	0.2682
$K_{\alpha 2}$	0 4 6	41.284	0.057	83.359	0.0224	0.3467
$K_{\alpha 2}$	0 5 1	27.266	0.252	83.600	0.0224	0.0025
$K_{\alpha 1}$	0 0 10	15.184	0.221	83.630	0.0271	0.0019
$K_{\alpha 2}$	0 0 10	44.485	0.175	83.861	0.0224	0.0079
$K_{\alpha 1}$	2 4 0	87.474	0.087	84.020	0.0252	0.0045
$K_{\alpha 2}$	2 4 0	55.361	0.024	84.263	0.0319	0.0058
$K_{\alpha 1}$	2 0 8	24.368	0.250	84.557	0.0224	0.0051
$K_{\alpha 1}$	3 2 1	14.624	0.186	84.575	0.0224	0.0170
$K_{\alpha 2}$	2 0 8	9.182	0.309	84.815	0.0239	0.0013
$K_{\alpha 2}$	3 2 1	38.979	0.038	84.879	0.0224	0.1481

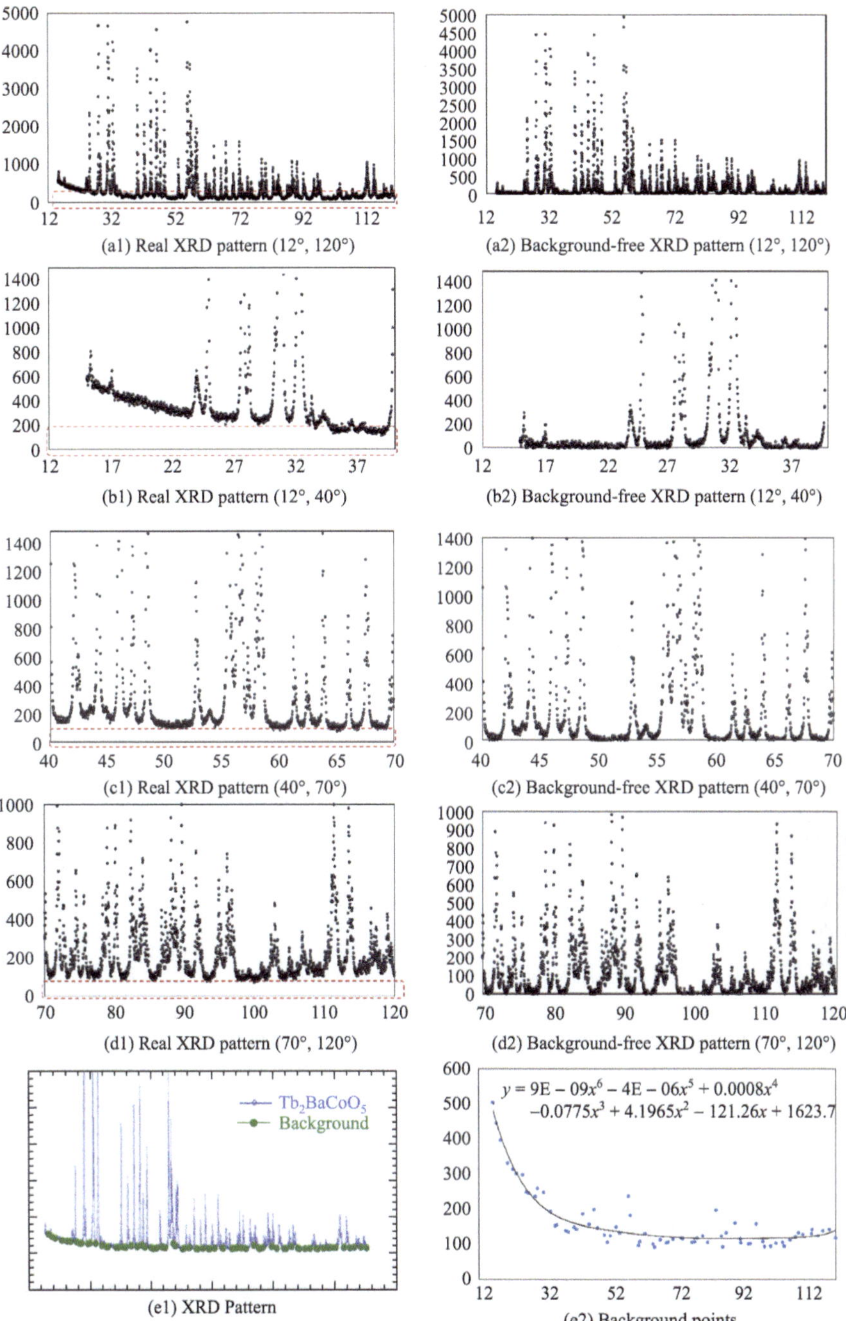

Fig. 7.15 **(a1)** The original whole powder XRD pattern. **(a2)** The background-free whole powder XRD pattern. **(b1)**, **(c1)**, **(d1)** Segments of the original XRD pattern within the 2θ ranges of (12°, 40°), (40°, 70°) and (70°, 120°). **(b2)**, **(c2)**, **(d2)** The corresponding background-free patterns. **(e1)** A set of base points are chosen by FullProf, highlighted by green points. **(e2)** The blue dots are discrete background points, and the blue curve is the smooth background fitted by the 6-order polynomial

$(w_i, A_i, \mu_i, \gamma_i, \sigma_i^2)$ for 16 XRD peaks, and Fig. 7.14d shows the original and EM XRD patterns. Replacing the constant $I_{\min}$ by the polynomial background, we have the weighted profile factor $R_{\text{wp}} = 5.72\%$ and profile factor $R_{\text{p}} = 4.46\%$ for the final EM pattern.

One-time background elimination is applied to alleviating the computational burden on the whole powder XRD pattern fitting. Figure 7.15a1 shows a whole profile of XRD pattern from $2\theta = 12°$ to $2\theta = 120°$ of crystal Tb_2BaCoO_5. To improve the signal/noise ratio, a polynomial fitting method is applied via the FullProf Suite software to eliminating the background, yielding the background-free pattern, as shown in Fig. 7.15a2. To clearly illustrate the background, the original whole pattern is divided into three segments, viz., $2\theta = (12°,\ 40°)$, $(40°,\ 70°)$ and $(70°,\ 120°)$, which are shown in Fig. 7.15b1, c1, d1, respectively, and the corresponding background-free patterns are shown in Figs. 7.15b2, c2, d2. Comparing the background-free patterns with the original one indicates that the background is almost eliminated. Among them, a set of base points are selected by FullProf Suite automatically, as shown by the green points in Fig. 7.15e1. These points are finally fitted by the 6-order polynomial

$$\begin{aligned} I_{\text{bac}} &= 9 \times 10^{-9}x^6 - 4 \times 10^{-6}x^5 + 0.0008x^4 - 0.0775x^3 \\ &\quad + 4.1965x^2 - 121.26x + 1623.7. \end{aligned} \tag{7.53}$$

The 6-order polynomial background is shown in Fig. 7.15e2 and removed from the original XRD pattern $(I_j - I_{\text{bac}})$.

7.6 Gaussian Process (GP) Regression

GPs for machine learning (Rasmussen & Williams, 2006; Williams & Rasmussen, 1996) are developed based on the Bayesian theorem and the Gaussian probability distribution. Unlike in other deterministic ML regressions, GP regression utilizes Gaussian probability distribution to regress data, and express regression results in terms of mean and covariance of the maximal posterior distribution. As described in Bayesian linear regression, the prior normal distribution $N(0,\ \boldsymbol{\Sigma}_p)$ is adopted in the coefficient space, where $\boldsymbol{\Sigma}_p$ is the covariance matrix of coefficient sets. The prior distribution suggests infinite linear functions, in which coefficient sets follow the normal distribution $N(0,\ \boldsymbol{\Sigma}_p)$, although each linear function has a unique set of coefficients. Input data provide the conditional likelihood, and the posterior distribution without normalization can be calculated by ignoring the total likelihood. After that, the maximal posterior distribution selects the best sets of coefficients from the infinite linear functions provided by the prior. The best sets of coefficients also follow a normal distribution, and the mean of the selected sets yields the linear prediction as "close" as possible to the data points. Like Bayesian linear regression, GPs are typically probabilistic ML methods based on Bayesian formalism and Gaussian probability distribution. Unlike Bayesian linear regression, GPs are developed in the sample function space, rather than in the coefficient space.

In the deterministic ML regression, only a function $f(\boldsymbol{x})$ is used to represent the observed response $y(\boldsymbol{x})$

$$y(\boldsymbol{x}) = f(\boldsymbol{x}) + \varepsilon, \tag{7.54a}$$

where $\boldsymbol{x} = \{x_j\}(j = 1,\ 2,\ \cdots,\ m)$ denotes the m features, and ε denotes the difference between $f(\boldsymbol{x})$ and $y(\boldsymbol{x})$, i.e., the error of predictions. It is usually accepted that the errors follow a normal distribution with zero mean and variance σ_n^2

$$p(\varepsilon) \sim N\left(0, \sigma_n^2\right). \tag{7.54b}$$

In GP regression, however, there are infinite sample functions $f(\boldsymbol{x})$, which follow a probability distribution $p(f(\boldsymbol{x}))$ at each given value of feature vector $\boldsymbol{x} = \{x_j\}(j = 1, 2, \cdots, m)$. The function mean $\overline{f}(\boldsymbol{x})$ and function covariance $K(\boldsymbol{x}, \boldsymbol{x}')$ of infinite sample functions $f(\boldsymbol{x})$ are defined, respectively, by

$$\overline{f}(\boldsymbol{x}) = E[f(\boldsymbol{x})] = \int f(\boldsymbol{x})p(f(\boldsymbol{x}))\mathrm{d}(f(\boldsymbol{x})), \tag{7.55a}$$

$$K(\boldsymbol{x}, \boldsymbol{x}') = E\left\{\left[f(\boldsymbol{x}) - \overline{f}(\boldsymbol{x})\right]\left[f(\boldsymbol{x}') - \overline{f}(\boldsymbol{x}')\right]\right\}, \tag{7.55b}$$

where $\boldsymbol{x}$ and $\boldsymbol{x}'$ represent two feature vectors. Covariance function $K(\boldsymbol{x}, \boldsymbol{x}')$ is also called kernel function $K(\boldsymbol{x}, \boldsymbol{x}')$. In the sample function space, rather than in the coefficient space, prior distribution is given by a normal distribution with zero mean and a preset kernel function, i.e.,

$$p(f(\boldsymbol{x})) \sim N(0, K(\boldsymbol{x}, \boldsymbol{x}')). \tag{7.55c}$$

The prior distribution indicates that without seeing any inputs, there are infinite potential functions $f(\boldsymbol{x})$ following such normal distribution $N(0, K(\boldsymbol{x}, \boldsymbol{x}'))$, although each function of $f(\boldsymbol{x})$ has a unique value at a given $\boldsymbol{x}$, and without seeing any inputs, $\boldsymbol{x}$ and $\boldsymbol{x}'$ in the prior just represent two arbitrary feature vectors. A prior distribution provides constraints on the following selection of $f(\boldsymbol{x})$, and the preset kernel function has an explicit mathematic form with involved parameters that are optimized during training. Obviously, the prior defined in the GP function space is consistent with that defined in the weight (coefficient) space in GP Bayesian linear regression.

It can be regarded that the function $f(\boldsymbol{x})$ is built up in a high-dimensional space, which is mapped from the original feature space. In the high-dimensional space, there are M features and each datum $X_i = (x_{i1}, x_{i2}, \cdots, x_{im})(i = 1, 2, \cdots, n)$ in the original m feature space is expressed by $D_i = (d_{i1}, d_{i2}, \cdots, d_{iM})(i = 1, 2, \cdots, n)$. with the new feature vector of $\boldsymbol{d} = (d_1, d_2, \cdots, d_M)$, and $f(\boldsymbol{x})$ is expressed as a linear function of $\boldsymbol{d}$, viz., $f(\boldsymbol{x}) = \boldsymbol{dW}$ with $\boldsymbol{W}$ being a $(M + 1) \times 1$ coefficient vector, which obeys the normal distribution $N(0, \boldsymbol{\Sigma}_p)$ in the prior, the same as that in GP linear regression. The input feature data $\boldsymbol{X}$ in the original feature space are mapped into input feature data $\boldsymbol{D}$ in the high-dimensional space, and the function $f(\boldsymbol{x})$ is linearly expressed by

$$f(\boldsymbol{x}) = \boldsymbol{D}\boldsymbol{W}. \tag{7.56a}$$

The input feature data $\boldsymbol{D}$ is called the design matrix, because it depends on the mapping function. The mapping trick has been briefly introduced in Chap. 4. Since $\boldsymbol{W}$ follows the normal distribution $N(0, \boldsymbol{\Sigma}_p)$, Eq. (7.56a) indicates that for each input X_i, which is mapped to D_i, we have

$$f(X_i) = D_i\boldsymbol{W}. \tag{7.56b}$$

If the error of predictions is not considered, we shall have $y(\boldsymbol{x}) = f(\boldsymbol{x})$. In this circumstance, the mean of the probabilistic prediction of input X_i is

$$\overline{f}(X_i) = 0. \tag{7.57a}$$

Furthermore, the covariance of the probabilistic prediction of input X_i and X_j is calculated with the definition of Eq. (7.55b) to be

$$K(X_i, X_j) = \boldsymbol{D}_i\boldsymbol{W}(\boldsymbol{D}_j\boldsymbol{W})^{\mathrm{T}} = \boldsymbol{D}_i\boldsymbol{\Sigma}_p\boldsymbol{D}_j^{\mathrm{T}} \;\; (i, j = 1, 2, \cdots, n). \tag{7.57b}$$

In the vector and matrix form, the mean and covariance of the probabilistic prediction of inputs $\boldsymbol{X}$ are, respectively

$$\overline{f}(\boldsymbol{x}) = \boldsymbol{0}, \tag{7.57c}$$

$$\boldsymbol{K} = \boldsymbol{D}\boldsymbol{\Sigma}_p\boldsymbol{D}^{\mathrm{T}}. \tag{7.57d}$$

The use of design matrix $\boldsymbol{D}$ aims at the heuristic description of GP regression. Most widely-used kernel functions are in spaces of infinite dimensions. For example, among the kernel functions listed in Table 4.1 in Chap. 4, Gaussian radial basis function (RBF) kernel, Laplace kernel, and Sigmoid kernel all have infinite and continuous dimensions, like Laplace space in Laplace transformation and Fourier space in Fourier transformation. In addition to the kernel functions listed in Table 4. 1, Table 7.12 lists another three kernel functions used in GPs. A preset kernel function plays the essential and fundamental role in GP regression, just as mapping functions. Particularly, parameters involved in a kernel function are called hyperparameters in SVM, which are optimized by cross-validation as described in Chap. 4, while the involved parameters are optimized during training in GP regression.

Table 7.12 Commonly used kernel functions in GPs

Names	Functions	Parameters
Rational quadratic kernel	$K(X_i, X_j) = \left(1 + \frac{\\|X_i - X_j\\|^2}{2\alpha l^2}\right)^{-\alpha}$	$\alpha > 0,\ l > 0$
ExpSine squared kernel	$K(X_i, X_j) = \exp\left(-\frac{2\sin\left(\frac{\pi\\|X_i - X_j\\|^2}{p}\right)}{l^2}\right)$	$p > 0,\ l > 0$
Dop product kernel	$K(X_i, X_j) = \sigma_0^2 + X_i \cdot X_j$	$\sigma_0^2 \geqslant 0$

Note The $\|\cdot\|$ represents the Euclidean distance of argument vector

With training data, GP regression takes $y(\boldsymbol{x}) = f(\boldsymbol{x})$ so that $y(\boldsymbol{x})$ follows the normal distribution $p(y(\boldsymbol{x})) \sim N(0, \boldsymbol{K})$, and employs feature inputs $\boldsymbol{X}$ to calculate the function covariance $\boldsymbol{K}(\boldsymbol{X}|\boldsymbol{\theta})$ with the preset mathematic form and involved parameter $\boldsymbol{\theta}$. The likelihood of response distribution probability is given by

$$p(\boldsymbol{y}|\boldsymbol{K}(\boldsymbol{X}|\boldsymbol{\theta})) = \frac{1}{(2\pi)^{n/2}|\boldsymbol{K}|^{1/2}} \exp\left(-\frac{1}{2}\boldsymbol{y}^{\mathrm{T}}\boldsymbol{K}^{-1}\boldsymbol{y}\right). \tag{7.58a}$$

Maximizing the likelihood or, equivalently, maximizing its natural logarithm determines the values of involved parameters, viz.,

$$\hat{\boldsymbol{\theta}} = \underset{\boldsymbol{\theta}}{\operatorname{argmax}}\left[-\frac{1}{2}(n\ln(2\pi) + \ln|\boldsymbol{K}| + \boldsymbol{y}^{\mathrm{T}}\boldsymbol{K}^{-1}\boldsymbol{y})\right]. \tag{7.58b}$$

After training, the function covariance $\boldsymbol{K}$ is optimal with optimal values of $\hat{\boldsymbol{\theta}}$. In prediction, a GP model with the trained covariance $\boldsymbol{K}$ predicts values of $\boldsymbol{y}_*(\boldsymbol{X}_*)$ when new feature input data $\boldsymbol{X}_*$ are available. The responses $\boldsymbol{y}$ used in training and responses $\boldsymbol{y}_*$ to be predicted form a joint normal distribution of $\begin{pmatrix}\boldsymbol{y}\\ \boldsymbol{y}_*\end{pmatrix}$ with zero mean and joint covariance:

$$p\begin{pmatrix}\boldsymbol{y}\\ \boldsymbol{y}_*\end{pmatrix} \sim N\left(\begin{pmatrix}0\\ 0\end{pmatrix}, \begin{bmatrix}\boldsymbol{K} & \boldsymbol{K}_*^{\mathrm{T}}\\ \boldsymbol{K}_* & \boldsymbol{K}_{**}\end{bmatrix}\right), \tag{7.59a}$$

where $\boldsymbol{K}$ and $\boldsymbol{K}_{**}$ are the covariance matrixes for data $\boldsymbol{X}$ and $\boldsymbol{X}_*$, respectively, and $\boldsymbol{K}_*$ is the covariance correlated between data $\boldsymbol{X}$ and $\boldsymbol{X}_*$. The joint distribution is expressed by

$$p\begin{pmatrix}\boldsymbol{y}\\ \boldsymbol{y}_*\end{pmatrix} = p(\boldsymbol{y}_*|\boldsymbol{y})p(\boldsymbol{y}). \tag{7.59b}$$

Clearly, the predication $p(\boldsymbol{y}_*|\boldsymbol{y})$ is also a conditional normal distribution

$$p(\boldsymbol{y}_*|\boldsymbol{y}) = \frac{p\begin{pmatrix} \boldsymbol{y} \\ \boldsymbol{y}_* \end{pmatrix}}{p(\boldsymbol{y})}. \tag{7.59c}$$

Calculating the exponential arguments gives

$$-\frac{1}{2}\begin{pmatrix} \boldsymbol{y} \\ \boldsymbol{y}_* \end{pmatrix}^{\mathrm{T}} \begin{bmatrix} \boldsymbol{K} & \boldsymbol{K}_*^{\mathrm{T}} \\ \boldsymbol{K}_* & \boldsymbol{K}_{**} \end{bmatrix}^{-1} \begin{pmatrix} \boldsymbol{y} \\ \boldsymbol{y}_* \end{pmatrix} + \frac{1}{2}\boldsymbol{y}^{\mathrm{T}} K^{-1} \boldsymbol{y}$$

$$= -\frac{1}{2}\begin{pmatrix} \boldsymbol{y} \\ \boldsymbol{y}_* \end{pmatrix}^{\mathrm{T}}$$

$$\begin{bmatrix} \boldsymbol{K}^{-1} + \boldsymbol{K}^{-1}\boldsymbol{K}_*(\boldsymbol{K}_{**} - \boldsymbol{K}_*\boldsymbol{K}^{-1}\boldsymbol{K}_*^{\mathrm{T}})^{-1}\boldsymbol{K}_*\boldsymbol{K}^{-1} & -\boldsymbol{K}^{-1}\boldsymbol{K}_*(\boldsymbol{K}_{**} - \boldsymbol{K}_*\boldsymbol{K}^{-1}\boldsymbol{K}_*^{\mathrm{T}})^{-1} \\ -(\boldsymbol{K}_{**} - \boldsymbol{K}_*\boldsymbol{K}^{-1}\boldsymbol{K}_*^{\mathrm{T}})^{-1}\boldsymbol{K}_*\boldsymbol{K}^{-1} & (\boldsymbol{K}_{**} - \boldsymbol{K}_*\boldsymbol{K}^{-1}\boldsymbol{K}_*^{\mathrm{T}})^{-1} \end{bmatrix}$$

$$\begin{pmatrix} \boldsymbol{y} \\ \boldsymbol{y}_* \end{pmatrix} + \frac{1}{2}\boldsymbol{y}^{\mathrm{T}} \boldsymbol{K}^{-1} \boldsymbol{y}$$

$$= -\frac{1}{2}(\boldsymbol{y}_* - \boldsymbol{K}_*\boldsymbol{K}^{-1}\boldsymbol{y})^{\mathrm{T}}(\boldsymbol{K}_{**} - \boldsymbol{K}_*\boldsymbol{K}^{-1}\boldsymbol{K}_*^{\mathrm{T}})^{-1}(\boldsymbol{y}_* - \boldsymbol{K}_*\boldsymbol{K}^{-1}\boldsymbol{y}). \tag{7.59d}$$

Equation (7.55d) indicates that the prediction follows the normal distribution $p(\boldsymbol{y}_*|\boldsymbol{y}) \sim N(\boldsymbol{K}_*\boldsymbol{K}^{-1}\boldsymbol{y}, \boldsymbol{K}_{**} - \boldsymbol{K}_*\boldsymbol{K}^{-1}\boldsymbol{K}_*^{\mathrm{T}})$. For each predicted sample, the GP model predicts a mean and variance. If the GP model is correct, repeatedly experimental tests or computations at a given X_* many times will generate many data $\boldsymbol{y}_*(X_*)$, which will follow the predicted normal distribution. The predicted mean and variance indicate how the normal distribution is the prediction of $\boldsymbol{y}_*$ at that point X_*. Both mean and variance play important roles in Bayesian global optimization, the active learning of materials informatics which will be introduced in Volume II.

Like in deterministic ML regression, we may use the function mean $\overline{f}(X_*)$ to calculate the prediction error with $y(X_*) = \overline{f}(X_*) + \varepsilon$, where $y(X_*)$ is the observed response. When an original training dataset is divided into a training subset and a validation subset, cross-validation will let the original whole data be tested. Therefore, the prediction errors of the original data can be calculated. As mentioned above, the prediction errors follow the normal distribution with zero mean and variance σ_{n}^2. Finally, the trained GP model, or the prior distribution, is $p(\boldsymbol{y}) \sim N(0, \overline{K} + \sigma_{\mathrm{n}}^2 \boldsymbol{I})$ with the consideration of the prediction errors, where $\overline{K}$ is the average of optimal $\boldsymbol{K}$s in cross validation. The prediction $p(\boldsymbol{y}_*|\boldsymbol{y})$ will obey the distribution of $N\big(\boldsymbol{K}_*(\overline{K} + \sigma_{\mathrm{n}}^2\boldsymbol{I})^{-1}\boldsymbol{y}, \boldsymbol{K}_{**} - \boldsymbol{K}_*(\overline{K} + \sigma_{\mathrm{n}}^2\boldsymbol{I})^{-1}\boldsymbol{K}_*^{\mathrm{T}}\big)$.

Similarly as that described above in Bayesian linear regression, the L_2 regularization can be added into the GP regression to enhance its generalization power. With the L_2 regularization, a hyperparameter λ replaces σ_n^2 in the prior distribution so that $p(\boldsymbol{y}) \sim N(0, \overline{K} + \lambda\boldsymbol{I})$. The optimal value of hyperparameter λ can be determined during training or by cross validation, which will be illustrated in detail in Volume II.

Example 7.10 Table 7.13 lists 13 Ferritic Martensitic (FM) steels' oxidation data in supercritical water (SCW) environment. The data include one feature of oxidation Cr equivalent concentration $\widetilde{[\mathrm{Cr}]} = [\mathrm{Cr}] + 40.3[\mathrm{V}] + 2.3[\mathrm{Si}] + 10.7[\mathrm{Ni}] - 1.5[\mathrm{Mn}](\mathrm{wt.\%})$ and one response of weight gains $\tilde{\omega} = \frac{\omega}{\omega_0}$, where ω is the experimental oxidation weight gain in the unit of $\mathrm{mg} \cdot \mathrm{cm}^{-2}$, ω_0 is the reference weight gain of $1\mathrm{mg} \cdot \mathrm{cm}^{-2}$. The 13 data are randomly divided into two subsets, a training subset of ten data and a testing subset of three data. Conduct GP regression on this dataset.

Table 7.13 Thirteen data of the oxidation of FM steels in SCW

Dataset	Data	$\widetilde{[\mathrm{Cr}]}$ (wt.%)	$\tilde{\omega}$
Training data	1	28.754	3.860938
	2	24.255	4.321611
	3	30.319	4.214262
	4	27.636	2.920559
	5	11.022	7.340569
	6	19.754	3.476809
	7	8.48	6.576576
	8	15.269	5.47903
	9	22.321	3.400953
	10	12.951	7.898581
Testing data	11	17.4596	5.708319
	12	9.73	9.517995
	13	29.8908	1.959625

A Gaussian kernel $\boldsymbol{K}(X_i, X_j) = a^2 \times \exp\left(-\frac{\|\boldsymbol{X}_i - \boldsymbol{X}_j\|^2}{l^2}\right)$ is used here. Since there is only one feature, the training feature matrix is reduced to a vector $\boldsymbol{X} = \begin{pmatrix} X_1 \\ \vdots \\ X_{10} \end{pmatrix} = \begin{pmatrix} 28.754 \\ \vdots \\ 12.951 \end{pmatrix}_{10\times 1}$. Thus, we have

$$\boldsymbol{K}(a^2, l^2) = \begin{bmatrix} a^2 \exp\left(-\frac{(\boldsymbol{X}_1 - \boldsymbol{X}_1)(\boldsymbol{X}_1 - \boldsymbol{X}_1)^{\mathrm{T}}}{l^2}\right) & \cdots & a^2 \exp\left(-\frac{(\boldsymbol{X}_1 - \boldsymbol{X}_{10})(\boldsymbol{X}_1 - \boldsymbol{X}_{10})^{\mathrm{T}}}{l^2}\right) \\ \vdots & & \vdots \\ a^2 \exp\left(-\frac{(\boldsymbol{X}_{10} - \boldsymbol{X}_1)(\boldsymbol{X}_{10} - \boldsymbol{X}_1)^{\mathrm{T}}}{l^2}\right) & \cdots & a^2 \exp\left(-\frac{(\boldsymbol{X}_{10} - \boldsymbol{X}_{10})^{\mathrm{T}}(\boldsymbol{X}_{10} - \boldsymbol{X}_{10})}{l^2}\right) \end{bmatrix}$$

$$= a^2 \begin{bmatrix} 1 & \cdots & \exp\left(-\frac{(28.754-12.951)^2}{l^2}\right) \\ \vdots & & \vdots \\ \exp\left(-\frac{(12.951-28.754)^2}{l^2}\right) & \cdots & 1 \end{bmatrix} = a^2 \tilde{K}(l^2).$$

The response vector in the training dataset is $\boldsymbol{y} = \begin{pmatrix} y_1 \\ \vdots \\ y_{10} \end{pmatrix}_{10\times 1} = \begin{pmatrix} 3.860938 \\ \vdots \\ 7.898581 \end{pmatrix}_{10\times 1}$.

Thus, the likelihood of $\boldsymbol{y}$ is given as $p(\boldsymbol{y}|\boldsymbol{K}(a^2, l^2)) = \frac{1}{(2\pi)^{n/2}|\boldsymbol{K}|^{1/2}} \exp\left(-\frac{1}{2}\boldsymbol{y}^{\mathrm{T}}\boldsymbol{K}^{-1}\boldsymbol{y}\right)$, and the natural logarithm of likelihood

$$L(\boldsymbol{\theta}) = -\frac{1}{2}\left(n \ln(2\pi) + \ln|\boldsymbol{K}(\boldsymbol{\theta})| + \boldsymbol{y}^{\mathrm{T}}\boldsymbol{K}(\boldsymbol{\theta})^{-1}\boldsymbol{y}\right).$$

The maximum of the natural logarithm of likelihood requires

$$\frac{\partial L(\boldsymbol{\theta})}{\partial \boldsymbol{\theta}} = \frac{\partial\left(-\frac{1}{2}\left(n \ln(2\pi) + \ln|\boldsymbol{K}(\boldsymbol{\theta})| + \boldsymbol{y}^{\mathrm{T}}\boldsymbol{K}(\boldsymbol{\theta})^{-1}\boldsymbol{y}\right)\right)}{\partial \boldsymbol{\theta}} = 0,$$

viz.,

$$\begin{aligned} \frac{\partial(\ln|\boldsymbol{K}(\boldsymbol{\theta})|)}{\partial \boldsymbol{\theta}} + \frac{\partial(\boldsymbol{y}^{\mathrm{T}}\boldsymbol{K}(\boldsymbol{\theta})^{-1}\boldsymbol{y})}{\partial \boldsymbol{\theta}} &= \frac{1}{|\boldsymbol{K}(\boldsymbol{\theta})|}|\boldsymbol{K}(\boldsymbol{\theta})|\mathrm{tr}\left(\boldsymbol{K}^{-1}(\boldsymbol{\theta})\frac{\partial \boldsymbol{K}(\boldsymbol{\theta})}{\partial \boldsymbol{\theta}}\right) \\ &\quad - \boldsymbol{y}^{\mathrm{T}}\boldsymbol{K}^{-1}(\boldsymbol{\theta})\frac{\partial \boldsymbol{K}(\boldsymbol{\theta})}{\partial \boldsymbol{\theta}}\boldsymbol{K}^{-1}(\boldsymbol{\theta})\boldsymbol{y} \\ &= \mathrm{tr}\left(\boldsymbol{K}^{-1}(\boldsymbol{\theta})\frac{\partial \boldsymbol{K}(\boldsymbol{\theta})}{\partial \boldsymbol{\theta}}\right) \\ &\quad - \boldsymbol{y}^{\mathrm{T}}\boldsymbol{K}^{-1}(\boldsymbol{\theta})\frac{\partial \boldsymbol{K}(\boldsymbol{\theta})}{\partial \boldsymbol{\theta}}\boldsymbol{K}^{-1}(\boldsymbol{\theta})\boldsymbol{y} = 0. \end{aligned} \tag{7.60}$$

In this example, $\boldsymbol{\theta} = \{a^2, l^2\}$. Thus, we have

$$\mathrm{tr}\left(\boldsymbol{K}^{-1}\frac{\partial \boldsymbol{K}}{\partial a^2}\right) - \boldsymbol{y}^{\mathrm{T}}\boldsymbol{K}^{-1}\frac{\partial \boldsymbol{K}}{\partial a^2}\boldsymbol{K}^{-1}\boldsymbol{y} = \mathrm{tr}(\boldsymbol{K}^{-1}\tilde{\boldsymbol{K}}) - \boldsymbol{y}^{\mathrm{T}}\boldsymbol{K}^{-1}\tilde{\boldsymbol{K}}\boldsymbol{K}^{-1}\boldsymbol{y} = f_1(a^2, l^2), \tag{7.61}$$

$$\mathrm{tr}\left(\boldsymbol{K}^{-1}\frac{\partial \boldsymbol{K}}{\partial l^2}\right) - \boldsymbol{y}^{\mathrm{T}}\boldsymbol{K}^{-1}\frac{\partial \boldsymbol{K}}{\partial l^2}\boldsymbol{K}^{-1}\boldsymbol{y} = \mathrm{tr}\left(\boldsymbol{K}^{-1}a^2\frac{\partial \tilde{\boldsymbol{K}}}{\partial l^2}\right) - \boldsymbol{y}^{\mathrm{T}}\boldsymbol{K}^{-1}a^2\frac{\partial \tilde{\boldsymbol{K}}}{\partial l^2}\boldsymbol{K}^{-1}\boldsymbol{y} = f_2(a^2, l^2), \tag{7.62}$$

with $\dfrac{\partial \tilde{\boldsymbol{K}}}{\partial l^2} =$

$$\begin{bmatrix} 0 & \cdots & \frac{(28.754-12.951)^2}{l^4}\exp\left(-\frac{(28.754-12.951)^2}{l^2}\right) \\ \vdots & & \vdots \\ \frac{(12.951-28.754)^2}{l^4}\exp\left(-\frac{(12.951-28.754)^2}{l^2}\right) & \cdots & 0 \end{bmatrix}.$$

To determine the values of a^2 and l^2, numerical calculation with iteration is usually employed to simultaneously solve Eqs. (7.61 and 7.62). In iteration, the gradients, the left sides of Eqs. (7.61 and 7.62), are not zero, but give the direction to approach zero. Letting $a^{2,\,(t)}$ and $l^{2,\,(t)}$ be the values of a^2 and l^2 after the t-th iteration, we have the log-likelihood value $L^{(t)} = L(a^{2,\,(t)}, l^{2,\,(t)})$. During the iteration, if $L^{(t+1)} < L^{(t)}$, $L^{(t)}$ will be the maximum of L (scikit-learn, package of optimizer), and the optimal values will be $a^{2,\,(t)}$ and $l^{2,\,(t)}$. Taking the initial value of $a^2 = 1$ and $l^2 = 1$, we conduct the iteration and obtain the optimal value of $\hat{a^2} = 4.13^2$ and $\hat{l^2} = 2.67^2$ with the log-likelihood value of 23.4.

The testing feature vector $\boldsymbol{X}_*$ and its response vector $\boldsymbol{y}_*$ are given as $\boldsymbol{X}_* = \begin{pmatrix} X_{11} \\ X_{12} \\ X_{13} \end{pmatrix} = \begin{pmatrix} 17.4596 \\ 9.73 \\ 29.8908 \end{pmatrix}_{3\times 1}$ and $\boldsymbol{y}_* = \begin{pmatrix} y_{11} \\ y_{12} \\ y_{13} \end{pmatrix}_{3\times 1} = \begin{pmatrix} 5.708319 \\ 9.517995 \\ 1.959625 \end{pmatrix}_{3\times 1}$, with the solved kernel $\boldsymbol{K}(\boldsymbol{X}_i, \boldsymbol{X}_j) = 4.13^2 \times \exp\left(-\dfrac{\left\|\boldsymbol{X}_i - \boldsymbol{X}_j\right\|^2}{2.67^2}\right)$, all kernels defined in Eq. (7.59b) can be calculated as

$$\boldsymbol{K} = 4.13^2 \begin{bmatrix} \exp\left(-\frac{(\boldsymbol{X}_1-\boldsymbol{X}_1)(\boldsymbol{X}_1-\boldsymbol{X}_1)^{\mathrm{T}}}{2.67^2}\right) & \cdots & \exp\left(-\frac{(\boldsymbol{X}_1-\boldsymbol{X}_{10})(\boldsymbol{X}_1-\boldsymbol{X}_{10})^{\mathrm{T}}}{2.67^2}\right) \\ \vdots & & \vdots \\ \exp\left(-\frac{(\boldsymbol{X}_{10}-\boldsymbol{X}_1)(\boldsymbol{X}_{10}-\boldsymbol{X}_1)^{\mathrm{T}}}{2.67^2}\right) & \cdots & \exp\left(-\frac{(\boldsymbol{X}_{10}-\boldsymbol{X}_{10})(\boldsymbol{X}_{10}-\boldsymbol{X}_{10})^{\mathrm{T}}}{2.67^2}\right) \end{bmatrix}_{10\times 10},$$

$$\boldsymbol{K}_* = 4.13^2 \begin{bmatrix} \exp\left(-\frac{(\boldsymbol{X}_{11}-\boldsymbol{X}_1)(\boldsymbol{X}_{11}-\boldsymbol{X}_1)^{\mathrm{T}}}{2.67^2}\right) & \cdots & \exp\left(-\frac{(\boldsymbol{X}_{11}-\boldsymbol{X}_{10})(\boldsymbol{X}_{11}-\boldsymbol{X}_{10})^{\mathrm{T}}}{2.67^2}\right) \\ \exp\left(-\frac{(\boldsymbol{X}_{12}-\boldsymbol{X}_1)(\boldsymbol{X}_{12}-\boldsymbol{X}_1)^{\mathrm{T}}}{2.67^2}\right) & \cdots & \exp\left(-\frac{(\boldsymbol{X}_{12}-\boldsymbol{X}_{10})(\boldsymbol{X}_{12}-\boldsymbol{X}_{10})^{\mathrm{T}}}{2.67^2}\right) \\ \exp\left(-\frac{(\boldsymbol{X}_{13}-\boldsymbol{X}_1)(\boldsymbol{X}_{13}-\boldsymbol{X}_1)^{\mathrm{T}}}{2.67^2}\right) & \cdots & \exp\left(-\frac{(\boldsymbol{X}_{13}-\boldsymbol{X}_{10})(\boldsymbol{X}_{13}-\boldsymbol{X}_{10})^{\mathrm{T}}}{2.67^2}\right) \end{bmatrix}_{3\times 10}$$

and

$$K_{**} = 4.13^2 \begin{bmatrix} 1 & \exp\left(-\frac{(X_{11}-X_{12})(X_{11}-X_{12})^{\mathrm{T}}}{2.67^2}\right) & \exp\left(-\frac{(X_{11}-X_{13})(X_{11}-X_{13})^{\mathrm{T}}}{2.67^2}\right) \\ \exp\left(-\frac{(X_{12}-X_{11})(X_{12}-X_{11})^{\mathrm{T}}}{2.67^2}\right) & 1 & \exp\left(-\frac{(X_{12}-X_{13})(X_{12}-X_{13})^{\mathrm{T}}}{2.67^2}\right) \\ \exp\left(-\frac{(X_{13}-X_{11})(X_{13}-X_{11})^{\mathrm{T}}}{2.67^2}\right) & \exp\left(-\frac{(X_{13}-X_{12})(X_{13}-X_{12})^{\mathrm{T}}}{2.67^2}\right) & 1 \end{bmatrix}_{3\times 3}.$$

Without considering the training data uncertainty for the GP model, i.e., $\varepsilon \equiv \sigma_n^2 = 0$, the prediction of $p(\boldsymbol{y}_*|\boldsymbol{y}) \sim N(\boldsymbol{K}_*\boldsymbol{K}^{-1}\boldsymbol{y}, \boldsymbol{K}_{**} - \boldsymbol{K}_*\boldsymbol{K}^{-1}\boldsymbol{K}_*^{\mathrm{T}})$ is shown in Fig. 7.16a and listed in Table 7.14.

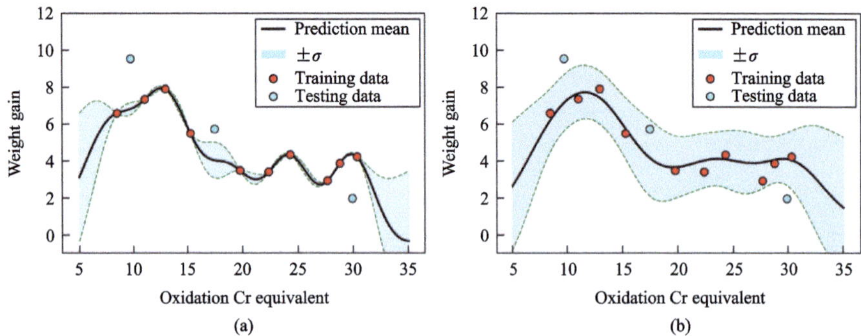

Fig. 7.16 Prediction mean and variance on each point of oxidation Cr equivalent from the GP models **(a)** without and **(b)** with prediction errors in the training data

Table 7.14 The prediction results on the 3 testing data

Data	Weight gain	Models	Prediction values	Deviations
11	5.708319	$\varepsilon = 0$	4.022 ± 0.97	1.69
		$\varepsilon \sim N(0.407, 2.067^2)$	4.154 ± 2.07	1.55
12	9.517995	$\varepsilon = 0$	6.816 ± 0.34	2.702
		$\varepsilon \sim N(0.407, 2.067^2)$	7.137 ± 1.53	2.381
13	1.959625	$\varepsilon = 0$	4.365 ± 0.07	–2.405
		$\varepsilon \sim N(0.407, 2.067^2)$	4.095 ± 1.42	–2.135

The LOOCV is applied to the original training dataset in which 10 data are divided into a training subset (9 data) and a validation subset (one datum). The cross validation gives 10 optimal values of a^2 and l^2, or 10 GP models, and 10 prediction means of the original training data. The 10 prediction errors follow the normal distribution $\varepsilon \sim N(0.407, 2.067^2)$. The mean deviated from zero is caused by the small dataset. Table 7.15 shows the results from the LOOCV. We take the average value of $\hat{a^2}$ and $\hat{l^2}$, i.e., $\overline{a^2} = 4.297^2$ and $\overline{l^2} = 2.961^2$, and the error variance of $\sigma^2 = 2.067^2$ to build up the GP model $p(\boldsymbol{y}) \sim N(0, \overline{K} + \sigma_n^2\boldsymbol{I})$. The prediction $p(\boldsymbol{y}_*|\boldsymbol{y}) \sim N\big(\boldsymbol{K}_*(\overline{K} + \sigma_n^2\boldsymbol{I})^{-1}\boldsymbol{y}, \boldsymbol{K}_{**} - \boldsymbol{K}_*(\overline{K} + \sigma_n^2\boldsymbol{I})^{-1}\boldsymbol{K}_*^{\mathrm{T}}\big)$ on the three testing data is conducted, and the prediction result is plotted in Fig. 7.16b and listed in Table 7.14. Without considering the prediction errors in the training data, the predictions pass the training data without any errors, and the prediction variance on each of the testing data is small so that the three testing data are located outside the corresponding variances, as shown in Fig. 7.16a. The results indicate that the GP model without considering the prediction errors in the training data is overfitting on the training data and has a poor prediction power on the testing data. If the prediction errors in the training data are included in the GP model, the model will have a larger variance on each of all training and testing data, as illustrated in Fig. 7.16b. In this case, the GP model improves the power of generalization, and enhances the prediction accuracy on the testing data, as shown in Table 7.14.

Table 7.15 The results of LOOCV on the original training dataset

Data	Weight gain y	Prediction values $\hat{y}$	Optimal parameters on the training subset	Error ε $(y-\hat{y})$
1	3.860938	3.047	$\widehat{a^2} = 4.34^2, \widehat{l^2} = 3.03^2$	0.814
2	4.321611	1.945	$\widehat{a^2} = 4.28^2, \widehat{l^2} = 3.5^2$	2.376
3	4.214262	5.424	$\widehat{a^2} = 4.46^2, \widehat{l^2} = 2.83^2$	–1.210
4	2.920559	4.050	$\widehat{a^2} = 4.32^2, \widehat{l^2} = 3.2^2$	–1.130
5	7.340569	8.459	$\widehat{a^2} = 4.31^2, \widehat{l^2} = 2.72^2$	–1.118
6	3.476809	0.010	$\widehat{a^2} = 4.26^2, \widehat{l^2} = 2.91^2$	3.467
7	6.576576	3.273	$\widehat{a^2} = 3.99^2, \widehat{l^2} = 2.68^2$	3.304
8	5.47903	7.192	$\widehat{a^2} = 4.42^2, \widehat{l^2} = 2.77^2$	–1.713
9	3.400953	5.756	$\widehat{a^2} = 4.5^2, \widehat{l^2} = 3.25^2$	–2.355
10	7.898581	6.263	$\widehat{a^2} = 4.09^2, \widehat{l^2} = 2.72^2$	1.636

Homework

1. Train a Naive Bayes classifier with data 1–16 from Table 7.1, and use it to test datum 17 in Table 7.1.
2. A set of data (7.68, 8.36, 9.47, 6.69, 3.98, 8.62, 9.26) follows the Weibull distribution, calculate the values of m and σ_0 by using the maximum likelihood.
3. Randomly generate a mixture of two Gaussian distributions, $0.3N(5, 1^2) + 0.7N(10, 1^2)$. Using the EM algorithm to estimate the latent variables of $(w_i, \mu_i, \sigma_i^2), i = 1, 2$.
4. Using the EM algorithm to fit a whole powder XRD pattern.
5. Discuss the relationship between GP regression and ridge regression with kernel functions.
6. Perform the GP regression with Rational Quadratic kernel and Dot Product kernel on the dataset in Table 7.13.

References

Berger, J. O. (2013). *Statistical decision theory and bayesian analysis*. Springer Science and Business Media.

Bishop, C. M., & Nasrabadi, N. M. (2006). *Pattern recognition and machine learning*. Springer.

Dempster, A. P., Laird, N. M., & Rubin, D. B. (1977). Maximum likelihood from incomplete data via the EM algorithm. *Journal of the Royal Statistical Society, 39*(1), 1–22.

Fu, R., & Zhang, T. Y. (1998). Effects of an applied electric field on the modulus of rupture of poled lead zirconate titanate ceramics. *Journal of the American Ceramic Society, 81*(4), 1058–1060.

Hoover, C. G., Bažant, Z. P., Vorel, J., et al. (2013). Comprehensive concrete fracture tests: Description and results. *Engineering Fracture Mechanics, 114*, 92–103.

John, G. H. (1995). Estimating continuous distributions in Bayesian classifiers. *Eleventh conference on uncertainty in artificial intelligence* (pp. 338–345). Morgan Kaufmann.

McCallum, A., & Nigam, K. (1998). A comparison of event models for Naive Bayes text classification. *AAAI-98 Workshop on Learning for Text Categorization 752*(1), 41–48.

Newcomb, S. (1886). A generalized theory of the combination of observations so as to obtain the best result. *American Journal of Mathematics, 8*(4), 343–366.

Patel, J. K., & Read, C. B. (1996). *Handbook of the normal distribution*. CRC Press.

Pedregosa, F., Varoquaux, G., Gramfort, A., et al. (2011). Scikit-learn: Machine learning in Python. *Journal of Machine Learning Research, 12*, 2825–2830.

Rasmussen, C. E., & Williams, C. K. (2006). *Gaussian processes for machine learning*. MIT Press.

Rish, I. (2001). An empirical study of the Naive Bayes classifier. *IJCAI Workshop on Empirical Methods in Artificial Intelligence 3*(2), 41–46.

Rodríguez-Carvajal, J. 2001. Introduction to the program FULLPROF: Refinement of crystal and magnetic structures from powder and single crystal data. Laboratoire Léon Brillouin (CEA-CNRS), Saclay, France.

Spackman, K. A. (1989). Signal detection theory: Valuable tools for evaluating inductive learning. In *Proceedings of the Sixth International Workshop on Machine Learning* (pp.160–163). Morgan Kaufmann.

Weibull, W. (1939). A statistical theory of the strength of materials. *Swed R Inst Eng Res, 151*, 1–45.

Williams, C. K. I., & Rasmussen, C. E. (1996). *Gaussian processes for regression advances in neural information processing systems 8* (pp. 514–520). MIT Press.

Xiong, J., Shi, S. Q., & Zhang, T. Y. (2021). Machine learning of phases and mechanical properties in complex concentrated alloys. *Journal of Materials Science and Technology, 87*, 133–142.

Zhang, T. Y., & Zheng, Y. R. (1998). Effects of absorption and desorption on hydrogen permeation—I. Theoretical modeling and room temperature verification. *Acta Materialia 46*(14), 5023–5033.

Chapter 8
Symbolic Regression

The scientific goal of materials informatics is to discover material science, technology and knowledge from data, where symbolic regression (SR) is playing a greater role, specifically in this era of exponential data growth. When combined with domain knowledge, SRs are able to discover scientifically meaningful formulas from data. SR is trying to find a globally optimal solution in a heuristic manner. Global optimization is a common problem in many areas. The main challenge of global optimization is that the space of interest for the search is usually too large to access all possible solutions within the current capability of computation, and thus highly efficient algorithms are required to locate the global optimum (or close to the global optimum) without the need of exhaustive search over the entire space. Heuristic algorithms, such as evolutionary computation (EC), are efficient for solving the global optimization problems (de Jong, 2006; Sette & Boullart, 2001).

8.1 Overview of Evolutionary Computation

EC is a population-based iterative heuristic algorithm inspired by the Darwin's theory of evolution by natural selection. Here "population" means a set of candidate solutions called individuals. The number of individuals in a population is termed as the size of the population. Figure 8.1 shows a typical flowchart of EC, where iterations are conducted until a satisfactory solution is found. Each EC iteration involves individual evaluation, individual selection, and breeding of individuals via crossover and mutation. After each iteration, a new population (or generation) is formed which comprises the selected and bred individuals. In the first iteration, EC starts with an initial population containing individuals sampled randomly from the search space of interest. Then, the process of evaluation is performed to select promising individuals based on their fitness to the data, e.g. with the metric of mean square error (MSE). Breeding of new individuals is carried out on the survived individuals (or

T. Zhang, *An Introduction to Materials Informatics*,
https://doi.org/10.1007/978-981-99-7992-9_8

parents) by crossover and mutation operations. These newly bred individuals are then put together with their parents to form a new population for the next iteration. The fitness of individuals will improve during iterations and the iteration continues until a satisfactory solution is found or the number of iterations reaches the prescribed maximal value. There are many algorithms under the EC umbrella, e.g., genetic algorithms (Holland, 1992), evolution strategies (Rechenberg, 1965), evolutionary programming (Fogel et al., 1966), and genetic programming (GP) (Koza, 1992; Poli et al., 2008). The GP also includes many methods, e.g., grammar-guided genetic programming (GGGP) (McKay et al., 2010), grammatical evolution (GE) (Ryan et al., 1998), statistical genetic programming (Amir Haeri et al., 2017), clone selection programming (Gan et al., 2009), and artificial immune system programming (Johnson, 2003). The methods of GP, GGGP, and GE are introduced below.

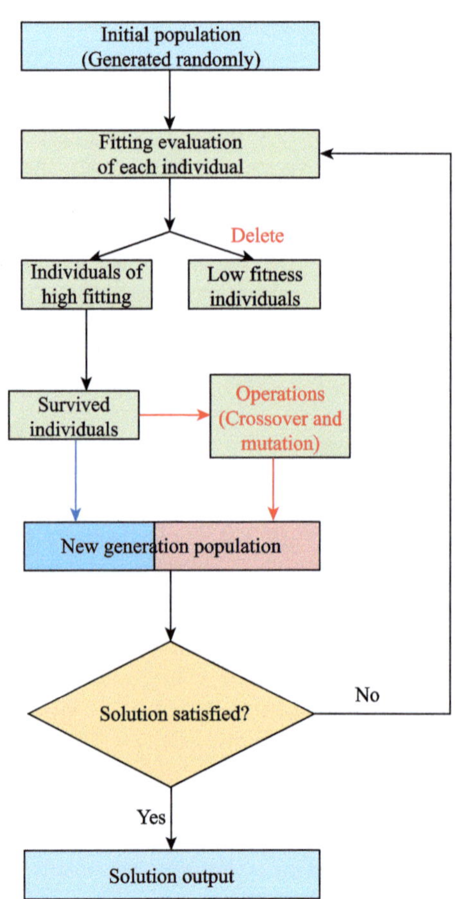

Fig. 8.1 The chart of EC working flow

8.2 Genetic Programming

In GP, one individual is represented by one tree, which is called the (mathematical) expression tree. For example, Fig. 8.2 shows an expression tree of $f(x) = \log(x) + \sqrt{x} + 3$. An expression tree comprises nodes and leaves, where one node is a mathematical operator, such as the " + ", "log" and "sqrt" in Fig. 8.2, and a leaf is a variable (or termed feature) or a constant, e.g., x and 3 in Fig. 8.2. Usually, a set of operators, e.g., $\left\{+,\ -,\ \times,\ \div,\ \cdots,\ \log,\ \sqrt{\ },\ \sin,\ \exp,\ \cdots\right\}$, is preset in GP. Initial trees are randomly generated, and many techniques can be used for such purpose, for example, the widely employed methods proposed by Koza (1992): "grow", "full" and "ramped-half-and-half" (RHH). Alternative methods for generating initial trees can be found in the work (Burke et al., 2003).

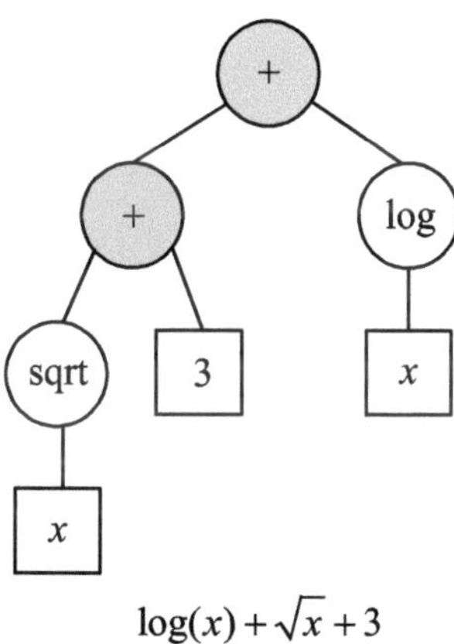

Fig. 8.2 An expression tree in GP

The fitness of a tree is evaluated by using a loss function. Many loss functions are described in Chap. 2. The lower the minimum of the loss function is, the better the GP formula will be. Trees are ranked according to their fitness (e.g., the one with the lowest MSE will be ranked at the first place). Highly scored trees will be selected for further evolution. There are many strategies to select trees, such as fitting-proportionate selection, greedy over-selection and tournament selection (Koza, 1992). Taking greedy over-selection as an example, the population is divided into two groups with the top-fittest 20% individuals in Group I and the remaining 80% individuals in Group II, and the 80% low fitness individuals will be eliminated or bred. A half of the top 20% fitness individuals are used for breeding. The other half of the top 20% fitness individuals with the newly bred individuals (called offspring) and virginal un-evaluated individuals form a new population for the next iteration.

Breeding takes place with the operations of crossover and mutation on the selected trees. For crossover, two individuals are selected at prescribed probability (an input parameter) for interchange between their sub trees or leaves, as shown in Fig. 8.3. For mutation, nodes in a tree can mutate to another operator, function or terminal, and an example is shown in Fig. 8.4. To start a GP evolution, several parameters need to be input: The size of the population, the maximal generation number, and the probabilities both for crossover and for mutation.

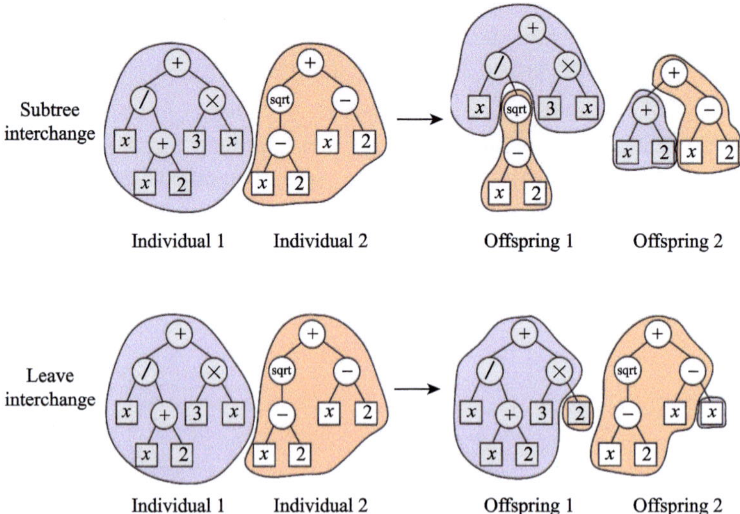

Fig. 8.3 An example of crossover between two expression trees in GP

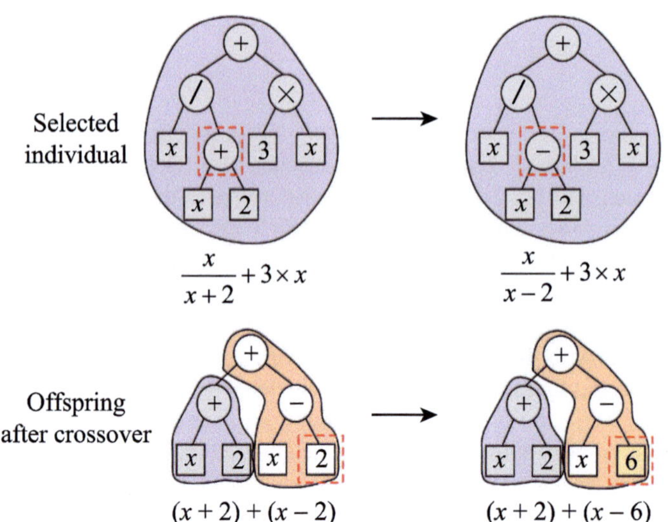

Fig. 8.4 An example of mutations in GP

The efficiency of GP algorithms is mainly determined by two aspects: the search space of trees, in which the best individuals will be searched, and the tree complexity. Since expression trees are generated via combination of all possible nodes and leaves, the search space of trees can be immense when the operator set, the terminal number, and the size (or length) of expression are large. Huge tree space and limited generations pose challenges regarding how to find the best solution. A simple direction to tackle this challenge is to construct appropriate search space of trees by using only small sets of operators and/or terminals with the aid of domain knowledge. Certainly, this requires a deep understanding of the studied problem and the desired final functional forms. Other strategies exist to reduce the tree space. For instance, the strong-typed GP (Montana, 1995) method allows the mathematical combinations to happen only between consistent data types, such as that float-type (integer-type) variables are only allowed to combine with float-type (integer-type) variables for linear operations. Grammar-guided-based methods reduce the space by constraining the types and the admissible forms of expressions, which will be described in detail in the following section. Lu et al. (2016) proposed a method to connect prior formula knowledge (PFK) and GP to reduce the tree space. In the PFK-GP method, data are analyzed firstly to determine their characteristics and also the corresponding expression candidates that can describe the data characteristics, thereby forming an initial set of expression candidates. Then, initial individuals for GP can be obtained from a small set of expression candidates to avoid huge tree space. The concern regarding the complexity of expressions in GP is called bloating (Luke & Panait, 2006), which means increasing complexity of expressions without further improving their quality. The simplest method to avoid bloating is to limit the depth of trees, or add a penalty term to the size of the expressions. Other methods include parsimony pressure (Luke & Panait, 2002; Poli & McPhee, 2008) and multi-objective GP (MOGP) (Fonseca & Fleming, 1995). The MOGP tries to find a Pareto optimal solution (Smits & Kotanchek, 2005) to achieve a balance between accuracy and complexity of expressions.

8.3 Grammar-Guided Genetic Programming and Grammatical Evolution

Grammar-guided genetic programming (GGGP) uses a grammar to derive the expression tree (McKay et al., 2010). Grammar, which originates from the field of human language processing, includes a set of production rules that define legal language models. Among others, the context free grammar (CFG) is one popular method of the GGGP family. Each production rule in CFG comprises two parts and is denoted as: A single nonterminal symbol (or variable):: = string of variables and/or terminal symbols. The mark "$::=$ " means that the symbol on the left-hand side can be replaced by any element on the right-hand side of the production. An example of CFG is shown in Table 8.1. A derivative process rewrites all variables, starting from the <start>, following the production rules until all leaves are terminals. Other types of grammars, such as logic grammars, semantic grammars and tree adjoining grammars, are reviewed in the review article (McKay et al., 2010). One advantage of the introduction of grammar to GP is that the size of the tree space can be greatly reduced because the type constraints and the admissible forms of expressions can be easily provided in the grammar setup. In addition, domain knowledge can be easily incorporated into the grammar setup to further reduce the space size and also drive the expressions toward meaningful formulas.

Table 8.1 An example of context free grammar

<start>	$::=$	<expr>
<expr>	$::=$	(<expr>) | (<expr><op><expr>) | <terminal>
<op>	$::=$	$+ \mid - \mid \times$
<terminal>	$::=$	a (some constant) | x

Note The symbol "|" means "OR"

Unlike that of GP, the breeding operations of crossover and mutation are not applied directly to the expression tree in GGGP. In the language of GGGP, an expression tree is called a phenotype obtained by transforming a genotype via the breeding operations. Genotype takes different forms in different kinds of GGGP. For example, it can be a tree representation (termed derivation tree) in tree-based GGGP (Hoai et al., 2002; Vergilio & Pozo, 2006), or a linear representation in grammatical evolution (Ryan et al., 1998).

Figure 8.5 shows an example of a derivation tree and how it grows into an expression in accordance with the grammar defined in Table 8.1. By reading out leaves of the derivation tree one by one from left to right, an expression tree can be constructed. In the evolution of a tree-based GGGP, compared with that in canonical GP, the operations of crossover and mutation are subject to additional constraints. For crossover, it must happen between two nodes, each from different trees, with the same nonterminal symbol of the grammar. The target nodes or subtrees can be randomly selected and exchanged. For example, Fig. 8.6 shows the crossover of two derivation trees. For mutation, a node in a derivation tree is selected randomly, and then all the "tails" (subtrees) connected to this node are deleted. Then, a new subtree, which is created according to the grammar, is appended to the selected node, as shown in Fig. 8.7.

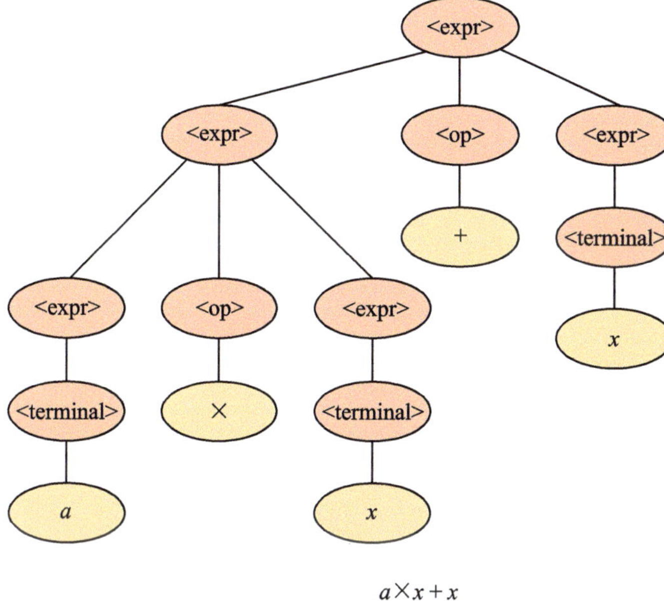

Fig. 8.5 An example of a derivation tree based on the grammar shown in Table 8.1

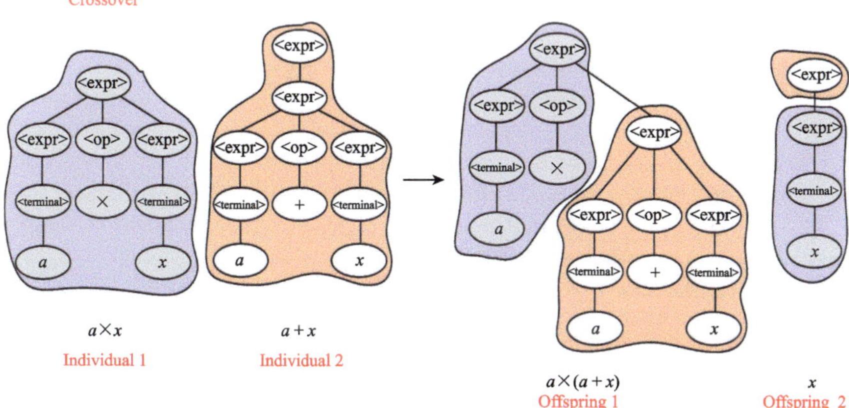

Fig. 8.6 An example of crossover between two derivation trees in GGGP

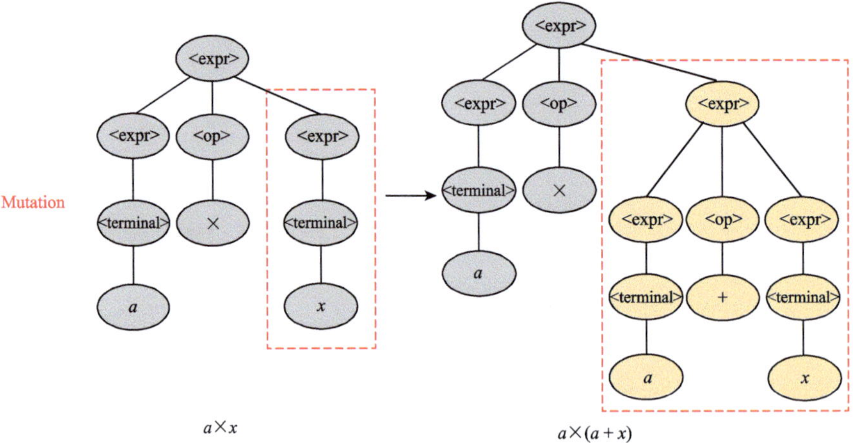

Fig. 8.7 An example of mutation in a derivation tree of GGGP

In grammatical evolution (GE), the genotype is a random number sequence, as shown in Fig. 8.8. A mathematical expression (phenotype) is obtained by translating the random number sequence according to the user defined grammar (Ryan et al., 1998). For example, with the grammar shown in Fig. 8.8, the number sequence [235, 458, 43, 45, 567, $\cdots$] can be translated into an expression by the following steps.

(1) The derivative process begins from the <start>, which can only be replaced by another non-terminal symbol <expr>. The <expr> represents one of the three candidate symbol expressions (see the second line of the grammar table): (<expr>), (<expr> <op> <expr>), or <terminal>, and let's index them by (0, 1, 2), respectively. Now, the first number 235 in the gene sequence is divided by the total number 3 of indices (0, 1, 2) in <expr>, giving the remainder of 1. So the symbol <expr> is substituted by the second symbol expression (<expr> <op> <expr>), whose index is 1.

(2) Repeating the above steps for the first <expr> in (<expr> <op> <expr>), we divide the second number of 458 in the gene sequence by 3 and obtain its remainder of 2, and so <expr> is substituted by <terminal>, whose index is 2, resulting in (<terminal> <op> <expr>).

(3) When the third number 43 is divided by the total number 2 of indices (0, 1) in <terminal>, we obtain the remainder of 1, and so <terminal> is substituted by the name of variable, denoted as x. Now the symbol expression is (x <op> <expr>).

(4) The <op>, which has three candidates (see the third line of the grammar table) " + ", " " and "*", are assigned indices (0, 1, 2), respectively. The remainder is 0 when the next random number 45 is divided by 3, and so the <op> is substituted by the symbol " + ". Now the symbol expression becomes ($x+$ <expr>). Repeating these steps, and the final <expr> can be translated to a corresponding expression. This type of translation methods is very flexible and different translation rules can be designed based on studied problems. For such linear representation of the genotype, operations of crossover and mutation are

easy to apply. As the example shown in Fig. 8.9, for crossover, it exchanges two selected number sequences from two different genotypes; for mutation, a randomly selected number is changed to a new number.

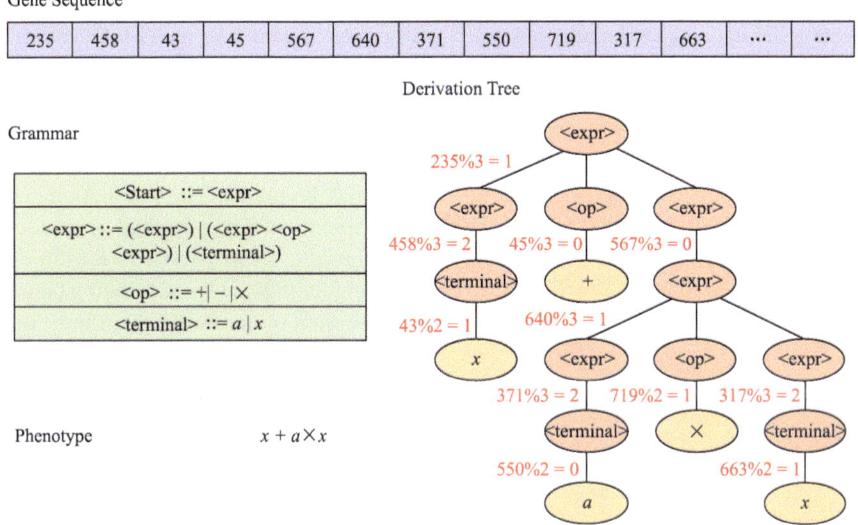

Fig. 8.8 An example of the process of GE to reach an expression tree (phenotype) from the linear expression of random integer sequences (genotype) by using the MOD mapping rule

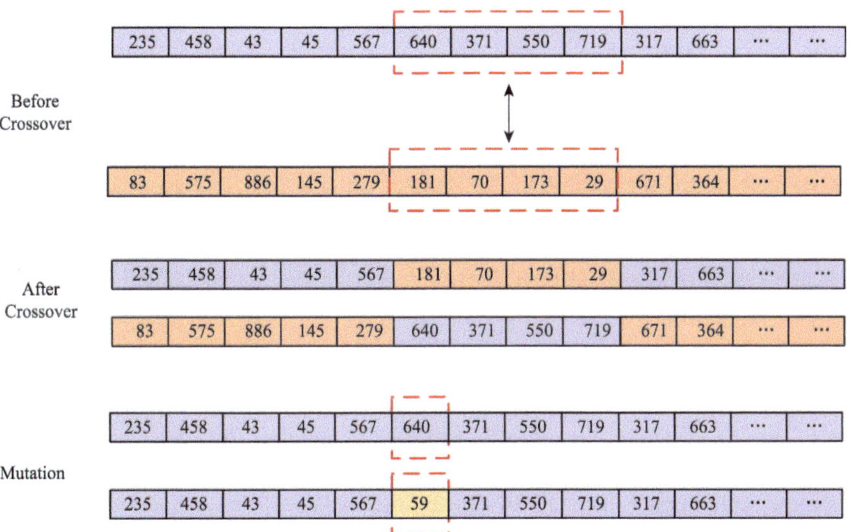

Fig. 8.9 Examples of crossover and mutation between/in genotypes in GE

Example 8.1 This example illustrates GP with tournament selection in detail, which is performed with the open software gplearn (https://gplearn.readthedocs.io/en/stable/).

Step 1: **Data and used algorithm parameters**

About 100 data are uniformly generated within the range of x from -1.0 to 1.0 with a step of 0.02 by using the following equation

$$f(x) = x^2 + \sin(x) + 1. \tag{8.1}$$

Table 8.2 lists the used parameters, which must be preset. The population size indicates how many individuals are there in a population. The maximal generation number is the maximum of iterations. Mean absolute error (MAE) is the fitness parameter to evaluate the performance of the fitting, with a lower value being better. A preset value of MAE is used as the stopping criterion of iteration, which indicates that the found formula has such accuracy of MAE. The function set includes a set of mathematic operators used in the GP example. The terminal set is composed of the variables and numbers used in terminal nodes. The shape and size of individual trees in the initial generation are controlled by two parameters, Init Depth and Init Method. The Init Depth sets a range of the tree depth for initial trees. For example, Init Depth of (2,6) allows a tree in the initial generation to randomly selects a tree depth to be an integer among 2, 3, 4, 5, and 6. The Init Method provides three options: “grow”, “full” and “ramped-half-and-half”. The option “full” requires that the depth of each branch in an initial individual tree must reach the same depth randomly selected from Init Depth; the “grow” does not require all branches of an initial tree having the same depth, although the tree depth is selected; the “half-and-half” yields a mixture of tree shapes in the initial generation with half “full” and half “grow” trees. Since crossover and mutation do not change the contents in the terminal nodes, the shape and size of formula trees in later generations are controlled by genetic operators.

Tournament selection is perhaps the most popular employed method to select individuals in GP for breeding. The tournament selection strategy draws a preset number of individuals from the current population each time (with replacement) and then selects the best individual among them to enter the next-generation population. This process is repeated until the size of the next-generation population reaches that of the current population. This preset number of individuals is referred to as the tournament size. In this example, the tournament size is two, and thus, the individual with a lower MAE in a subset is called the winner and selected for breeding. To operate crossover, two winners from two subsets, one from each, are needed, while only one winner from one subset is needed for mutation. The best winners have the chance to be individuals in the next generation, which is implemented by the reproduction operator. The crossover rate, the mutation rate, and the reproduction rate correspondingly indicate the probabilities of generating the next population of individuals.

Table 8.2 The preset values of algorithm parameters

Parameters	Values
Population size	100
Maximal generation number	2000
Metric (lower is better)	$\frac{\sum_{i=1}^{N} \lvert y_i - \hat{y}_i \rvert}{N}$
Stopping criterion	$\frac{\sum_{i=1}^{N} \lvert y_i - \hat{y}_i \rvert}{N} < 10^{-5}$
Function set	$+,\ \times,\ \sin$
Terminal set	$x,\ 1$
Init depth	(3, 3)
Init method	Full
Tournament size	2
Crossover rate	0.70
Mutation rate	0.25
Reproduction rate	0.05

Step 2: **Initial generation**

One hundred individuals are randomly generated from the terminal set and the function set to form the initial generation (generation 1). As an example, Fig. 8.10 shows only the first three individuals and their fitness (or errors), where the abbreviations “add”, “mul”, and “sin”, represent “add (+)”, “mul (×)” and “sin function (sin)”, respectively, and will be used hereafter.

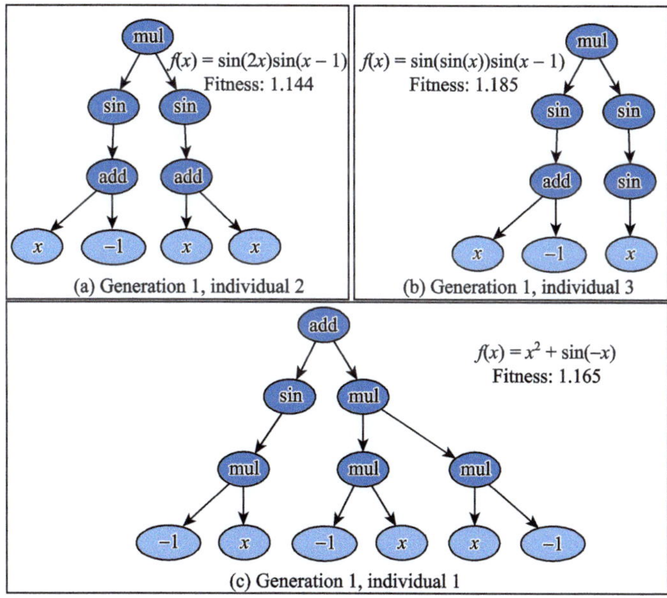

Fig. 8.10 Three individuals in the initial generation (generation 1)

Step 3: **Breeding**

Individuals 85, 73, 64, and 52 in generation 1 beat their competitors of individuals 97, 89, 7, and 49, respectively, in tournament selection with the winner MAE/loser MAE of 1.32/2.95, 1.02/6.03, 0.92/1.34, and 1.33/3.36, and thus become parents for breeding generation 2. Figure 8.11 shows that parents 85 and 73 are randomly selected from the winners and crossed over each other to breed two offspring. In tournament selection, only one offspring is randomly selected to become an individual in generation 2 and the other is discarded. Through mutation, parent 64 varies to individual 3 in generation 2, as shown in Fig. 8.12. Parent 52 is lucky to survive to generation 2, as shown in Fig. 8.13. To have 100 individuals in generation 2, 89 subsets, including 44 repeated subsets, have been randomly selected by Gplearn from generation 1, and hence yielding only 45 winners. Even in different subsets, some individuals are also the same so that only 82 individuals in generation 1 are calculated their MAEs. The 45 winners, with replacement sampling, are used for breeding via crossover, mutation, and reproduction on predefined probabilities of 0.7, 0.25, and 0.05, respectively, to generate 100 individuals in generation 2.

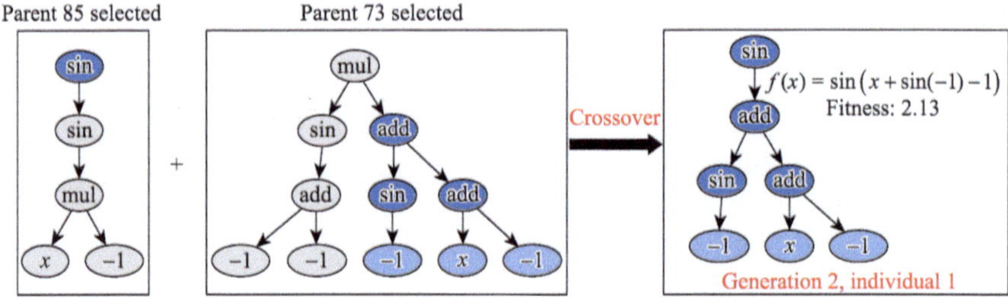

Fig. 8.11 The breeding of individual 1 in generation 2

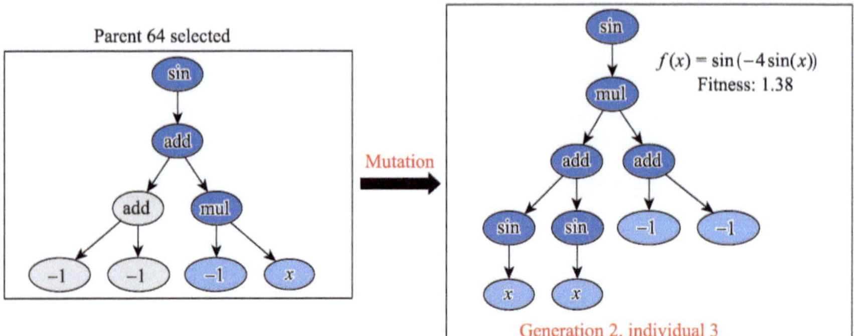

Fig. 8.12 The breeding of individual 3 in generation 2

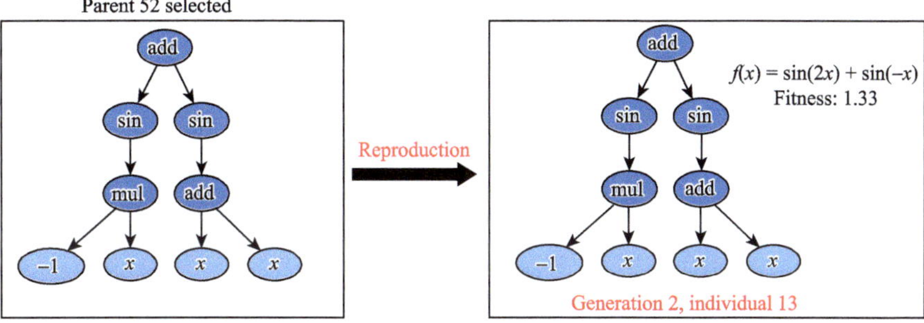

Fig. 8.13 Individual 52 in generation 1 is copied as individual 13 in generation 2

Step 4: **Iteration until meeting the stopping criterion**

The iteration goes on one generation after another. The converged formula is founded in generation 6, which has the tree shape shown in Fig. 8.14.

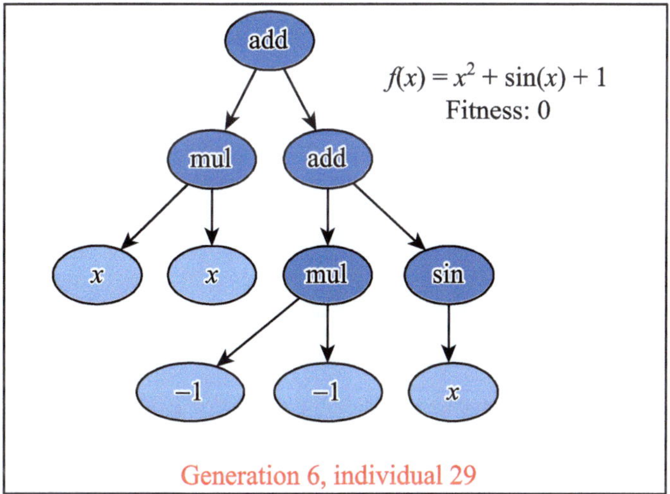

Fig. 8.14 The formula is founded in generation 6

It should be noted that the exactly correct result may not be found within 6 generations due to the fact that GP is a stochastic optimization algorithm. Another two independent runs found the formula with different numbers of generations and tree shapes (Fig. 8.15).

As described above, tournament selection has the advantage of generating large diversity in the offspring and thus, the individual with the highest goodness of fitting in a parent generation may not be selected to survive in its child generation. In this example, all data are generated from the same formula without any noise so that greedy selection might be more appropriate.

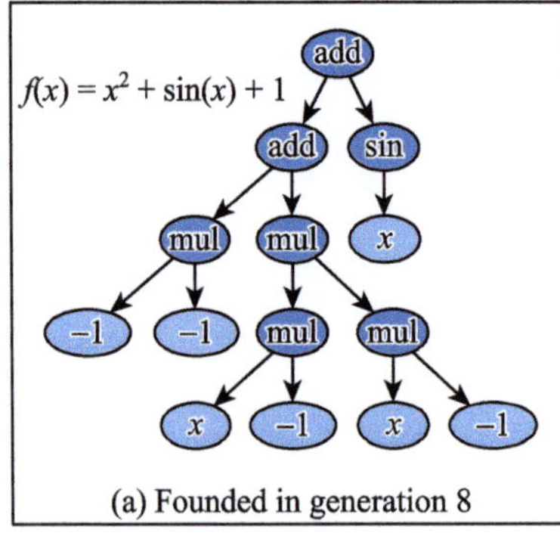

(a) Founded in generation 8

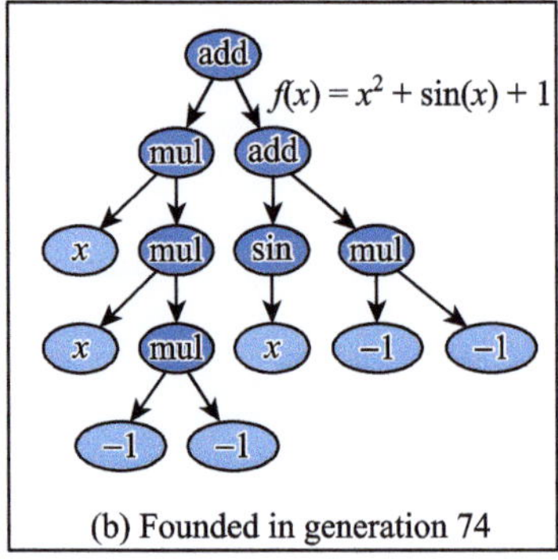

(b) Founded in generation 74

Fig. 8.15 Another two independent runs found the formula in generations 8 and 74, respectively

---

8.4 The Application of LASSO in Symbolic Regression

LASSO (Tibshirani, 1996) is described in Chap. 2, which is based on the theory of compressed sensing (Candes & Wakin, 2008) originated from the field of signal processing and image compression. From a large number of candidate trees in SR, LASSO can also select candidate trees, as selecting features described in Chap. 2. There have been many recent applications of LASSO (or modified LASSO) for finding explicit analytical expressions for materials properties in materials science (Ghiringhelli et al., 2015, 2017; Nelson et al., 2013).

Homework

Let us create some synthetic data based on the relationship $f(x) = x_1^2 - x_2^2 + x_1 - 1$ with the x_1 and x_2 in the range from -1.0 to 1.0 with a step of 0.1. Based on the man-made data, using the package of gplearn to perform GP. Please use the preset values for the algorithm parameters given in the table below, and the default values of gplearn for others.

Parameters	Values
Population size	4
Maximal generations number	2000
Metric (lower is better)	$\frac{\sum_{i=1}^{N} \lvert y_i - \hat{y}_i \rvert}{N}$
Stopping criterion	$\frac{\sum_{i=1}^{N} \lvert y_i - \hat{y}_i \rvert}{N} < 10^{-5}$

(continued)

(continued)

Parameters	Values
Function set	$+, -, \times$
Terminal set	$x_1, x_2, -1$

1. Please give the formula trees of the four individuals in the initial generation.
2. Please select some individuals with the tournament-selection method from the initial generation, and perform the genetic operation of crossover, mutation and reproduction between the winners.
3. Please give the final regression formula obtained from GP.

References

Amir Haeri, M., Ebadzadeh, M. M., & Folino, G. (2017). Statistical genetic programming for symbolic regression. *Applied Soft Computing, 60*, 447–469.

Burke, E., Gustafson, S., & Kendall, G. (2003). Ramped half-n-half initialisation bias in GP. In E. Cantú-Paz, et al. (Eds.), *Genetic and evolutionary computation—GECCO* 2003. *GECCO* 2003 (pp. 1800–1801). Springer.

Candes, E. J., & Wakin, M. B. (2008). An introduction to compressive sampling. *IEEE Signal Processing Magazine, 25*(2), 21–30.

de Jong, K. A. (2006). *Evolutionary computation: A unified approach*. MIT Press.

Fogel, L. J., Owens, A. J., & Walsh, M. J. (1966). *Artificial intelligence through simulated evolution*. Wiley-IEEE Press.

Fonseca, C. M., & Fleming, P. J. (1995). An overview of evolutionary algorithms in multiobjective optimization. *Evolutionary Computation, 3*(1), 1–16.

Gan, Z., Chow, T. W. S., & Chau, W. N. (2009). Clone selection programming and its application to symbolic regression. *Expert Systems with Applications, 36*(2), 3996–4005.

Ghiringhelli, L. M., Vybiral, J., & Ahmetcik, E. (2017). Learning physical descriptors for materials science by compressed sensing. *New Journal of Physics, 19*(2), 023017.

Ghiringhelli, L. M., Vybiral, J., & Levchenko, S. V. (2015). Big data of materials science: Critical role of the descriptor. *Physical Review Letters, 114*(10), 105503.

Hoai, N. X., McKay, R. I., & Essam, D. 2002. Solving the symbolic regression problem with tree-adjunct grammar guided genetic programming: The comparative results. *Proceedings of the* 2002 *Congress on Evolutionary Computation*, 2: 1326–1331.

Holland, J. H. (1992). Genetic algorithms. *Scientific American, 267*(1), 66–73.

Johnson, C. G. (2003). Artificial immune system programming for symbolic regression. In C. Ryan, T. Soule, M. Keijzer, et al. (Eds.), *Genetic programming. EeroGP* 2033 (pp. 345–353). Springer.

Koza, J. R. (1992). *Genetic programming: On the programming of computers by means of natural selection*. MIT Press.

Lu, Q., Ren, J., & Wang, Z. G. (2016). Using genetic programming with prior formula knowledge to solve symbolic regression problem. *Computational Intelligence and Neuroscience, 2*, 1–17.

Luke, S., & Panait, L. (2002). Fighting bloat with nonparametric parsimony pressure. In J. J. M. Guervós P. Adamidis, H. -G. Beyer, et al. (Eds.), *Parallel problem solving from nature—PPSN VII. PPSN* 2002 (pp. 411–421). Springer.

Luke, S., & Panait, L. (2006). A comparison of bloat control methods for genetic programming. *Evolutionary Computation, 14*(3), 309–344.

McKay, R. I., Hoai, N. X., & Whigham, P. A. (2010). Grammar-based genetic programming: A survey. *Genetic Programming and Evolvable Machines, 11*(3), 365–396.

Montana, D. J. (1995). Strongly typed genetic programming. *Evolutionary Computation, 3*(2), 199–230.

Nelson, L. J., Hart, G. L. W., & Zhou, F. (2013). Compressive sensing as a paradigm for building physics models. *Physical Review B, 87*(3), 035125.

Poli, R., & McPhee, N. F. (2008). *Parsimony pressure made easy*. ACM Press.

Poli, R., Langdon, W. B., & McPhee, N. F. (2008). *A field guide to genetic programming*. Lulu Press.

Rechenberg, I. (1965). Cybernetic solution path of an experimental problem. In Royal Aircraft Establishment Library Translation 1122.

Ryan, C., Collins, J. J. & Neill, M. O. (1998). Grammatical evolution: Evolving programs for an arbitrary language. In W. Banzhaf, R. Poli, M. Schoenauer, & T.C. Fogarty (Eds.), *Genetic programming. EuroGP* 1998. *Lecture notes in computer science*, vol 1391 (pp. 83–96). Springer.

Sette, S., & Boullart, L. (2001). Genetic programming: Principles and applications. *Engineering Applications of Artificial Intelligence, 14*(6), 727–736.

Smits, G. F., Kotanchek, M. (2005). Pareto-front exploitation in symbolic regression. In U. M. O'Reilly, T. Yu, & R. Riolo, et al. (Eds.), Genetic programming theory and practice II (pp. 283-299). Springer.

Tibshirani, R. (1996). Regression shrinkage and selection via the LASSO. *Journal of the Royal Statistical Society Series B (methodological), 73*(1), 273–282.

Vergilio, S. R., & Pozo, A. (2006). A grammar-guided genetic programming framework configured for data mining and software testing. *International Journal of Software Engineering and Knowledge Engineering, 16*, 245–267.

Chapter 9
Neural Networks

9.1 Neural Networks and Perceptron

The neural network (NN) or artificial neural network (ANN) (Hopfield, 1982; Ackley et al., 1985; McCulloch & Pitts, 1990; Tickle et al., 1998) in the field of artificial intelligence, database and machine learning is developed and inspired by the biological NNs. NNs have a layer structure, as shown in Fig. 9.1. The first layer is the input layer, the last layer is the output layer, and the layers between the input and output layers are the middle layers, called the hidden layers as well. There are neurons (nodes) in each layer and the numbers of neurons in the input and output layers are equal to the numbers of features and responses, respectively. Typically, synapses link neurons in one layer to neurons in its closest neighboring layer(s). Usually, no synapses link neurons among the same layer and no synapses link neurons in one layer to other layers, rather than its closest neighbor layer. NNs with such topology structure are called the multilayer feedforward NNs, which are introduced here. Neurons in hidden layers are mathematical (logical) operators to generate predictions. A mathematical (logical) operator uses an activation function to implement the mathematical (logical) operation and then passes the results in a weighted manner to the next layer of neurons. Neurons in the output layer gather the prediction and make the final result. If there are $(L - 1)$ hidden layers in an NN, the total number of layers of the NN will be L, meaning that the smallest NN will have only one layer even there must exist the input layer and the output layer. This might be attributed to the fact that the neurons in the input layer do not have the mathematical (logical) operation functions, while the neurons in the output layer might have or might not have the mathematical (logical) operation functions.

Let's start with some simple examples of NNs based on the perceptron. As described in Chap. 3, a perceptron binary classifier is able to find a separating hyperplane that separates data into two classes. Thus, a perceptron binary classifier can directly perform logical operations of AND, OR, and AND NOT = NAND. Figure 9.2a1 shows that logical operations AND, OR and NAND employ two neurons

T. Zhang, *An Introduction to Materials Informatics*,
https://doi.org/10.1007/978-981-99-7992-9_9

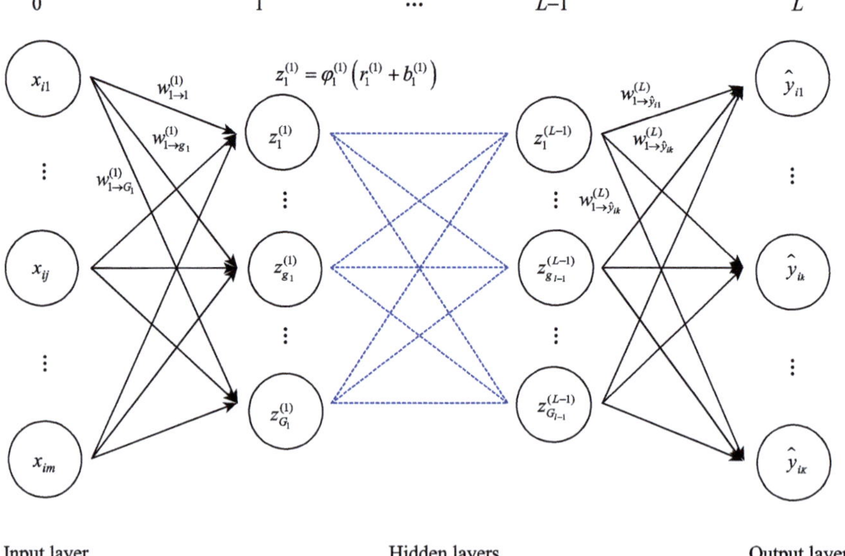

Fig. 9.1 The structure of an NN

in the input layer and one neuron in the output layer. Logical exclusive OR = XOR combines logical AND and NAND to find two separating hyperplanes simultaneously and thus has two neurons in the input layer and two neurons in the output layer, as shown in Fig. 9.2a2. To illustrate the logical operations, we use zero and one to represent false and true. Now consider a simple case of four types of events with two features, where each feature can be false (0) or true (1). The logical AND of true ($y = 1$) class is defined as that an event is true only if both features are true so that there is only one true point (1, 1) for $y = 1$. Accordingly, the logical AND of false ($y = 0$) class is defined as that one feature is false so that there are three false points of (0, 0), (1, 0) and (0, 1) in $y = 0$ class. The four points in logical AND are three false points of (0, 0), (1, 0), and (0, 1) with $y = 0$ and one true point (1, 1) with $y = 1$. The logical true AND operation is equivalent to finding a separating hyperplane $\boldsymbol{XW} + b = 0$ to separate the true point (1, 1) from the other three false points, where $\boldsymbol{X}$ is the data, $\boldsymbol{W} = (w_1, w_2)^{\mathrm{T}}$ is the normal vector of the separating hyperplane, and b is the interception, which is called the bias as well. Any points on the positive side of the separating hyperplane represent true events, i.e., $\boldsymbol{X}_i\boldsymbol{W} + b > 0$ and $y = 1$, where $\boldsymbol{X}_i = (x_{i1}, x_{i2}),\ i = 1,\ 2,\ 3,\ 4$. Any points on the negative side of the separating hyperplane represent false events, $\boldsymbol{X}_i\boldsymbol{W} + b < 0$ and $y = 0$, where $\boldsymbol{X}_i = (x_{i1}, x_{i2}),\ i = 1,\ 2,\ 3,\ 4$. Clearly, there exist many separating hyperplanes to separate the true point (1, 1) with $y = 1$ from the false points of (0, 0), (1, 0), and (0, 1) with $y = 0$, meaning that the solution is not unique, as shown in Fig. 9.2b1. The logical OR of the true ($y = 1$) class is defined as that one feature is true. Under the logical OR, the three data of (1, 0), (0, 1), and (1, 1) are in the true class $y = 1$ and the datum (0, 0) is in the false class $y = 0$. The logical OR operation is equivalent to finding a separating hyperplane $\boldsymbol{XW} + b = 0$ with three data of (1, 0), (0, 1), and (1, 1) on the right side and the datum of (0, 0) on the left side, as shown in Fig. 9.2b2. The logical NAND of true ($y = 1$) and false ($y = 0$) classes are defined as the corresponding opposites of logical OR classes and hence the logical NAND can be implemented by changing the sign of the normal vector $\boldsymbol{W}$ in the logical OR, as shown in Fig. 9.2b3. The logical XOR of true ($y = 1$) occurs if and only if one feature is true, viz., data of (1, 0) and (0, 1) are in the true class ($y = 1$) and data of (0, 0) and (1, 1) are in the false class ($y = 0$). The logical XOR operation is equivalent to finding two separating hyperplanes to establish a true region between them with all other regions identified as false. Similarly, there exist many separating hyperplanes for each of the logical operations, as shown in Fig. 9.2b4.

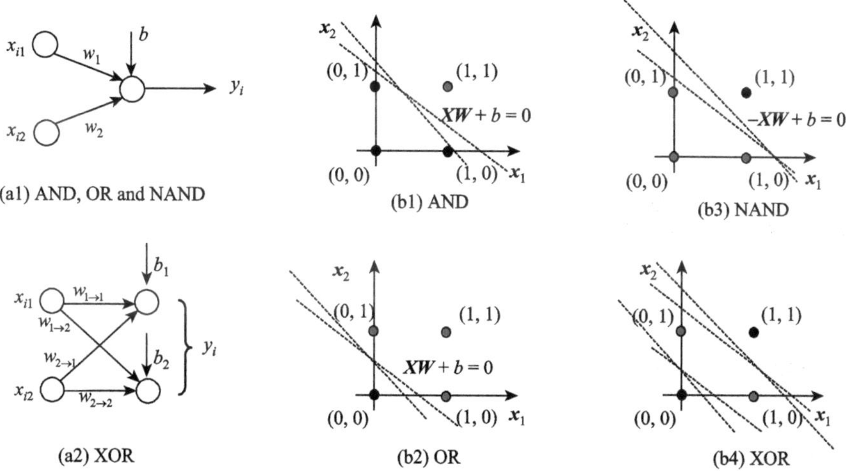

Fig. 9.2 NN structure **(a1)** for AND, OR and NAND, and **(a2)** for XOR. The illustration of logical operations of **(b1)** AND, **(b2)** OR, **(b3)** NAND, and **(b4)** XOR, where the red and black dots denote the true and false events labelled by $y = 1$ and $y = 0$, respectively

9.2 Back Propagation Algorithm

Consider a dataset $D = \{(\boldsymbol{X}_1, \boldsymbol{Y}_1), (\boldsymbol{X}_2, \boldsymbol{Y}_2), \cdots, (\boldsymbol{X}_n, \boldsymbol{Y}_n)\}$ of n data, where $\boldsymbol{X}_i = (x_{i1}, x_{i2}, \cdots, x_{im})(i = 1, 2, \cdots, n) \in \mathbf{R}^m$ denotes a $1 \times m$ input feature vector and $\boldsymbol{Y}_i = (y_{i1}, y_{i2}, \cdots, y_{iK}) \in \mathbf{R}^K$ is a $1 \times K$ response vector. Figure 9.1 schematically illustrates an NN with L layers. The first and last hidden layers are denoted by 1 and $L - 1$, respectively, and the output layer is the L-th layer. The neural position in a hidden layer is denoted by $g_l (g_l = 1, 2, \cdots, G_l)$ for layer l ($l = 1, 2, \cdots, L - 1$).

Note that the number G_l of neurons in layer l may vary or not vary from layer to layer. The received latent variable by neuron g_l is denoted by $r_{g_l}^{(l)}$. Neuron g_l, acting as an activation function or transfer function, performs the mathematical (logical) operation to the received variable $r_{g_l}^{(l)}$ and generates the latent variable $z_{g_l}^{(l)}$, which will be sent forward, through synapses in the weighted manner, to "connected" neurons in the next layer $l+1$. The forward synapses connect all neurons $g_{l-1}(g_{l-1} = 1, 2, \cdots, G_{l-1})$ in layer $l-1$, in the weighted manner, to neuron g_l in layer l. Denoting the weights by $w_{g_{l-1}\to g_l}^{(l)}$, the received variable will be

$$r_{g_l}^{(l)} = \sum_{g_{l-1}=1}^{G_{l-1}} z_{g_{l-1}}^{(l-1)} w_{g_{l-1}\to g_l}^{(l)} = Z_{g_{l-1}}^{(l-1)} W_{g_{l-1}\to g_l}^{(l)}, \tag{9.1}$$

where $Z_{g_{l-1}}^{(l-1)} = \left(z_1^{(l-1)}, z_2^{(l-1)}, \cdots, z_{G_{l-1}}^{(l-1)}\right)$ and $W_{g_{l-1}\to g_l}^{(l)} = \left(w_{1\to g_l}^{(l)}, w_{2\to g_l}^{(l)}, \cdots, w_{G_{l-1}\to g_l}^{(l)}\right)^{\mathrm{T}}$. If the activation function including a constant of $b_{g_l}^{(l)}$ in neuron g_l is expressed by $\varphi_{g_l}^{(l)}$, the generated latent variable $z_{g_l}^{(l)}$ will be

$$z_{g_l}^{(l)} = \varphi_{g_l}^{(l)}\left(r_{g_l}^{(l)} + b_{g_l}^{(l)}\right) = \varphi_{g_l}^{(l)}\left(Z_{g_{l-1}}^{(l-1)} W_{g_{l-1}\to g_l}^{(l)} + b_{g_l}^{(l)}\right) \quad (g_l = 1, 2, \cdots, G_l). \tag{9.2a}$$

The constant $b_{g_l}^{(l)}$ may function as a threshold to control the generated latent variable of neuron g_l. Each neuron in the first hidden layer receives information from all input neurons, performs the mathematical (logical) operation, and then delivers information. Thus, we have, in a similar treatment, the latent variable $z_{g_1}^{(1)}$ for neurons in the first hidden layer

$$z_{g_1}^{(1)} = \varphi_{g_1}^{(1)}\left(\boldsymbol{X} W_{x\to g_1}^{(1)} + b_{g_1}^{(1)}\right) \quad (g_1 = 1, 2, \cdots, G_1). \tag{9.2b}$$

As mentioned above, the output layer might have or not have the mathematical (logical) operation functions. For apparent consistency, the output layer is still expressed in the similar term

$$\hat{y}_k = \varphi_{y_k}^{(L-1)}\left(Z_{g_{L-1}}^{(L-1)} W_{g_{L-1}\to \hat{y}_k}^{(L)} + b_{\hat{y}_k}^{(L)}\right). \tag{9.2c}$$

In the above description, each neuron in the hidden layer and the output layer may have its own activation function. If so, too many activation functions will be used in such an NN and the neuron network will be too complicated. In practice, neurons in each layer use one common activation function with different received and generated latent variables so that $\varphi_{g_l}^{(l)}$ will be replaced by $\varphi^{(l)}(l = 1, 2, \cdots, L-1)$. The most used activation functions in NNs are listed in Table 9.1, and the plot of activation functions are shown in Fig. 9.3.

The error back propagation (BP) method (Biron, 1997; Buscema, 1998) is the common method to train an NN and determines the values of all parameters of $W_{g_{L-1}\to \hat{y}_k}^{(L)}$, $b_{\hat{y}_k}^{(L)}$, $W_{g_{l-1}\to g_l}^{(l)}$ and $b_{g_l}^{(l)}(g_l = 1, 2, \cdots, G_l,\ l = 1, 2, \cdots, L-1)$ with

Table 9.1 Commonly used activation functions

Names	Functions	Derivatives	Plots
Sigmoid	$\varphi = \frac{1}{1+\exp(-u)}$	$\frac{d\varphi}{du} = \varphi(1-\varphi)$	Figure 9.3a1, b1
tanh	$\varphi = \frac{\exp(u)-\exp(-u)}{\exp(u)+\exp(-u)}$	$\frac{d\varphi}{du} = 1-\varphi^2$	Figure 9.3a2, b2
ReLU	$\varphi = \max(0, u)$	$\frac{d\varphi}{du} = \begin{cases} 0, & u < 0 \\ \text{non}, & u = 0 \\ 1, & u > 0 \end{cases}$	Figure 9.3a3, b3

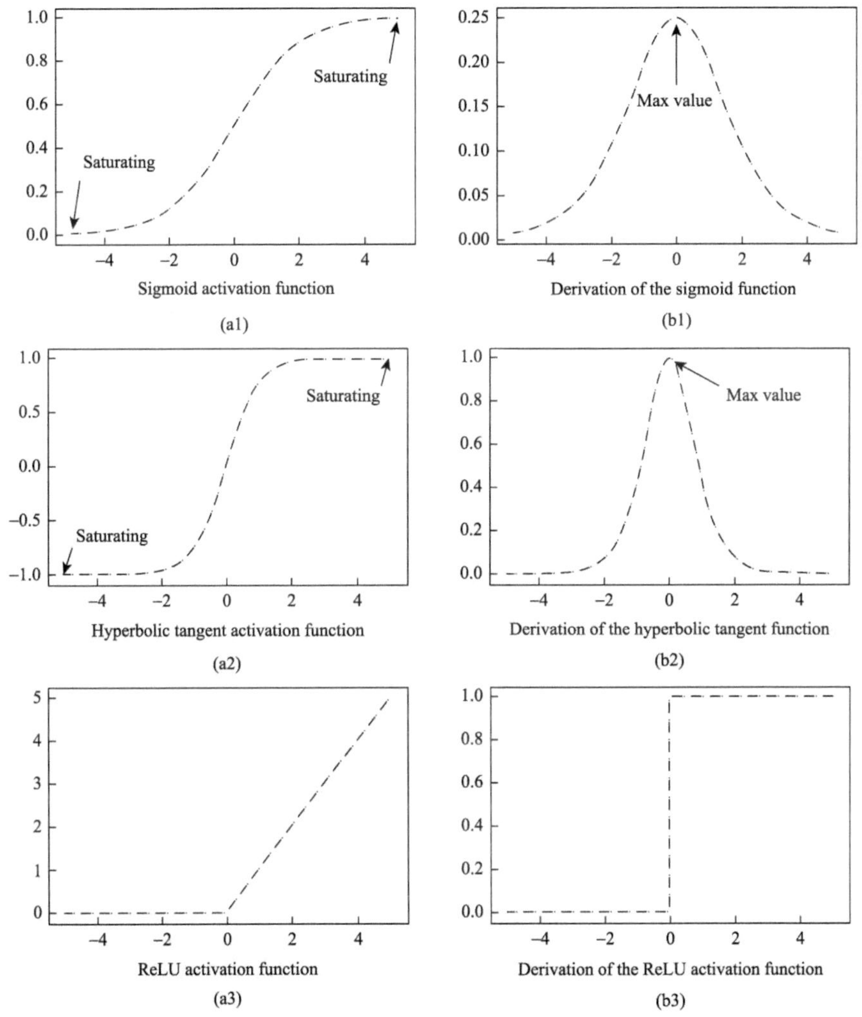

Sigmoid activation function

(a1)

Derivation of the sigmoid function

(b1)

Hyperbolic tangent activation function

(a2)

Derivation of the hyperbolic tangent function

(b2)

ReLU activation function

(a3)

Derivation of the ReLU activation function

(b3)

Fig. 9.3 Plots of commonly used activation functions and their derivation functions

a loss function. As mentioned in Chap. 2, the RSS $\mathcal{L}(\boldsymbol{W})$ is the widely used loss function for regression,

$$\mathcal{L}(\boldsymbol{W}) = \frac{1}{2}\sum_{i=1}^{n}\left[\hat{\boldsymbol{Y}}_i(\boldsymbol{X}_i, \boldsymbol{W}) - \boldsymbol{Y}_i(\boldsymbol{X}_i)\right]^2, \tag{9.3a}$$

where $\boldsymbol{W} = \left\{W^{(L)}_{g_{L-1}\to\hat{y}_k}, W^{(l)}_{g_{l-1}\to g_l}, b^{(l)}_{g_l}, b^{(L)}_{\hat{y}_k}\right\}$ $(g_l = 1, 2, \cdots, G_l; l = 1, 2, \cdots, L-1)$. Minimizing the loss function determines the values of all trainable variables,

$$\hat{\boldsymbol{W}} = \underset{\boldsymbol{W}}{\operatorname{argmin}}\, \mathcal{L}(\boldsymbol{W}). \tag{9.3b}$$

The minimization of $\mathcal{L}(\boldsymbol{W})$ requires that the gradient of $\mathcal{L}(\boldsymbol{W})$ with respect to $\boldsymbol{W}$ vanishes, i.e.,

$$\nabla\mathcal{L}(\boldsymbol{W}) = 0\,. \tag{9.4a}$$

Equation (9.3a) indicates that $\mathcal{L}(\boldsymbol{W}) = \frac{1}{2}\sum_{i=1}^{n}\mathcal{L}_i(\boldsymbol{W})$, $\mathcal{L}_i(\boldsymbol{W}) = [\hat{\boldsymbol{Y}}_i(\boldsymbol{X}_i, \boldsymbol{W}) - \boldsymbol{Y}_i(\boldsymbol{X}_i)]^2$, and hence

$$\nabla\mathcal{L}(\boldsymbol{W}) = \sum_{i=1}^{n}\nabla\mathcal{L}_i(\boldsymbol{W}) = 0\,. \tag{9.4b}$$

Solving Eq. (9.4b) yields the values of $\boldsymbol{W}$. Usually, the gradient descent method is used to solve for $\boldsymbol{W}$,

$$\boldsymbol{W}^{\text{new}} = \boldsymbol{W} - \eta\nabla\mathcal{L}(\boldsymbol{W}) = \boldsymbol{W} - \eta\sum_{i=1}^{n}\nabla\mathcal{L}_i(\boldsymbol{W}), \tag{9.5}$$

where $\eta \in (0, 1]$ is called the learning rate or the step size. Obviously, the gradient descent at a point $\boldsymbol{W}$ shows the direction how to approach the optimal $\hat{\boldsymbol{W}}$ from $\boldsymbol{W}$ and the learning rate gives how large the step in the iteration. An appropriate η value should be chosen, because a too big η value may cause oscillation, while a too small η value makes the convergence very slow. Although each step in the gradient descent method performs a linear iteration, it can handle nonlinear iterations. In the BP method, the coefficient $W^{(L)}_{g_{L-1}\to\hat{y}_k}$ and the threshold $b^{(L)}_{\hat{y}_k}$ are iterated first. Thus, for each input datum, we have

$$W^{(L)\text{new}}_{g_{L-1}\to\hat{y}_k} = W^{(L)}_{g_{L-1}\to\hat{y}_k} - \eta\frac{\partial\mathcal{L}_i}{\partial W^{(L)}_{g_{L-1}\to\hat{y}_k}}, \tag{9.6a}$$

$$b^{(L)\text{new}}_{\hat{y}_k} = b^{(L)}_{\hat{y}_k} - \eta\frac{\partial\mathcal{L}_i}{\partial b^{(L)}_{\hat{y}_k}}, \tag{9.6b}$$

with

$$\frac{\partial \mathcal{L}_i}{\partial W^{(L)}_{g_{L-1}\to \hat{y}_k}} = \sum_{k=1}^{K} \frac{\partial \mathcal{L}_i}{\partial \hat{y}_k} \frac{\partial \hat{y}_k}{\partial \varphi^{(L-1)}} \frac{\partial \varphi^{(L-1)}}{\partial W^{(L)}_{g_{L-1}\to \hat{y}_k}}, \tag{9.6c}$$

$$\frac{\partial \mathcal{L}_i}{\partial b^{(L)}_{\hat{y}_k}} = \sum_{k=1}^{K} \frac{\partial \mathcal{L}_i}{\partial \hat{y}_k} \frac{\partial \hat{y}_k}{\partial \varphi^{(L-1)}} \frac{\partial \varphi^{(L-1)}}{\partial b^{(L)}_{\hat{y}_k}}. \tag{9.6d}$$

Similarly, the coefficient $W^{(l)}_{g_{l-1}\to g_l}$ and the threshold $b^{(l)}_{g_l}(g_l = 1, 2, \cdots, G_l;\ l = 1, 2, \cdots, L-1)$ will be iterated in the BP manner with the chain rule of derivatives so that

$$W^{(l)\text{new}}_{g_{l-1}\to g_l} = W^{(l)}_{g_{l-1}\to g_l} - \eta \frac{\partial \mathcal{L}_i}{\partial W^{(l)}_{g_{l-1}\to g_l}}, \tag{9.7a}$$

$$b^{(l)\text{new}}_{g_l} = b^{(l)}_{g_l} - \eta \frac{\partial \mathcal{L}_i}{\partial b^{(l)}_{g_l}}, \tag{9.7b}$$

with

$$\frac{\partial \mathcal{L}_i}{\partial W^{(l)}_{g_{l-1}\to g_l}} = \sum_{k=1}^{K} \frac{\partial \mathcal{L}_i}{\partial \hat{y}_k} \frac{\partial \hat{y}_k}{\partial \varphi^{(L-1)}} \cdots \frac{\partial \varphi^{(l)}}{\partial W^{(l)}_{g_{l-1}\to g_l}}, \tag{9.7c}$$

$$\frac{\partial \mathcal{L}_i}{\partial b^{(l)}_{g_l}} = \sum_{k=1}^{K} \frac{\partial \mathcal{L}_i}{\partial \hat{y}_k} \frac{\partial \hat{y}_k}{\partial \varphi^{(L-1)}} \cdots \frac{\partial \varphi^{(l)}}{\partial b^{(l)}_{g_l}}. \tag{9.7d}$$

After all training data have been used once to update $W^{(L)}_{g_{L-1}\to \hat{y}_k}$, $b^{(L)}_{\hat{y}_k}$, $W^{(l)}_{g_{l-1}\to g_l}$ and $b^{(l)}_{g_l}$ $(g_l = 1, 2, \cdots, G_l;\ l = 1, 2, \cdots, L-1)$, the iteration process is conducted one round and called one epoch. The iteration process continues until convergence of all trainable variables or until reaching a preset limit of the epoch number. Obviously, the BP method is trying to find the global minimum in the space of trainable variables. It is great challenging to find the global minimum and the BP method might very often find just a local minimum or a saddle point in the worst case. There are different methods used in the iteration approach. In the whole batch method, $\nabla \mathcal{L}_i(\boldsymbol{W})(i = 1, 2, \cdots, n)$ is calculated for all n data and then $\nabla \mathcal{L}(\boldsymbol{W}) = \sum_{i=1}^{n} \nabla \mathcal{L}_i(\boldsymbol{W})$ is used in the iteration process of $\boldsymbol{W}$. The iteration accuracy of the whole batch method depends highly on the initial values of trainable variables and the whole batch method may find just a local minimum. In the mini-batch method, a random subset S of the n training data is used to calculate $\nabla \mathcal{L}(\boldsymbol{W}) = \sum_{\hat{i}=1}^{\hat{n}} \nabla \mathcal{L}_{\hat{i}}(\boldsymbol{W})$, where the subset S is composed of $\hat{n}$ data $(\hat{n} < n)$ and $\hat{n}$ is called the batch size. Various subsets, randomly formed from the training dataset, will be used in different iteration steps. A special case of this mini-batch method is when the subset S is an individual datum. If each iteration step for one trainable variable is conducted with one individual datum and different iteration steps employ different data randomly selected from the training dataset, the method is stochastic and the batch size is 1. The advantage of the mini-batch method and the stochastic method lies in the high possibility of finding the global minimum.

Example 9.1 Four data of lead-free solder alloys listed in Table 9.2 include four chemical elements of silver (Ag), copper (Cu) and bismuth (Bi) in wt.% and one property of the yield strength σ_y in MPa. The regression is conducted by the NN with the BP algorithm.

Table 9.2 Data of lead-free solder alloys

No.	Ag (wt.%)	Cu (wt.%)	Bi (wt.%)	σ_y (MPa)	Normalized $\hat{\sigma}_y$
1	3.8	0.7	0	42	0
2	3	0	2	48	1
3	3	2	0	46	0.67
4	3	4	0	43	0.17

Figure 9.4 shows the NN structure for this example with three features and one target, where there is only one hidden layer with two neurons. The detailed neural structure in the hidden layer is illustrated in Fig. 9.4 to show the neural function clearly. Neurons 1 and 2 in the hidden layer add constants $b_1^{(1)}$ and $b_2^{(1)}$ on the received variables of $r_1^{(1)}$ and $r_2^{(1)}$, respectively. Then, the same sigmoid function $\varphi(u) = \frac{1}{1+\exp(-u)}$ is used for the activation functions $\varphi_1^{(1)}$ and $\varphi_2^{(1)}$, viz., $\varphi = \varphi_1^{(1)} = \varphi_2^{(1)}$ so that $\varphi(u)$ will work on $\left(r_1^{(1)} + b_1^{(1)}\right)$ and $\left(r_2^{(1)} + b_2^{(1)}\right)$, respectively, and generate $z_1^{(1)}$ and $z_2^{(1)}$. The neuron in the output layer will receive the received variable $r_1^{(2)}$, add constant $b_1^{(2)}$, implement the linear function and deliver the output $\hat{y}_i = r_1^{(2)} + b_1^{(2)}$.

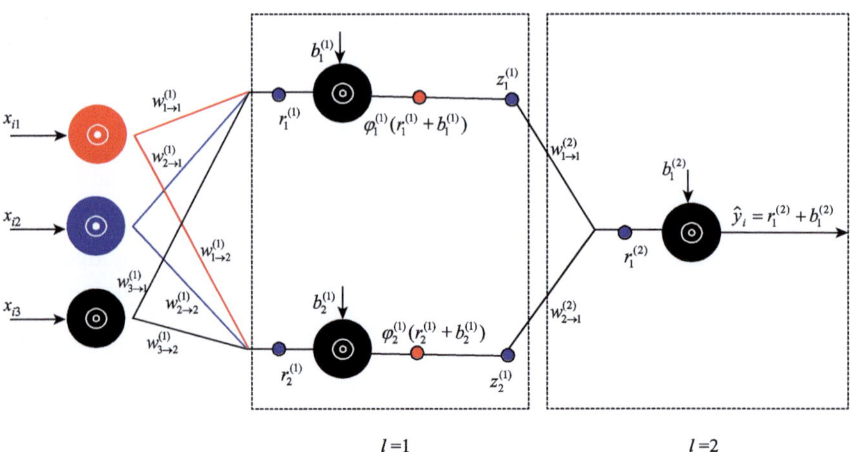

Fig. 9.4 The NN structure used in Example 9.1 ($i = 1, 2, 3, 4$)

There are 11 trainable parameters in the NN, including 8 parameters $\left\{w_{1\to1}^{(1)}, w_{1\to2}^{(1)}; w_{2\to1}^{(1)}, w_{2\to2}^{(1)}; w_{3\to1}^{(1)}, w_{3\to2}^{(1)}; b_1^{(1)}, b_2^{(1)}\right\}$ in the hidden layer and 3 parameters $\left\{w_{1\to1}^{(2)}, w_{2\to1}^{(2)}; b_1^{(2)}\right\}$ in the output layer. From Eq. (9.2b), we have

$$z_{i1}^{(1)} = \varphi_1^{(1)}\left(r_{i1}^{(1)} + b_1^{(1)}\right) \text{ and } z_{i2}^{(1)} = \varphi_2^{(1)}\left(r_{i2}^{(1)} + b_2^{(1)}\right),$$

with

$$\begin{aligned} r_{i1}^{(1)} &= w_{1\to1}^{(1)}x_{i1} + w_{2\to1}^{(1)}x_{i2} + w_{3\to1}^{(1)}x_{i3} \quad (i = 1, 2, 3, 4), \\ r_{i2}^{(1)} &= w_{1\to2}^{(1)}x_{i1} + w_{2\to2}^{(1)}x_{i2} + w_{3\to2}^{(1)}x_{i3} \quad (i = 1, 2, 3, 4). \end{aligned}$$

From Eq. (9.2c), we have

$$\hat{y}_i = w_{1\to1}^{(2)}z_{i1}^{(1)} + w_{2\to1}^{(2)}z_{i2}^{(1)} + b_1^{(2)} \quad (i = 1, 2, 3, 4).$$

The loss function $\mathcal{L}_i = \frac{1}{2}(\hat{y}_i - y_i)^2$ $(i = 1,\ 2,\ 3,\ 4)$, where $\hat{y}_i$ and y_i are the predicted and actual values, respectively. The minimization of the loss function requires

$$\nabla\mathcal{L}_i(\mathbf{W}) = (\hat{y}_i - y_i)\nabla\hat{y}_i = 0.$$

With Eqs. (9.7c) and (9.7d), we have

$$\begin{aligned} \frac{\partial\mathcal{L}_i}{\partial w_{1\to1}^{(2)}} &= (\hat{y}_i - y_i)\frac{\partial\hat{y}_i}{\partial w_{1\to1}^{(2)}} = (\hat{y}_i - y_i)z_{i1}^{(1)}, \\ \frac{\partial\mathcal{L}_i}{\partial w_{2\to1}^{(2)}} &= (\hat{y}_i - y_i)\frac{\partial\hat{y}_i}{\partial w_{2\to1}^{(2)}} = (\hat{y}_i - y_i)z_{i2}^{(1)}, \\ \frac{\partial\mathcal{L}_i}{\partial w_{j\to1}^{(1)}} &= (\hat{y}_i - y_i)\frac{\partial\hat{y}_i}{\partial z_{i1}^{(1)}}\frac{\partial z_{i1}^{(1)}}{\partial\varphi_1^{(1)}}\frac{\partial\varphi_1^{(1)}}{\partial r_{i1}^{(1)}}\frac{\partial r_{i1}^{(1)}}{\partial w_{j\to1}^{(1)}} = (\hat{y}_i - y_i)w_{1\to1}^{(2)}\frac{\partial z_{i1}^{(1)}}{\partial\varphi_1^{(1)}}\frac{\partial\varphi_1^{(1)}}{\partial r_{i1}^{(1)}}x_{ij}, \\ \frac{\partial\mathcal{L}_i}{\partial w_{j\to2}^{(1)}} &= (\hat{y}_i - y_i)\frac{\partial\hat{y}_i}{\partial z_{i2}^{(1)}}\frac{\partial z_{i2}^{(1)}}{\partial\varphi_2^{(2)}}\frac{\partial\varphi_2^{(1)}}{\partial r_{i2}^{(1)}}\frac{\partial r_{i2}^{(1)}}{\partial w_{j\to2}^{(1)}} = (\hat{y}_i - y_i)w_{2\to1}^{(2)}\frac{\partial z_{i2}^{(1)}}{\partial\varphi_2^{(1)}}\frac{\partial\varphi_2^{(1)}}{\partial r_{i2}^{(1)}}x_{ij}. \\ &(i = 1, 2, 3, 4;\ j = 1, 2, 3). \end{aligned}$$

The same sigmoid $\varphi = \frac{1}{1+\exp(-u)}$ is used here for the two activation functions and $\frac{\partial z_{i1}^{(1)}}{\partial\varphi_1^{(1)}} = 1$ and $\frac{\partial z_{i2}^{(1)}}{\partial\varphi_2^{(1)}} = 1$ due to $z_{i1}^{(1)} = \varphi_1^{(1)}\left(r_{i1}^{(1)}\right)$ and $z_{i2}^{(1)} = \varphi_2^{(1)}\left(r_{i2}^{(1)}\right)$. Thus, the gradients of the loss function are

$$\begin{cases} \frac{\partial\mathcal{L}_i}{\partial w_{1\to1}^{(1)}} = (\hat{y}_i - y_i)w_{1\to1}^{(2)}\varphi\left(r_{i1}^{(1)} + b_1^{(1)}\right)\left(1 - \varphi\left(r_{i1}^{(1)} + b_1^{(1)}\right)\right) \times x_{i1}, \\ \frac{\partial\mathcal{L}_i}{\partial w_{2\to1}^{(1)}} = (\hat{y}_i - y_i)w_{1\to1}^{(2)}\varphi\left(r_{i1}^{(1)} + b_1^{(1)}\right)\left(1 - \varphi\left(r_{i1}^{(1)} + b_1^{(1)}\right)\right) \times x_{i2}, \\ \frac{\partial\mathcal{L}_i}{\partial w_{3\to1}^{(1)}} = (\hat{y}_i - y_i)w_{1\to1}^{(2)}\varphi\left(r_{i1}^{(1)} + b_1^{(1)}\right)\left(1 - \varphi\left(r_{i1}^{(1)} + b_1^{(1)}\right)\right) \times x_{i3}, \end{cases}$$

$$\begin{cases} \frac{\partial \mathcal{L}_i}{\partial w_{1\to 2}^{(2)}} = (\hat{y}_i - y_i) w_{2\to 1}^{(2)} \varphi\left(r_{i2}^{(1)} + b_2^{(1)}\right)\left(1 - \varphi\left(r_{i2}^{(1)} + b_2^{(1)}\right)\right) \times x_{i1}, \\ \frac{\partial \mathcal{L}_i}{\partial w_{2\to 2}^{(2)}} = (\hat{y}_i - y_i) w_{2\to 1}^{(2)} \varphi\left(r_{i2}^{(1)} + b_2^{(1)}\right)\left(1 - \varphi\left(r_{i2}^{(1)} + b_2^{(1)}\right)\right) \times x_{i2}, \\ \frac{\partial \mathcal{L}_i}{\partial w_{3\to 2}^{(2)}} = (\hat{y}_i - y_i) w_{2\to 1}^{(2)} \varphi\left(r_{i2}^{(1)} + b_2^{(1)}\right)\left(1 - \varphi\left(r_{i2}^{(1)} + b_2^{(1)}\right)\right) \times x_{i3}, \end{cases}$$

$$\frac{\partial \mathcal{L}_i}{\partial b_1^{(2)}} = (\hat{y}_i - y_i)\frac{\partial \hat{y}_i}{\partial b_1^{(2)}} = \hat{y}_i - y_i,$$

$$\begin{aligned} \frac{\partial \mathcal{L}_i}{\partial b_1^{(1)}} &= (\hat{y}_i - y_i)\frac{\partial \hat{y}_i}{\partial z_{i1}^{(1)}}\frac{\partial z_{i1}^{(1)}}{\partial \varphi_1^{(1)}}\frac{\partial \varphi_1^{(1)}}{\partial b_1^{(1)}} \\ &= (\hat{y}_i - y_i) w_{1\to 1}^{(2)} \varphi\left(r_{i1}^{(1)} + b_1^{(1)}\right)\left(1 - \varphi\left(r_{i1}^{(1)} + b_1^{(1)}\right)\right), \end{aligned}$$

$$\begin{aligned} \frac{\partial \mathcal{L}_i}{\partial b_2^{(1)}} &= (\hat{y}_i - y_i)\frac{\partial \hat{y}_i}{\partial z_{i1}^{(1)}}\frac{\partial z_{i2}^{(1)}}{\partial \varphi_2^{(1)}}\frac{\partial \varphi_2^{(1)}}{\partial b_2^{(1)}} \\ &= (\hat{y}_i - y_i) w_{2\to 1}^{(2)} \varphi\left(r_{i2}^{(1)} + b_2^{(1)}\right)\left(1 - \varphi\left(r_{i2}^{(1)} + b_2^{(1)}\right)\right), \\ &(i = 1, 2, 3, 4). \end{aligned}$$

The 11 trainable parameters are updated iteratively by the gradient descent method and the mini-batch method with $\hat{n} = 1$ being employed in this example. The target data are normalized to the range of [0, 1], which are listed in Table 9.2. The normalized target data of $\hat{\sigma}_y$ might reduce the variation increment $\eta \nabla \mathcal{L}(\boldsymbol{W})$ in the iteration of $\boldsymbol{W}$, because $\nabla \mathcal{L}(\boldsymbol{W})$ contains the term $\frac{\mathrm{d}\mathcal{L}_i}{\mathrm{d}\hat{y}_i} = (\hat{y}_i - y_i)$ and the term becomes smaller. The sigmoid activation function is used in this example. As mentioned in the text, the result of training parameters depends on their initial values. If the initial values are randomly selected, the trained values of them might be of some randomness. In this example, we set the initial values manually by $w_{1\to 1}^{(1)} = w_{1\to 2}^{(1)} = w_{2\to 1}^{(1)} = w_{2\to 2}^{(1)} = w_{3\to 1}^{(1)} = w_{3\to 2}^{(1)} = b_1^{(1)} = b_2^{(1)} = w_{1\to 1}^{(2)} = w_{2\to 1}^{(2)} = b_1^{(2)} = 1$, and take the value of the learning rate to be 0.01.

As shown above, these values of $\hat{y}_i$, $\varphi'\left(r_{i1}^{(1)} + b_1^{(1)}\right)$, and $\varphi'\left(r_{i2}^{(1)} + b_2^{(1)}\right)$ must be used in the BP method, which means that they must be calculated in a manner of forward propagation (FP) with the old (initial) values in order to calculate the new values in the BP method. Feeding data x_1 =[3.8, 0.7, 0] and $y_1 = 0$ into the NN yields:

$$\begin{aligned} &w_{1\to 1}^{(1)} x_{11} + w_{2\to 1}^{(1)} x_{12} + w_{3\to 1}^{(1)} x_{13} + b_1^{(1)} = 1 \times 3.8 + 1 \times 0.7 + 1 \times 0 + 1 = 5.5, \\ &w_{1\to 2}^{(1)} x_{11} + w_{2\to 2}^{(1)} x_{12} + w_{3\to 2}^{(1)} x_{13} + b_2^{(1)} = 1 \times 3.8 + 1 \times 0.7 + 1 \times 0 + 1 = 5.5, \\ &\hat{y}_1 = w_{1\to 1}^{(2)} z_1^{(1)} + w_{2\to 1}^{(2)} z_2^{(1)} + b_1^{(2)} = 1 \times \varphi(5.5) + 1 \times \varphi(5.5) + 1 = 2.992, \end{aligned}$$

which leads to the value of the loss function

$$\mathcal{L}_1 = \frac{1}{2}(\hat{y}_1 - y_1) \cdot (\hat{y}_1 - y_1) = 4.476.$$

Take $w_{2\to2}^{(1)}$ as an example to illustrate the BP process and the others will be conducted in the same way. The gradient is calculated to be

$$\begin{aligned}\frac{\partial\mathcal{L}_1}{\partial w_{2\to2}^{(1)}} &= (\hat{y}_1 - y_1)w_{2\to1}^{(2)}\varphi\left(r_{12}^{(1)} + b_2^{(1)}\right)\left(1 - \varphi\left(r_{12}^{(1)} + b_2^{(1)}\right)\right) \times x_{12} \\ &= (2.992 - 0) \times \varphi(2.992) \times (1 - \varphi(2.992)) \times 0.7 = 0.0953.\end{aligned}$$

Then, the parameter is adjusted according to the gradient, and the new value of $w_{2\to2}^{(1)}$ is obtained,

$$w_{2\to2}^{(1)\text{new}} = w_{2\to2}^{(1)} - \frac{\partial\mathcal{L}_1}{\partial w_{2\to2}^{(1)}} \times 0.01 = 0.99905.$$

Similarly, all values of the 11 parameters are updated and the first cycle of the BP is finished. For each step, parameters are updated dependent on one individual datum. After four BP steps, all data in the dataset are used once and such process that utilized all data once is denoted as one epoch.

The iteration processes of the FP and the BP continue until convergence. In this example, the ANN model is trained by 40 epochs, and its results are listed in Table 9.3. The value of loss versus the epoch number is plotted in Fig. 9.5, which indicates that every epoch training decreases the value of the loss function and after training 40 epochs, the value of the loss function maintains almost a constant. Table 9.4 lists the trained values of variables, which change greatly from initial values.

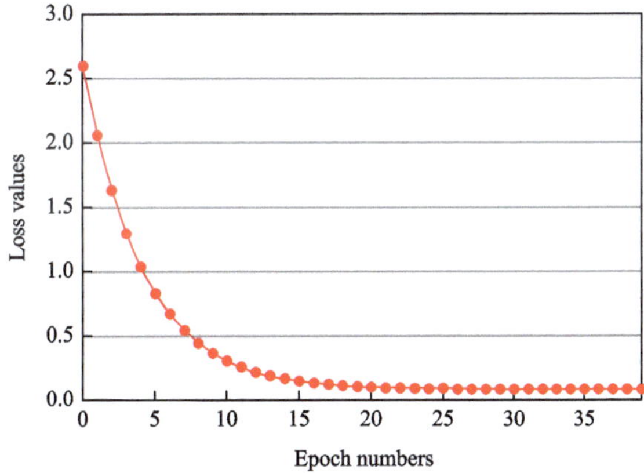

Fig. 9.5 The loss values versus epoch numbers

Table 9.3 The values of loss in each epoch

Epoch	**0**	**1**	**2**	**3**	**4**	**5**	**6**	**7**	**8**	**9**
$\mathcal{L}$	2.599230	2.055386	1.628953	1.294584	1.032399	0.826808	0.665587	0.539152	0.439992	0.362216
Epoch	**10**	**11**	**12**	**13**	**14**	**15**	**16**	**17**	**18**	**19**
$\mathcal{L}$	0.301207	0.253347	0.215799	0.186339	0.163223	0.145083	0.130847	0.119675	0.110906	0.104022
Epoch	**20**	**21**	**22**	**23**	**24**	**25**	**26**	**27**	**28**	**29**
$\mathcal{L}$	0.098619	0.094378	0.091048	0.088433	0.086381	0.084769	0.083503	0.082509	0.081728	0.081114
Epoch	**30**	**31**	**32**	**33**	**34**	**35**	**36**	**37**	**38**	**39**
$\mathcal{L}$	0.080632	0.080254	0.079956	0.079722	0.079538	0.079393	0.07928	0.07919	0.07912	0.079064

Table 9.4 The trained values of parameters

Parameters	Values	Parameters	Values
$w^{(1)}_{1\to 1}$	0.822	$w^{(2)}_{1\to 1}$	0.162
$w^{(1)}_{2\to 1}$	0.889	$w^{(2)}_{2\to 1}$	0.162
$w^{(1)}_{3\to 1}$	0.997	$b^{(1)}_{1}$	0.947
$w^{(1)}_{1\to 2}$	0.822	$b^{(1)}_{2}$	0.947
$w^{(1)}_{2\to 2}$	0.889	$b^{(2)}_{1}$	0.159
$w^{(1)}_{3\to 2}$	0.997	–	–

9.3 Regularization in NNs

9.3.1 L₁ Regularization

To avoid overfitting, many types of regularizations are added into NNs as penalty terms. Chapter 2 has provided a detailed introduction to the widely used L_1 regularization and L_2 regularization. In NNs, an L_1 regularization-like penalty term takes the form of

$$\Omega_{L_1}(\boldsymbol{W}) = \sum_{l=1}^{L} \sum_{g_{l-1}=1}^{G_{l-1}} \sum_{g_l=1}^{G_l} \left| w^{(l)}_{g_{l-1}\to g_l} \right| . \tag{9.8a}$$

With the L_1 penalty terms, the values of trainable variables are determined from the minimum of the regularized loss function

$$\hat{\boldsymbol{W}} = \underset{\boldsymbol{W}}{\operatorname{argmin}}\left[\mathcal{L}(\boldsymbol{W}) + \lambda \Omega_{L_1}(\boldsymbol{W})\right], \tag{9.8b}$$

where λ is a positive hyperparameter. The value of the hyperparameter λ determines the weight of the L_1 penalty term in the loss function. Obviously, sufficiently large values of the hyperparameter λ make all trainable variables zero and as the λ value decreases, components of $\hat{\boldsymbol{W}}$ will appear one by one. In practice, values of $\hat{\boldsymbol{W}}$ are determined from a training dataset with a preset value of λ and the optimal value of the hyperparameter λ is determined by cross-validation. The advantage of L_1 regularization lies in that a few and important trainable variables are selected from a large pool of variables so that an NN with L_1 regularization performs better than the same NN without L_1 regularization due to the redundant trainable variables. An NN with L_1 regularization also performs better than the most similar NNs with the same number of trainable variables as that in the NN with L_1 regularization, because the NN with L_1 regularization selects and uses the most important trainable variables.

Example 9.2 Take the 20 NIMS fatigue data in Table 2.4 of Chap. 2, as an example, to illustrate the NN with L_1 regularization. Only consider one property of tensile strength and five features of Cr, Mo, TT, dB, and S in this example. Figure 9.6 shows the NN structure for this example with five features and one target, and only one hidden layer of two neurons. Neurons 1 and 2 in the later hidden layer add constants $b_1^{(1)}$ and $b_2^{(1)}$ on the received variables of $r_1^{(1)}$ and $r_2^{(1)}$, respectively. The same sigmoid function $\varphi = \frac{1}{1+\exp(-u)}$ is used for the activation functions $\varphi_1^{(1)}$ and $\varphi_2^{(1)}$, viz., $\varphi = \varphi_1^{(1)} = \varphi_2^{(1)}$ so that $\varphi(u)$ will work on $\left(r_1^{(1)} + b_1^{(1)}\right)$ and $\left(r_2^{(1)} + b_2^{(1)}\right)$ and generate $z_1^{(1)}$ and $z_2^{(1)}$, respectively. The neuron in the output layer will receive the variable $r_1^{(2)}$, add constant $b_1^{(2)}$, implement the linear function and deliver the output $\hat{y}_i = r_1^{(2)} + b_1^{(2)}$.

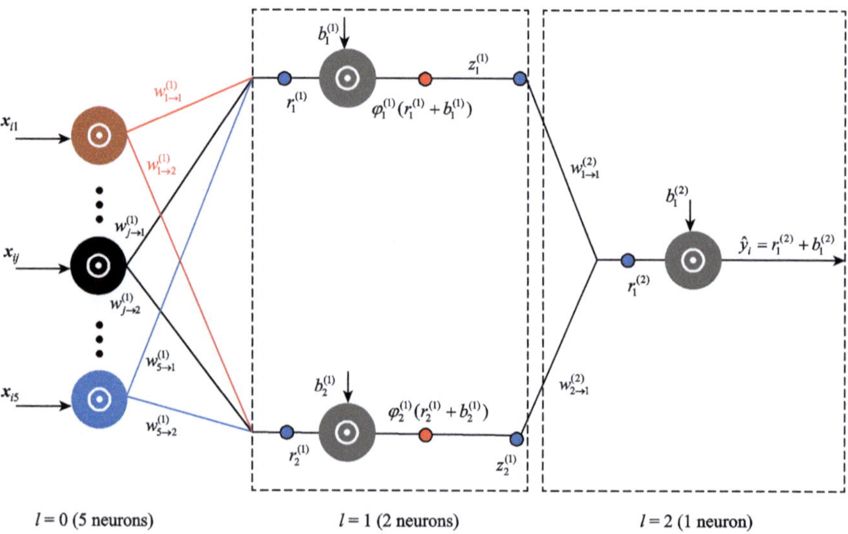

Fig. 9.6 The NN structure used in Example 9.2 ($i = 1, 2, \ldots, 20; j = 1, 2, \ldots, 5$)

There are 15 trainable variables in the NN, including 12 trainable variables $\left\{\left(w_{1\to1}^{(1)}, w_{1\to2}^{(1)}\right), \left(w_{2\to1}^{(1)}, \cdots, w_{5\to2}^{(1)}\right); b_1^{(1)}, b_2^{(1)}\right\}$ in the hidden layer and 3 trainable variables $\left\{w_{1\to1}^{(2)}, w_{2\to1}^{(2)}; b_1^{(2)}\right\}$ in the output layer. From Eq. (9.2b), we have

$$z_{i1}^{(1)} = \varphi_1^{(1)}\left(r_{i1}^{(1)} + b_1^{(1)}\right) \text{ and } z_{i2}^{(1)} = \varphi_2^{(1)}\left(r_{i2}^{(1)} + b_2^{(1)}\right),$$

with

$$r_{i1}^{(1)} = w_{1\to1}^{(1)}x_{i1} + \cdots + w_{5\to1}^{(1)}x_{i5} \quad (i = 1, 2, \cdots, 20),$$
$$r_{i2}^{(1)} = w_{1\to2}^{(1)}x_{i1} + \cdots + w_{5\to2}^{(1)}x_{i5} \quad (i = 1, 2, \cdots, 20).$$

From Eq. (9.2c), we have

$$\hat{y}_i = w^{(2)}_{1\to 1} z^{(1)}_{i1} + w^{(2)}_{2\to 1} z^{(1)}_{i2} + b^{(2)}_1 \quad (i = 1, 2, \cdots, 20).$$

The loss function $\mathcal{L}_i = \frac{1}{2}(\hat{y}_i - y_i)^2 + \lambda \sum_{l=1}^{L} \sum_{g_{l-1}=1}^{G_{l-1}} \sum_{g_l=1}^{G_l} \left| w^{(l)}_{g_{l-1}\to g_l} \right|$, where $\hat{y}_i$ and y_i are the predicted and actual values, respectively, and $w^{(l)}_{g_{l-1}\to g_l}$ is the weight of forward synapses between the neuron g_{l-1} of layer $l-1$ and the neuron g_l of layer l. The minimization of the loss function requires

$$\sum \nabla \mathcal{L}_i(\boldsymbol{W}) = \sum\left((\hat{y}_i - y_i)\nabla \hat{y}_i + \nabla\left(\lambda \sum_{l=1}^{L} \sum_{g_{l-1}=1}^{G_{l-1}} \sum_{g_l=1}^{G_l} \left| w^{(l)}_{g_{l-1}\to g_l} \right| \right)\right) = 0. \quad (9.9)$$

The penalty term Eq. (9.8a) is the sum of 12 absolute weights, i.e.,

$$\begin{aligned} \Omega_{L_1}(\boldsymbol{W}) &= \sum_{l=1}^{L} \sum_{g_{l-1}=1}^{G_{l-1}} \sum_{g_l=1}^{G1} \left| w^{(l)}_{g_{l-1}\to g_l} \right| \\ &= \left| w^{(1)}_{1\to 1} \right| + \left| w^{(1)}_{1\to 2} \right| + \left| w^{(1)}_{2\to 1} \right| + \left| w^{(1)}_{2\to 2} \right| + \left| w^{(1)}_{3\to 1} \right| \\ &\quad + \left| w^{(1)}_{3\to 2} \right| + \left| w^{(1)}_{4\to 1} \right| + \left| w^{(1)}_{4\to 2} \right| + \left| w^{(1)}_{5\to 1} \right| + \left| w^{(1)}_{5\to 2} \right| \\ &\quad + \left| w^{(2)}_{1\to 1} \right| + \left| w^{(2)}_{2\to 1} \right|. \end{aligned}$$

Let's define the parameter $I\left(w^{(l)}_{g_{l-1}\to g_l}\right) = \frac{\partial \Omega_{L_1}}{\partial w^{(l)}_{g_{l-1}\to g_l}}$. Obviously, if $w^{(l)}_{g_{l-1}\to g_l} \geq 0$, $I = 1$, and if $w^{(l)}_{g_{l-1}\to g_l} < 0$, $I = -1$. With Eqs. (9.7c) and (9.7d), we have

$$\begin{aligned} \frac{\partial \mathcal{L}_i}{\partial w^{(2)}_{1\to 1}} &= (\hat{y}_i - y_i)\frac{\partial \hat{y}_i}{\partial w^{(2)}_{1\to 1}} + I \times \lambda = (\hat{y}_i - y_i) z^{(1)}_{i1} + I \times \lambda, \\ \frac{\partial \mathcal{L}_i}{\partial w^{(2)}_{2\to 1}} &= (\hat{y}_i - y_i)\frac{\partial \hat{y}_i}{\partial w^{(2)}_{2\to 1}} + I \times \lambda = (\hat{y}_i - y_i) z^{(1)}_{i2} + I \times \lambda, \\ \frac{\partial \mathcal{L}_i}{\partial w^{(1)}_{j\to g_l}} &= (\hat{y}_i - y_i)\frac{\partial \hat{y}_i}{\partial z^{(1)}_{ig_l}} \frac{\partial z^{(1)}_{ig_l}}{\partial \varphi^{(1)}_{g_l}} \frac{\partial \varphi^{(1)}_{g_l}}{\partial r^{(1)}_{ig_l}} \frac{\partial r^{(1)}_{ig_l}}{\partial w^{(1)}_{j\to g_l}} + I \times \lambda \\ &= (\hat{y}_i - y_i) w^{(2)}_{g_1\to 1} \frac{\partial z^{(1)}_{ig_l}}{\partial \varphi^{(1)}_{g_l}} \frac{\partial \varphi^{(1)}_{g_l}}{\partial r^{(1)}_{ig_l}} x_{ij} + I \times \lambda \\ &\quad (j = 1, 2, \cdots, 5;\ g_l = 1, 2). \end{aligned}$$

The same sigmoid function $\varphi(u) = \frac{1}{1+\exp(-u)}$ is used here for all activation functions, and hence again $\frac{\partial z^{(1)}_{ig_1}}{\partial \varphi^{(1)}_{g_1}} = 1$ due to $z^{(1)}_{ig_1} = \varphi^{(1)}_{g_1}\left(r^{(1)}_{ig_1}\right)$. Thus, the gradients of the loss function are

$$\frac{\partial \mathcal{L}_i}{\partial w_{j\to g_1}^{(1)}} = (\hat{y}_i - y_i) w_{g_1\to 1}^{(2)} \varphi\left(r_{ig_1}^{(1)} + b_{g_1}^{(1)}\right)\left(1 - \varphi\left(r_{ig_1}^{(1)} + b_{g_1}^{(1)}\right)\right) x_{ij} + I \times \lambda,$$

$$\frac{\partial \mathcal{L}_i}{\partial b_1^{(2)}} = (\hat{y}_i - y_i)\frac{\partial \hat{y}_i}{\partial b_1^{(2)}} = (\hat{y}_i - y_i),$$

$$\begin{aligned}\frac{\partial \mathcal{L}_i}{\partial b_{g_1}^{(1)}} &= (\hat{y}_i - y_i)\frac{\partial \hat{y}_i}{\partial z_{ig_1}^{(1)}}\frac{\partial z_{ig_1}^{(1)}}{\partial \varphi_{g_1}^{(1)}}\frac{\partial \varphi_{g_1}^{(1)}}{\partial b_{g_1}^{(1)}} \\ &= (\hat{y}_i - y_i) w_{g_1\to 1}^{(2)} \varphi\left(r_{ig_1}^{(1)} + b_{g_1}^{(1)}\right)\left(1 - \varphi\left(r_{ig_1}^{(1)} + b_{g_1}^{(1)}\right)\right) \\ &\quad (i = 1, 2, \cdots, 20).\end{aligned} \tag{9.10}$$

The 15 trainable variables are solved iteratively with the gradient descent method. The data are normalized to the range of (0, 1), which are listed in Table 9.5. In this example, we use leave-one-out cross-validation (LOOCV) to optimize the regularization hyperparameter λ with the initial value of $\lambda = 1$. In this example, the learning rate η is 0.01, the epoch number is set to be 2000, and the mini-batch method is employed, with five batches each containing about four data.

Table 9.5 Twenty NIMS data after being normalized to the range of (0, 1) with one property and five features

No.	Cr (x_1)	Mo (x_2)	TT (x_3)	dB (x_4)	S (x_5)	Tensile strength (y)
1	0.66	0.00	0.38	0.25	0.29	0.48
2	0.92	0.89	0.00	0.00	0.71	0.98
3	0.92	1.00	0.38	0.00	0.19	0.65
4	0.73	1.00	0.62	0.00	0.29	0.67
5	0.00	0.00	0.77	1.00	0.90	0.00
6	0.18	0.00	0.00	0.00	0.52	0.33
7	0.88	0.94	0.77	0.00	0.00	0.57
8	0.05	0.00	0.00	0.00	0.24	0.46
9	0.92	0.94	0.00	0.00	0.05	1.00
10	0.08	0.00	0.00	0.00	0.71	0.42
11	0.08	0.00	0.77	0.00	0.81	0.21
12	0.03	0.00	0.77	0.50	1.00	0.22
13	0.94	0.89	0.38	0.00	0.24	0.69
14	0.72	0.00	0.00	0.00	0.24	0.72
15	0.92	0.89	0.38	0.00	0.71	0.70
16	0.99	1.00	0.00	0.25	0.52	0.97
17	0.73	1.00	1.00	0.00	0.29	0.45
18	0.20	0.00	0.00	0.25	0.19	0.49
19	1.00	0.89	0.38	0.00	0.38	0.82
20	0.11	0.00	0.00	0.00	0.57	0.55

The initial values of weights are randomly selected from a series of numbers conforming to the normal distribution $N(0,\ 1)$ so that

$$\begin{bmatrix} w_{1\to 1}^{(1)} & w_{2\to 1}^{(1)} & w_{3\to 1}^{(1)} & w_{4\to 1}^{(1)} & w_{5\to 1}^{(1)} \\ w_{1\to 2}^{(1)} & w_{2\to 2}^{(1)} & w_{3\to 2}^{(1)} & w_{4\to 2}^{(1)} & w_{5\to 2}^{(1)} \end{bmatrix}_{2\times 5} = \begin{bmatrix} 0.53 & -0.89 & -0.13 & 0.65 & 0.62 \\ -0.51 & 0.83 & 0.03 & -0.03 & -0.91 \end{bmatrix}_{2\times 5},$$

$$\begin{bmatrix} w_{1\to 1}^{(2)} & w_{2\to 1}^{(2)} \end{bmatrix}_{1\times 2} = [0.53\ -0.51]_{1\times 2},$$

$$\begin{bmatrix} b_1^{(1)} \\ b_2^{(1)} \end{bmatrix}_{2\times 1} = \begin{bmatrix} 0.53 \\ -0.51 \end{bmatrix}_{2\times 1},$$

$$\begin{bmatrix} b_1^{(2)} \end{bmatrix}_{1\times 1} = [0.53]_{1\times 1}.$$

As an example, the calculation steps are shown below for the first cycle of LOOCV in epoch one of the first batch with data 2–5 for training and datum 1 for validation. Letting $\hat{\boldsymbol{y}} = (\hat{y}_2, \hat{y}_3, \hat{y}_4, \hat{y}_5)^{\mathrm{T}}$, $\boldsymbol{u}_{g_1}^{(1)} = \left(r_{2g_1}^{(1)} + b_{g_1}^{(1)}, \cdots, r_{5g_1}^{(1)} + b_{g_1}^{(1)}\right)^{\mathrm{T}}$, $\boldsymbol{z}_{g_1}^{(1)} = \varphi\left(\boldsymbol{u}_{g_1}^{(1)}\right)$, $\boldsymbol{x}_j = (x_{2j}, x_{3j}, x_{4j}, x_{5j})^{\mathrm{T}}$ $(g_1 = 1, 2;\ j = 1, 2, \cdots, 5)$, we conduct the calculations in a matrix and vector manner.

The data matrix of data 2–5 is $(\boldsymbol{X}, \boldsymbol{y}) = \begin{bmatrix} x_{2,1} & \cdots & x_{2,5} & y_2 \\ x_{3,1} & \cdots & x_{3,5} & y_3 \\ x_{4,1} & \cdots & x_{4,5} & y_4 \\ x_{5,1} & \cdots & x_{5,5} & y_5 \end{bmatrix} = \begin{bmatrix} 0.92 & \cdots & 0.71 & 0.98 \\ 0.92 & \cdots & 0.19 & 0.65 \\ 0.73 & \cdots & 0.29 & 0.67 \\ 0.00 & \cdots & 0.90 & 0.00 \end{bmatrix}.$

With the initial values of weights, we have

$$w_{1\to 1}^{(1)} x_1 + \cdots + w_{5\to 1}^{(1)} x_5 + b_1^{(1)} = 0.53 \times \begin{bmatrix} 0.92 \\ 0.92 \\ 0.73 \\ 0.00 \end{bmatrix} + \cdots + 0.62 \times \begin{bmatrix} 0.71 \\ 0.19 \\ 0.29 \\ 0.90 \end{bmatrix}$$

$$+ 0.53 \times \begin{bmatrix} 1 \\ 1 \\ 1 \\ 1 \end{bmatrix} = \begin{bmatrix} 0.68 \\ 0.21 \\ 0.13 \\ 1.65 \end{bmatrix},$$

$$w_{1\to 2}^{(1)}x_1 + \cdots + w_{5\to 2}^{(1)}x_5 + b_2^{(1)} = -0.51 \times \begin{bmatrix} 0.92 \\ 0.92 \\ 0.73 \\ 0.00 \end{bmatrix} + \cdots - 0.91 \times \begin{bmatrix} 0.71 \\ 0.19 \\ 0.29 \\ 0.90 \end{bmatrix} - 0.51 \times \begin{bmatrix} 1 \\ 1 \\ 1 \\ 1 \end{bmatrix} = \begin{bmatrix} -0.26 \\ -0.15 \\ -0.04 \\ -0.48 \end{bmatrix}.$$

$$\hat{y} = w_{1\to 1}^{(2)} z_1^{(1)} + w_{2\to 1}^{(2)} z_2^{(1)} + b_1^{(2)}$$
$$= 0.53 \times \varphi\left(\begin{bmatrix} 0.68 \\ 0.21 \\ 0.13 \\ 1.65 \end{bmatrix}\right) - 0.51 \times \varphi\left(\begin{bmatrix} -0.26 \\ -0.15 \\ -0.04 \\ -0.48 \end{bmatrix}\right) + 0.53 \times \begin{bmatrix} 1 \\ 1 \\ 1 \\ 1 \end{bmatrix} = \begin{bmatrix} 0.66 \\ 0.59 \\ 0.56 \\ 0.78 \end{bmatrix},$$

which leads to the value of the loss function

$$\mathcal{L}^0 = \frac{1}{2}(\hat{y} - y)^{\mathrm{T}}(\hat{y} - y) + \lambda \sum_{l=1}^{L} \sum_{g_{l-1}=1}^{G_{l-1}} \sum_{g_l=1}^{G_l} \left| w_{g_{l-1}\to g_l}^{(l)} \right|$$
$$= \frac{1}{2} \begin{bmatrix} 0.66 - 0.98 \\ 0.59 - 0.65 \\ 0.56 - 0.67 \\ 0.78 - 0.00 \end{bmatrix}^{\mathrm{T}} \begin{bmatrix} 0.66 - 0.98 \\ 0.59 - 0.65 \\ 0.56 - 0.67 \\ 0.78 - 0.00 \end{bmatrix} + 1 \times 6.17 = 6.53.$$

Take $w_{1\to 1}^{(1)}$ as an example to illustrate the BP process and the others will be conducted in the same way. The average loss is given by $\overline{\mathcal{L}} = \frac{1}{4}\sum_{i=2}^{5} \mathcal{L}_i$. According to Eq. (9.10), the gradient of average loss is calculated to be

$$\begin{bmatrix} \frac{\partial \mathcal{L}_2}{\partial w_{1\to 1}^{(1)}} \\ \frac{\partial \mathcal{L}_3}{\partial w_{1\to 1}^{(1)}} \\ \frac{\partial \mathcal{L}_4}{\partial w_{1\to 1}^{(1)}} \\ \frac{\partial \mathcal{L}_5}{\partial w_{1\to 1}^{(1)}} \end{bmatrix} = (\hat{y}_i - y_i) w_{1\to 1}^{(2)} \varphi\left(u_1^{(1)}\right)\left(1 - \varphi\left(u_1^{(1)}\right)\right) \times x_1 \times 1_4 + I \times \lambda$$

$$
\begin{aligned}
&= \begin{bmatrix} 0.66-0.98 \\ 0.59-0.65 \\ 0.56-0.67 \\ 0.78-0.00 \end{bmatrix} \times 0.53 \odot \varphi\left(\begin{bmatrix} 0.68 \\ 0.21 \\ 0.13 \\ 1.65 \end{bmatrix}\right) \odot \left(\begin{bmatrix} 1 \\ 1 \\ 1 \\ 1 \end{bmatrix} - \varphi\left(\begin{bmatrix} 0.68 \\ 0.21 \\ 0.13 \\ 1.65 \end{bmatrix}\right)\right) \\
&\quad \odot \begin{bmatrix} 0.92 \\ 0.92 \\ 0.73 \\ 0.00 \end{bmatrix} + 1 \times \lambda \times \begin{bmatrix} 1 \\ 1 \\ 1 \\ 1 \end{bmatrix} \\
&= \begin{bmatrix} -0.009+\lambda \\ -0.002+\lambda \\ -0.003+\lambda \\ -0.000+\lambda \end{bmatrix},
\end{aligned}
$$

$$
\begin{aligned}
\frac{\partial \overline{\mathcal{L}}}{\partial w_{1\to 1}^{(1)}} &= \frac{\frac{\partial \mathcal{L}_2}{\partial w_{1\to 1}^{(1)}} + \frac{\partial \mathcal{L}_3}{\partial w_{1\to 1}^{(1)}} + \frac{\partial \mathcal{L}_4}{\partial w_{1\to 1}^{(1)}} + \frac{\partial \mathcal{L}_5}{\partial w_{1\to 1}^{(1)}}}{4} \\
&= \frac{(-0.009+\lambda)+(-0.002+\lambda)+(-0.003+\lambda)+(-0.000+\lambda)}{4} \\
&= -0.0035+\lambda.
\end{aligned}
$$

For the mini-batch method, the parameters are adjusted according to the average gradient of data 2–5 in the first iteration. Then, the new value of $w_{1\to 1}^{(1)}$ is obtained,

$$
w_{1\to 1}^{(1)\text{new}} = w_{1\to 1}^{(1)} - \eta \times \frac{\partial \overline{\mathcal{L}}}{\partial w_{1\to 1}^{1}} = 0.53 - 0.01 \times (-0.0035+\lambda) = 0.53 - 0.01\lambda.
$$

Similarly, all values of the 15 trainable variables are updated and the first epoch of training is finished.

Figure 9.7 shows the variations of weights with the value of λ decrease. When the hyperparameter λ is sufficiently large, the weights of all synapse links in the NN are zero. As the value of λ decreases (from left to right on the horizontal axis), the weights appear one by one. When none of the weights of synapse links is zero, the NN changes from a non-fully connected network to a fully connected network. When λ is smaller than a critical value, the values of weights do not change any more, indicating that the L_1 penalty term stops working and the regularized NN degrades into an ordinary NN. Table 9.6 shows the appearing orders of weights as the value of λ decreases from 10^{-1} to 10^{-6}.

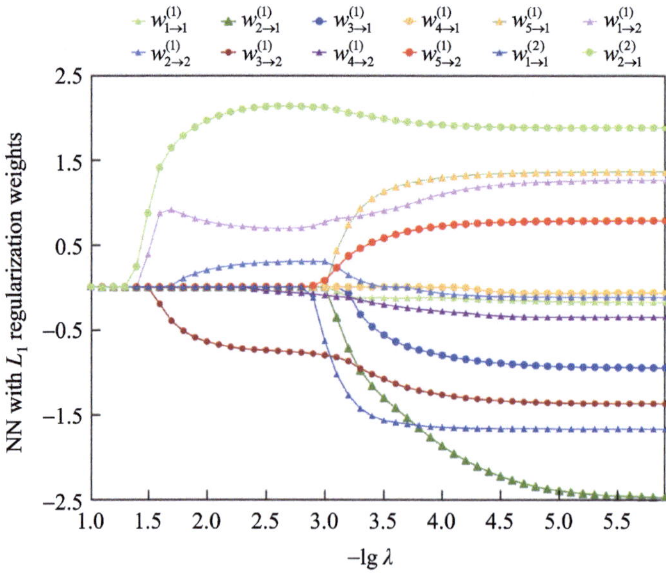

Fig. 9.7 The variations of weights with the value of λ in the L_1 NN

In this example, we use LOOCV to optimize the regularization hyperparameter λ, and the optimal value of λ is determined by the maximum LOOCV-*R*, as shown in Fig. 9.8a (blue line). Figure 9.8a (blue line) shows the LOOCV-*R* value of the tensile properties varies with the value of λ, where the Pearson correlation coefficient *R* is calculated by $R = \frac{\sum_{i=1}^{n}(y_i - \overline{y})(\hat{y}_i - \overline{\hat{y}})}{\sqrt{\sum_{i=1}^{n}(y_i - \overline{y})^2 \sum_{i=1}^{n}(\hat{y}_i - \overline{\hat{y}})^2}}$. Figure 9.8a (red line) shows the number of zero weights of synapse links (average number of LOOCV) varies with the value of λ. Figure 9.8b shows the measured values of tensile strength against the predicted values of LOOCV with the optimal λ.

Table 9.6 The appearing orders of weights as the value of λ decreases from 10^{-1} to 10^{-6}

Orders	1	2	3	4	5	6	7	8	9	10	11	12
Weights	$w^{(2)}_{2\to1}$	$w^{(1)}_{1\to2}$	$w^{(1)}_{3\to2}$	$w^{(1)}_{2\to2}$	$w^{(1)}_{4\to2}$	$w^{(1)}_{5\to2}$	$w^{(2)}_{1\to1}$	$w^{(1)}_{5\to1}$	$w^{(1)}_{2\to1}$	$w^{(1)}_{1\to1}$	$w^{(1)}_{3\to1}$	$w^{(1)}_{4\to1}$

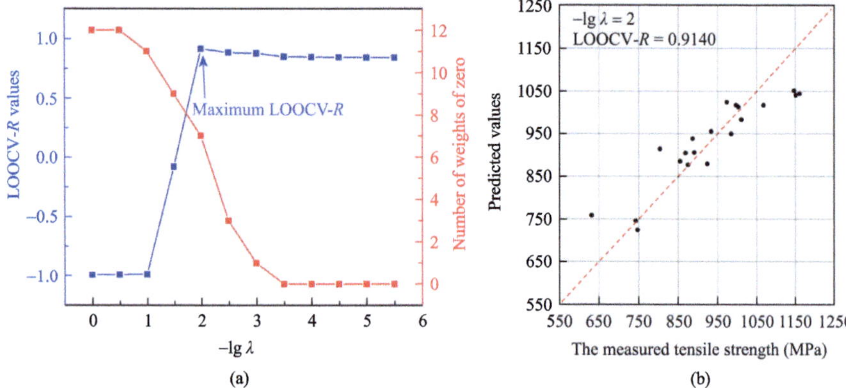

Fig. 9.8 **(a)** The NN model with L_1 regularization for different λ values; **(b)** Predicted values, with the optimal value of λ against the measured tensile strength

---

9.3.2 L₂ *Regularization*

A L_2 regularization-like penalty term can also be added into an NN, which takes the form of

$$\Omega_{L_2}(\boldsymbol{W}) = \sum_{l=1}^{L} \sum_{g_{l-1}=1}^{G_{l-1}} \sum_{g_l=1}^{G_l} \left(w_{g_{l-1}\to g_l}^{(l)}\right)^2. \tag{9.11a}$$

With the L_2 penalty terms, the values of trainable variables are determined from the minimum of the regularized loss function

$$\hat{\boldsymbol{W}} = \underset{\boldsymbol{W}}{\operatorname{argmin}}\left[\mathcal{L}(\boldsymbol{W}) + \lambda \Omega_{L_2}(\boldsymbol{W})\right], \tag{9.11b}$$

where λ is a positive hyperparameter. Larger values of λ will tend to shrink the trainable variables toward zero, and cross-validation is used to optimize λ.

Example 9.3 Take the 20 normalized NIMS data in Table 9.5 to conduct NN with L_2 regularization. The NN structure and the model hyperparameter (excluding λ) of this example are the same as those used in Example 9.2. The minimization of the loss function requires

$$\sum \nabla \mathcal{L}_i(\boldsymbol{W}) = \sum\left((\hat{y}_i - y_i)\nabla \hat{y}_i + \nabla\left(\lambda \sum_{l=1}^{L} \sum_{g_{l-1}=1}^{G_{l-1}} \sum_{g_l=1}^{G_l} \left(w_{g_{l-1}\to g_l}^{(l)}\right)^2 \right)\right) = 0. \tag{9.12}$$

The penalty term Eq. (9.11a) is the sum of 12 square weights, i.e.,

$$\begin{aligned}\Omega_{L_2}(\boldsymbol{W}) &= \sum_{l=1}^{L}\sum_{g_{l-1}=1}^{G_{l-1}}\sum_{g_1=1}^{G_1}\left(w_{g_{l-1}\to g_1}^{(l)}\right)^2 \\ &= \left(w_{1\to1}^{(1)}\right)^2+\left(w_{1\to2}^{(1)}\right)^2+\left(w_{2\to1}^{(1)}\right)^2+\left(w_{2\to2}^{(1)}\right)^2 \\ &\quad+\left(w_{3\to1}^{(1)}\right)^2+\left(w_{3\to2}^{(1)}\right)^2+\left(w_{4\to1}^{(1)}\right)^2+\left(w_{4\to2}^{(1)}\right)^2+\left(w_{5\to1}^{(1)}\right)^2 \\ &\quad+\left(w_{5\to2}^{(1)}\right)^2+\left(w_{1\to1}^{(2)}\right)^2+\left(w_{2\to1}^{(2)}\right)^2.\end{aligned}$$

With Eqs. (9.7c) and (9.7d), we have the gradient expressions of

$$\frac{\partial\Omega_{L_2}}{\partial w_{g_{l-1}\to g_l}^{(l)}} = 2\times w_{g_{l-1}\to g_l}^{(l)},$$

$$\begin{aligned}
\frac{\partial\mathcal{L}_i}{\partial w_{1\to1}^{(2)}} &= (\hat{y}_i-y_i)\frac{\partial\hat{y}_i}{\partial w_{1\to1}^{(2)}}+2\times w_{1\to1}^{(2)}\times\lambda = (\hat{y}_i-y_i)Z_{i1}^{(1)}+2\times w_{1\to1}^{(2)}\times\lambda,\\
\frac{\partial\mathcal{L}_i}{\partial w_{2\to1}^{(2)}} &= (\hat{y}_i-y_i)\frac{\partial\hat{y}_i}{\partial w_{2\to1}^{(2)}}+2\times w_{2\to1}^{(2)}\times\lambda = (\hat{y}_i-y_i)Z_{i2}^{(1)}+2\times w_{2\to1}^{(2)}\times\lambda,\\
\frac{\partial\mathcal{L}_i}{\partial w_{j\to g_1}^{(1)}} &= (\hat{y}_i-y_i) = \frac{\partial\hat{y}_i}{\partial z_{ig_1}^{(1)}}\frac{\partial z_{ig_1}^{(1)}}{\partial\varphi_{g_1}^{(1)}}\frac{\partial\varphi_{g_1}^{(1)}}{\partial r_{ig_1}^{(1)}}\frac{\partial r_{ig_1}^{(1)}}{\partial w_{j\to g_1}^{(1)}}+2\times w_{j\to g_1}^{(1)}\times\lambda\\
&= (\hat{y}_i-y_i)w_{g_1\to1}^{(2)}\frac{\partial z_{ig_1}^{(1)}}{\partial\varphi_{g_1}^{(1)}}\frac{\partial\varphi_{g_1}^{(1)}}{\partial r_{ig_1}^{(1)}}x_{ij}+2\times w_{j\to g_1}^{(1)}\times\lambda\\
&= (\hat{y}_i-y_i)w_{g_1\to1}^{(2)}\varphi\left(r_{ig_1}^{(1)}+b_{g_1}^{(1)}\right)\left(1-\varphi\left(r_{ig_1}^{(1)}+b_{g_1}^{(1)}\right)\right)x_{ij}+2\times w_{j\to g_1}^{(1)}\times\lambda,\\
\frac{\partial\mathcal{L}_i}{\partial b_1^{(2)}} &= (\hat{y}_i-y_i)\frac{\partial\hat{y}_i}{\partial b_1^{(2)}} = \hat{y}_i-y_i,\\
\frac{\partial\mathcal{L}_i}{\partial b_{g_1}^{(1)}} &= (\hat{y}_i-y_i)\frac{\partial\hat{y}_i}{\partial z_{ig_1}^{(1)}}\frac{\partial z_{ig_1}^{(1)}}{\partial\varphi_{g_1}^{(1)}}\frac{\partial\varphi_{g_1}^{(1)}}{\partial b_{g_1}^{(1)}}\\
&= (\hat{y}_i-y_i)w_{g_1\to1}^{(2)}\varphi\left(r_{ig_1}^{(1)}+b_{g_1}^{(1)}\right)\left(1-\varphi\left(r_{ig_1}^{(1)}+b_{g_1}^{(1)}\right)\right)\\
&\quad(i=1,2,\cdots,20;\ j=1,2,\cdots,5;\ g_1=1,2).
\end{aligned}$$

The 15 trainable variables can be solved iteratively by the gradient descent method and the BP process, as that used in Example 9.2. Figure 9.9 shows the variations of the weights with the λ decrease, for the L_2 regularized NN with the sigmoid activation function. With the value of λ decreasing from $10^{-0.5}$ to $10^{-5.9}$, the 12 weights appear in two batches when λ reaches two different values. When λ reaches about 10^{-1}, the weights of $w_{1\to2}^{(1)}$, $w_{2\to2}^{(1)}$, $w_{3\to2}^{(1)}$, $w_{4\to2}^{(1)}$, $w_{5\to2}^{(1)}$ and $w_{2\to2}^{(2)}$ appear; when λ reaches about $10^{-2.1}$, the remaining weights of $w_{1\to1}^{(1)}$, $w_{2\to1}^{(1)}$, $w_{3\to1}^{(1)}$, $w_{4\to1}^{(1)}$, $w_{5\to1}^{(1)}$ and $w_{1\to1}^{(2)}$ appear. The weights appear separately, which is different from the whole batch appearance of weights observed in the ridge regression shown in Chap. 2. The reason for such difference may be that the sigmoid activation function is used in the NN, which may lead to the fact that the least squares error function of the NN is a non-convex function, while the least squares error function of the ridge regression is a strictly convex function. To demonstrate this, an L_2 regularized NN model without any nonlinear activation function is constructed, and Fig. 9.10 shows the variations of weights with the λ decrease of this model, where the weights appear simultaneously. The optimal λ value of the L_2 NN (with the sigmoid activation) model is determined by the maximum LOOCV-R^2, as shown in Fig. 9.11.

In addition to regression, NNs can also accomplish classification tasks, thereby having the terminologies of regression NNs and classification NNs. The classification NNs are briefly introduced in Sect. 9.4.

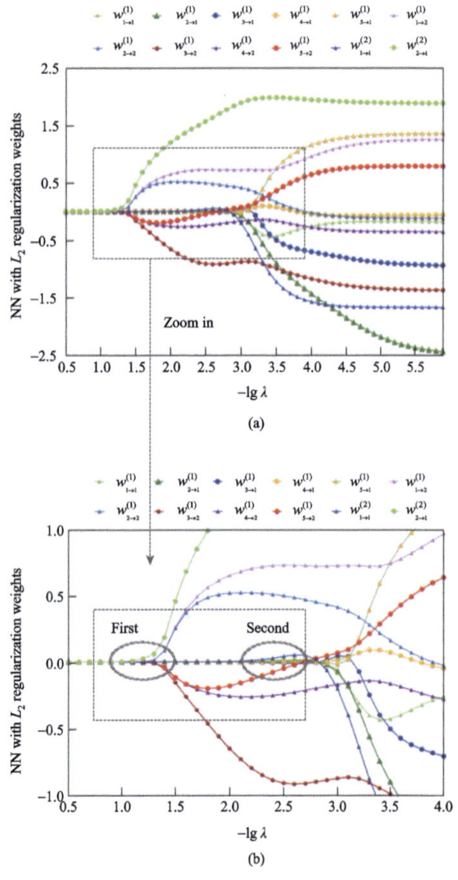

Fig. 9.9 The variations of weights with the value of λ in the L_2 NN with the sigmoid activation function

Fig. 9.10 The variations of weights with the value of λ in the L_2 NN without any nonlinear activation function

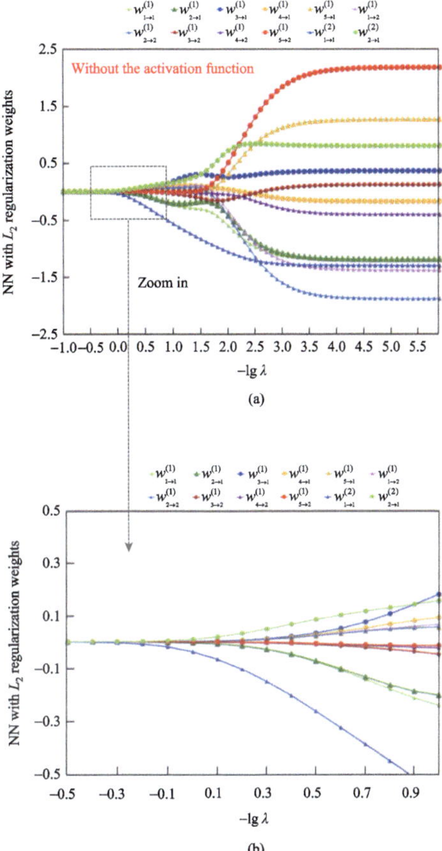

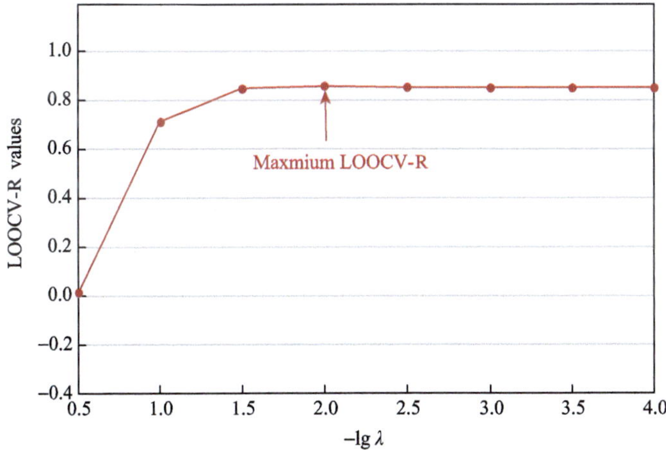

Fig. 9.11 The LOOCV-R versus λ in the NN model

9.4 Classification NNs

9.4.1 Binary Classification

In addition to regression, NNs are also used in classification. The structure of an NN for classification is the same as that shown in Fig. 9.1, where the neurons in the output layer are equipped with the sigmoid or softmax function as the activation function. The widely used loss function in binary classification is based on the cross entropy, as given by Eq. (3.7d) in Chap. 3. The training of trainable variables in a NN is performed by minimizing the loss function.

For binary classification, each of the output neurons delivers a prediction variable, which is set to be 0 or 1, i.e., $\hat{y}_p \in \{0, 1\}$ $(p = 1, 2, \cdots, P)$, where P is the number of the neurons in the output layer. Each of the neurons in the output layer receives forward information variable $r_{g_L}^{(L)}$, puts a constant $b_{g_L}^{(L)}$, executes the activation function, sets the threshold Θ, and then classifies the datum. Let $u_{g_L}^{(L)} \equiv r_{g_L}^{(L)} + b_{g_L}^{(L)}$. When the sigmoid function is employed as the activation function φ, it will act on $u_{g_L}^{(L)}$ and produce $\varphi\left(u_{g_L}^{(L)}\right)$. If $\varphi\left(u_{g_L}^{(L)}\right) < \Theta$, the datum is classified to zero, or otherwise to one. Usually, the activation function is the sigmoid function, and the threshold is set to be $\Theta = 0.5$, as described in Chap. 3.

If there is only one output variable, $P = 1$, there will be only one neuron in the output layer. In classification, the loss function is based on cross entropy, as shown by Eq. (3.7d) in Chap. 3, which is used here and given by

$$\mathcal{L}(\boldsymbol{W}) = -\sum_{i=1}^{n}\left[y_i u_i^{(L)} - \ln\left(1 + \mathrm{e}^{u_i^{(L)}}\right)\right]. \tag{9.13}$$

with $u_i^{(L)} = Z_{ig_{L-1}}^{(L-1)} W_{g_{L-1}\to\hat{y}}^{(L)} + b_{\hat{y}}^{(L)}$ and a minus sign is added here. The loss function links, step by step, input feature data to all parameters $W_{g_{L-1}\to\hat{y}}^{(L)}$, $b_{\hat{y}}^{(L)}$, $W_{g_{l-1}\to g_l}^{(l)}$, and $b_{g_l}^{(l)}$ $(g_l = 1, 2, \cdots, G_l;\quad l = 1, 2, \cdots, L-1)$, i.e., $\boldsymbol{W} = \left\{W_{g_{L-1}\to\hat{y}}^{(L)}, b_{\hat{y}}^{(L)}, W_{g_{l-1}\to g_l}^{(l)}, b_{g_l}^{(l)}\right\}$. Minimizing the loss function determines the values of, as stated by Eq. (9.3b). The minimization of $\mathcal{L}(\boldsymbol{W})$ requires that the gradient of $\mathcal{L}(\boldsymbol{W})$ with respect to $\boldsymbol{W}$ nulls, as stated by Eq. (9.4b). Again, the gradient descent method is used to solve the equation set of $\sum_{i=1}^{n} \nabla\mathcal{L}_i(\boldsymbol{W}) = 0$. Following Eq. (3.8b) in Chap. 3, we have the gradients with the chain rule of derivatives,

$$\frac{\partial\mathcal{L}}{\partial W_{g_{L-1}\to\hat{y}}^{(L)}} = \sum_{i=1}^{n} Z_{ig_{L-1}}^{(L-1)}\left(\frac{\mathrm{e}^{u_i^{(L)}}}{1+\mathrm{e}^{u_i^{(L)}}} - y_i\right) = \sum_{i=1}^{n} Z_{ig_{L-1}}^{(L-1)}\left(\varphi\left(u_i^{(L)}\right) - y_i\right), \tag{9.14a}$$

$$\frac{\partial\mathcal{L}}{\partial b_{\hat{y}}^{(L)}} = \sum_{i=1}^{n}\left(\frac{\mathrm{e}^{u_i^{(L)}}}{1+\mathrm{e}^{u_i^{(L)}}} - y_i\right) = \sum_{i=1}^{n}\left(\varphi\left(u_i^{(L)}\right) - y_i\right), \tag{9.14b}$$

$$\frac{\partial \mathcal{L}}{\partial W_{g_{l-1}\to g_l}^{(l)}} = \sum_{i=1}^{n}\left[\left(\varphi\left(u_i^{(L)}\right) - y_i\right)\frac{\partial Z_{ig_{L-1}}^{(L-1)}}{\partial W_{g_{l-1}\to g_l}^{(l)}}\right], \tag{9.14c}$$

$$\frac{\partial \mathcal{L}}{\partial b_{g_l}^{(l)}} = \sum_{i=1}^{n}\left[\left(\varphi\left(u_i^{(L)}\right) - y_i\right)\frac{\partial Z_{ig_{L-1}}^{(L-1)}}{\partial b_{g_l}^{(l)}}\right]. \tag{9.14d}$$

Then, the values of trainable variables are iterated until convergence, as described in Eqs. (9.6a)–(9.6b) and Eqs. (9.7a)–(9.7b).

Example 9.4 The catalyst data in Example 5.3 of Chap. 5 are used here again to illustrate neural network binary classification and the role of the activation function. The data are normalized by $\hat{x}_{ij} = \frac{x_{ij}-x_{\min,\,j}}{x_{\max,\,j}-x_{\min,\,j}}$, x_{ij} $(i = 1, 2, \cdots, n;\ j = 1, 2, \cdots, m)$ denotes the value feature x_j in sample i, and $x_{\max,\,j}$ and $x_{\min,\,j}$ are the maximum and minimum among the feature data x_{ij} $(i = 1, 2, \cdots, n)$. The normalized data are dimensionless and range from zero to one, as listed in Table 9.7, where nitrogen oxide (NOx) conversion efficiency is classed into low and high grades.

Because of the small amount of data, we only use one hidden layer to achieve the classification. The number of neurons in the hidden layer is determined by fivefold cross- validation. Figure 9.12 shows the classification accuracy versus the number of

Table 9.7 The normalized feature data of 22 catalysts

Catalysts	GHSV (x_1)	T (x_2)	NO_x conversion (class)
Mn/Glass-fiber	0.27	0.36	Low (0)
Mn_8Ce_1/Glass-fiber	0.27	0.27	Low (0)
Mn_5Ce_1/Glass-fiber	0.27	0.32	Low (0)
Mn_1Ce_2/Graphene	0.00	0.00	Low (0)
$Mn_{20}Gd_1/MnO_x$	0.13	0.18	Low (0)
$Mn_{10}Gd_1/MnO_x$	0.79	0.43	Low (0)
$V_1Mn_4Fe_1$/Attapulgite	0.17	0.09	Low (0)
Mn_5Ce_1/TiO_2	1.00	0.64	Low (0)
Mn_5Eu_1/TiO_2	0.88	0.32	Low (0)
Mn/ZSM-5	0.13	0.45	High (1)
Mn_5Ce_2/Titanate nanotubes	0.79	1.00	High (1)
Fe_1Mn_8/MnO_x	0.06	0.25	High (1)
Fe_1Mn_5/MnO_x	0.06	0.18	High (1)
Mn/TiO_2	0.88	1.00	High (1)
Mn_2Ce_1/Graphene	0.00	0.27	High (1)
Mn_4Ce_1/Graphene	0.00	0.09	High (1)
Mn_8Ti_2	0.38	0.77	High (1)
$Fe_1Mn_8Ti_1$	0.38	0.55	High (1)
$Mn_{10}Gd_3/MnO_x$	0.13	0.73	High (1)
$Mn_{10}Gd_1/MnO_x$	0.13	0.55	High (1)
MnO_x	0.79	0.55	High (1)
Ce/MnO_x	0.17	0.55	High (1)

neurons. If no activation function or only the linear function is used in the neurons, the classification accuracy from fivefold cross-validation is independent of the number of neurons. The classification accuracy is the highest for 2 neurons with the tan*h* activation function, the second highest for 2 neurons with the sigmoid function, and the third highest 5 neurons with the ReLU activation function. Therefore, the tanh activation function is used in the neural network classifier for detailed illustration.

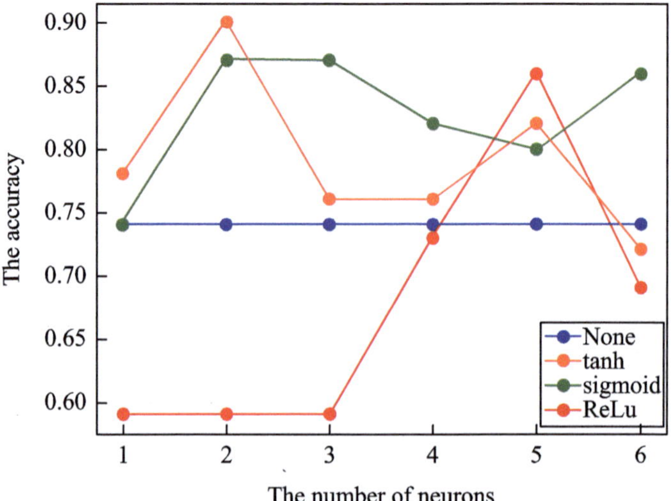

Fig. 9.12 Classification accuracy versus the number of neurons in the one hidden layer

Figure 9.13 illustrates the structure of the NN classifier with only one hidden layer. There are in total 9 latent variables in the NN classifier, in which 6 are in the hidden layer, $\left\{w_{1\to1}^{(1)}, w_{2\to1}^{(1)}, w_{1\to2}^{(1)}, w_{2\to2}^{(1)}; b_1^{(1)}, b_2^{(1)}\right\}$, and 3 are in the output layer

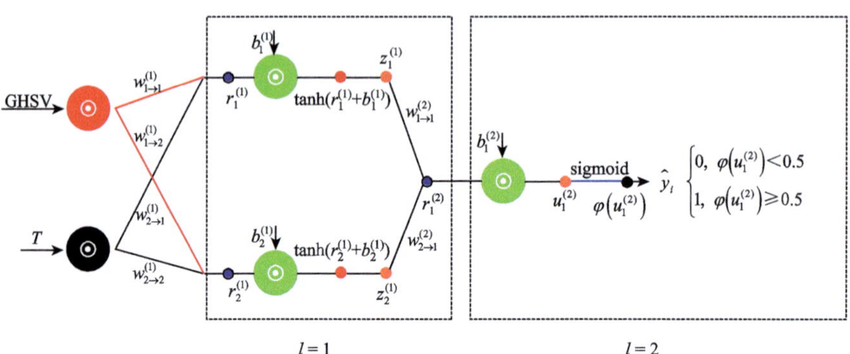

Fig. 9.13 The NN structure used in Example 9.4

$\left\{w^{(2)}_{1\to1}, w^{(2)}_{2\to1}; b^{(2)}_1\right\}$. Besides, a value of the threshold is given in the output layer, which is 0.5 in the present work.

Letting $x_{i1} = \text{GHSV}_i$ and $x_{i2} = T_i$, we have

$$z^{(1)}_{i1} = \tanh\left(\boldsymbol{X}_i\begin{bmatrix} w^{(1)}_{1\to1} \\ w^{(1)}_{2\to1}\end{bmatrix} + b^{(1)}_1\right) = \tanh\left(x_{i1}w^{(1)}_{1\to1} + x_{i2}w^{(1)}_{2\to1} + b^{(1)}_1\right),$$

$$z^{(1)}_{i2} = \tanh\left(\boldsymbol{X}_i\begin{bmatrix} w^{(1)}_{1\to2} \\ w^{(1)}_{2\to2}\end{bmatrix} + b^{(1)}_2\right) = \tanh\left(x_{i1}w^{(1)}_{1\to2} + x_{i2}w^{(1)}_{2\to2} + b^{(1)}_2\right),$$

$$u^{(2)}_i = \boldsymbol{Z}_i\begin{bmatrix} w^{(2)}_{1\to1} \\ w^{(2)}_{2\to1}\end{bmatrix} + b^{(2)}_1 = z^{(1)}_{i1}w^{(2)}_{1\to1} + z^{(1)}_{i2}w^{(2)}_{2\to1} + b^{(2)}_1.$$

According to Eq. (9.13), the sigmoid cross entropy loss function is

$$\mathcal{L}(\boldsymbol{W}) = -\sum_{i=1}^{22}\left[y_iu^{(2)}_i - \ln\left(1 + \mathrm{e}^{u^{(2)}_i}\right)\right].$$

Using Eqs. (9.14a)–(9.14b) gives

$$\begin{cases} \dfrac{\partial\mathcal{L}_i}{\partial w^{(2)}_{1\to1}} = \dfrac{\partial\mathcal{L}_i}{\partial u^{(2)}_i}\dfrac{\partial u^{(2)}_i}{\partial w^{(2)}_{1\to1}} = \left(\text{sigmoid}\left(u^{(2)}_i\right) - y_i\right)z^{(1)}_{i1}, \\ \dfrac{\partial\mathcal{L}_i}{\partial w^{(2)}_{2\to1}} = \dfrac{\partial\mathcal{L}_i}{\partial u^{(2)}_i}\dfrac{\partial u^{(2)}_i}{\partial w^{(2)}_{2\to1}} = \left(\text{sigmoid}\left(u^{(2)}_i\right) - y_i\right)z^{(1)}_{i2}, \quad i = 1, 2, \cdots, 22. \\ \dfrac{\partial\mathcal{L}_i}{\partial b^{(2)}_1} = \dfrac{\partial\mathcal{L}_i}{\partial u^{(2)}_i}\dfrac{\partial u^{(2)}_i}{\partial b^{(2)}_1} = \left(\text{sigmoid}\left(u^{(2)}_i\right) - y_i\right), \end{cases}$$

Using Eqs. (9.14c)–(9.14d) gives

$$\begin{cases} \dfrac{\partial\mathcal{L}_i}{\partial w^{(1)}_{j\to1}} = \dfrac{\partial\mathcal{L}_i}{\partial u^{(2)}_i}\dfrac{\partial u^{(2)}_i}{\partial z^{(1)}_{i1}}\dfrac{\partial z^{(1)}_{i1}}{\partial w^{(1)}_{j\to1}}, \\ \dfrac{\partial\mathcal{L}_i}{\partial w^{(1)}_{j\to2}} = \dfrac{\partial\mathcal{L}_i}{\partial u^{(2)}_i}\dfrac{\partial u^{(2)}_i}{\partial z^{(1)}_{i2}}\dfrac{\partial z^{(1)}_{i1}}{\partial w^{(1)}_{j\to2}}, \quad i = 1, 2, \cdots, 22; \; j = 1, 2. \\ \dfrac{\partial\mathcal{L}_i}{\partial b^{(1)}_j} = \dfrac{\partial\mathcal{L}_i}{\partial u^{(2)}_i}\dfrac{\partial u^{(2)}_i}{\partial \boldsymbol{Z}^{(1)}_i}\dfrac{\partial \boldsymbol{Z}_i}{\partial b^{(1)}_j}, \end{cases}$$

Thus, we have

$$\begin{cases} \dfrac{\partial \mathcal{L}_i}{\partial w_{1\to 1}^{(1)}} = \left(\text{sigmoid}\left(u_i^{(2)}\right) - y_i\right) w_{1\to 1}^{(2)} \left(1 - \tanh^2\left(x_{i1} w_{1\to 1}^{(1)} + x_{i2} w_{2\to 1}^{(1)} + b_1^{(1)}\right)\right) x_{i1}, \\ \dfrac{\partial \mathcal{L}_i}{\partial w_{2\to 1}^{(1)}} = \left(\text{sigmoid}\left(u_i^{(2)}\right) - y_i\right) w_{1\to 1}^{(2)} \left(1 - \tanh^2\left(x_{i1} w_{1\to 1}^{(1)} + x_{i2} w_{2\to 1}^{(1)} + b_1^{(1)}\right)\right) x_{i2}, \\ \dfrac{\partial \mathcal{L}_i}{\partial w_{1\to 2}^{(1)}} = \left(\text{sigmoid}\left(u_i^{(2)}\right) - y_i\right) w_{2\to 1}^{(2)} \left(1 - \tanh^2\left(x_{i1} w_{1\to 2}^{(1)} + x_{i2} w_{2\to 2}^{(1)} + b_2^{(1)}\right)\right) x_{i1}, \\ \dfrac{\partial \mathcal{L}_i}{\partial w_{2\to 2}^{(1)}} = \left(\text{sigmoid}\left(u_i^{(2)}\right) - y_i\right) w_{2\to 1}^{(2)} \left(1 - \tanh^2\left(x_{i1} w_{1\to 2}^{(1)} + x_{i2} w_{2\to 2}^{(1)} + b_2^{(1)}\right)\right) x_{i2}, \end{cases}$$

$$\begin{cases} \dfrac{\partial \mathcal{L}_i}{\partial b_1^{(1)}} = \left(\text{sigmoid}\left(u_i^{(2)}\right) - y_i\right) w_{1\to 1}^{(2)} \left(1 - \tanh^2\left(x_{i1} w_{1\to 1}^{(1)} + x_{i2} w_{2\to 1}^{(1)} + b_1^{(1)}\right)\right), \\ \dfrac{\partial \mathcal{L}_i}{\partial b_2^{(1)}} = \left(\text{sigmoid}\left(u_i^{(2)}\right) - y_i\right) w_{2\to 1}^{(2)} \left(1 - \tanh^2\left(x_{i1} w_{1\to 2}^{(1)} + x_{i2} w_{2\to 2}^{(1)} + b_2^{(1)}\right)\right), \end{cases}$$

$i = 1, 2, \cdots, 22.$

To start the iteration in the gradient descent method, we should assign initial values to the 9 trainable variables. In Example 9.1, the same value of one is assigned to all trainable variables, which might cause the problem of gradient symmetry. In this example, the initial values are randomly sampled from the normal distribution $N(0, 1)$ so that

$$\boldsymbol{W}(\text{initial}) = \boldsymbol{W}(\text{old}) = \begin{bmatrix} w_{1\to 1}^{(1)} & w_{1\to 2}^{(1)} \\ w_{2\to 1}^{(1)} & w_{2\to 2}^{(1)} \\ w_{1\to 1}^{(2)} & w_{2\to 1}^{(2)} \end{bmatrix} = \begin{bmatrix} -0.8113 & 1.4846 \\ 0.0653 & -2.4427 \\ -0.2089 & 0.0054 \end{bmatrix},$$

$$\boldsymbol{b}(\text{initial}) = \boldsymbol{b}(\text{old}) = \begin{pmatrix} b_1^{(1)} \\ b_2^{(1)} \\ b_1^{(2)} \end{pmatrix} = \begin{pmatrix} -0.6759 \\ 0.2867 \\ -1.6338 \end{pmatrix}.$$

The whole batch method is employed here again and the learning rate is set to be 1. Summing up $\sum_{i=1}^{22} \nabla \mathcal{L}_i(\boldsymbol{W})$ yields

$$\begin{bmatrix} \dfrac{\partial \mathcal{L}}{\partial w_{1\to 1}^{(1)}} & \dfrac{\partial \mathcal{L}}{\partial w_{1\to 2}^{(1)}} \\ \dfrac{\partial \mathcal{L}}{\partial w_{2\to 1}^{(1)}} & \dfrac{\partial \mathcal{L}}{\partial w_{2\to 2}^{(1)}} \\ \dfrac{\partial \mathcal{L}}{\partial w_{1\to 1}^{(2)}} & \dfrac{\partial \mathcal{L}}{\partial w_{2\to 1}^{(2)}} \end{bmatrix} = \begin{bmatrix} 0.0087 & -0.0003 \\ 0.0227 & -0.0007 \\ 0.2762 & 0.2337 \end{bmatrix}, \quad \begin{pmatrix} \dfrac{\partial \mathcal{L}}{\partial b_1^{(1)}} \\ \dfrac{\partial \mathcal{L}}{\partial b_2^{(1)}} \\ \dfrac{\partial \mathcal{L}}{\partial b_1^{(2)}} \end{pmatrix} = \begin{pmatrix} 0.0449 \\ -0.0014 \\ -0.4064 \end{pmatrix}.$$

With the expression of matrixes and vectors, one iteration updates the values of all trainable variables. The first iteration leads to the new values of trainable variables,

$$\boldsymbol{W}(\text{new}) = \boldsymbol{W}(\text{old}) - 1 \times \begin{bmatrix} \dfrac{\partial \mathcal{L}}{\partial w_{1\to 1}^{(1)}} & \dfrac{\partial \mathcal{L}}{\partial w_{1\to 2}^{(1)}} \\ \dfrac{\partial \mathcal{L}}{\partial w_{2\to 1}^{(1)}} & \dfrac{\partial \mathcal{L}}{\partial w_{2\to 2}^{(1)}} \\ \dfrac{\partial \mathcal{L}}{\partial w_{1\to 1}^{(2)}} & \dfrac{\partial \mathcal{L}}{\partial w_{2\to 1}^{(2)}} \end{bmatrix} = \begin{bmatrix} -0.82 & 1.4849 \\ 0.0426 & -2.442 \\ -0.4851 & -0.2283 \end{bmatrix},$$

$$\boldsymbol{b}(\text{new}) = \boldsymbol{b}(\text{old}) - 1 \times \begin{pmatrix} \frac{\partial \mathcal{L}}{\partial b_1^{(1)}} \\ \frac{\partial \mathcal{L}}{\partial b_2^{(1)}} \\ \frac{\partial \mathcal{L}}{\partial b_1^{(2)}} \end{pmatrix} = \begin{pmatrix} -0.7208 \\ 0.2881 \\ -1.2274 \end{pmatrix}.$$

After 538 iterations, the loss function reaches the minimum with the values of trainable variables as

$$\boldsymbol{W}(538) = \begin{bmatrix} -0.4807 & 5.5275 \\ -0.6431 & -5.2322 \\ -3.0465 & -5.5594 \end{bmatrix}, \quad \boldsymbol{b}(538) = \begin{pmatrix} -0.0383 \\ 0.8807 \\ 1.4763 \end{pmatrix}.$$

Table 9.8 lists the value of the sigmoid function and the classification of each sample by the NN classifier and Table 9.9 shows the classification confusion matrix.

Table 9.8 The value of the sigmoid function and the classification of each sample

Samples	$\varphi\left(u_1^{(2)}\right)$	$\hat{y}_i$
1	0.55	1
2	0.16	0
3	0.3	0
4	0.088	0
5	0.28	0
6	0.098	0
7	0.056	0
8	0.14	0
9	0.09	0
10	0.998	1
11	0.98	1
12	0.93	1
13	0.63	1
14	0.8	1
15	0.99	1
16	0.41	0
17	0.9996	1
18	0.92	1
19	0.9998	1
20	0.9995	1
21	0.12	0
22	0.9992	1

Table 9.9 The confusion matrix of the NN classification

Actual classes	Prediction classes	
	Class 0	Class 1
Class 0	8	1
Class 1	2	11

9.4.2 *Multiclassification of Multiply Grades in a Category*

Materials possess various properties and characteristics, each of which can form a category. For example, the nitrogen oxide (NO_x) conversion ability studied in Example 9.4 is one of the catalyst properties. Clearly, catalysts can be classified by other properties, chemical compositions, crystalline structures, etc., which leads to many categories of catalysts. In addition, there may exist different grades in one category. Obviously, classifying multiple grades in a category requires multiple corresponding thresholds, which means that the category must be quantitatively described. In this sense, the discrete grades will be more and more, approaching a spectrum, along with the increase in the number of thresholds. As an extreme result, the classification will convert to regression. In multiclassification of multiple grades in a category, the one-hot coding is widely used, which expresses all K classes in a vector as $[c_1, c_2, \cdots, c_K]$ and uses 0 or 1 to express $c_k (k = 1, 2, \cdots, K)$. If a datum belongs to c_k, the component $c_k = 1$ and other components $c_l = 0,\ l \neq k$. Taking three classes for example, we will have the one-hot coding expressions [1, 0, 0], [0, 1, 0], and [0, 0, 1] for a datum belonging to classes c_1,c_2, and c_3, respectively. The following paragraphs introduce how an NN model classifies multiple grades of one category with the one-hot coding and this kind of NN classification is also called multiclassification for convenience.

The basic structure of an NN to classify multiple grades of one category is the same as that shown in Fig. 9.1, except that the softmax function and multiple neurons are used in the output layer, where the number of neurons in the output layer is the same as the number of classes (grades). The output of the softmax function is a vector of $[\hat{c}_1, \hat{c}_2, \cdots, c_K]$ and each of the component $\hat{c}_k (k = 1, 2, \cdots, K)$ is a normalized exponential function,

$$\hat{c}_k = \text{softmax}\left(u_k^{(L)}\right) = \frac{\mathrm{e}^{u_k^{(L)}}}{\sum\limits_{k=1}^{K} \mathrm{e}^{u_k^{(L)}}}, \tag{9.15}$$

where K represents the number of classes and

$$u_k^{(L)} = \varphi_{g_l}^{(L-1)}\left(Z_{g_{L-1}}^{(L-1)} W_{g_{L-1}\to g_l}^{(L)} + b_{g_l}^{(L)}\right),$$

is the information received by neuron k in the output layer. The softmax function transforms the output value of the NN into a probability value $\hat{c}_k$ in (0, 1).

The cross-entropy function is also used as the loss function. The total loss function for the entire data is given by

$$\mathcal{L}(\boldsymbol{W}) = -\sum_{i=1}^{n}\sum_{k=1}^{K} c_{ik} \ln\left(\text{softmax}\left(u_{ik}^{(L)}\right)\right) = -\sum_{i=1}^{n}\sum_{k=1}^{K} c_{ik} \ln(\hat{c}_{ik}) \tag{9.16a}$$

or

$$\mathcal{L}(\boldsymbol{W}) = \sum_{i=1}^{n} \mathcal{L}_i\left(u_{ik}^{(L)}\right). \tag{9.16b}$$

With the loss function for datum i

$$\mathcal{L}_i(\boldsymbol{W}) = -\sum_{k=1}^{K} c_{ik} \ln\left(\text{softmax}\left(u_{ik}^{(L)}\right)\right) = -\sum_{k=1}^{K} c_{ik} \ln(\hat{c}_{ik}), \tag{9.16c}$$

where c_{ik} is the real label of class for datum i expressed in the one-hot coding and $\boldsymbol{W}$ represents all trainable variables via $u_{ik}^{(L)}$. There are K neurons in the output layer and each neuron will produce a component of $[\hat{c}_1, \hat{c}_2, \cdots, \hat{c}_K]$. The minimization of the loss function determines the output of the NN classifier. The one-hot coding of the real label of class brings the advantage that Eq. (9.16c) can be reduced to

$$\mathcal{L}_i(\boldsymbol{W}) = -\ln\left(\text{softmax}\left(u_{ik}^{(L)}\right)\right) = -\ln(\hat{c}_{ik}), \tag{9.16d}$$

because only one $c_{ik} = 1$ and all others $c_{il} = 0$ $(c_{il} \neq c_{ik})$. Since $\hat{c}_{ik} \leq 1$,the minimization of $\mathcal{L}_i(\boldsymbol{W})$ is equivalent to the maximization of $\hat{c}_{ik}$, viz., the maximization of $\text{softmax}\left(u_{ik}^{(L)}\right)$. The whole batch method minimizes the total loss function or the average of total loss $\frac{1}{n}\mathcal{L}(\boldsymbol{W})$ and hence the $\nabla\mathcal{L}(\boldsymbol{W})$ or $\frac{1}{n}\nabla\mathcal{L}(\boldsymbol{W})$ is used in the iteration process of $\boldsymbol{W}$, where n is the number of data.

Example 9.5 We divided the NO_x conversion in the above catalyst data into three classes: high, medium and low (labelled by 0, 1 and 2 respectively). Three data shown in Table 9.10 are used to illustrate how the NN performs multiclassification.

Table 9.10 Data of catalysts

Catalysts	GHSV	T	Classes	One-hot coding ($[c_1 \quad c_2 \quad c_3]$)
Mn_5Eu_1/TiO_2	1	0	0	[1, 0, 0]
Mn_8Ti_2	0.33	1	1	[0, 1, 0]
$Mn_{10}Gd_1/MnO_x$	0	0.9	2	[0, 0, 1]

Figure 9.14 illustrates the structure of the NN classifier with only one hidden layer and the hidden layer does not use any activation function. There are total 15 trainable variables in the NN classifier, in which 6 are in the hidden layer, $\left\{w_{1\to1}^{(1)}, w_{1\to2}^{(1)}, w_{2\to1}^{(1)}, w_{2\to2}^{(1)}, b_1^{(1)}, b_2^{(1)}\right\}$, and 9 are in the output layer $\left\{w_{1\to1}^{(2)}, w_{1\to2}^{(2)}, w_{1\to3}^{(2)}, w_{2\to1}^{(2)}, w_{2\to2}^{(2)}, w_{2\to3}^{(2)}, b_1^{(2)}, b_2^{(2)}, b_3^{(2)}\right\}$.

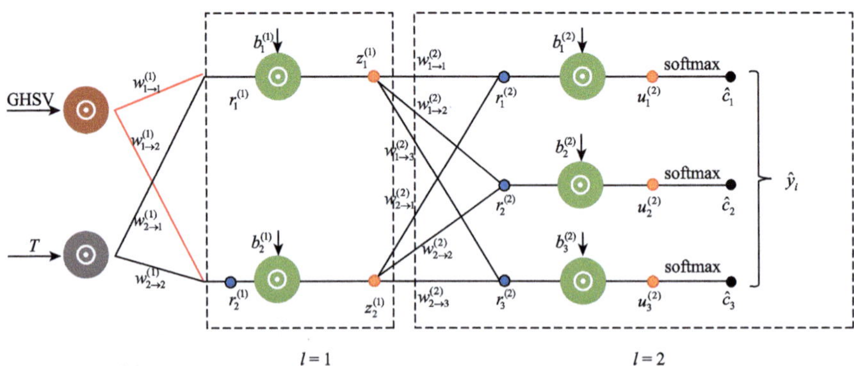

Fig. 9.14 The NN structure used in Example 9.5

We randomly select the initial parameters in a series of numbers conforming to the normal distribution $N(0,\ 1)$.

$$\begin{bmatrix} w_{1\to1}^{(1)} & w_{1\to2}^{(1)} \\ w_{2\to1}^{(1)} & w_{2\to2}^{(1)} \end{bmatrix} = \begin{bmatrix} -0.81 & 1.48 \\ 0.065 & -2.44 \end{bmatrix}, \quad \begin{bmatrix} b_1^{(1)} & b_2^{(1)} \end{bmatrix} = [-0.666\ -0.42],$$

$$\begin{bmatrix} w_{1\to1}^{(2)} & w_{1\to2}^{(2)} & w_{1\to3}^{(2)} \\ w_{2\to1}^{(2)} & w_{2\to2}^{(2)} & w_{2\to3}^{(2)} \end{bmatrix} = \begin{bmatrix} -0.21 & 0.0054 & 0.31 \\ 0.5999 & -0.303 & 0.685 \end{bmatrix},$$

$$\begin{bmatrix} b_1^{(2)} & b_2^{(2)} & b_3^{(2)} \end{bmatrix} = [0.429\ 0.0321\ 0.119].$$

The feature data matrix is $\boldsymbol{X} = \begin{bmatrix} 1 & 0 \\ 0.33 & 1 \\ 0 & 0.9 \end{bmatrix}$. The whole batch method is employed here again,

$$\begin{bmatrix} z_{11}^{(1)} & z_{12}^{(1)} \\ z_{21}^{(1)} & z_{22}^{(1)} \\ z_{31}^{(1)} & z_{32}^{(1)} \end{bmatrix} = \boldsymbol{X} \begin{bmatrix} w_{1\to1}^{(1)} & w_{1\to2}^{(1)} \\ w_{2\to1}^{(1)} & w_{2\to2}^{(1)} \end{bmatrix} + \begin{pmatrix} 1 \\ 1 \\ 1 \end{pmatrix} \left(b_1^{(1)}\ b_2^{(1)} \right)$$

$$= \begin{bmatrix} -1.476 & 1.06 \\ -0.6277 & -2.8112 \\ -0.6075 & -2.616 \end{bmatrix},$$

$$\begin{bmatrix} u_{11}^{(2)} & u_{12}^{(2)} & u_{13}^{(2)} \\ u_{21}^{(2)} & u_{22}^{(2)} & u_{23}^{(2)} \\ u_{31}^{(2)} & u_{32}^{(2)} & u_{33}^{(2)} \end{bmatrix} = \begin{bmatrix} z_{11}^{(1)} & z_{12}^{(1)} \\ z_{21}^{(1)} & z_{22}^{(1)} \\ z_{31}^{(1)} & z_{32}^{(1)} \end{bmatrix} \begin{bmatrix} w_{1\to1}^{(2)} & w_{1\to2}^{(2)} & w_{1\to3}^{(2)} \\ w_{2\to1}^{(2)} & w_{2\to2}^{(2)} & w_{2\to3}^{(2)} \end{bmatrix}$$

$$+ \begin{pmatrix} 1 \\ 1 \\ 1 \end{pmatrix} \begin{pmatrix} b_1^{(2)} & b_2^{(2)} & b_3^{(2)} \end{pmatrix}$$

$$= \begin{bmatrix} 1.37484 & -0.29705 & 0.38754 \\ -1.12559 & 0.88049 & -2.001241 \\ -1.01276 & 0.82147 & -1.861285 \end{bmatrix}^{\mathrm{T}},$$

$$\begin{bmatrix} \hat{c}_{i1} & \hat{c}_{i2} & \hat{c}_{i3} \end{bmatrix} = \begin{bmatrix} \mathrm{softmax}\left(u_{i1}^{(2)}\right) & \mathrm{softmax}\left(u_{i2}^{(2)}\right) & \mathrm{softmax}\left(u_{i3}^{(2)}\right) \end{bmatrix}$$

$$= \begin{bmatrix} 0.64 & 0.12 & 0.24 \\ 0.11 & 0.84 & 0.05 \\ 0.13 & 0.81 & 0.06 \end{bmatrix}^{\mathrm{T}}.$$

Calculate the initial loss according to Eq. (9.14b),

$$\begin{aligned} \mathcal{L} &= -\sum_{i=1}^{3} [c_{i1} \ln \hat{c}_{i1} + c_{i2} \ln \hat{c}_2 + c_{i3} \ln \hat{c}_{i3}] = \mathcal{L}_1 + \mathcal{L}_2 + \mathcal{L}_3 \\ &= -[1 \times \ln 0.64 + 0 \times \ln 0.12 \\ &\quad + 0 \times \ln 0.24] - [0 \times \ln 0.11 + 1 \times \ln 0.84 + 0 \times \ln 0.05] \\ &\quad - [0 \times \ln 0.13 + 0 \times \ln 0.81 + 1 \times \ln 0.06] \\ &= 0.4463 + 0.1737 + 2.8882 = 3.5082. \end{aligned}$$

The NN model is trained by minimizing the average loss of

$$\frac{1}{3} \sum_{i=1}^{3} \nabla \mathcal{L}(\boldsymbol{W}).$$

The values of trainable variables are determined by iterations and the minimization is finally achieved, as what we have done above for binary classification. The iteration process is conducted by

$$\boldsymbol{W}(\text{new}) = \boldsymbol{W}(\text{old}) - \eta \times \frac{1}{3} \nabla \mathcal{L}(\boldsymbol{W})$$

until convergence, where η is the learning rate and $\eta = 1$ is used here. The chain rule of derivatives is used in the calculation of the gradient $\frac{1}{3}\nabla \mathcal{L}(\boldsymbol{W})$. The final results are shown in Tables 9.11 and 9.12.

Table 9.11 The prediction results after 100 iterations

$\hat{c}_1$	$\hat{c}_2$	$\hat{c}_3$	$\hat{y}_i$
0.8248	0.175	0.0002	0
0.0185	0.8666	0.01149	1
0.0065	0.1932	0.8003	2

Table 9.12 The trained values of trainable variables

Trainable variables	Values	Trainable variables	Values
$w_{1\to1}^{(1)}$	−4.082	$w_{2\to2}^{(2)}$	0.115
$w_{1\to2}^{(1)}$	3.613	$w_{2\to3}^{(2)}$	−2.266
$w_{2\to1}^{(1)}$	−0.3164	$b_1^{(1)}$	0.065
$w_{2\to2}^{(1)}$	−2.448	$b_2^{(1)}$	0.082
$w_{1\to1}^{(2)}$	−0.562	$b_1^{(2)}$	0.055
$w_{1\to2}^{(2)}$	−1.845	$b_2^{(2)}$	0.238
$w_{1\to3}^{(2)}$	2.516	$b_3^{(2)}$	0.288
$w_{2\to1}^{(2)}$	3.132		

9.5 Autoencoders

9.5.1 Introduction

Autoencoders are one kind of NNs, in which the number of output neurons equals the number of input neurons, viz., $K = m$, in order to let the output reconstructs the input, i.e., $\hat{y}_{ik} \approx x_{ij}$ or $\hat{y}_{ij} \approx x_{ij}(i = 1, 2, \cdots, n, \ j = 1, 2, \cdots, m) \in \mathbf{R}^m$ for clarity. In an autoencoder, the front half part has the structure shown in Fig. 9.1 and encodes the input data, as described above for the NN regressor and the NN classifier, and the back half part has a structure with mirror-symmetry to the front half part and decodes the encoded data. Thus, the layer number in an autoencoder is even and the number of hidden layers is odd. When an NN is trained, the characteristic information in the input data is encoded and stored in the trained trainable variables. Notice that real data in autoencoders might not have explicit responses and labels. In the front half part of an autoencoder, an encoder function ϕ mathematically maps the input $\boldsymbol{X}_i = (x_{i1}, x_{i2}, \cdots, x_{im})(i = 1, 2, \cdots, n)$ to $\left\{z_{ig_{L/2}}^{(L/2)}, b_{g_{L/2}}^{(L/2)}\right\}(g_{L/2} = 1, 2, \cdots, G_{L/2})$ of neurons in the hidden layer $L/2$. The job of the back half part uses a decoder function θ to decode the encoded input or to map back $\left\{z_{ig_{L/2}}^{(L/2)}, b_{g_{L/2}}^{(L/2)}\right\}(g_{L/2} = 1, 2, \cdots, G_{L/2})$

to $\hat{y}_{ij} \approx x_{ij}$ $(i = 1, 2, \cdots, n;\ j = 1, 2, \cdots, m)$. The basic structure of an autoencoder is shown in Fig. 9.15, where the encoder function ϕ and the decoder function θ can be any functions, and the structure of an autoencoder NN is shown in Fig. 9.16. The mean squared error is the typical loss function for autoencoders

$$\mathcal{L}(\boldsymbol{W}) = \frac{1}{n}\sum_{i=1}^{n}[\theta(\phi(\boldsymbol{X}_i)) - \boldsymbol{X}_i]^2 . \tag{9.17}$$

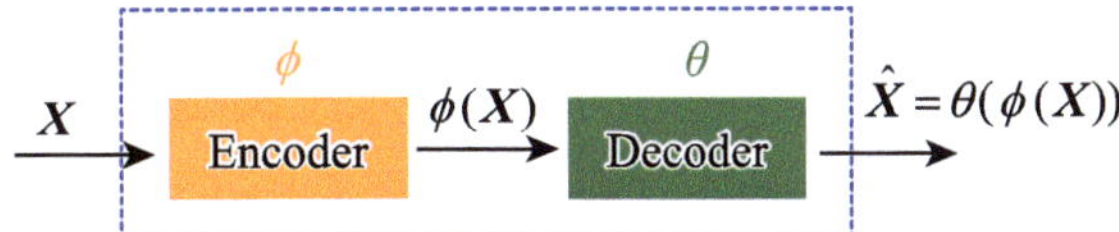

Fig. 9.15 The basic structure of an autoencoder

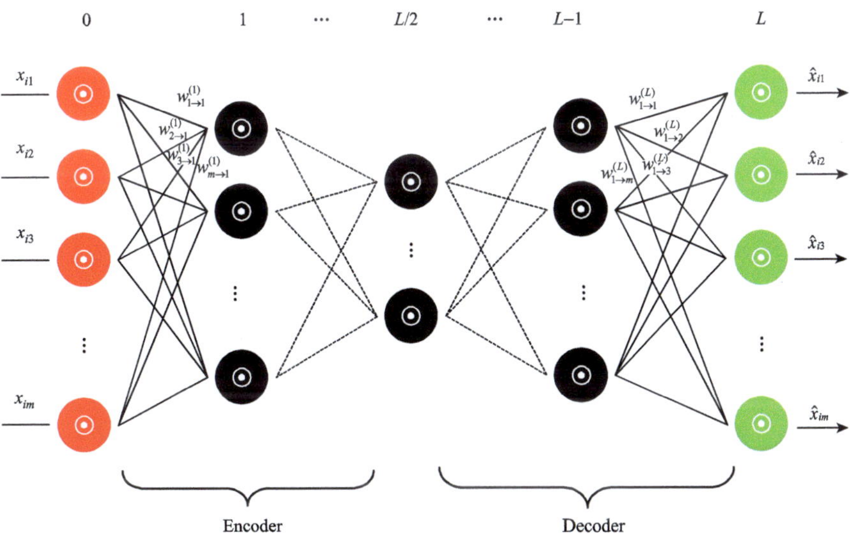

Fig. 9.16 The structure of autoencoder NNs used in Example 9.5

9.5.2 *Denoising Autoencoder*

Humans have the ability to recognize damaged images that are blocked, which stems from our advanced associative memory and sensory functions. We can remember in many ways (such as images, sounds, odours, etc.), so even if the data is partially damaged or lost, we can recall it. The denoising autoencoder is a type of autoencoders that accepts corrupted data as input and is trained to predict the original uncorrupted data as output. The idea is to make the learned representation robust to partial damage of the input pattern. A good representation is expected to capture stable structures in the form of dependencies and regularities of the (unknown) distribution in the input data (Vincent et al., 2008).

In denoising autoencoders, the original input data $\boldsymbol{X}_i = (x_{i1}, x_{i2}, \cdots, x_{im})$ $(i = 1, 2, \cdots, n)$ first undergo probabilistic corruption, transiting into corrupted data $\hat{\boldsymbol{X}}_i = (\hat{x}_{i1}, \hat{x}_{i2}, \cdots, \hat{x}_{im})$ $(i = 1, 2, \cdots, n)$. The denoising autoencoders aim to recover the original input $\boldsymbol{X}_i$ and hence the loss function takes the form of

$$\mathcal{L}(\boldsymbol{W}) = \frac{1}{n}\sum_{i=1}^{n}\left[\theta(\phi(\hat{\boldsymbol{X}}_i)) - \boldsymbol{X}_i\right]^2. \tag{9.18}$$

Example 9.6 Take the 17 catalyst data, which have five features, in Table 9.13 as an example to illustrate the denoising autoencoder, where the probabilistic corruption (noise) conforming to the Gaussian distribution N (0,1) is added into the inputs. Figure 9.17 shows the structure of used denoising autoencoder NN, including five input neurons, five output neurons and one hidden layer. Neurons 1 and 2 in the hidden layer add constants $b_1^{(1)}$ and $b_2^{(1)}$ on the received variables of $r_1^{(1)}$ and $r_2^{(1)}$, respectively. The same tan$h$ function $\varphi = \frac{\exp(u)-\exp(-u)}{\exp(u)+\exp(-u)}$ is used for the activation functions $\varphi_1^{(1)}$ and

Table 9.13 Normalized data of catalysts

No	Mn (x_1)	Ce (x_2)	Fe (x_3)	GHSV (x_4)	T (x_5)
1	1	0.8451	0	0.25	0
2	0.2269	0	0	1	0.4
3	0.8067	0	0.6216	1	0.4
4	0.7661	0	0.8919	1	0.4
5	0.8880	0	0	1	0.6
6	0.8590	0	0.3243	1	0.4
7	0.2879	0	0	1	0.3
8	0	0.5610	0	0.7917	0.8
9	0	0.6463	0	0.7917	0.8
10	0	0.6463	0	0.375	0.6
11	0	0.6463	0	0	0.4
12	0	1	0	0.375	0.7
13	0	0.9634	0.1622	0.375	0.6
14	0	0.9146	0.2973	0.375	0.6
15	0	0.9024	0.4054	0.375	0.5
16	0	0.8293	1	0.375	0.5
17	0	0.6463	0	0.7917	1

Notes (1) The maximum value of gas hour space velocity (GHSV) is 60000 h^{-1}, and the minimum value is 12000 h^{-1}

(2) The maximum value of temperature (T) is 350 °C, and the minimum value is 100 °C

$\varphi_2^{(1)}$, viz., $\varphi = \varphi_1^{(1)} = \varphi_2^{(1)}$ so that $\varphi(u)$ will work on $\left(r_1^{(1)} + b_1^{(1)}\right)$ and $\left(r_2^{(1)} + b_2^{(1)}\right)$, and generate $z_1^{(1)}$ and $z_2^{(1)}$, respectively. The output layer will receive the variable $\boldsymbol{W}^{(2)}$, add constant $\boldsymbol{b}^{(2)}$, implement the linear function and deliver the output $\hat{\boldsymbol{X}} = \boldsymbol{W}^{(2)} + \boldsymbol{b}^{(2)}$.

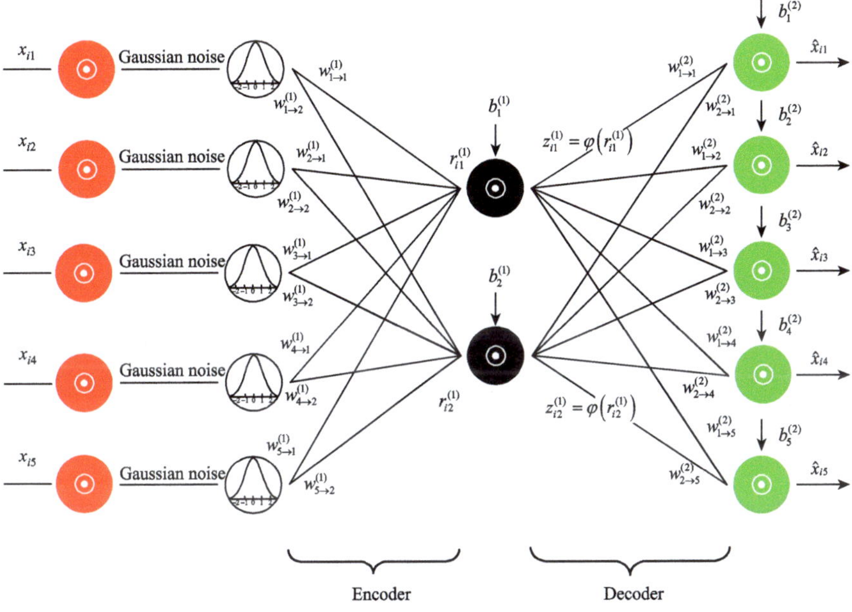

Fig. 9.17 The structure of the denoising autoencoder NN used in Example 9.5

There are 27 trainable variables in the autoencoder NN, including 12 in the hidden layer, $\left\{w_{1\to1}^{(1)}, w_{1\to2}^{(1)}, w_{2\to1}^{(1)}, w_{2\to2}^{(1)}, w_{3\to1}^{(1)}, w_{3\to2}^{(1)}, w_{4\to1}^{(1)}, w_{4\to2}^{(1)}, w_{5\to1}^{(1)}, w_{5\to2}^{(1)}, b_1^{(1)}, b_2^{(1)}\right\}$ and 15 in the output layer, $\left\{w_{1\to1}^{(2)}, w_{2\to1}^{(2)}, w_{1\to2}^{(2)}, w_{2\to2}^{(2)}, w_{1\to3}^{(2)}, w_{2\to3}^{(2)}, w_{1\to4}^{(2)}, w_{2\to4}^{(2)}, w_{1\to5}^{(2)}, w_{2\to5}^{(2)}, b_1^{(2)}, b_2^{(2)}, b_3^{(2)}, b_4^{(2)}, b_5^{(2)}\right\}$. The initial values of the 27 trainable variables are randomly selected from a series of numbers conforming to the normal distribution N (0, 1) so that

$$\boldsymbol{W}^{(1)} = \begin{bmatrix} w_{1\to1}^{(1)} & w_{1\to2}^{(1)} \\ w_{2\to1}^{(1)} & w_{2\to2}^{(1)} \\ w_{3\to1}^{(1)} & w_{3\to2}^{(1)} \\ w_{4\to1}^{(1)} & w_{4\to2}^{(1)} \\ w_{5\to1}^{(1)} & w_{5\to2}^{(1)} \end{bmatrix} = \begin{bmatrix} -0.8113 & 1.4846 \\ 0.0653 & -2.4427 \\ 0.0992 & 0.5912 \\ 0.5928 & -2.1229 \\ -0.7229 & -0.0563 \end{bmatrix},$$

$$\boldsymbol{b}^{(1)} = \left[b_1^{(1)}\ b_2^{(1)}\right] = [-0.7682\ -0.4501],$$

$$\boldsymbol{W}^{(2)} = \begin{bmatrix} w_{1\to1}^{(2)} & w_{1\to2}^{(2)} & w_{1\to3}^{(2)} & w_{1\to4}^{(2)} & w_{1\to5}^{(2)} \\ w_{2\to1}^{(2)} & w_{2\to2}^{(2)} & w_{2\to3}^{(2)} & w_{2\to4}^{(2)} & w_{2\to5}^{(2)} \end{bmatrix}$$
$$= \begin{bmatrix} -0.666 & -0.4168 & 1.8211 & 0.6804 & -0.2614 \\ 0.8771 & 0.3983 & -0.888 & -1.453 & -1.0287 \end{bmatrix},$$
$$\boldsymbol{b}^{(2)} = \begin{bmatrix} b_1^{(2)} & b_2^{(2)} & b_3^{(2)} & b_4^{(2)} & b_5^{(2)} \end{bmatrix}$$
$$= [\,0.5364\ 0.3593\ 0.2706\ -0.1529\ 0.3305\,].$$

In each iteration step, one datum is used to update the 27 trainable variables through the BP process, and when all the 17 data are used once, the 17 BP processes are called one epoch. In the first iteration step, datum 1 is used to update the trainable variables, as described below.

With a randomly generated Gaussian noise (GN) of $\mathbf{GN} = \begin{bmatrix} 1.0617 & -0.8786 & 0.1119 & 0.1851 & 0.9092 \end{bmatrix}$ and $\boldsymbol{X}_1 = [\,x_{11}\ x_{12}\ x_{13}\ x_{14}\ x_{15}\,] = [\,1\ 0.8451\ \ 0\ 0.25\ 0\,]$, we have

$$\boldsymbol{X}_1' = \boldsymbol{X}_1 + \mathbf{GN} = [\,x_{11}'\ x_{12}'\ x_{13}'\ x_{14}'\ x_{15}'\,]$$
$$= [2.0617\ \ -0.0335\ \ 0.1119\ \ 0.4351\ \ 0.9092].$$

Subsequently, we have

$$\boldsymbol{r}^{(1)} = \begin{bmatrix} r_{11}^{(1)} & r_{12}^{(1)} \end{bmatrix} = \boldsymbol{X}_1'\boldsymbol{W}^{(1)} + \boldsymbol{b}^{(1)}$$
$$= [\,2.0617\ -0.0335\ 0.1119\ 0.4351\ 0.9092\,]$$
$$\begin{bmatrix} -0.8113 & 1.4846 \\ 0.0653 & -2.4427 \\ 0.0992 & 0.5912 \\ 0.5928 & -2.1229 \\ -0.7229 & -0.0563 \end{bmatrix} + [\,-0.7682\ -0.4501\,]$$
$$= [\,-2.8313\ 1.7838\,].$$

Then, implementing the activation function $\varphi = \tanh$ leads to

$$\boldsymbol{z}^{(1)} = \varphi\left(\boldsymbol{r}^{(1)}\right) = \left[\tanh\left(r_{11}^{(1)}\right)\ \tanh\left(r_{12}^{(1)}\right)\right] = [\,-0.9931\ 0.9451\,].$$

Thus, the predicted value for datum 1 from the denoising autoencoder is

$$\hat{\boldsymbol{X}}_1 = [\,\hat{x}_{11}\ \hat{x}_{12}\ \hat{x}_{13}\ \hat{x}_{14}\ \hat{x}_{15}\,] = \boldsymbol{z}^{(1)}\boldsymbol{W}^{(2)} + \boldsymbol{b}^{(2)}$$
$$= [\,-0.9931\ 0.9451\,]\begin{bmatrix} -0.666 & -0.4168 & 1.8211 & 0.6804 & -0.2614 \\ 0.8771 & 0.3983 & -0.888 & -1.453 & -1.0287 \end{bmatrix}$$
$$+ [\,0.5364\ 0.3593\ 0.2706\ -0.1529\ 0.3305\,]$$

$$= [\,2.0267\ 1.1496\ 1.2398\ -2.2018\ -0.3821\,].$$

The loss function is

$$\mathcal{L} = \frac{1}{n}\sum_{i=1}^{n}(\hat{\boldsymbol{X}}_i - \boldsymbol{X}_i)^2. \tag{9.19}$$

The loss function for $\boldsymbol{X}_1$ is

$$\mathcal{L}_1 = (\hat{\boldsymbol{X}}_1 - \boldsymbol{X}_1)^2 = (\hat{x}_{11} - x_{11})^2 + (\hat{x}_{12} - x_{12})^2 + (\hat{x}_{13} - x_{13})^2 + (\hat{x}_{14} - x_{14})^2 + (\hat{x}_{15} - x_{15})^2 = 11.2943.$$

The gradients of the loss function are

$$\begin{cases} \dfrac{\partial \mathcal{L}_1}{\partial w_{j\to 1}^{(1)}} = 2(\hat{\boldsymbol{X}}_1 - \boldsymbol{X}_1)\dfrac{\partial \hat{\boldsymbol{X}}_1}{\partial z_{11}^{(1)}}\dfrac{\partial z_{11}^{(1)}}{\partial r_{11}^{(1)}}\dfrac{\partial r_{11}^{(1)}}{\partial w_{j\to 1}^{(1)}}, \\ \dfrac{\partial \mathcal{L}_1}{\partial w_{j\to 2}^{(1)}} = 2(\hat{\boldsymbol{X}}_1 - \boldsymbol{X}_1)\dfrac{\partial \hat{\boldsymbol{X}}_1}{\partial z_{12}^{(1)}}\dfrac{\partial z_{12}^{(1)}}{\partial r_{12}^{(1)}}\dfrac{\partial r_{12}^{(1)}}{\partial w_{j\to 2}^{(1)}}, \end{cases} \tag{9.20a}$$

$$\begin{cases} \dfrac{\partial \mathcal{L}_1}{\partial w_{1\to k}^{(2)}} = 2(\hat{\boldsymbol{X}}_1 - \boldsymbol{X}_1)\dfrac{\partial \hat{\boldsymbol{X}}_1}{\partial w_{1\to k}^{(2)}}, \\ \dfrac{\partial \mathcal{L}_1}{\partial w_{2\to k}^{(2)}} = 2(\hat{\boldsymbol{X}}_1 - \boldsymbol{X}_1)\dfrac{\partial \hat{\boldsymbol{X}}_1}{\partial w_{2\to k}^{(2)}}, \end{cases} \tag{9.20b}$$

$$\frac{\partial \mathcal{L}_1}{\partial b_1^{(1)}} = 2(\hat{\boldsymbol{X}}_1 - \boldsymbol{X}_1)\frac{\partial \hat{\boldsymbol{X}}_1}{\partial z_{11}^{(1)}}\frac{\partial z_{11}^{(1)}}{\partial r_{11}^{(1)}}\frac{\partial r_{11}^{(1)}}{\partial b_1^{(1)}}, \tag{9.20c}$$

$$\frac{\partial \mathcal{L}_1}{\partial b_2^{(1)}} = 2(\hat{\boldsymbol{X}}_1 - \boldsymbol{X}_1)\frac{\partial \hat{\boldsymbol{X}}_1}{\partial z_{12}^{(1)}}\frac{\partial z_{12}^{(1)}}{\partial r_{12}^{(1)}}\frac{\partial r_{12}^{(1)}}{\partial b_2^{(1)}}, \tag{9.20d}$$

$$\frac{\partial \mathcal{L}_1}{\partial b_k^{(2)}} = 2(\hat{\boldsymbol{X}}_1 - \boldsymbol{X}_1)\frac{\partial \hat{\boldsymbol{X}}_1}{\partial b_k^{(2)}}$$
$$(j = 1, 2, 3, 4, 5;\ k = 1, 2, 3, 4, 5). \tag{9.20e}$$

The derivative of the activation function $\varphi = \frac{\exp(u)-\exp(-u)}{\exp(u)+\exp(-u)}$ is

$$\frac{\mathrm{d}\varphi}{\mathrm{d}u} = 1 - \varphi^2.$$

In this example, the $z_{11}^{(1)} = \varphi_1^{(1)}\left(r_{11}^{(1)}\right)$ and $z_{12}^{(1)} = \varphi_2^{(1)}\left(r_{12}^{(1)}\right)$, and thus

$$\frac{\partial \boldsymbol{z}_{11}^{(1)}}{\partial \boldsymbol{r}_{11}^{(1)}} = 1 - \left(\frac{\exp\left(\boldsymbol{r}_{11}^{(1)}\right) - \exp\left(-\boldsymbol{r}_{11}^{(1)}\right)}{\exp\left(\boldsymbol{r}_{11}^{(1)}\right) + \exp\left(-\boldsymbol{r}_{11}^{(1)}\right)}\right)^2, \tag{9.20f}$$

$$\frac{\partial z_{12}^{(1)}}{\partial \boldsymbol{r}_{12}^{(1)}} = 1 - \left(\frac{\exp\left(\boldsymbol{r}_{12}^{(1)}\right) - \exp\left(-\boldsymbol{r}_{12}^{(1)}\right)}{\exp\left(\boldsymbol{r}_{12}^{(1)}\right) + \exp\left(-\boldsymbol{r}_{12}^{(1)}\right)}\right)^2. \tag{9.20g}$$

Then, the gradients of the loss function, i.e., Eqs. (9.20a)-(9.20g) are

$$\frac{\partial \mathcal{L}_1}{\partial w_{j\to 1}^{(1)}} = 2(\hat{\boldsymbol{X}}_1 - \boldsymbol{X}_1)\left(1 - \left(\frac{\exp\left(\boldsymbol{r}_{11}^{(1)}\right) - \exp\left(-\boldsymbol{r}_{11}^{(1)}\right)}{\exp\left(\boldsymbol{r}_{11}^{(1)}\right) + \exp\left(-\boldsymbol{r}_{11}^{(1)}\right)}\right)^2\right)\begin{pmatrix} w_{1\to 1}^{(2)} \\ w_{1\to 2}^{(2)} \\ w_{1\to 3}^{(2)} \\ w_{1\to 4}^{(2)} \\ w_{1\to 5}^{(2)} \end{pmatrix} x'_{1j}, \tag{9.21a}$$

$$\frac{\partial \mathcal{L}_1}{\partial w_{j\to 2}^{(1)}} = 2(\hat{\boldsymbol{X}}_1 - \boldsymbol{X}_1)\left(1 - \left(\frac{\exp\left(\boldsymbol{r}_{12}^{(1)}\right) - \exp\left(-\boldsymbol{r}_{12}^{(1)}\right)}{\exp\left(\boldsymbol{r}_{12}^{(1)}\right) + \exp\left(-\boldsymbol{r}_{12}^{(1)}\right)}\right)^2\right)\begin{pmatrix} w_{2\to 1}^{(2)} \\ w_{2\to 2}^{(2)} \\ w_{2\to 3}^{(2)} \\ w_{2\to 4}^{(2)} \\ w_{2\to 5}^{(2)} \end{pmatrix} x'_{1j}, \tag{9.21b}$$

$$\frac{\partial \mathcal{L}_1}{\partial w_{1\to k}^{(2)}} = 2(\hat{\boldsymbol{X}}_1 - \boldsymbol{X}_1)\begin{pmatrix} z_{11}^{(1)} \\ z_{11}^{(1)} \\ z_{11}^{(1)} \\ z_{11}^{(1)} \\ z_{11}^{(1)} \end{pmatrix}, \tag{9.21c}$$

$$\frac{\partial \mathcal{L}_1}{\partial w_{2\to k}^{(2)}} = 2(\hat{\boldsymbol{X}}_1 - \boldsymbol{X}_1)\begin{pmatrix} z_{12}^{(1)} \\ z_{12}^{(1)} \\ z_{12}^{(1)} \\ z_{12}^{(1)} \\ z_{12}^{(1)} \end{pmatrix}, \tag{9.21d}$$

$$\frac{\partial \mathcal{L}_1}{\partial b_1^{(1)}} = 2(\hat{\boldsymbol{X}}_1 - \boldsymbol{X}_1)\left(1 - \left(\frac{\exp\left(\boldsymbol{r}_{11}^{(1)}\right) - \exp\left(-\boldsymbol{r}_{11}^{(1)}\right)}{\exp\left(\boldsymbol{r}_{11}^{(1)}\right) + \exp\left(-\boldsymbol{r}_{11}^{(1)}\right)}\right)^2\right)\begin{pmatrix} w_{1\to 1}^{(2)} \\ w_{1\to 2}^{(2)} \\ w_{1\to 3}^{(2)} \\ w_{1\to 4}^{(2)} \\ w_{1\to 5}^{(2)} \end{pmatrix}, \tag{9.21e}$$

$$\frac{\partial \mathcal{L}_1}{\partial b_2^{(1)}} = 2(\hat{\boldsymbol{X}}_1 - \boldsymbol{X}_1)\left(1 - \left(\frac{\exp\left(\boldsymbol{r}_{12}^{(1)}\right) - \exp\left(-\boldsymbol{r}_{12}^{(1)}\right)}{\exp\left(\boldsymbol{r}_{12}^{(1)}\right) + \exp\left(-\boldsymbol{r}_{12}^{(1)}\right)}\right)^2\right)\begin{pmatrix} w_{2\to 1}^{(2)} \\ w_{2\to 2}^{(2)} \\ w_{2\to 3}^{(2)} \\ w_{2\to 4}^{(2)} \\ w_{2\to 5}^{(2)} \end{pmatrix}, \tag{9.21f}$$

$$\begin{pmatrix} \frac{\partial \mathcal{L}_1}{\partial b_1^{(2)}} \\ \frac{\partial \mathcal{L}_1}{\partial b_2^{(2)}} \\ \frac{\partial \mathcal{L}_1}{\partial b_3^{(2)}} \\ \frac{\partial \mathcal{L}_1}{\partial b_4^{(2)}} \\ \frac{\partial \mathcal{L}_1}{\partial b_5^{(2)}} \end{pmatrix} = 2(\hat{\boldsymbol{X}}_1 - \boldsymbol{X}_1) \begin{bmatrix} 1 & 0 & 0 & 0 & 0 \\ 0 & 1 & 0 & 0 & 0 \\ 0 & 0 & 1 & 0 & 0 \\ 0 & 0 & 0 & 1 & 0 \\ 0 & 0 & 0 & 0 & 1 \end{bmatrix}$$

$$(k = 1, 2, 3, 4, 5). \tag{9.21g}$$

Take $w_{1\to 1}^{(1)}$ as an example to illustrate the BP process and the others will be conducted in the same way. According to Eq. (9.20f), we have

$$\frac{\partial z_{11}^{(1)}}{\partial r_{11}^{(1)}} = 1 - \left(\frac{\exp\left(\boldsymbol{r}_{11}^{(1)}\right) - \exp\left(-\boldsymbol{r}_{11}^{(1)}\right)}{\exp\left(\boldsymbol{r}_{11}^{(1)}\right) + \exp\left(-\boldsymbol{r}_{11}^{(1)}\right)} \right)^2$$

$$= 1 - \left(\frac{\exp(-2.8312) - \exp(2.8312)}{\exp(-2.8312) + \exp(2.8312)} \right)^2 = 0.0138,$$

thus,

$$\frac{\partial \mathcal{L}_1}{\partial w_{1\to 1}^{(1)}} = 2(\hat{\boldsymbol{X}}_1 - \boldsymbol{X}_1) \left(1 - \left(\frac{\exp\left(\boldsymbol{r}_{11}^{(1)}\right) - \exp\left(\boldsymbol{r}_{11}^{(1)}\right)}{\exp\left(\boldsymbol{r}_{11}^{(1)}\right) - \exp\left(\boldsymbol{r}_{11}^{(1)}\right)} \right)^2 \right) \begin{pmatrix} w_{1\to 1}^{(2)} \\ w_{1\to 2}^{(2)} \\ w_{1\to 3}^{(2)} \\ w_{1\to 4}^{(2)} \\ w_{1\to 5}^{(2)} \end{pmatrix} x'_{11}$$

$$= 2 \begin{bmatrix} 2.1010 & -1 \\ 1.0698 & -0.8451 \\ -1.3134 & -0 \\ -2.7448 & -0.25 \\ -1.5045 & -0 \end{bmatrix}^{\mathrm{T}} \times 0.0138 \times \begin{pmatrix} -0.666 \\ -0.4168 \\ -1.8211 \\ 0.6804 \\ -0.2614 \end{pmatrix}$$

$$\times 2.0617 = -0.2858.$$

The learning rate η is set to be 0.01 and the parameters are adjusted according to the gradients, which yield the new value of $w_{1\to 1}^{(1)}$

$$w_{1\to 1}^{(1)\text{new}} = w_{1\to 1}^{(1)} - \frac{\partial \mathcal{L}_1}{w_{1\to 1}^{(1)}} \times \eta = -0.8113 + 0.2858 \times 0.01 = -0.808442.$$

Similarly, all values of the 27 trainable variables are updated in the first iteration step with $\boldsymbol{X}_1$. When all the 17 data are used once, i.e., 17 iteration steps have been completed, one epoch is finished.

After 400 epochs, the loss is $\mathcal{L} = 0.03$ with added noise. For comparison, we also calculate the output results without noise in the same way. Table 9.14 shows the predicted values of each sample from the two models and Table 9.15 shows the R^2 values of the predicted results of the two models. The average prediction accuracy is slightly improved after adding noise.

Table 9.14 Predicted values of each sample by the two models

After adding noise of $N(0,1)$					No noise				
Mn	Ce	Fe	GHSV	T	Mn	Ce	Fe	GHSV	T
0.85	0.78	0.44	0.18	0.09	0.92	0.76	0.38	0.17	0.05
0.47	0.03	0.25	1.06	0.55	0.31	0.09	0.22	1.04	0.62
0.80	−0.05	0.36	1.05	0.38	0.84	−0.01	0.35	0.98	0.33
0.87	0.00	0.39	0.98	0.32	0.87	−0.01	0.35	0.98	0.32
0.73	−0.02	0.34	1.04	0.41	0.69	−0.04	0.31	1.06	0.44
0.79	0.04	0.36	0.96	0.36	0.83	−0.01	0.34	0.99	0.34
0.25	0.16	0.18	0.99	0.64	0.39	0.10	0.24	1.01	0.57
−0.15	0.51	0.08	0.74	0.76	−0.07	0.48	0.14	0.76	0.72
−0.08	0.56	0.10	0.67	0.70	−0.08	0.53	0.14	0.71	0.72
−0.08	0.78	0.12	0.44	0.63	0.01	0.74	0.16	0.47	0.60
0.06	0.96	0.18	0.22	0.49	0.11	0.93	0.19	0.25	0.48
−0.03	0.85	0.14	0.36	0.58	−0.09	0.92	0.14	0.31	0.60
0.01	0.88	0.16	0.32	0.55	−0.04	0.91	0.16	0.31	0.57
0.07	0.81	0.17	0.37	0.54	−0.01	0.87	0.16	0.34	0.57
0.01	0.84	0.16	0.36	0.56	0.04	0.88	0.17	0.32	0.54
0.26	0.82	0.24	0.31	0.42	0.14	0.80	0.20	0.37	0.50
−0.12	0.47	0.08	0.77	0.75	−0.13	0.50	0.13	0.75	0.76

Table 9.15 The R^2 values of the two models

Features	After adding noise of $N(0,1)$	No noise
Mn	0.9406	0.9019
Ce	0.9367	0.9098
Fe	0.1061	0.1608
GHSV	0.9425	0.9487
T	0.6942	0.6391
The average R^2 values of five features	0.7240	0.7121

9.5.3 Sparse Autoencoder

When the layer number and/or neuron number increase, the number of trainable variables could be extremely large, which demands big data and large computation capacity. Sparse autoencoders (Andrew, 2011) aim to make some trainable variables zero and others non-zero. If k-trainable variables are non-zero in a sparse autoencoder and the others are zero, it is named the k-sparse autoencoder. Like in L_1 regularization, a sparsity-regularization-penalty term takes the form of

$$\boldsymbol{\Omega}_{\mathrm{s}}(\boldsymbol{W}) = \sum_{l=1}^{L-1} \sum_{g_l=1}^{G_l} \left(p\ln\frac{p}{p_{g_l}^{(l)}} + (1-p)\ln\frac{1-p}{1-p_{g_l}^{(l)}} \right), \tag{9.22a}$$

where p is called the sparsity proportion of the sparse autoencoder, which is a hyperparameter among (0, 1) and optimized by cross-validation, and $p_{g_l}^{(l)}$ is defined by

$$p_{g_l}^{(l)} = \frac{1}{n} \sum_{i=1}^{n} z_{ig_l}^{(l)}. \tag{9.22b}$$

The sparsity-regularization-penalty term Eq. (9.22a) actually sums up the Kullback-Leibler (KL) divergence between p and $p_{g_l}^{(l)}$,

$$D_{\mathrm{KL}}\left(p \| p_{g_l}^{(l)}\right) = \left(p\ln\frac{p}{p_{g_l}^{(l)}} + (1-p)\ln\frac{1-p}{1-p_{g_l}^{(l)}} \right),$$

and hence Eq. (9.22a) is rewritten as

$$\Omega_{\mathrm{s}}(\boldsymbol{W}) = \sum_{l=1}^{L-1} \sum_{g_l=1}^{G_l} D_{\mathrm{KL}}\left(p \| p_{g_l}^{(l)}\right). \tag{9.22c}$$

Obviously, if $p = p_{g_l}^{(l)}$, the $D_{\mathrm{KL}}\big(p \| p_{g_l}^{(l)}\big)$ is null and the neuron's output $z_{g_l}^{(l)}$ will not be penalized, otherwise it will be penalized. An L_2 regularization-like penalty term can also be added into an NN, which takes the form of

$$\Omega_{\mathrm{w}}(\boldsymbol{W}) = \frac{1}{2}\sum_{l=1}^{L}\sum_{g_{l-1}=1}^{G_{l-1}}\sum_{g_l=1}^{G_l}\left(w_{g_{l-1}\to g_l}^{(l)}\right)^2 . \tag{9.23}$$

If $\phi(\hat{\boldsymbol{X}}_i) = \big(\phi_1(\hat{\boldsymbol{X}}_i)\ \phi_2(\hat{\boldsymbol{X}}_i)\ \cdots\ \phi_{G_{L/2}}(\hat{\boldsymbol{X}}_i)\big)$ in the middle layer $L/2$, its Jacobian matrix is given by

$$J_\phi(\hat{\boldsymbol{X}}_i) = \begin{bmatrix} \frac{\partial \phi_1}{\partial x_1} & \cdots & \frac{\partial \phi_1}{\partial x_m} \\ \vdots & & \vdots \\ \frac{\partial \phi_{G_{L/2}}}{\partial x_1} & \cdots & \frac{\partial \phi_{G_{L/2}}}{\partial x_m} \end{bmatrix} . \tag{9.24a}$$

The Frobenius norm is defined by

$$\left(J_\phi(\hat{\boldsymbol{X}}_i)\right)_{\mathrm{F}} = \sqrt{\mathrm{tr}((J_\phi(\hat{\boldsymbol{X}}_i))^{\mathrm{T}} J_\phi(\hat{\boldsymbol{X}}_i))} = \sqrt{\sum_{g_{L/2}=1}^{G_{L/2}}\sum_{j=1}^{m}(J_\phi(\hat{\boldsymbol{X}}_i))_{g_{L/2}j}^2} . \tag{9.24b}$$

The Frobenius penalty term is given by

$$\Omega_{\mathrm{F}}(\boldsymbol{W}) = \frac{1}{2}\sum_{i=1}^{n}\left(J_\phi(\hat{\boldsymbol{X}}_i)\right)_{\mathrm{F}}^2 = \frac{1}{2}\sum_{i=1}^{n}\sum_{g_{L/2}=1}^{G_{L/2}}\sum_{j=1}^{m}(J_\phi(\hat{\boldsymbol{X}}_i))_{g_{L/2}j}^2 . \tag{9.24c}$$

With one of the penalty terms, the values of trainable variables are determined from the minimum of the regularized loss function

$$\hat{\boldsymbol{W}} = \underset{\boldsymbol{W}}{\mathrm{argmin}}[\mathcal{L}(\boldsymbol{W}) + \lambda\Omega_{\mathrm{s}}(\boldsymbol{W})] , \tag{9.25a}$$

$$\hat{\boldsymbol{W}} = \underset{\boldsymbol{W}}{\mathrm{argmin}}[\mathcal{L}(\boldsymbol{W}) + \lambda\Omega_{\mathrm{w}}(\boldsymbol{W})] , \tag{9.25b}$$

$$\hat{\boldsymbol{W}} = \underset{\boldsymbol{W}}{\mathrm{argmin}}[\mathcal{L}(\boldsymbol{W}) + \lambda\Omega_{\mathrm{F}}(\boldsymbol{W})] , \tag{9.25c}$$

where λ is a positive hyperparameter, as discussed before in L_1 regularization, L_2 regularization, and Frobenius regularization. The value of the hyperparameter λ is optimized by cross-validation.

Example 9.7 The data listed in Table 9.13 are used in this example, Fig. 9.18 shows the neural structure used in this example, and the L_2 regularization-like penalty term $\Omega_{\mathrm{s}}(\boldsymbol{W})$is adopted here.

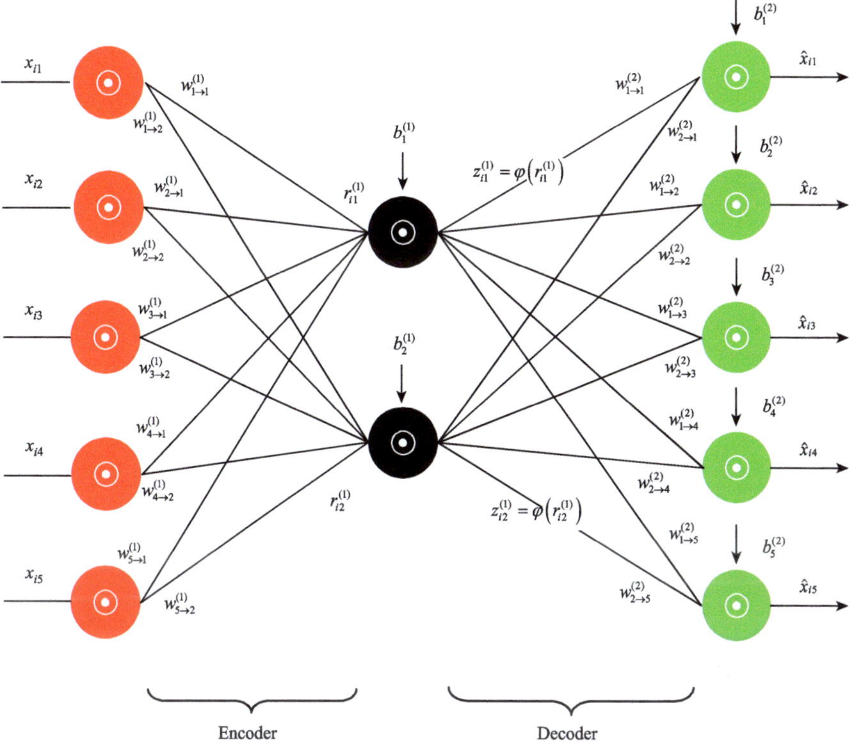

Fig. 9.18 The structure of the sparse autoencoder NN used in Example 9.7

The initial values of the trainable variables are set as a series of numbers conforming to the normal distribution N (0, 1) so that

$$\boldsymbol{W}^{(1)} = \begin{bmatrix} w_{1\to1}^{(1)} & w_{1\to2}^{(1)} \\ w_{2\to1}^{(1)} & w_{2\to2}^{(1)} \\ w_{3\to1}^{(1)} & w_{3\to2}^{(1)} \\ w_{4\to1}^{(1)} & w_{4\to2}^{(1)} \\ w_{5\to1}^{(1)} & w_{5\to2}^{(1)} \end{bmatrix} = \begin{bmatrix} -0.7258 & 1.0124 \\ -0.7520 & -0.8886 \\ -2.0785 & 0.1374 \\ 0.2474 & -0.4235 \\ -1.5115 & -0.3998 \end{bmatrix},$$

$$\boldsymbol{b}^{(1)} = \left(b_1^{(1)}\ b_2^{(1)}\right) = (-0.1233\ 0.3371),$$

$$\boldsymbol{W}^{(2)} = \begin{bmatrix} w_{1\to1}^{(2)} & w_{1\to2}^{(2)} & w_{1\to3}^{(2)} & w_{1\to4}^{(2)} & w_{1\to5}^{(2)} \\ w_{2\to1}^{(2)} & w_{2\to2}^{(2)} & w_{2\to3}^{(2)} & w_{2\to4}^{(2)} & w_{2\to5}^{(2)} \end{bmatrix}$$

$$= \begin{bmatrix} -0.1954 & -0.9807 & -0.4593 & 2.2266 & -1.0782 \\ -0.5019 & -0.0340 & 1.7136 & -0.2694 & 0.9063 \end{bmatrix},$$

$$\boldsymbol{b}^{(2)} = \left(b_1^{(2)}\ b_2^{(2)}\ b_3^{(2)}\ b_4^{(2)}\ b_5^{(2)}\right)$$

$$= \begin{pmatrix} -0.9500 & -1.0599 & 0.4022 & -0.2024 & -0.4045 \end{pmatrix}.$$

Data 1–4 are employed in the first iteration step, which are

$$\begin{bmatrix} \boldsymbol{X}_1 \\ \boldsymbol{X}_2 \\ \boldsymbol{X}_3 \\ \boldsymbol{X}_4 \end{bmatrix} = \begin{bmatrix} 1 & 0.8451 & 0 & 0.25 & 0 \\ 0.2269 & 0 & 0 & 1 & 0.4 \\ 0.8067 & 0 & 0.6216 & 1 & 0.4 \\ 0.7661 & 0 & 0.8919 & 1 & 0.4 \end{bmatrix}.$$

In each iteration, different samples use the same values of trainable variables in the FP. The following shows the detailed calculation in the FP.

$$\boldsymbol{r}_1^{(1)} = \begin{bmatrix} r_{11}^{(1)} & r_{12}^{(1)} \\ r_{21}^{(1)} & r_{22}^{(1)} \\ r_{31}^{(1)} & r_{32}^{(1)} \\ r_{41}^{(1)} & r_{42}^{(1)} \end{bmatrix} = \begin{bmatrix} \boldsymbol{X}_1 \\ \boldsymbol{X}_2 \\ \boldsymbol{X}_3 \\ \boldsymbol{X}_4 \end{bmatrix} \boldsymbol{W}^{(1)} + \boldsymbol{b}^{(1)}$$

$$= \begin{bmatrix} \boldsymbol{X}_1 \\ \boldsymbol{X}_2 \\ \boldsymbol{X}_3 \\ \boldsymbol{X}_4 \end{bmatrix} \begin{bmatrix} w_{1\to1}^{(1)} & w_{1\to2}^{(1)} \\ w_{2\to1}^{(1)} & w_{2\to2}^{(1)} \\ w_{3\to1}^{(1)} & w_{3\to2}^{(1)} \\ w_{4\to1}^{(1)} & w_{4\to2}^{(1)} \\ w_{5\to1}^{(1)} & w_{5\to2}^{(1)} \end{bmatrix} + \begin{bmatrix} b_1^{(1)} & b_2^{(1)} \\ b_1^{(1)} & b_2^{(1)} \\ b_1^{(1)} & b_2^{(1)} \\ b_1^{(1)} & b_2^{(1)} \end{bmatrix}$$

$$= \begin{bmatrix} 1 & 0.8451 & 0 & 0.25 & 0 \\ 0.2269 & 0 & 0 & 1 & 0.4 \\ 0.8067 & 0 & 0.6216 & 1 & 0.4 \\ 0.7661 & 0 & 0.8919 & 1 & 0.4 \end{bmatrix} \begin{bmatrix} -0.7258 & 1.0124 \\ -0.7520 & -0.8886 \\ -2.0785 & 0.1374 \\ 0.2474 & -0.4235 \\ -1.5115 & -0.3998 \end{bmatrix}$$

$$+ \begin{bmatrix} -0.1233 & 0.3371 \\ -0.1233 & 0.3371 \\ -0.1233 & 0.3371 \\ -0.1233 & 0.3371 \end{bmatrix} = \begin{bmatrix} -1.4228 & 0.4926 \\ -0.6452 & -0.0166 \\ -2.3580 & 0.6560 \\ -2.8903 & 0.6518 \end{bmatrix}.$$

Taking the tanh function as the activation function in this example, we have

$$\boldsymbol{z}_1^{(1)} = \begin{bmatrix} z_{11}^{(1)} & z_{12}^{(1)} \\ z_{21}^{(1)} & z_{22}^{(1)} \\ z_{31}^{(1)} & z_{22}^{(1)} \\ z_{41}^{(1)} & z_{22}^{(1)} \end{bmatrix} = \varphi\left(\boldsymbol{r}_1^{(1)}\right) = \begin{bmatrix} \tanh(r_{11}^{(1)}) & \tanh(r_{12}^{(1)}) \\ \tanh(r_{21}^{(1)}) & \tanh(r_{22}^{(1)}) \\ \tanh(r_{31}^{(1)}) & \tanh(r_{32}^{(1)}) \\ \tanh(r_{41}^{(1)}) & \tanh(r_{42}^{(1)}) \end{bmatrix}$$

$$= \begin{bmatrix} -0.8902 & 0.4563 \\ -0.5684 & -0.0166 \\ -0.9823 & 0.5756 \\ -0.9938 & 0.5729 \end{bmatrix}.$$

Then, decoding the $z_1^{(1)}$ to produce the prediction of $\hat{\boldsymbol{X}}_1$, viz.,

$$\begin{bmatrix} \hat{\boldsymbol{X}}_1 \\ \hat{\boldsymbol{X}}_2 \\ \hat{\boldsymbol{X}}_3 \\ \hat{\boldsymbol{X}}_4 \end{bmatrix} = \boldsymbol{z}_1^{(1)} \boldsymbol{W}^{(2)} + \boldsymbol{b}^{(2)}$$

$$= \begin{bmatrix} z_{11}^{(1)} & z_{12}^{(1)} \\ z_{21}^{(1)} & z_{22}^{(1)} \\ z_{31}^{(1)} & z_{22}^{(1)} \\ z_{41}^{(1)} & z_{22}^{(1)} \end{bmatrix} \begin{bmatrix} w_{1\to1}^{(2)} & w_{1\to2}^{(2)} & w_{1\to3}^{(2)} & w_{1\to4}^{(2)} & w_{1\to5}^{(2)} \\ w_{2\to1}^{(2)} & w_{2\to2}^{(2)} & w_{2\to3}^{(2)} & w_{2\to4}^{(2)} & w_{2\to5}^{(2)} \end{bmatrix}$$

$$+ \begin{bmatrix} b_1^{(2)} & b_2^{(2)} & b_3^{(2)} & b_4^{(2)} & b_5^{(2)} \\ b_1^{(2)} & b_2^{(2)} & b_3^{(2)} & b_4^{(2)} & b_5^{(2)} \\ b_1^{(2)} & b_2^{(2)} & b_3^{(2)} & b_4^{(2)} & b_5^{(2)} \\ b_1^{(2)} & b_2^{(2)} & b_3^{(2)} & b_4^{(2)} & b_5^{(2)} \end{bmatrix}$$

$$= \begin{bmatrix} -0.8902 & 0.4563 \\ -0.5684 & -0.0166 \\ -0.9823 & 0.5756 \\ -0.9938 & 0.5729 \end{bmatrix}$$

$$\begin{bmatrix} -0.1954 & -0.9807 & -0.4593 & 2.2266 & -1.0782 \\ -0.5019 & -0.0340 & 1.7136 & -0.2694 & 0.9063 \end{bmatrix}$$

$$+ \begin{bmatrix} -0.9500 & -1.0599 & 0.4022 & -0.2024 & -0.4045 \\ -0.9500 & -1.0599 & 0.4022 & -0.2024 & -0.4045 \\ -0.9500 & -1.0599 & 0.4022 & -0.2024 & -0.4045 \\ -0.9500 & -1.0599 & 0.4022 & -0.2024 & -0.4045 \end{bmatrix}$$

$$= \begin{bmatrix} -1.005 & -0.2024 & 1.5930 & -2.3074 & 0.9689 \\ -0.8306 & -0.5019 & 0.6348 & -1.4636 & 0.1933 \\ -1.0469 & -0.1162 & 1.8396 & -2.5446 & 1.1762 \\ -1.0433 & -0.1047 & 1.8404 & -2.5696 & 1.1863 \end{bmatrix}.$$

Therefore, the average loss function $\overline{\mathcal{L}}$ of samples is

$$\overline{\mathcal{L}} = \frac{1}{4} \sum_{i=1}^{4} (\hat{\boldsymbol{X}}_i - \boldsymbol{X}_i)^2 + \frac{1}{4} \lambda \Omega_{\mathrm{w}}(\boldsymbol{W}).$$

The first term $\overline{\mathcal{L}}$ is the mean square error, i.e.,

$$\frac{1}{4}\sum_{i=1}^{4}(\hat{\boldsymbol{X}}_i-\boldsymbol{X}_i)^2=\frac{1}{4}\sum_{i=1}^{4}((\hat{x}_{i1}-x_{i1})^2+(\hat{x}_{i2}-x_{i2})^2+(\hat{x}_{i3}-x_{i3})^2 + (\hat{x}_{i4}-x_{i4})^2+(\hat{x}_{i5}-x_{i5})^2).$$

The second term $\overline{\mathcal{L}}$ is the sum of 20 square weights, i.e.,

$$\begin{aligned}\Omega_{\mathrm{w}}(\boldsymbol{W})=\frac{1}{2}\Big(&\left|w_{1\to1}^{(1)}\right|^2+\left|w_{1\to2}^{(1)}\right|^2+\left|w_{2\to1}^{(1)}\right|^2+\left|w_{2\to2}^{(1)}\right|^2+\left|w_{3\to1}^{(1)}\right|^2\\&+\left|w_{3\to2}^{(1)}\right|^2+\left|w_{4\to1}^{(1)}\right|^2+\left|w_{4\to2}^{(1)}\right|^2+\left|w_{5\to1}^{(1)}\right|^2+\left|w_{5\to2}^{(1)}\right|^2+\left|w_{1\to1}^{(2)}\right|^2\\&+\left|w_{1\to2}^{(2)}\right|^2+\left|w_{1\to3}^{(2)}\right|^2+\left|w_{1\to4}^{(2)}\right|^2+\left|w_{1\to5}^{(2)}\right|^2\\&+\left|w_{2\to1}^{(2)}\right|^2+\left|w_{2\to2}^{(2)}\right|^2+\left|w_{2\to3}^{(2)}\right|^2+\left|w_{2\to4}^{(2)}\right|^2+\left|w_{2\to5}^{(2)}\right|^2\Big).\end{aligned}$$

Setting $\lambda=0.1$ gives the average loss

$$\overline{\mathcal{L}}=\frac{\mathcal{L}_1+\mathcal{L}_2+\mathcal{L}_3+\mathcal{L}_4}{4}=14.9328.$$

For data $\boldsymbol{X}_1$, the gradients of the loss function are

$$\begin{cases}\dfrac{\partial\mathcal{L}_1}{\partial w_{j\to1}^{(1)}}=2(\hat{\boldsymbol{X}}_1-\boldsymbol{X}_1)\dfrac{\partial\hat{\boldsymbol{X}}_1}{\partial z_{11}^{(1)}}\dfrac{\partial z_{11}^{(1)}}{\partial r_{11}^{(1)}}\dfrac{\partial r_{11}^{(1)}}{\partial w_{j\to1}^{(1)}}+\lambda w_{j\to1}^{(1)},\\ \dfrac{\partial\mathcal{L}_1}{\partial w_{j\to2}^{(1)}}=2(\hat{\boldsymbol{X}}_1-\boldsymbol{X}_1)\dfrac{\partial\hat{\boldsymbol{X}}_1}{\partial z_{12}^{(1)}}\dfrac{\partial z_{12}^{(1)}}{\partial r_{12}^{(1)}}\dfrac{\partial r_{12}^{(1)}}{\partial w_{j\to2}^{(1)}}+\lambda w_{j\to2}^{(1)},\end{cases}$$

$$\begin{cases}\dfrac{\partial\mathcal{L}_1}{\partial w_{1\to k}^{(2)}}=2(\hat{\boldsymbol{X}}_1-\boldsymbol{X}_1)\dfrac{\partial\hat{\boldsymbol{X}}_1}{\partial w_{1\to k}^{(2)}}+\lambda w_{1\to k}^{(2)},\\ \dfrac{\partial\mathcal{L}_1}{\partial w_{2\to k}^{(2)}}=2(\hat{\boldsymbol{X}}_1-\boldsymbol{X}_1)\dfrac{\partial\hat{\boldsymbol{X}}_1}{\partial w_{2\to k}^{(2)}}+\lambda w_{2\to k}^{(2)},\end{cases}$$

$$\frac{\partial\mathcal{L}_1}{\partial b_1^{(1)}}=2(\hat{\boldsymbol{X}}_1-\boldsymbol{X}_1)\frac{\partial\hat{\boldsymbol{X}}_1}{\partial z_{11}^{(1)}}\frac{\partial z_{11}^{(1)}}{\partial r_{11}^{(1)}}\frac{\partial r_{11}^{(1)}}{\partial b_1^{(1)}},$$

$$\frac{\partial\mathcal{L}_1}{\partial b_2^{(1)}}=2(\hat{\boldsymbol{X}}_1-\boldsymbol{X}_1)\frac{\partial\hat{\boldsymbol{X}}_1}{\partial z_{12}^{(1)}}\frac{\partial z_{12}^{(1)}}{\partial r_{12}^{(1)}}\frac{\partial r_{12}^{(1)}}{\partial b_2^{(1)}},$$

$$\frac{\partial\mathcal{L}_1}{\partial b_k^{(2)}}=2(\hat{\boldsymbol{X}}_1-\boldsymbol{X}_1)\frac{\partial\hat{\boldsymbol{X}}_1}{\partial b_k^{(2)}}$$

$$(j=1,2,3,4,5;\ k=1,2,3,4,5).$$

For data $\boldsymbol{X}_1$, the gradients of loss are calculated to be

$$\frac{\partial \mathcal{L}_1}{\partial w^{(1)}_{j\to 1}} = 2(\hat{\boldsymbol{X}}_1 - \boldsymbol{X}_1)\left(1 - \left(\frac{\exp\left(\boldsymbol{r}^{(1)}_{11}\right) - \exp\left(-\boldsymbol{r}^{(1)}_{11}\right)}{\exp\left(\boldsymbol{r}^{(1)}_{11}\right) + \exp\left(-\boldsymbol{r}^{(1)}_{11}\right)}\right)^2\right)\begin{bmatrix} w^{(2)}_{1\to 1} \\ w^{(2)}_{1\to 2} \\ w^{(2)}_{1\to 3} \\ w^{(2)}_{1\to 4} \\ w^{(2)}_{1\to 5} \end{bmatrix} x_{1j} + \lambda w^{(1)}_{j\to 1},$$

$$\frac{\partial \mathcal{L}_1}{\partial w^{(1)}_{j\to 2}} = 2(\hat{\boldsymbol{X}}_1 - \boldsymbol{X}_1)\left(1 - \left(\frac{\exp\left(\boldsymbol{r}^{(1)}_{12}\right) - \exp\left(-\boldsymbol{r}^{(1)}_{12}\right)}{\exp\left(\boldsymbol{r}^{(1)}_{12}\right) + \exp\left(-\boldsymbol{r}^{(1)}_{12}\right)}\right)^2\right)\begin{bmatrix} w^{(2)}_{1\to 1} \\ w^{(2)}_{1\to 2} \\ w^{(2)}_{1\to 3} \\ w^{(2)}_{1\to 4} \\ w^{(2)}_{1\to 5} \end{bmatrix} x_{1j} + \lambda w^{(1)}_{j\to 2},$$

$$\frac{\partial \mathcal{L}_1}{\partial w^{(2)}_{1\to k}} = 2(\hat{\boldsymbol{X}}_1 - \boldsymbol{X}_1)\begin{bmatrix} z^{(1)}_{11} \\ z^{(1)}_{11} \\ z^{(1)}_{11} \\ z^{(1)}_{11} \\ z^{(1)}_{11} \end{bmatrix} + \lambda w^{(2)}_{1\to k},$$

$$\frac{\partial \mathcal{L}_1}{\partial w^{(2)}_{2\to k}} = 2(\hat{\boldsymbol{X}}_1 - \boldsymbol{X}_1)\begin{bmatrix} z^{(1)}_{12} \\ z^{(1)}_{12} \\ z^{(1)}_{12} \\ z^{(1)}_{12} \\ z^{(1)}_{12} \end{bmatrix} + \lambda w^{(2)}_{2\to k},$$

$$\frac{\partial \mathcal{L}_1}{\partial b^{(1)}_1} = 2(\hat{\boldsymbol{X}}_1 - \boldsymbol{X}_1)\left(1 - \left(\frac{\exp\left(\boldsymbol{r}^{(1)}_{11}\right) - \exp\left(-\boldsymbol{r}^{(1)}_{11}\right)}{\exp\left(\boldsymbol{r}^{(1)}_{11}\right) + \exp\left(-\boldsymbol{r}^{(1)}_{11}\right)}\right)^2\right)\begin{bmatrix} w^{(2)}_{1\to 1} \\ w^{(2)}_{1\to 2} \\ w^{(2)}_{1\to 3} \\ w^{(2)}_{1\to 4} \\ w^{(2)}_{1\to 5} \end{bmatrix},$$

$$\frac{\partial \mathcal{L}_1}{\partial b^{(1)}_2} = 2(\hat{\boldsymbol{X}}_1 - \boldsymbol{X}_1)\left(1 - \left(\frac{\exp\left(\boldsymbol{r}^{(1)}_{12}\right) - \exp\left(-\boldsymbol{r}^{(1)}_{12}\right)}{\exp\left(\boldsymbol{r}^{(1)}_{12}\right) + \exp\left(-\boldsymbol{r}^{(1)}_{12}\right)}\right)^2\right)\begin{bmatrix} w^{(2)}_{2\to 1} \\ w^{(2)}_{2\to 2} \\ w^{(2)}_{2\to 3} \\ w^{(2)}_{2\to 4} \\ w^{(2)}_{2\to 5} \end{bmatrix},$$

$$\begin{bmatrix} \frac{\partial \mathcal{L}_1}{\partial b^{(2)}_1} \\ \frac{\partial \mathcal{L}_1}{\partial b^{(2)}_2} \\ \frac{\partial \mathcal{L}_1}{\partial b^{(2)}_3} \\ \frac{\partial \mathcal{L}_1}{\partial b^{(2)}_4} \\ \frac{\partial \mathcal{L}_1}{\partial b^{(2)}_5} \end{bmatrix} = 2(\hat{\boldsymbol{X}}_1 - \boldsymbol{X}_1)\begin{bmatrix} 1 & 0 & 0 & 0 & 0 \\ 0 & 1 & 0 & 0 & 0 \\ 0 & 0 & 1 & 0 & 0 \\ 0 & 0 & 0 & 1 & 0 \\ 0 & 0 & 0 & 0 & 1 \end{bmatrix}.$$

Take $w^{(1)}_{1\to 1}$ as an example to illustrate the BP process and the other trainable variables will be conducted in the same way. Thus

$$\frac{\partial \mathcal{L}_1}{\partial w_{1\to 1}^{(1)}} = 2(\hat{\boldsymbol{X}}_1 - \boldsymbol{X}_1)\left(1 - \left(\frac{\exp(\boldsymbol{r}_{11}^1) - \exp(-\boldsymbol{r}_{11}^1)}{\exp(\boldsymbol{r}_{11}^1) + \exp(-\boldsymbol{r}_{11}^1)}\right)^2\right)\begin{bmatrix} w_{1\to 1}^{(2)} \\ w_{1\to 2}^{(2)} \\ w_{1\to 3}^{(2)} \\ w_{1\to 4}^{(2)} \\ w_{1\to 5}^{(2)} \end{bmatrix} x_{11} + \lambda w_{1\to 1}^{(1)}$$

$$= 2\begin{bmatrix} -1.1972 & -1 \\ -1.0766 & -0.8451 \\ 1.2464 & -0 \\ -0.3351 & -0.25 \\ 0.0420 & -0 \end{bmatrix}^{\mathrm{T}} \times \left(1 - \left(\frac{\exp(-1.4228) - \exp(1.4228)}{\exp(-1.4228) + \exp(-1.4228)}\right)^2\right)$$

$$\times \begin{bmatrix} -0.1954 \\ -0.9807 \\ -0.4593 \\ 2.2266 \\ -1.0782 \end{bmatrix} \times 1 + 0.1 \times (-0.7258) = -2.5850.$$

Since in the example we use data 1–4 in the first iteration for updating trainable variables, we have

$$\frac{\partial \mathcal{L}_2}{\partial w_{1\to 1}^{(1)}} = -2.0884,$$

$$\frac{\partial \mathcal{L}_3}{\partial w_{1\to 1}^{(1)}} = -3.7313,$$

$$\frac{\partial \mathcal{L}_4}{\partial w_{1\to 1}^{(1)}} = -3.7159,$$

$$\frac{\partial \overline{\mathcal{L}}}{\partial w_{1\to 1}^{(1)}} = \frac{\frac{\partial \mathcal{L}_1}{\partial w_{1\to 1}^{(1)}} + \frac{\partial \mathcal{L}_2}{\partial w_{1\to 1}^{(1)}} + \frac{\partial \mathcal{L}_3}{\partial w_{1\to 1}^{(1)}} + \frac{\partial \mathcal{L}_4}{\partial w_{1\to 1}^{(1)}}}{4} = -3.0302.$$

Let $\eta = 0.01$, the value of $w_{1\to 1}^{(1)}$ is updated by

$$w_{1\to 1(\text{new})}^{(1)} = w_{1\to 1(\text{old})}^{(1)} - \eta \frac{\partial \overline{\mathcal{L}}}{\partial w_{1\to 1}^{(1)}} = -0.7258 - 0.01 \times (-3.0302) = -0.695498.$$

The other trainable variables are updated in the same way, and the iteration stops until the average loss function $\overline{\mathcal{L}}$ reaches convergence.

Figure 9.19 shows the variations of weights with λ decreasing from 1 to $10^{-3.5}$ in the L_2 regularized autoencoder, where the weight appears simultaneously. The optimal λ value of the L_2 regularized autoencoder is determined by the maximum LOOCV-R^2, as shown in Fig. 9.20.

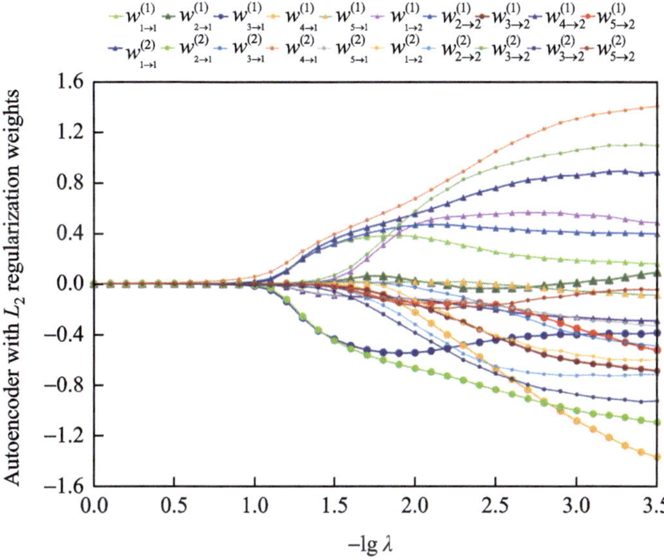

Fig. 9.19 The variation of trainable variables with hyperparameter λ in the L_2 regularized autoencoder on the training dataset

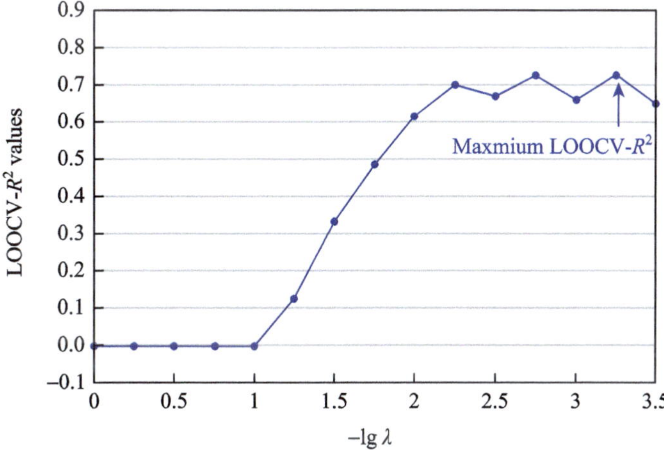

Fig. 9.20 The LOOCV-R^2 versus the λ value in the L_2 regularized autoencoder

---

9.5.4 Variational Autoencoder

The variational autoencoder (Kingma & Welling, 2013) is developed based on the consideration that data are generated by some random process, involving unobserved continuous random variables. Therefore, the variational autoencoder (Kingma & Welling, 2013) takes the same probabilistic prediction nature based on the normal distribution (or some "nice" distribution family) and the Bayesian theorem as that in Gaussian processes, which is briefly described in Chap. 7. Unlike other autoencoders, which encode an input sample as a single point, the variational autoencoder (VAE) encodes an input sample as many (infinite) normal distributions in the latent space, as shown in Fig. 9.21. Frankly speaking, the encoder in VAE is a Gaussian process, which regresses input data with variational virtual tasks in VAE. Kingma and Welling (2013) called the decoder function θ the generative model and involved parameters the generative model parameters, and parameters involved in the encoder function ϕ the variational parameters.

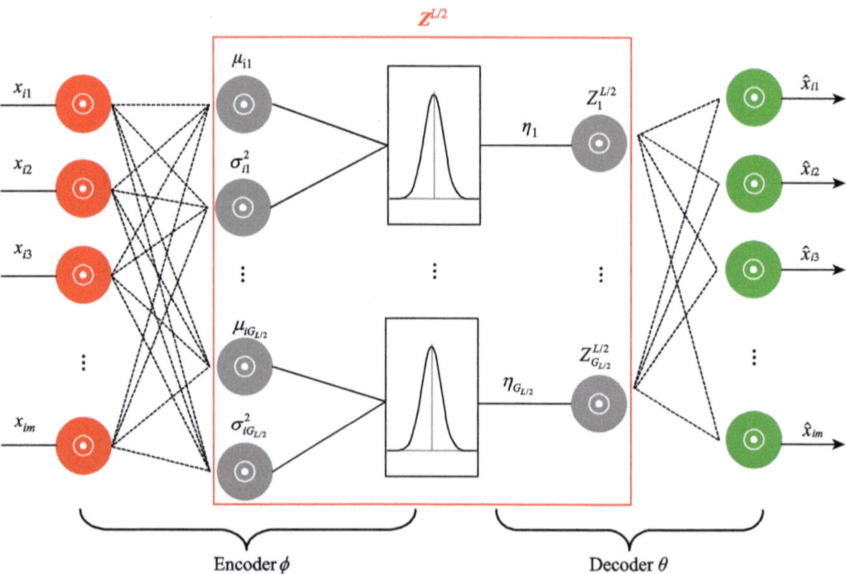

Fig. 9.21 The topology structure of a variational autoencoder, where only the input layer, the middle layer, and the output layer are highlighted and other hidden layers are ignored with dashed lines

For simplicity, the latent space called here is the latent variable space in the middle layer. For other autoencoders, the latent space is composed of $\boldsymbol{Z}^{L/2} = \left\{z_{ig_{L/2}}^{L/2}, b_{g_{L/2}}^{L/2}\right\}(g_{L/2} = 1, 2, \cdots, G_{L/2})$ of neurons in hidden layer $L/2$. In VAE, the latent space is composed of the infinite normal distribution $\boldsymbol{Z}^{L/2}$. Training data $\boldsymbol{X} = \{x_{i1}, x_{i2}, \cdots, x_{im}\}$ $(i = 1, 2, \cdots, n)$ of m features are mapped from the data space to the latent space and expressed in normal distributions of $\boldsymbol{Z}^{L/2}$, viz.,

$$\phi(\boldsymbol{X}) = \boldsymbol{Z}^{L/2}(\boldsymbol{X}) \sim N((\boldsymbol{\mu}(\boldsymbol{Z}^{L/2})|\boldsymbol{X}), (\boldsymbol{\Sigma}(\boldsymbol{Z}^{L/2})|\boldsymbol{X})), \tag{9.26}$$

where ϕ denotes the encoder function involved in the first half part, $\boldsymbol{\mu}(\boldsymbol{Z}^{L/2})$ is the mean vector and $\boldsymbol{\Sigma}(\boldsymbol{Z}^{L/2}) = \begin{bmatrix} \sigma_1^2 & & \\ & \ddots & \\ & & \sigma_{G_{L/2}}^2 \end{bmatrix}$ is the diagonal covariance matrix of multivariate normal distributions. In terms of probability, encode $\phi(\boldsymbol{X})$ is expressed by $q_\phi(\boldsymbol{Z}^{L/2}|\boldsymbol{X})$. The decoder function θ involved in the second half part gives the output $\hat{\boldsymbol{X}} = \theta(\phi(\boldsymbol{X})) = \theta(\boldsymbol{Z}^{L/2})$. In other words, the encoder function maps the input data to virtual tasks $q_\phi(\boldsymbol{Z}^{L/2}|\boldsymbol{X})$ and is expressed by normal distributions of multivariable mean vector and covariance matrix and the number of neurons in the middle layer $G_{L/2}$ might be regarded as the number of multiple virtual tasks. Based on the consideration that data are generated by some random process, the decoder should sample data randomly from the distributions of virtual tasks so that noises must be stochastically added on.

With the consideration that data are generated by some random process, the output datum points $\hat{\boldsymbol{X}}$ in VAE are from a distribution function of $p_\theta(\hat{\boldsymbol{X}})$, which is actually the marginal probability of the joint distribution $p_\theta\left(\hat{\boldsymbol{X}}, \boldsymbol{Z}^{L/2}\right)$ over $\boldsymbol{Z}^{(L/2)}$, i.e.,

$$p_\theta(\hat{\boldsymbol{X}}) = \int p_\theta\left(\hat{\boldsymbol{X}}, \boldsymbol{Z}^{L/2}\right)\mathrm{d}\boldsymbol{Z}^{L/2} = \int p_\theta\left(\hat{\boldsymbol{X}}|\boldsymbol{Z}^{L/2}\right)p_\theta(\boldsymbol{Z}^{L/2})\mathrm{d}\boldsymbol{Z}^{L/2}, \tag{9.27a}$$

where the subscript "θ" denotes the decoder part, i.e., the generative model. The marginal probability is called the evidence (Blei et al., 2017). Kingma and Welling (2013) emphasize that the integral of the marginal likelihood is intractable. The Bayesian theorem gives

$$p_\theta\left(\boldsymbol{Z}^{L/2}|\hat{\boldsymbol{X}}\right) = \frac{p_\theta\left(\hat{\boldsymbol{X}}|\boldsymbol{Z}^{L/2}\right)p_\theta(\boldsymbol{Z}^{L/2})}{p_\theta(\hat{\boldsymbol{X}})}, \tag{9.27b}$$

where the evidence (marginal probability) $p_\theta(\hat{\boldsymbol{X}})$ is called the total likelihood $p_\theta(\hat{\boldsymbol{X}})$, the conditional probability $p_\theta\left(\boldsymbol{Z}^{L/2}|\hat{\boldsymbol{X}}\right)$ is termed as the posterior probability of $\boldsymbol{Z}^{L/2}$, $p_\theta(\boldsymbol{Z}^{L/2})$ is the prior probability of $\boldsymbol{Z}^{L/2}$, and the conditional probability $p_\theta\left(\hat{\boldsymbol{X}}|\boldsymbol{Z}^{L/2}\right)$ is called conditional likelihood. As described in Chap. 7, the Bayesian theorem is usually used to calculate the posterior probability by ignoring total likelihood $p_\theta(\hat{\boldsymbol{X}})$, because the total likelihood is regarded as a normalizing constant to normalize the posterior probability (or density) and can be ignored if without considering normalization. The same approach is adopted in the VAE, as described by Blei et al. (2017). Kingma and Welling (2013) also emphasize that the posterior is intractable so that the EM algorithm cannot be used. In this circumstance, it becomes a challenge how to apply the error BP method with the loss function of the mean squared error of Eq. (9.17).

To solve these problems, Kingma and Welling (2013) introduce a recognition model $q_\phi(\boldsymbol{Z}^{L/2}|\boldsymbol{X})$ to approximately replace the intractable true posterior $p_\theta\left(\boldsymbol{Z}^{L/2}|\hat{\boldsymbol{X}}\right)$. The goal of the VAE to minimize the loss function of the mean squared error of $\mathcal{L}(\boldsymbol{W}) = \frac{1}{n}\sum_{i=1}^{n}\left[\left|\hat{\boldsymbol{X}}_i - \boldsymbol{X}_i\right|\right]^2$ can be achieved by minimizing the KL divergence between distribution $q_\phi(\boldsymbol{Z}^{L/2}|\boldsymbol{X})$ and $p_\theta\left(\boldsymbol{Z}^{L/2}|\hat{\boldsymbol{X}}\right)$, i.e.,

$$\min \mathrm{KL}\left(q_\phi(\boldsymbol{Z}^{L/2}|\boldsymbol{X}) \| p_\theta\left(\boldsymbol{Z}^{L/2}|\hat{\boldsymbol{X}}\right)\right), \tag{9.28a}$$

$$\mathrm{KL}\left(q_\phi\left(\boldsymbol{Z}^{L/2}|\boldsymbol{X}\right) \| p_\theta\left(\boldsymbol{Z}^{L/2}|\hat{\boldsymbol{X}}\right)\right) \equiv \int q_\phi\left(\boldsymbol{Z}^{L/2}|\boldsymbol{X}\right) \ln \frac{q_\phi\left(\boldsymbol{Z}^{L/2}|\boldsymbol{X}\right)}{p_\theta\left(\boldsymbol{Z}^{L/2}|\hat{\boldsymbol{X}}\right)} \mathrm{d}\boldsymbol{Z}^{L/2}. \tag{9.28b}$$

The KL divergence is always ≥ 0 and obviously, if $\boldsymbol{X} = \hat{\boldsymbol{X}}$, the KL divergence will be zero. Therefore, minimizing the KL divergence lets $\hat{\boldsymbol{X}}$ approach $\boldsymbol{X}$. The KL divergence, however, is also intractable. Taking the invariable nature of $\ln p_\theta(\hat{\boldsymbol{X}})$ with respect to $q_\phi\left(\boldsymbol{Z}^{L/2}|\boldsymbol{X}\right)$, Kingma and Welling (2013) have

$$\begin{aligned}\ln p_\theta(\hat{\boldsymbol{X}}) = &\int q_\phi\left(\boldsymbol{Z}^{L/2}|\boldsymbol{X}\right) \ln \frac{p_\theta\left(\hat{\boldsymbol{X}}, \boldsymbol{Z}^{L/2}\right)}{q_\phi\left(\boldsymbol{Z}^{L/2}|\boldsymbol{X}\right)} \mathrm{d}\boldsymbol{Z}^{L/2} \\ &+ \mathrm{KL}\left(q_\phi\left(\boldsymbol{Z}^{L/2}|\boldsymbol{X}\right) \| p_\theta\left(\boldsymbol{Z}^{L/2}|\hat{\boldsymbol{X}}\right)\right).\end{aligned} \tag{9.29a}$$

Since the KL divergence between distribution $q_\phi\left(\boldsymbol{Z}^{L/2}|\boldsymbol{X}\right)$ and $p_\theta\left(\boldsymbol{Z}^{L/2}|\hat{\boldsymbol{X}}\right)$ is always ≥ 0, the term $\int q_\phi\left(\boldsymbol{Z}^{L/2}|\boldsymbol{X}\right) \ln \frac{p_\theta\left(\hat{\boldsymbol{X}}, \boldsymbol{Z}^{L/2}\right)}{q_\phi\left(\boldsymbol{Z}^{L/2}|\boldsymbol{X}\right)} \mathrm{d}\boldsymbol{Z}^{L/2}$ is a lower bound for the evidence $\ln p_\theta(\hat{\boldsymbol{X}})$ and called as the evidence lower bound (ELBO), denoted by L_b. The ELBO is the negative KL divergence plus $\ln p_\theta(\hat{\boldsymbol{X}})$, which is a constant with respect to $q_\phi\left(\boldsymbol{Z}^{L/2}|\boldsymbol{X}\right)$. Thus, maximizing the ELBO is equivalent to minimizing the KL divergence (Blei et al., 2017). The ELBO can be rearranged to

$$\begin{aligned}L_b &= \int q_\phi(\boldsymbol{Z}^{L/2}|\boldsymbol{X}) \ln \frac{p_\theta(\hat{\boldsymbol{X}}, \boldsymbol{Z}^{L/2})}{q_\phi(\boldsymbol{Z}^{L/2}|\boldsymbol{X})} \mathrm{d}\boldsymbol{Z}^{L/2} \\ &= \int q_\phi(\boldsymbol{Z}^{L/2}|\boldsymbol{X}) \ln \frac{p_\theta\left(\hat{\boldsymbol{X}}|\boldsymbol{Z}^{L/2}\right) p_\theta(\boldsymbol{Z}^{L/2})}{q_\phi(\boldsymbol{Z}^{L/2}|\boldsymbol{X})} \mathrm{d}\boldsymbol{Z}^{L/2}\end{aligned}$$

$$
\begin{aligned}
&= \int q_\phi(\mathbf{Z}^{L/2}|\boldsymbol{X}) \ln \frac{p_\theta(\mathbf{Z}^{L/2})}{q_\phi(\mathbf{Z}^{L/2}|\boldsymbol{X})} \mathrm{d}\mathbf{Z}^{L/2} \\
&\quad + \int q_\phi(\mathbf{Z}^{L/2}|\boldsymbol{X}) \ln\ p_\theta\left(\hat{\boldsymbol{X}}\middle|\mathbf{Z}^{L/2}\right) \mathrm{d}\mathbf{Z}^{L/2} \\
&= - \int q_\phi(\mathbf{Z}^{L/2}|\boldsymbol{X}) \ln \frac{q_\phi(\mathbf{Z}^{L/2}|\boldsymbol{X})}{p_\theta(\mathbf{Z}^{L/2})} \mathrm{d}\mathbf{Z}^{L/2} \\
&\quad + \int q_\phi(\mathbf{Z}^{L/2}|\boldsymbol{X}) \ln\ p_\theta\left(\hat{\boldsymbol{X}}\middle|\mathbf{Z}^{L/2}\right) \mathrm{d}\mathbf{Z}^{L/2} \\
&= -\mathrm{KL}\big(q_\phi(\mathbf{Z}^{L/2}|\boldsymbol{X}) \| p_\theta(\mathbf{Z}^{L/2})\big) + \int q_\phi(\mathbf{Z}^{L/2}|\boldsymbol{X}) \ln\ p_\theta\left(\hat{\boldsymbol{X}}\middle|\mathbf{Z}^{L/2}\right) \mathrm{d}\mathbf{Z}^{L/2}.
\end{aligned}
\tag{9.29b}
$$

Equation (9.29b) indicates that ELBO is composed of two terms. The first term is the minus KL divergence between distribution $q_\phi\left(\mathbf{Z}^{L/2}\middle|\boldsymbol{X}\right)$ and $p_\theta(\mathbf{Z}^{L/2})$ and the prior $p_\theta(\mathbf{Z}^{L/2})$ is preset to follow the standard normal distribution $N(0, \boldsymbol{I})$. Thus, maximizing $-\mathrm{KL}\left(q_\phi\left(\mathbf{Z}^{L/2}\middle|\boldsymbol{X}\right) \| p_\theta(\mathbf{Z}^{L/2})\right)$ or minimizing $\mathrm{KL}\left(q_\phi\left(\mathbf{Z}^{L/2}\middle|\boldsymbol{X}\right) \| p_\theta(\mathbf{Z}^{L/2})\right)$ makes the encoder conditional probability as close as possible to the decoder prior so that the virtual tasks $q_\phi(\mathbf{Z}^{L/2}|\boldsymbol{X})$ will be under the constraint of decoder prior. The negative KL divergence can be explicitly expressed by

$$
\begin{aligned}
&-\mathrm{KL}\big(q_\phi(\mathbf{Z}^{L/2}|\boldsymbol{X}) \| p_\theta(\mathbf{Z}^{L/2})\big) \\
&= -\int q_\phi(\mathbf{Z}^{L/2}|\boldsymbol{X}) \ln q_\phi(\mathbf{Z}^{L/2}|\boldsymbol{X}) \mathrm{d}\mathbf{Z}^{L/2} \\
&\quad + \int q_\phi(\mathbf{Z}^{L/2}|\boldsymbol{X}) \ln p_\theta(\mathbf{Z}^{L/2}) \mathrm{d}\mathbf{Z}^{L/2} \\
&= -\int N\big(\boldsymbol{\mu}(\mathbf{Z}^{L/2}), \boldsymbol{\Sigma}(\mathbf{Z}^{L/2})\big) \ln\ N \\
&= \big(\boldsymbol{\mu}(\mathbf{Z}^{L/2}), \Sigma(\mathbf{Z}^{L/2})\big) \mathrm{d}\mathbf{Z}^{L/2} \\
&\quad + \int N\big(\boldsymbol{\mu}(\mathbf{Z}^{L/2}), \boldsymbol{\Sigma}(\mathbf{Z}^{L/2})\big) \ln\ N(\mathbf{0},\ \boldsymbol{I}) \mathrm{d}\mathbf{Z}^{L/2} \\
&= -\left[-\frac{G_{L/2}}{2} \ln\ 2\pi - \frac{1}{2} \sum_{g_{L/2}=1}^{G_{L/2}} \left(1 + \ln\ \sigma_{g_{L/2}}^2\right)\right] \\
&\quad + \left[-\frac{G_{L/2}}{2} \ln\ 2\pi - \frac{1}{2} \sum_{g_{L/2}=1}^{G_{L/2}} \left(\mu_{g_{L/2}}^2 + \sigma_{g_{L/2}}^2\right)\right] \\
&= \frac{1}{2} \sum_{g_{L/2}=1}^{G_{L/2}} \left(1 + \ln \sigma_{g_{L/2}}^2 - \mu_{g_{L/2}}^2 - \sigma_{g_{L/2}}^2\right),
\end{aligned}
\tag{9.30}
$$

where $\mu_{g_{L/2}}$ and $\sigma^2_{g_{L/2}}$ $(g_{L/2} = 1, 2, \cdots, G_{L/2})$ are the mean and the variance of the virtual task $Z^{L/2}_{g_{L/2}}$, respectively. Data fed into the decoder are sampled from the mean vector $\boldsymbol{\mu}(\boldsymbol{Z}^{L/2})$ and the diagonal covariance matrix $\boldsymbol{\Sigma}(\boldsymbol{Z}^{L/2})$ with a diagonal noise matrix, i.e.,

$$\boldsymbol{Z}^{L/2} = \begin{bmatrix} Z_1^{L/2} \\ \vdots \\ Z_{G_{L/2}}^{L/2} \end{bmatrix} = \begin{bmatrix} \mu_1 + \eta_1\sigma_1 \\ \vdots \\ \mu_{G_{L/2}} + \eta_{G_{L/2}}\sigma_{G_{L/2}} \end{bmatrix}, \tag{9.31}$$

where $\boldsymbol{\eta}$ is an auxiliary noise $\boldsymbol{\eta} \sim N(0, \boldsymbol{I})$. The second term the ELBO is the expectation of the generative model, $\ln p_\theta\left(\hat{\boldsymbol{X}}\middle|\boldsymbol{Z}^{L/2}\right)$, with respect to $q_\phi\left(\boldsymbol{Z}^{L/2}|\boldsymbol{X}\right)$, i.e.,

$$\mathrm{E}_{q_\phi(\boldsymbol{Z}^{L/2}|\boldsymbol{X})}\left(\ln p_\theta\left(\hat{\boldsymbol{X}}\middle|\boldsymbol{Z}^{L/2}\right)\right) = \int q_\phi\left(\boldsymbol{Z}^{L/2}|\boldsymbol{X}\right)\ln p_\theta\left(\hat{\boldsymbol{X}}\middle|\boldsymbol{Z}^{L/2}\right)\mathrm{d}\boldsymbol{Z}^{L/2}. \tag{9.32a}$$

Clearly, virtual tasks $\boldsymbol{Z}^{L/2}$ are variational and the optimal virtual tasks should meet the goal of the VAE that the loss function of the mean squared error of $\mathcal{L}(\boldsymbol{W}) = \frac{1}{n}\sum_{i=1}^{n}\left[\hat{\boldsymbol{X}}_i - \boldsymbol{X}_i\right]^2$ should be minimized. The output data are generated by the decoder, i.e.,

$$\hat{\boldsymbol{X}}_i = \boldsymbol{\theta}\left(\boldsymbol{\mu}\left(\boldsymbol{Z}^{L/2}|\ \boldsymbol{X}_i\right), \eta_i\boldsymbol{\Sigma}\left(\boldsymbol{Z}^{L/2}|\boldsymbol{X}_i\right)\right). \tag{9.32b}$$

Kingma and Welling (2013) use the reparameterization trick and, for normal distribution, have

$$\mathrm{E}_{q_\phi(\boldsymbol{Z}^{L/2}|\boldsymbol{X})}\left(\ln p_\theta\left(\hat{\boldsymbol{X}}\middle|\boldsymbol{Z}^{L/2}\right)\right) = -\frac{1}{2}\sum_{i=1}^{n}\left[\hat{\boldsymbol{X}}_i - \boldsymbol{X}_i\right]^2. \tag{9.32c}$$

Finally, $\max_{\phi,\theta} L_b$ is given by

$$\max_{\phi,\theta} L_b \sim \max_{\phi,\theta}\left\{\frac{1}{2}\sum_{g=1}^{G_{L/2}}\left(1 + \ln\boldsymbol{\Sigma}\left(Z_g^{L/2}\right) - \mu^2\left(Z_g^{L/2}\right) - \boldsymbol{\Sigma}\left(Z_g^{L/2}\right)\right) - \frac{1}{2}\left\|\hat{\boldsymbol{X}} - \boldsymbol{X}\right\|^2\right\}. \tag{9.33a}$$

Equivalently, $\max_{\phi,\theta} L_b$ can be converted to $\min_{\phi,\theta}(-L_b)$, i.e.,

$$\min_{\phi,\theta}\left\{\frac{1}{2}\sum_{g=1}^{G_{L/2}}\left(-\ln\boldsymbol{\Sigma}\left(Z_g^{L/2}\right) + \mu^2\left(Z_g^{L/2}\right) + \boldsymbol{\Sigma}\left(Z_g^{L/2}\right)\right) + \frac{1}{2}\left\|\hat{\boldsymbol{X}} - \boldsymbol{X}\right\|^2\right\}. \tag{9.33b}$$

Example 9.8 Take the 20 NIMS data of Table 2.4 from Chap. 2 as an example to illustrate the variational autoencoder. In this example, we consider only five variables, including the weight percentage of carbon, fatigue strength, ultimate tensile strength, true fracture strength, and hardness. The data are standardized and Table 9.16 shows the original and standardized data.

Table 9.16 Twenty NIMS data

Original data					Standardized data				
C	F	T	Fr	H	C	F	T	Fr	H
0.40	542	990	1761	310	–0.073	0.796	0.646	0.851	0.516
0.40	543	1037	1746	328	–0.073	0.809	0.978	0.742	0.905
0.47	471	911	1669	288	1.386	–0.147	0.090	0.181	0.040
0.40	410	772	1531	242	–0.073	–0.957	–0.890	–0.823	–0.955
0.35	344	635	1384	201	–1.115	–1.833	–1.856	–1.894	–1.841
0.41	610	1139	1874	372	0.135	1.699	1.697	1.674	1.856
0.40	524	985	1716	309	–0.073	0.557	0.611	0.524	0.494
0.43	582	1052	1756	328	0.552	1.327	1.083	0.815	0.905
0.42	505	938	1693	298	0.344	0.305	0.280	0.356	0.256
0.36	524	964	1742	306	–0.906	0.557	0.463	0.713	0.429
0.35	496	898	1759	290	–1.115	0.185	–0.002	0.837	0.083
0.40	543	1013	1750	328	–0.073	0.809	0.809	0.771	0.905
0.40	425	758	1562	246	–0.073	–0.758	–0.989	–0.598	–0.868
0.56	492	918	1517	291	3.261	0.132	0.139	–0.925	0.105
0.37	389	721	1462	226	–0.698	–1.236	–1.250	–1.326	–1.300
0.44	378	696	1424	219	0.761	–1.382	–1.426	–1.603	–1.452
0.40	396	715	1498	229	–0.073	–1.143	–1.292	–1.064	–1.236
0.34	476	891	1725	287	–1.323	–0.080	–0.051	0.589	0.018
0.36	587	1118	1784	362	–0.906	1.394	1.549	1.019	1.640
0.41	404	815	1529	263	0.135	–1.037	–0.587	–0.838	–0.501

Note C, F, T, Fr, and H denote the weight percentage of carbon, fatigue strength, ultimate tensile strength, true fracture strength, and hardness, respectively

Figure 9.22 illustrates the VAE structure used in this example, where the input layer and one hidden layer form the encoder and one hidden layer and the output layer construct the decoder without any activation function. The mean vector $\boldsymbol{\mu}(\mathbf{Z}^{L/2})$ and diagonal covariance matrix $\boldsymbol{\Sigma}(\mathbf{Z}^{L/2})$ are expressed by

$$\boldsymbol{\mu}(\mathbf{Z}^{L/2}) = \left[\mu_1^{(1)}, \mu_2^{(1)}\right], \tag{9.34}$$

$$\boldsymbol{\Sigma}(\mathbf{Z}^{L/2}) = \begin{bmatrix} \sigma_1^{(1)} & \\ & \sigma_2^{(1)} \end{bmatrix}. \tag{9.35}$$

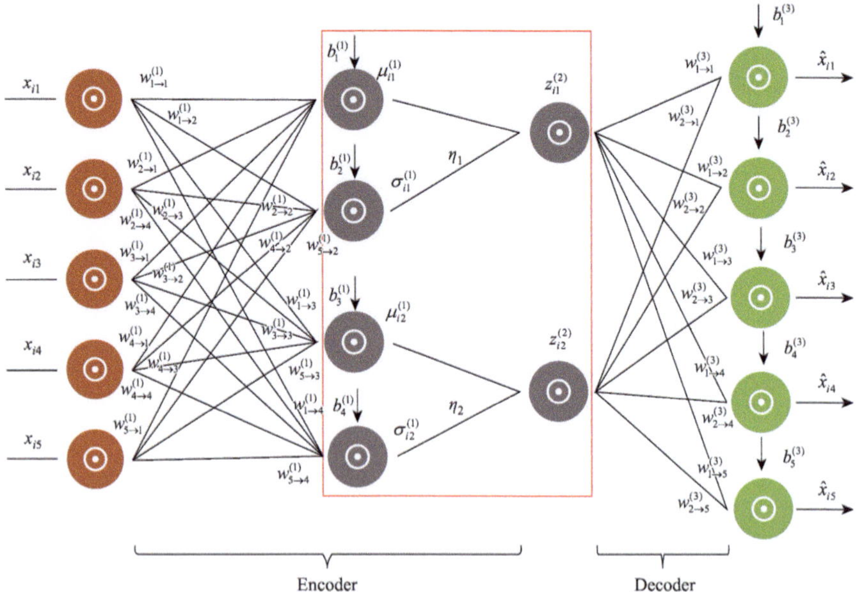

Fig. 9.22 The structure of the VAE model used in Example 9.8

For convenience of calculation, we mix $\boldsymbol{\mu}(\mathbf{Z}^{L/2})$ and $\ln \boldsymbol{\Sigma}(\mathbf{Z}^{L/2})$ into a vector form and express by

$$\mathbf{Z}^{(1)} = \left[\boldsymbol{\mu}_1^{(1)} \ \ln \boldsymbol{\sigma}_1^{(1)} \ \boldsymbol{\mu}_2^{(1)} \ \ln \boldsymbol{\sigma}_2^{(1)} \right] = \boldsymbol{X}\boldsymbol{W}^{(1)} + \boldsymbol{b}^{(1)}, \tag{9.36}$$

where the $\boldsymbol{W}^{(1)}$ and $\boldsymbol{b}^{(1)}$ are the weights and biases (trainable variables), i.e., the variational parameters ϕ. With the auxiliary noise $\boldsymbol{\eta} = (\eta_1, \eta_2)$, we have

$$\mathbf{Z}^{(2)} = \left[\boldsymbol{\mu}_1^{(1)} + \eta_1 \boldsymbol{\sigma}_1^{(1)}, \boldsymbol{\mu}_2^{(1)} + \eta_2 \boldsymbol{\sigma}_2^{(1)} \right]. \tag{9.37}$$

Then, the generative model is expressed by

$$\hat{X} = \mathbf{Z}^{(3)} = \mathbf{Z}^{(2)}\boldsymbol{W}^{(3)} + \boldsymbol{b}^{(3)}, \tag{9.38}$$

where the $\boldsymbol{W}^{(3)}$ and $\boldsymbol{b}^{(3)}$ are the weights and biases (trainable variables), i.e., the generative model parameters θ.

Totally, there are 39 trainable variables in the VAE, including 24 trainable variables $\left\{ w_{1\to1}^{(1)}, w_{1\to2}^{(1)}, \cdots, w_{5\to3}^{(1)}, w_{5\to4}^{(1)}; b_1^{(1)}, b_2^{(1)}, b_3^{(1)}, b_4^{(1)} \right\}$ in the encoder and 15 trainable variables $\left\{ w_{1\to1}^{(3)}, w_{1\to2}^{(3)}, \cdots, w_{2\to4}^{(3)}, w_{2\to5}^{(3)}; b_1^{(3)}, b_2^{(3)}, b_3^{(3)}, b_4^{(3)}, b_5^{(3)} \right\}$ in the decoder. We use the mini-batch method in this example and set the bitch size to be 4, taking data 1–4 for the first iteration. The trainable variables are adjusted according to the average gradient of data 1–4 in the first iteration. The data matrix of data 1–4 is

$$\begin{bmatrix} \boldsymbol{X}_1 \\ \boldsymbol{X}_2 \\ \boldsymbol{X}_3 \\ \boldsymbol{X}_4 \end{bmatrix} = \begin{bmatrix} x_{11} \ x_{12} \cdots x_{15} \\ x_{21} \ x_{22} \cdots x_{25} \\ x_{31} \ x_{32} \cdots x_{35} \\ x_{41} \ x_{42} \cdots x_{45} \end{bmatrix} = \begin{bmatrix} -0.073 & 0.796 & 0.646 & 0.851 & 0.516 \\ -0.073 & 0.809 & 0.978 & 0.742 & 0.905 \\ 1.386 & -0.147 & 0.090 & 0.181 & 0.040 \\ -0.073 & -0.957 & -0.890 & -0.823 & -0.955 \end{bmatrix}.$$

The initial values of the 39 trainable variables are randomly selected from a normal distribution N (0,1), which are shown below

$$\boldsymbol{W}^{(1)} = \begin{bmatrix} w_{1\to1}^{(1)} & w_{1\to2}^{(1)} & w_{1\to3}^{(1)} & w_{1\to4}^{(1)} \\ w_{2\to1}^{(1)} & w_{2\to2}^{(1)} & w_{2\to3}^{(1)} & w_{2\to4}^{(1)} \\ w_{3\to1}^{(1)} & w_{3\to2}^{(1)} & w_{3\to3}^{(1)} & w_{3\to4}^{(1)} \\ w_{4\to1}^{(1)} & w_{4\to2}^{(1)} & w_{4\to3}^{(1)} & w_{4\to4}^{(1)} \\ w_{5\to1}^{(1)} & w_{5\to2}^{(1)} & w_{5\to3}^{(1)} & w_{5\to4}^{(1)} \end{bmatrix} = \begin{bmatrix} -0.6511 & 0.6718 & -0.0376 & 0.1456 \\ 0.5249 & -0.8216 & 0.0264 & 0.5645 \\ -0.2575 & 0.6771 & 0.7962 & 0.2235 \\ -0.6167 & -0.9133 & -0.1165 & 0.4257 \\ 0.4174 & -0.3368 & -0.8323 & -0.4840 \end{bmatrix},$$

$$\boldsymbol{b}^{(1)} = \begin{bmatrix} b_1^{(1)} & b_2^{(1)} & b_3^{(1)} & b_4^{(1)} \end{bmatrix} = [-0.7682 \ \ -0.4501 \ 0.2585 \ \ 0.3698],$$

$$\boldsymbol{W}^{(3)} = \begin{bmatrix} w_{1\to1}^{(3)} & w_{1\to2}^{(3)} & w_{1\to3}^{(3)} & w_{1\to4}^{(3)} & w_{1\to5}^{(3)} \\ w_{2\to1}^{(3)} & w_{2\to2}^{(3)} & w_{2\to3}^{(3)} & w_{2\to4}^{(3)} & w_{2\to5}^{(3)} \end{bmatrix} = \begin{bmatrix} -0.4582 & -0.6855 & 1.8542 & 0.5842 & -0.2102 \\ 1.5877 & 0.4654 & -0.9889 & -0.1222 & -0.3637 \end{bmatrix},$$

$$\boldsymbol{b}^{(3)} = \begin{bmatrix} b_1^{(3)} & b_2^{(3)} & b_3^{(3)} & b_4^{(3)} & b_5^{(3)} \end{bmatrix} = [0.5364 \ 0.3593 \ 0.2706 \ \ -0.1529 \ 0.3305].$$

According to Eq. (9.36), with data 1–4, the encoder generates

$$\boldsymbol{Z}^{(1)} = \begin{bmatrix} z_1^{(1)} \\ z_2^{(1)} \\ z_3^{(1)} \\ z_4^{(1)} \end{bmatrix} = \begin{bmatrix} \mu_{11}^{(1)} & \ln\sigma_{11}^{(1)} & \mu_{12}^{(1)} & \ln\sigma_{12}^{(1)} \\ \mu_{21}^{(1)} & \ln\sigma_{21}^{(1)} & \mu_{22}^{(1)} & \ln\sigma_{22}^{(1)} \\ \mu_{31}^{(1)} & \ln\sigma_{31}^{(1)} & \mu_{32}^{(1)} & \ln\sigma_{32}^{(1)} \\ \mu_{41}^{(1)} & \ln\sigma_{41}^{(1)} & \mu_{42}^{(1)} & \ln\sigma_{42}^{(1)} \end{bmatrix} = \begin{bmatrix} \boldsymbol{X}_1 \\ \boldsymbol{X}_2 \\ \boldsymbol{X}_3 \\ \boldsymbol{X}_4 \end{bmatrix} \boldsymbol{W}^{(1)} + \begin{bmatrix} 1 \\ 1 \\ 1 \\ 1 \end{bmatrix} \boldsymbol{b}^{(1)}$$

$$= \begin{bmatrix} -0.073 & 0.796 & 0.646 & 0.851 & 0.516 \\ -0.073 & 0.809 & 0.978 & 0.742 & 0.905 \\ 1.386 & -0.147 & 0.090 & 0.181 & 0.040 \\ -0.073 & -0.957 & -0.890 & -0.823 & -0.955 \end{bmatrix} \begin{bmatrix} -0.8113 & 1.4846 & 0.5277 & 2.1456 \\ 0.0653 & -2.4427 & 0.5275 & 0.5645 \\ 0.0992 & 0.5912 & -0.0202 & 0.2555 \\ 0.5920 & -2.1229 & 1.7962 & 1.0257 \\ -0.7229 & -0.0563 & 1.2073 & 1.7425 \end{bmatrix}$$

$$
+\begin{bmatrix}1\\1\\1\\1\end{bmatrix}\begin{bmatrix}-0.7682 & -0.4501 & 0.2585 & 0.3698\\-0.7682 & -0.4501 & 0.2585 & 0.3698\\-0.7682 & -0.4501 & 0.2585 & 0.3698\\-0.7682 & -0.4501 & 0.2585 & 0.3698\end{bmatrix}
$$

$$
=\begin{bmatrix}-0.7786 & -1.6667 & 0.2679 & 1.0654\\-0.6277 & 1.4840 & 0.2216 & 0.9122\\-1.8658 & 0.4839 & 0.2197 & 0.5664\\-0.8848 & 0.7578 & 0.4180 & -0.2681\end{bmatrix}.
$$

Then, we have

$$
\begin{bmatrix}\mu_{11}^{(1)} & \sigma_{11}^{(1)} & \mu_{12}^{(1)} & \sigma_{12}^{(1)}\\\mu_{21}^{(1)} & \sigma_{21}^{(1)} & \mu_{22}^{(1)} & \sigma_{22}^{(1)}\\\mu_{31}^{(1)} & \sigma_{31}^{(1)} & \mu_{32}^{(1)} & \sigma_{32}^{(1)}\\\mu_{41}^{(1)} & \sigma_{41}^{(1)} & \mu_{42}^{(1)} & \sigma_{42}^{(1)}\end{bmatrix}=\begin{bmatrix}\mu_{11}^{(1)} & \mathrm{e}^{\ln\sigma_{11}^{(1)}} & \mu_{12}^{(1)} & \mathrm{e}^{\ln\sigma_{12}^{(1)}}\\\mu_{21}^{(1)} & \mathrm{e}^{\ln\sigma_{21}^{(1)}} & \mu_{22}^{(1)} & \mathrm{e}^{\ln\sigma_{22}^{(1)}}\\\mu_{31}^{(1)} & \mathrm{e}^{\ln\sigma_{31}^{(1)}} & \mu_{32}^{(1)} & \mathrm{e}^{\ln\sigma_{32}^{(1)}}\\\mu_{41}^{(1)} & \mathrm{e}^{\ln\sigma_{41}^{(1)}} & \mu_{42}^{(1)} & \mathrm{e}^{\ln\sigma_{42}^{(1)}}\end{bmatrix}
$$

$$
=\begin{bmatrix}-0.7786 & 0.1888 & 0.2679 & 2.9020\\-0.6277 & 0.2267 & 0.2216 & 2.4900\\-1.8658 & 1.6224 & 0.2197 & 1.7619\\-0.8848 & 2.1335 & 0.4180 & 0.7648\end{bmatrix}.
$$

When vector $\boldsymbol{\eta}$ takes the value of $[\,\eta_1\ \eta_2\,]=[\,0.0425\ \ -0.3659\,]$ by randomly sampled from the standard normal distribution, we can obtain the value of $z_1^{(2)}$ for data $\boldsymbol{X}_1$, as below

$$
z_1^{(2)}=\begin{bmatrix}z_{11}^{(2)}\\z_{12}^{(2)}\end{bmatrix}^{\mathrm{T}}=\begin{bmatrix}\mu_{11}^{(1)}+\eta_1\times\sigma_{11}^{(1)}\\\mu_{12}^{(1)}+\eta_2\times\sigma_{12}^{(1)}\end{bmatrix}^{\mathrm{T}}
$$

$$
=\begin{bmatrix}-0.7786+0.0425\times 0.1888\\0.2679+(-0.3659)\times 2.9020\end{bmatrix}^{\mathrm{T}}=\begin{bmatrix}-0.7706\\-0.7938\end{bmatrix}^{\mathrm{T}}.
$$

Similarly, we have

$$
\boldsymbol{Z}^{(2)}=\begin{bmatrix}z_1^{(2)}\\z_2^{(2)}\\z_3^{(2)}\\z_4^{(2)}\end{bmatrix}=\begin{bmatrix}-0.7706 & -0.7938\\-0.6180 & -0.6894\\-1.7969 & -0.4249\\-0.7942 & 0.1382\end{bmatrix}.
$$

Finally, we have the initial output

$$
\begin{aligned}
\boldsymbol{Z}^{(3)} &= \begin{bmatrix} \hat{\boldsymbol{X}}_1 \\ \hat{\boldsymbol{X}}_2 \\ \hat{\boldsymbol{X}}_3 \\ \hat{\boldsymbol{X}}_4 \end{bmatrix} = \begin{bmatrix} z_1^{(2)} \\ z_2^{(2)} \\ z_3^{(2)} \\ z_4^{(2)} \end{bmatrix} \boldsymbol{W}^{(3)} + \begin{bmatrix} 1 \\ 1 \\ 1 \\ 1 \end{bmatrix} \boldsymbol{b}^{(3)} \\
&= \begin{bmatrix} -0.7706 & -0.7938 \\ -0.6180 & -0.6894 \\ -1.7969 & -0.4249 \\ -0.7942 & 0.1382 \end{bmatrix} \begin{bmatrix} -0.4582 & -0.6855 & 1.8542 & 0.5842 & -0.2102 \\ 1.5877 & 0.4654 & -0.9889 & -0.1222 & -0.3637 \end{bmatrix} \\
&+ \begin{bmatrix} 0.5364 & 0.3593 & 0.2706 & -0.1529 & 0.3305 \\ 0.5364 & 0.3593 & 0.2706 & -0.1529 & 0.3305 \\ 0.5364 & 0.3593 & 0.2706 & -0.1529 & 0.3305 \\ 0.5364 & 0.3593 & 0.2706 & -0.1529 & 0.3305 \end{bmatrix} \\
&= \begin{bmatrix} -0.3709 & 0.5180 & -0.3731 & -0.5060 & 0.7812 \\ -0.2750 & 0.4621 & -0.1935 & -0.4297 & 0.7111 \\ 0.6851 & 1.3933 & -2.6410 & -1.1507 & 0.8627 \\ 1.1197 & 0.9680 & -1.3387 & -0.6337 & 0.4471 \end{bmatrix}.
\end{aligned}
$$

With Eq. (9.33b), we obtain the average loss of mini-batch in the first iteration,

$$
\begin{aligned}
\overline{\mathcal{L}} &= \frac{1}{4} \sum_{i=1}^{4} \mathcal{L}_i \\
&= \frac{1}{4} \sum_{i=1}^{4} \left(\frac{1}{2} \sum_{j=1}^{2} \left(-\ln \sum \left(Z_{ij}^{(1)} \right) + \mu^2 \left(Z_{ij}^{(1)} \right) + \sum \left(Z_{ij}^{(1)} \right) \right) + \frac{1}{2} \left\| \hat{\boldsymbol{X}}_i - \boldsymbol{X}_i \right\|^2 \right) \\
&= \frac{1}{4} \sum_{i=1}^{4} \left(\frac{1}{2} \sum_{j=1}^{2} \left(-\ln \left(\sigma_{ij}^{(1)} \right)^2 + \left(\mu_{ij}^{(1)} \right)^2 + \left(\sigma_{ij}^{(1)} \right)^2 \right) + \frac{1}{2} \sum_{j=1}^{5} (\hat{x}_{ij} - x_{ij})^2 \right) \\
&= 7.07802.
\end{aligned}
$$

For data $\boldsymbol{X}_1$, the gradients of loss are calculated to be

$$
\begin{aligned}
\frac{\partial \mathcal{L}_1}{\partial w_{j \to 1}^{(1)}} &= \frac{\partial \mathcal{L}_1}{\partial \hat{\boldsymbol{X}}_1} \frac{\partial \hat{\boldsymbol{X}}_1}{\partial z_{11}^{(2)}} \frac{\partial z_{11}^{(2)}}{\partial \mu_{11}^{(1)}} \frac{\partial \mu_{11}^{(1)}}{\partial w_{j \to 1}^{(1)}} + \frac{\partial \mathcal{L}_1}{\partial \mu_{11}^{(1)}} \frac{\partial \mu_{11}^{(1)}}{\partial w_{j \to 1}^{(1)}} \\
&= (\hat{\boldsymbol{X}}_1 - \boldsymbol{X}_1) \begin{bmatrix} w_{1 \to 1}^{(3)} \\ w_{1 \to 2}^{(3)} \\ w_{1 \to 3}^{(3)} \\ w_{1 \to 4}^{(3)} \\ w_{1 \to 5}^{(3)} \end{bmatrix} [x_{1j}] + \mu_{11}^{(1)} [x_{1j}],
\end{aligned}
$$

$$\frac{\partial \mathcal{L}_1}{\partial w_{j\to 2}^{(1)}} = \frac{\partial \mathcal{L}_1}{\partial \hat{\boldsymbol{X}}_1}\frac{\partial \hat{\boldsymbol{X}}_1}{\partial z_{11}^{(2)}}\frac{\partial z_{11}^{(2)}}{\partial \sigma_{11}^{(1)}}\frac{\partial \sigma_{11}^{(1)}}{\partial w_{j\to 2}^{(1)}} + \frac{\partial \mathcal{L}_1}{\partial \sigma_{11}^{(1)}}\frac{\partial \sigma_{11}^{(1)}}{\partial w_{j\to 1}^{(1)}}$$

$$= (\hat{\boldsymbol{X}}_1 - \boldsymbol{X}_1)\begin{bmatrix} w_{1\to 1}^{(3)} \\ w_{1\to 2}^{(3)} \\ w_{1\to 3}^{(3)} \\ w_{1\to 4}^{(3)} \\ w_{1\to 5}^{(3)} \end{bmatrix}\eta_1[x_{1j}]$$

$$+ \left(\sigma_{11}^{(1)} - \frac{1}{\sigma_{11}^{(1)}}\right)[x_{1j}],$$

$$\frac{\partial \mathcal{L}_1}{\partial w_{j\to 3}^{(1)}} = \frac{\partial \mathcal{L}_1}{\partial \hat{\boldsymbol{X}}_1}\frac{\partial \hat{\boldsymbol{X}}_1}{\partial z_{12}^{(2)}}\frac{\partial z_{12}^{(2)}}{\partial \mu_{12}^{(1)}}\frac{\partial \mu_{12}^{(1)}}{\partial w_{j\to 3}^{(1)}} + \frac{\partial \mathcal{L}_1}{\partial \mu_{12}^{(1)}}\frac{\partial \mu_{12}^{(1)}}{\partial w_{j\to 3}^{(1)}}$$

$$= (\hat{\boldsymbol{X}}_1 - \boldsymbol{X}_1)\begin{bmatrix} w_{2\to 1}^{(3)} \\ w_{2\to 2}^{(3)} \\ w_{2\to 3}^{(3)} \\ w_{2\to 4}^{(3)} \\ w_{2\to 5}^{(3)} \end{bmatrix}[x_{1j}] + \mu_{12}^{(1)}[x_{1j}],$$

$$\frac{\partial \mathcal{L}_1}{\partial w_{j\to 4}^{(1)}} = \frac{\partial \mathcal{L}_1}{\partial \hat{\boldsymbol{X}}_1}\frac{\partial \hat{\boldsymbol{X}}_1}{\partial z_{12}^{(2)}}\frac{\partial z_{12}^{(2)}}{\partial \sigma_{12}^{(1)}}\frac{\partial \sigma_{12}^{(1)}}{\partial w_{j\to 4}^{(1)}} + \frac{\partial \mathcal{L}_1}{\partial \sigma_{12}^{(1)}}\frac{\partial \sigma_{12}^{(1)}}{\partial w_{j\to 4}^{(1)}}$$

$$= (\hat{\boldsymbol{X}}_1 - \boldsymbol{X}_1)\begin{bmatrix} w_{2\to 1}^{(3)} \\ w_{2\to 2}^{(3)} \\ w_{2\to 3}^{(3)} \\ w_{2\to 4}^{(3)} \\ w_{2\to 5}^{(3)} \end{bmatrix}\eta_2[x_{1j}] + \left(\sigma_{12}^{(1)} - \frac{1}{\sigma_{12}^{(1)}}\right)[x_{1j}],$$

$$\frac{\partial \mathcal{L}_1}{\partial b_1^{(1)}} = \frac{\partial \mathcal{L}_1}{\partial \hat{\boldsymbol{X}}_1}\frac{\partial \hat{\boldsymbol{X}}_1}{\partial z_{11}^{(2)}}\frac{\partial z_{11}^{(2)}}{\partial \mu_{11}^{(1)}}\frac{\partial \mu_{11}^{(1)}}{\partial b_1^{(1)}} + \frac{\partial \mathcal{L}_1}{\partial \mu_{11}^{(1)}}\frac{\partial \mu_{11}^{(1)}}{\partial b_1^{(1)}}$$

$$= (\hat{\boldsymbol{X}}_1 - \boldsymbol{X}_1)\begin{bmatrix} w_{1\to 1}^{(3)} \\ w_{1\to 2}^{(3)} \\ w_{1\to 3}^{(3)} \\ w_{1\to 4}^{(3)} \\ w_{1\to 5}^{(3)} \end{bmatrix} + \mu_{11}^{(1)}[x_{1j}] \quad (j = 1, 2, 3, 4, 5),$$

$$\frac{\partial \mathcal{L}_1}{\partial b_2^{(1)}} = \frac{\partial \mathcal{L}_1}{\partial \hat{\boldsymbol{X}}_1}\frac{\partial \hat{\boldsymbol{X}}_1}{\partial z_{11}^{(2)}}\frac{\partial z_{11}^{(2)}}{\partial \sigma_{11}^{(1)}}\frac{\partial \sigma_{11}^{(1)}}{\partial b_2^{(1)}} + \frac{\partial \mathcal{L}_1}{\partial \sigma_{11}^{(1)}}\frac{\partial \sigma_{11}^{(1)}}{\partial b_2^{(1)}}$$

$$= (\hat{\boldsymbol{X}}_1 - \boldsymbol{X}_1)\begin{bmatrix} w_{1\to1}^{(3)} \\ w_{1\to2}^{(3)} \\ w_{1\to3}^{(3)} \\ w_{1\to4}^{(3)} \\ w_{1\to5}^{(3)} \end{bmatrix}\eta_1 + \left(\sigma_{11}^{(1)} - \frac{1}{\sigma_{11}^{(1)}}\right),$$

$$\frac{\partial \mathcal{L}_1}{\partial b_3^{(1)}} = \frac{\partial \mathcal{L}_1}{\partial \hat{\boldsymbol{X}}_1}\frac{\partial \hat{\boldsymbol{X}}_1}{\partial z_{12}^{(2)}}\frac{\partial z_{12}^{(2)}}{\partial \mu_{12}^{(1)}}\frac{\partial \mu_{12}^{(1)}}{\partial b_3^{(1)}} + \frac{\partial \mathcal{L}_1}{\partial \mu_{12}^{(1)}}\frac{\partial \mu_{12}^{(1)}}{\partial b_3^{(1)}}$$

$$= (\hat{\boldsymbol{X}}_1 - \boldsymbol{X}_1)\begin{bmatrix} w_{2\to1}^{(3)} \\ w_{2\to2}^{(3)} \\ w_{2\to3}^{(3)} \\ w_{2\to4}^{(3)} \\ w_{2\to5}^{(3)} \end{bmatrix} + \mu_{12}^{(1)},$$

$$\frac{\partial \mathcal{L}_1}{\partial b_4^{(1)}} = \frac{\partial \mathcal{L}_1}{\partial \hat{\boldsymbol{X}}_1}\frac{\partial \hat{\boldsymbol{X}}_1}{\partial z_{12}^{(2)}}\frac{\partial z_{12}^{(2)}}{\partial \sigma_{12}^{(1)}}\frac{\partial \sigma_{12}^{(1)}}{\partial b_4^{(1)}} + \frac{\partial \mathcal{L}_1}{\partial \sigma_{12}^{(1)}}\frac{\partial \sigma_{12}^{(1)}}{\partial b_4^{(1)}}$$

$$= (\hat{\boldsymbol{X}}_1 - \boldsymbol{X}_1)\begin{bmatrix} w_{2\to1}^{(3)} \\ w_{2\to2}^{(3)} \\ w_{2\to3}^{(3)} \\ w_{2\to4}^{(3)} \\ w_{2\to5}^{(3)} \end{bmatrix}\eta_2 + \left(\sigma_{12}^{(1)} - \frac{1}{\sigma_{12}^{(1)}}\right),$$

$$\begin{bmatrix} \frac{\partial \mathcal{L}_1}{\partial w_{k\to1}^{(3)}} \\ \frac{\partial \mathcal{L}_1}{\partial w_{k\to2}^{(3)}} \\ \frac{\partial \mathcal{L}_1}{\partial w_{k\to3}^{(3)}} \\ \frac{\partial \mathcal{L}_1}{\partial w_{k\to4}^{(3)}} \\ \frac{\partial \mathcal{L}_1}{\partial w_{k\to5}^{(3)}} \end{bmatrix} = \begin{bmatrix} \frac{\partial \mathcal{L}_1}{\partial \hat{\boldsymbol{X}}_1}\frac{\partial \hat{\boldsymbol{X}}_1}{\partial w_{k\to1}^{(3)}} \\ \frac{\partial \mathcal{L}_1}{\partial \hat{\boldsymbol{X}}_1}\frac{\partial \hat{\boldsymbol{X}}_1}{\partial w_{k\to2}^{(3)}} \\ \frac{\partial \mathcal{L}_1}{\partial \hat{\boldsymbol{X}}_1}\frac{\partial \hat{\boldsymbol{X}}_1}{\partial w_{k\to3}^{(3)}} \\ \frac{\partial \mathcal{L}_1}{\partial \hat{\boldsymbol{X}}_1}\frac{\partial \hat{\boldsymbol{X}}_1}{\partial w_{k\to4}^{(3)}} \\ \frac{\partial \mathcal{L}_1}{\partial \hat{\boldsymbol{X}}_1}\frac{\partial \hat{\boldsymbol{X}}_1}{\partial w_{k\to5}^{(3)}} \end{bmatrix}$$

$$= (\hat{\boldsymbol{X}}_1 - \boldsymbol{X}_1)\begin{bmatrix} z_k^{(2)} & 0 & 0 & 0 & 0 \\ 0 & z_k^{(2)} & 0 & 0 & 0 \\ 0 & 0 & z_k^{(2)} & 0 & 0 \\ 0 & 0 & 0 & z_k^{(2)} & 0 \\ 0 & 0 & 0 & 0 & z_k^{(2)} \end{bmatrix} \quad (k = 1, 2),$$

$$\begin{bmatrix} \frac{\partial \mathcal{L}_1}{\partial b_1^{(3)}} \\ \frac{\partial \mathcal{L}_1}{\partial b_2^{(3)}} \\ \frac{\partial \mathcal{L}_1}{\partial b_3^{(3)}} \\ \frac{\partial \mathcal{L}_1}{\partial b_4^{(3)}} \\ \frac{\partial \mathcal{L}_1}{\partial b_5^{(3)}} \end{bmatrix} = \begin{bmatrix} \frac{\partial \mathcal{L}_1}{\partial \hat{X}_1} \frac{\partial \hat{X}_1}{\partial b_1^{(3)}} \\ \frac{\partial \mathcal{L}_1}{\partial \hat{X}_1} \frac{\partial \hat{X}_1}{\partial b_2^{(3)}} \\ \frac{\partial \mathcal{L}_1}{\partial \hat{X}_1} \frac{\partial \hat{X}_1}{\partial b_3^{(3)}} \\ \frac{\partial \mathcal{L}_1}{\partial \hat{X}_1} \frac{\partial \hat{X}_1}{\partial b_4^{(3)}} \\ \frac{\partial \mathcal{L}_1}{\partial \hat{X}_1} \frac{\partial \hat{X}_1}{\partial b_5^{(3)}} \end{bmatrix} = (\hat{\boldsymbol{X}}_1 - \boldsymbol{X}_1) \begin{bmatrix} 1\,0\,0\,0\,0 \\ 0\,1\,0\,0\,0 \\ 0\,0\,1\,0\,0 \\ 0\,0\,0\,1\,0 \\ 0\,0\,0\,0\,1 \end{bmatrix}.$$

Take $w_{1\to 1}^{(1)}$ as an example to illustrate the BP process.

$$\frac{\partial \mathcal{L}_1}{\partial w_{1\to 1}^{(1)}} = (\hat{\boldsymbol{X}}_1 - \boldsymbol{X}_1) \begin{bmatrix} w_{1\to 1}^{(3)} \\ w_{1\to 2}^{(3)} \\ w_{1\to 3}^{(3)} \\ w_{1\to 4}^{(3)} \\ w_{1\to 5}^{(3)} \end{bmatrix} [x_{11}] + \mu_{11}^{(1)} [x_{11}]$$

$$= \begin{bmatrix} -0.3709 + 0.073 \\ 0.5180 - 0.796 \\ -0.3731 - 0.646 \\ -0.5060 - 0.851 \\ 0.7812 - 0.516 \end{bmatrix}^{\mathrm{T}} \cdot \begin{bmatrix} -0.4582 \\ -0.6855 \\ 1.8542 \\ 0.5842 \\ -0.2102 \end{bmatrix} [-0.073]$$

$$- 0.7786 \times (-0.073) = 0.2328.$$

Similarly, for data $\boldsymbol{X}_2$, $\boldsymbol{X}_2$, and $\boldsymbol{X}_4$, we can obtain $\frac{\partial \mathcal{L}_2}{\partial w_{1\to 1}^{(1)}} = 0.2382$, $\frac{\partial \mathcal{L}_3}{\partial w_{1\to 1}^{(1)}} = -10.4341$, and $\frac{\partial \mathcal{L}_4}{\partial w_{1\to 1}^{(1)}} = 0.2672$, and thus

$$\frac{\partial \overline{\mathcal{L}}}{\partial w_{1\to 1}^{(1)}} = \frac{\frac{\partial \mathcal{L}_1}{\partial w_{1\to 1}^{(1)}} + \frac{\partial \mathcal{L}_2}{\partial w_{1\to 1}^{(1)}} + \frac{\partial \mathcal{L}_3}{\partial w_{1\to 1}^{(1)}} + \frac{\partial \mathcal{L}_4}{\partial w_{1\to 1}^{(1)}}}{4} = -2.4240.$$

With the learning rate $\eta = 0.01$, we update the value of $w_{1\to 1}^{(1)}$, viz.,

$$w_{1\to 1(\text{new})}^{(1)} = w_{1\to 1(\text{old})}^{(1)} - \eta \frac{\partial \overline{\mathcal{L}}}{\partial w_{1\to 1}^{(1)}} = -0.6511 - 0.01 \times (-2.4240) = -0.62686.$$

The other 38 trainable variables are updated in the same way.

After 3000 epochs, the loss $\mathcal{L}$ is 0.61. Table 9.17 shows the VAE outputs corresponding to the twenty input data. Table 9.18 shows the R^2 value of the predicted results.

Table 9.17 The VAE outputs and inputs

Outputs					Inputs				
C	F	T	Fr	H	C	F	T	Fr	H
−0.345	0.342	0.332	0.417	0.344	−0.073	0.796	0.646	0.851	0.516
0.194	0.835	0.840	0.762	0.837	−0.073	0.809	0.978	0.742	0.905
1.457	−0.017	0.023	−0.374	−0.022	1.386	−0.147	0.090	0.181	0.040
0.257	−0.724	−0.716	−0.764	−0.727	−0.073	−0.957	−0.890	−0.823	−0.955
−1.924	−1.958	−2.009	−1.424	−1.958	−1.115	−1.833	−1.856	−1.894	−1.841
−0.422	1.211	1.199	1.278	1.216	0.135	1.699	1.697	1.674	1.856
−0.250	0.198	0.191	0.254	0.199	−0.073	0.557	0.611	0.524	0.494
0.492	0.707	0.720	0.564	0.707	0.552	1.327	1.083	0.815	0.905
−0.091	0.224	0.221	0.240	0.224	0.344	0.305	0.280	0.356	0.256
−0.360	0.636	0.626	0.706	0.639	−0.906	0.557	0.463	0.713	0.429
−0.783	−0.322	−0.343	−0.119	−0.321	−1.115	0.185	−0.002	0.837	0.083
0.159	0.970	0.974	0.901	0.972	−0.073	0.809	0.809	0.771	0.905
−0.338	−0.426	−0.434	−0.328	−0.426	−0.073	−0.758	−0.989	−0.598	−0.868
2.346	−0.290	−0.225	−0.857	−0.297	3.261	0.132	0.139	−0.925	0.105
−0.624	−0.810	−0.826	−0.631	−0.811	−0.698	−1.236	−1.250	−1.326	−1.300
−0.227	−1.292	−1.297	−1.195	−1.295	0.761	−1.382	−1.426	−1.603	−1.452
0.339	−0.560	−0.550	−0.625	−0.562	−0.073	−1.143	−1.292	−1.064	−1.236
−1.258	0.211	0.177	0.515	0.215	−1.323	−0.080	−0.051	0.589	0.018
−0.723	0.932	0.911	1.081	0.936	−0.906	1.394	1.549	1.019	1.640
−0.190	−1.208	−1.212	−1.124	−1.211	0.135	−1.037	−0.587	−0.838	−0.501

Table 9.18 The R^2 of the VAE model

The variable	C	F	T	Fr	H	The average value
R^2	0.8830	0.9392	0.8453	0.7845	0.9086	0.8721

Example 9.9 This example illustrates how the model complexity affects the prediction accuracy in the VAE. Figure 9.23 shows the structure of the VAE model used in Example 9.9. Compared with the VAE model in Example 9.8, we add one hidden layer in the encoder and one hidden layer in the decoder and use the activation function $\varphi = \tanh(x)$. The same input data listed in Table 9.17 are used in this example.

Totally, there are 81 trainable variables in the VAE model, including 44 trainable variables $\left\{w_{1\to 1}^{(1)}, \cdots, w_{5\to 4}^{(1)}; b_1^{(1)}, \cdots, b_4^{(1)}; w_{1\to 1}^{(2)}, \cdots, w_{4\to 4}^{(2)}; b_1^{(2)}, \cdots, b_4^{(2)}\right\}$ in the encoder and 37 trainable variables $\left\{w_{1\to 1}^{(4)}, \cdots, w_{2\to 4}^{(4)}; b_1^{(4)}, \cdots, b_4^{(4)}; w_{1\to 1}^{(5)}, \cdots, \right.$

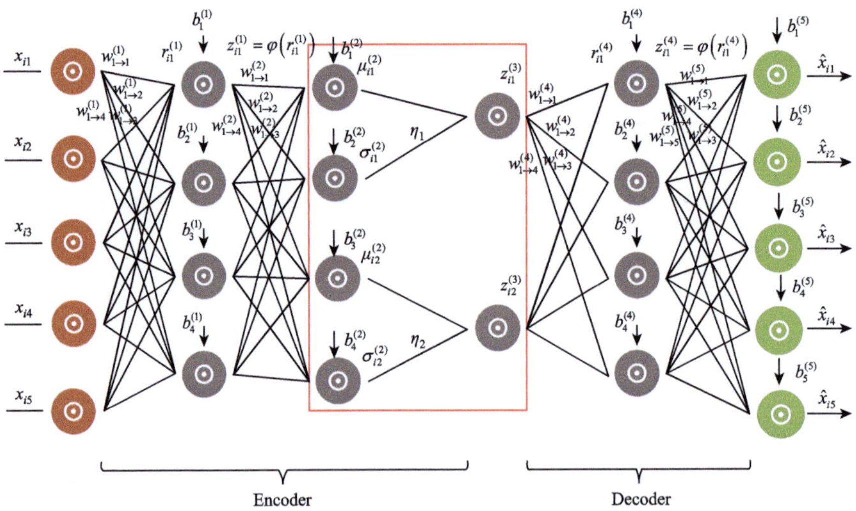

Fig. 9.23 The structure of the VAE model used in Example 9.9

$w^{(5)}_{4\to5}; b^{(5)}_1, \cdots, b^{(5)}_5\}$ in the decoder. Again, the mini-batch method is used with the bitch size of 4. The data matrix of data 1–4 is

$$\begin{bmatrix} \boldsymbol{X}_1 \\ \boldsymbol{X}_2 \\ \boldsymbol{X}_3 \\ \boldsymbol{X}_4 \end{bmatrix} = \begin{bmatrix} x_{11} & x_{12} & \cdots & x_{15} \\ x_{21} & x_{22} & \cdots & x_{25} \\ x_{31} & x_{32} & \cdots & x_{35} \\ x_{41} & x_{42} & \cdots & x_{45} \end{bmatrix}$$
$$= \begin{bmatrix} -0.073 & 0.796 & 0.646 & 0.851 & 0.516 \\ -0.073 & 0.809 & 0.978 & 0.742 & 0.905 \\ 1.386 & -0.147 & 0.090 & 0.181 & 0.040 \\ -0.073 & -0.957 & -0.890 & -0.823 & -0.955 \end{bmatrix}.$$

The initial values of the 81 trainable variables are randomly selected from a normal distribution N (0,1), which are shown below

$$\boldsymbol{W}^{(1)} = \begin{bmatrix} w^{(1)}_{1\to1} & w^{(1)}_{1\to2} & w^{(1)}_{1\to3} & w^{(1)}_{1\to4} \\ w^{(1)}_{2\to1} & w^{(1)}_{2\to2} & w^{(1)}_{2\to3} & w^{(1)}_{2\to4} \\ w^{(1)}_{3\to1} & w^{(1)}_{3\to2} & w^{(1)}_{3\to3} & w^{(1)}_{3\to4} \\ w^{(1)}_{4\to1} & w^{(1)}_{4\to2} & w^{(1)}_{4\to3} & w^{(1)}_{4\to4} \\ w^{(1)}_{5\to1} & w^{(1)}_{5\to2} & w^{(1)}_{5\to3} & w^{(1)}_{5\to4} \end{bmatrix}$$

$$= \begin{bmatrix} 0.0497 & 0.0592 & 0.5291 & -0.1718 \\ -0.1615 & -0.1819 & -0.4208 & -0.0198 \\ 0.0973 & 0.2147 & 0.5148 & 0.0232 \\ 0.2528 & 0.1029 & 0.6541 & -0.3643 \\ 0.2712 & -0.2460 & 0.4342 & -0.0533 \end{bmatrix},$$

$$\boldsymbol{b}^{(1)} = \left[b_1^{(1)}\ b_2^{(1)}\ b_3^{(1)}\ b_4^{(1)} \right] = [\,-0.2336\ 0.3193\ -0.1694\ -0.0064\,],$$

$$\boldsymbol{W}^{(2)} = \begin{bmatrix} w_{1\to1}^{(2)} & w_{1\to2}^{(2)} & w_{1\to3}^{(2)} & w_{1\to4}^{(2)} \\ w_{2\to1}^{(2)} & w_{2\to2}^{(2)} & w_{2\to3}^{(2)} & w_{2\to4}^{(2)} \\ w_{3\to1}^{(2)} & w_{3\to2}^{(2)} & w_{3\to3}^{(2)} & w_{3\to4}^{(2)} \\ w_{4\to1}^{(2)} & w_{4\to2}^{(2)} & w_{4\to3}^{(2)} & w_{4\to4}^{(2)} \end{bmatrix}$$

$$= \begin{bmatrix} -0.7286 & -1.0228 & -0.0376 & -0.0430 \\ -0.3639 & -0.5126 & -0.0264 & -0.0176 \\ 0.2743 & 0.1969 & 0.0050 & 0.0235 \\ 0.5367 & -0.7873 & -0.0165 & -0.0306 \end{bmatrix},$$

$$\boldsymbol{b}^{(2)} = \left[b_1^{(2)}\ b_2^{(2)}\ b_3^{(2)}\ b_4^{(2)} \right] = [\,-0.0179\ -0.0239\ -0.0407\ -0.0677\,],$$

$$\boldsymbol{W}^{(4)} = \begin{bmatrix} w_{1\to1}^{(4)} & w_{1\to2}^{(4)} & w_{1\to3}^{(4)} & w_{1\to4}^{(4)} \\ w_{2\to1}^{(4)} & w_{2\to2}^{(4)} & w_{2\to3}^{(4)} & w_{2\to4}^{(4)} \end{bmatrix}$$

$$= \begin{bmatrix} -0.7496 & -0.4817 & 2.1771 & 1.8239 \\ 3.1253 & -2.6128 & 1.8581 & -0.6759 \end{bmatrix},$$

$$\boldsymbol{b}^{(4)} = \left[b_1^{(4)}\ b_2^{(4)}\ b_3^{(4)}\ b_4^{(4)} \right] = [\,0.9628\ 0.4477\ 1.1277\ -0.1531\,],$$

$$\boldsymbol{W}^{(5)} = \begin{bmatrix} w_{1\to1}^{(5)} & w_{1\to2}^{(5)} & w_{1\to3}^{(5)} & w_{1\to4}^{(5)} & w_{1\to5}^{(5)} \\ w_{2\to1}^{(5)} & w_{2\to2}^{(5)} & w_{2\to3}^{(5)} & w_{2\to4}^{(5)} & w_{2\to5}^{(5)} \\ w_{3\to1}^{(5)} & w_{3\to2}^{(5)} & w_{3\to3}^{(5)} & w_{3\to4}^{(5)} & w_{3\to5}^{(5)} \\ w_{4\to1}^{(5)} & w_{4\to2}^{(5)} & w_{4\to3}^{(5)} & w_{4\to4}^{(5)} & w_{4\to5}^{(5)} \end{bmatrix}$$

$$= \begin{bmatrix} 0.1963 & -0.5255 & 0.2610 & 1.7003 & -0.6352 \\ 0.4165 & -0.0911 & -0.7577 & -1.2179 & -1.7871 \\ -0.7742 & -1.3710 & -0.5602 & -0.6589 & 0.4474 \\ -0.5544 & 0.0430 & 1.2429 & -0.8601 & -0.8536 \end{bmatrix},$$

$$\boldsymbol{b}^{(5)} = \left[b_1^{(5)}\ b_2^{(5)}\ b_3^{(5)}\ b_4^{(5)}\ b_5^{(5)} \right] = [\,-0.4543\ 0.8152\ 1.0417\ -0.6501\ 1.0287\,]\,.$$

With data 1–4, the encoder generates

$$\boldsymbol{r}^{(1)} = \begin{bmatrix} \boldsymbol{r}_1^{(1)} \\ \boldsymbol{r}_2^{(1)} \\ \boldsymbol{r}_3^{(1)} \\ \boldsymbol{r}_4^{(1)} \end{bmatrix} = \begin{bmatrix} \boldsymbol{X}_1 \\ \boldsymbol{X}_2 \\ \boldsymbol{X}_3 \\ \boldsymbol{X}_4 \end{bmatrix} \boldsymbol{W}^{(1)} + \begin{bmatrix} 1 \\ 1 \\ 1 \\ 1 \end{bmatrix} \boldsymbol{b}^{(1)}$$

$$= \begin{bmatrix} -0.073 & 0.796 & 0.646 & 0.851 & 0.516 \\ -0.073 & 0.809 & 0.978 & 0.742 & 0.905 \\ 1.386 & -0.147 & 0.090 & 0.181 & 0.040 \\ -0.073 & -0.957 & -0.890 & -0.823 & -0.955 \end{bmatrix} \begin{bmatrix} 0.0497 & 0.0592 & 0.5291 & -0.1718 \\ -0.1615 & -0.1819 & -0.4208 & -0.0198 \\ 0.0973 & 0.2147 & 0.5148 & 0.0232 \\ 0.2528 & 0.1029 & 0.6541 & -0.3643 \\ 0.2712 & -0.2460 & 0.4342 & -0.0533 \end{bmatrix}$$

$$+ \begin{bmatrix} 1 \\ 1 \\ 1 \\ 1 \end{bmatrix} \begin{bmatrix} -0.2336 & 0.3193 & -0.1694 & -0.0064 \end{bmatrix}$$

$$= \begin{bmatrix} 0.0521 & 0.2695 & 0.5703 & -0.3322 \\ 0.1603 & 0.2315 & 0.8333 & -0.3057 \\ -0.0756 & 0.4562 & 0.8079 & -0.3076 \\ -0.6363 & 0.4482 & -1.2165 & 0.3552 \end{bmatrix}.$$

Then, we have

$$\mathbf{Z}^{(1)} = \begin{bmatrix} z_1^{(1)} \\ z_2^{(1)} \\ z_3^{(1)} \\ z_4^{(1)} \end{bmatrix} = \varphi(\boldsymbol{r}^{(1)})$$

$$= \tanh\left(\begin{bmatrix} 0.0521 & 0.2695 & 0.5703 & -0.3322 \\ 0.1603 & 0.2315 & 0.8333 & -0.3057 \\ -0.0756 & 0.4562 & 0.8079 & -0.3076 \\ -0.6363 & 0.4482 & -1.2165 & 0.3552 \end{bmatrix} \right)$$

$$= \begin{bmatrix} 0.0521 & 0.2631 & 0.5155 & -0.3204 \\ 0.1589 & 0.2274 & 0.6822 & -0.2965 \\ -0.0754 & 0.4269 & 0.6684 & -0.2982 \\ -0.5623 & 0.4204 & -0.8386 & 0.3409 \end{bmatrix},$$

$$\mathbf{Z}^{(2)} = \begin{bmatrix} z_1^{(2)} \\ z_2^{(2)} \\ z_3^{(2)} \\ z_4^{(2)} \end{bmatrix} = \begin{bmatrix} \mu_{11}^{(2)} & \ln \sigma_{11}^{(2)} & \mu_{12}^{(2)} & \ln \sigma_{12}^{(2)} \\ \mu_{21}^{(2)} & \ln \sigma_{21}^{(2)} & \mu_{22}^{(2)} & \ln \sigma_{22}^{(2)} \\ \mu_{31}^{(2)} & \ln \sigma_{31}^{(2)} & \mu_{32}^{(2)} & \ln \sigma_{32}^{(2)} \\ \mu_{41}^{(2)} & \ln \sigma_{41}^{(2)} & \mu_{42}^{(2)} & \ln \sigma_{42}^{(2)} \end{bmatrix}$$

$$= \mathbf{Z}^{(1)} \boldsymbol{W}^{(2)} + \begin{bmatrix} 1 \\ 1 \\ 1 \\ 1 \end{bmatrix} \boldsymbol{b}^{(2)}$$

$$= \begin{bmatrix} 0.0521 & 0.2631 & 0.5155 & -0.3204 \\ 0.1589 & 0.2274 & 0.6822 & -0.2965 \\ -0.0754 & 0.4269 & 0.6684 & -0.2982 \\ -0.5623 & 0.4204 & -0.8386 & 0.3409 \end{bmatrix} \cdot \begin{bmatrix} -0.7286 & -1.0228 & -0.0376 & -0.0430 \\ -0.3639 & -0.5126 & -0.0264 & -0.0176 \\ 0.2743 & 0.1969 & 0.0050 & 0.0235 \\ 0.5367 & -0.7873 & -0.0165 & -0.0306 \end{bmatrix}$$

$$+\begin{bmatrix}1\\1\\1\\1\end{bmatrix}[-0.0179\ -0.0239\ -0.0407\ -0.0677]$$

$$=\begin{bmatrix}-0.1821 & 0.1417 & -0.0417 & -0.0526\\ -0.1884 & 0.0647 & -0.0443 & -0.0534\\ -0.0950 & 0.2008 & -0.0408 & -0.0471\\ 0.1918 & -0.0977 & -0.0404 & -0.0810\end{bmatrix}.$$

Then, we have

$$\begin{bmatrix}\mu_{11}^{(2)} & \sigma_{11}^{(2)} & \mu_{12}^{(2)} & \sigma_{12}^{(2)}\\ \mu_{21}^{(2)} & \sigma_{21}^{(2)} & \mu_{22}^{(2)} & \sigma_{22}^{(2)}\\ \mu_{31}^{(2)} & \sigma_{31}^{(2)} & \mu_{32}^{(2)} & \sigma_{32}^{(2)}\\ \mu_{41}^{(2)} & \sigma_{41}^{(2)} & \mu_{42}^{(2)} & \sigma_{42}^{(2)}\end{bmatrix}=\begin{bmatrix}\mu_{11}^{(2)} & e^{\ln\sigma_{11}^{(2)}} & \mu_{12}^{(2)} & e^{\ln\sigma_{12}^{(2)}}\\ \mu_{21}^{(2)} & e^{\ln\sigma_{21}^{(2)}} & \mu_{22}^{(2)} & e^{\ln\sigma_{22}^{(2)}}\\ \mu_{31}^{(2)} & e^{\ln\sigma_{31}^{(2)}} & \mu_{32}^{(2)} & e^{\ln\sigma_{32}^{(2)}}\\ \mu_{41}^{(2)} & e^{\ln\sigma_{41}^{(2)}} & \mu_{42}^{(2)} & e^{\ln\sigma_{42}^{(2)}}\end{bmatrix}$$

$$=\begin{bmatrix}-0.1821 & 1.1522 & -0.0417 & 0.9487\\ -0.1884 & 1.0668 & -0.0443 & 0.9479\\ -0.0950 & 1.2224 & -0.0408 & 0.9539\\ 0.1918 & 0.9068 & -0.0404 & 0.9221\end{bmatrix}.$$

When vector $\boldsymbol{\eta}$ takes the value of $[\eta_1\ \eta_2]=[0.0425\ -0.3659]$ by sampling from the standard normal distribution, we can obtain the value of $z_1^{(3)}$ for data $\boldsymbol{X}_1$, as below

$$z_1^{(3)}=\begin{bmatrix}z_{11}^{(3)}\\ z_{12}^{(3)}\end{bmatrix}^{\mathrm{T}}=\begin{bmatrix}\mu_{11}^{(2)}+\eta_1\times\sigma_{11}^{(2)}\\ \mu_{12}^{(2)}+\eta_2\times\sigma_{12}^{(2)}\end{bmatrix}^{\mathrm{T}}$$

$$=\begin{bmatrix}-0.1821+0.0425\times 1.1522\\ -0.0417+(-0.3659)\times 0.9487\end{bmatrix}^{\mathrm{T}}=\begin{bmatrix}-0.1332\\ -0.3888\end{bmatrix}^{\mathrm{T}}.$$

Similarly, for data $\boldsymbol{X}_2$, $\boldsymbol{X}_3$, and $\boldsymbol{X}_4$, we have

$$\boldsymbol{Z}^{(3)}=\begin{bmatrix}z_1^{(3)}\\ z_2^{(3)}\\ z_3^{(3)}\\ z_4^{(3)}\end{bmatrix}=\begin{bmatrix}-0.1332 & -0.3888\\ -0.1431 & -0.3912\\ -0.0430 & -0.3899\\ 0.2303 & -0.3778\end{bmatrix}.$$

Then,

$$\boldsymbol{r}^{(4)} = \begin{bmatrix} \boldsymbol{r}_1^{(4)} \\ \boldsymbol{r}_2^{(4)} \\ \boldsymbol{r}_3^{(4)} \\ \boldsymbol{r}_4^{(4)} \end{bmatrix} = \boldsymbol{Z}^{(3)} \boldsymbol{W}^{(4)} + \begin{bmatrix} 1 \\ 1 \\ 1 \\ 1 \end{bmatrix} \boldsymbol{b}^{(4)}$$

$$= \begin{bmatrix} -0.1332 & -0.3888 \\ -0.1431 & -0.3912 \\ -0.0430 & -0.3899 \\ 0.2303 & -0.3778 \end{bmatrix} \begin{bmatrix} -0.7496 & -0.4817 & 2.1771 & 1.8239 \\ 3.1253 & -2.6128 & 1.8581 & -0.6759 \end{bmatrix}$$

$$+ \begin{bmatrix} 1 \\ 1 \\ 1 \\ 1 \end{bmatrix} [\, 0.9628 \; 0.4477 \; 1.1277 \; -0.1531 \,]$$

$$= \begin{bmatrix} -0.1526 & 1.5279 & 0.1150 & -0.1332 \\ -0.1526 & 1.5388 & 0.0890 & -0.1497 \\ -0.2235 & 1.4872 & 0.3094 & 0.0319 \\ -0.3908 & 1.3240 & 0.9270 & 0.5224 \end{bmatrix},$$

$$\boldsymbol{Z}^{(4)} = \begin{bmatrix} z_1^{(4)} \\ z_2^{(4)} \\ z_3^{(4)} \\ 4z_4^{(4)} \end{bmatrix} = \varphi(\boldsymbol{r}^{(4)})$$

$$= \varphi\left(\begin{bmatrix} -0.1526 & 1.5279 & 0.1150 & -0.1332 \\ -0.1526 & 1.5388 & 0.0890 & -0.1497 \\ -0.2235 & 1.4872 & 0.3094 & 0.0319 \\ -0.3908 & 1.3240 & 0.9270 & 0.5224 \end{bmatrix} \right)$$

$$= \begin{bmatrix} -0.1515 & 0.9100 & 0.1145 & -0.1324 \\ -0.1074 & 0.9165 & -0.0382 & -0.1486 \\ -0.2199 & 0.9028 & 0.2999 & 0.0319 \\ -0.3721 & 0.8677 & 0.7292 & 0.4795 \end{bmatrix}.$$

Finally,

$$\begin{bmatrix} \hat{\boldsymbol{X}}_1 \\ \hat{\boldsymbol{X}}_2 \\ \hat{\boldsymbol{X}}_3 \\ \hat{\boldsymbol{X}}_4 \end{bmatrix} = \boldsymbol{Z}^{(4)} \boldsymbol{W}^{(5)} + \begin{bmatrix} 1 \\ 1 \\ 1 \\ 1 \end{bmatrix} \boldsymbol{b}^{(5)}$$

$$
= \begin{bmatrix} -0.1515 & 0.9100 & 0.1145 & -0.1324 \\ -0.1074 & 0.9165 & -0.0382 & -0.1486 \\ -0.2199 & 0.9028 & 0.2999 & 0.0319 \\ -0.3721 & 0.8677 & 0.7292 & 0.4795 \end{bmatrix} \begin{bmatrix} 0.1963 & -0.5255 & 0.2610 & 1.7003 & -0.6352 \\ 0.4165 & -0.0911 & -0.7577 & -1.2179 & -1.7871 \\ -0.7742 & -1.3710 & -0.5602 & -0.6589 & 0.4474 \\ -0.5544 & 0.0430 & 1.2429 & -0.8601 & -0.8536 \end{bmatrix}
$$

$$
+ \begin{bmatrix} 1 \\ 1 \\ 1 \\ 1 \end{bmatrix} \begin{bmatrix} -0.4543 & 0.8152 & 1.0417 & -0.6501 & 1.0287 \end{bmatrix}
$$

$$
= \begin{bmatrix} -0.1202 & 0.6491 & 0.0837 & -1.9776 & -0.3371 \\ -0.0905 & 0.6835 & 0.0766 & -1.9489 & -0.3381 \\ -0.3713 & 0.4386 & 0.1718 & -2.3486 & -0.3380 \\ -0.9963 & -0.0474 & 0.4746 & -3.2326 & -0.3688 \end{bmatrix}.
$$

With Eq. (9.33b), we obtain the average loss after the first iteration

$$
\begin{aligned}
\overline{\mathcal{L}} &= \frac{1}{4} \sum_{i=1}^{4} \mathcal{L}_i \\
&= \frac{1}{4} \sum_{i=1}^{4} \left(\frac{1}{2} \sum_{j=1}^{2} \left(-\ln \sum \left(Z_{ij}^{(1)} \right) + \mu^2 \left(Z_{ij}^{(1)} \right) + \sum \left(Z_{ij}^{(1)} \right) \right) + \frac{1}{2} \left\| \hat{\boldsymbol{X}}_i - \boldsymbol{X}_i \right\|^2 \right) \\
&= \frac{1}{4} \sum_{i=1}^{4} \left(\frac{1}{2} \sum_{j=1}^{2} \left(-\ln \left(\sigma_{ij}^{(1)} \right)^2 + \left(\mu_{ij}^{(1)} \right)^2 + \left(\sigma_{ij}^{(1)} \right)^2 \right) + \frac{1}{2} \sum_{j=1}^{5} (\hat{x}_{ij} - x_{ij})^2 \right) \\
&= 5.8336.
\end{aligned}
$$

For data $\boldsymbol{X}_1$, the gradients of loss are calculated to be

$$
\begin{aligned}
\frac{\partial \mathcal{L}_1}{\partial w_{j \to k}^{(1)}} &= \frac{\partial \mathcal{L}_1}{\partial \hat{\boldsymbol{X}}_1} \frac{\partial \hat{\boldsymbol{X}}_1}{\partial z_1^{(4)}} \frac{\partial z_1^{(4)}}{\partial r_1^{(4)}} \frac{\partial r_1^{(4)}}{\partial z_1^{(3)}} \frac{\partial z_1^{(3)}}{\partial z_1^{(2)}} \frac{\partial z_1^{(2)}}{\partial z_{1k}^{(1)}} \frac{\partial z_{1k}^{(1)}}{\partial r_{1k}^{(1)}} \frac{\partial r_{1k}^{(1)}}{\partial w_{j \to k}^{(1)}} \\
&= \Bigg((\hat{\boldsymbol{X}}_1 - \boldsymbol{X}_1) \cdot (\boldsymbol{W}^{(5)})^{\mathrm{T}} \\
&\quad \begin{bmatrix} 1 - \tanh^2\left(r_{11}^{(4)}\right) & 0 & 0 & 0 \\ 0 & 1 - \tanh^2\left(r_{12}^{(4)}\right) & 0 & 0 \\ 0 & 0 & 1 - \tanh^2\left(r_{13}^{(4)}\right) & 0 \\ 0 & 0 & 0 & 1 - \tanh^2\left(r_{14}^{(4)}\right) \end{bmatrix} \\
&\quad \cdot (\boldsymbol{W}^{(4)})^T \begin{bmatrix} 1 & \eta_1 & 0 & 0 \\ 0 & 0 & 1 & \eta_2 \end{bmatrix} + \left[-\mu_{11}^{(2)} \left(\sigma_{11}^{(2)} - \frac{1}{\sigma_{11}^{(2)}} \right) - \mu_{12}^{(2)} \left(\sigma_{12}^{(2)} - \frac{1}{\sigma_{12}^{(2)}} \right) \right] \Bigg) \\
&\quad \cdot \begin{bmatrix} w_{k \to 1}^{(2)} \\ w_{k \to 2}^{(2)} \\ w_{k \to 3}^{(2)} \\ w_{k \to 4}^{(2)} \end{bmatrix} \cdot \left(1 - \tanh^2\left(r_{1k}^{(1)} \right) \right) \cdot x_{1j} \quad (j = 1, 2, 3, 4, 5;\ k = 1, 2, 3, 4),
\end{aligned}
$$

$$\frac{\partial \mathcal{L}_1}{\partial b_k^{(1)}} = \frac{\partial \mathcal{L}_1}{\partial \hat{\boldsymbol{X}}_1}\frac{\partial \hat{\boldsymbol{X}}_1}{\partial z_1^{(4)}}\frac{\partial z_1^{(4)}}{\partial r_1^{(4)}}\frac{\partial r_1^{(4)}}{\partial z_1^{(3)}}\frac{\partial z_1^{(3)}}{\partial z_1^{(2)}}\frac{\partial z_1^{(2)}}{\partial z_{1k}^{(1)}}\frac{\partial z_{1k}^{(1)}}{\partial r_{1k}^{(1)}}\frac{\partial r_{1k}^{(1)}}{\partial b_k^{(1)}}$$

$$= \Bigg((\hat{\boldsymbol{X}}_1 - \boldsymbol{X}_1)\cdot(\boldsymbol{W}^{(5)})^{\mathrm{T}} \begin{bmatrix} 1-\tanh^2\left(r_{11}^{(4)}\right) & 0 & 0 & 0 \\ 0 & 1-\tanh^2\left(r_{12}^{(4)}\right) & 0 & 0 \\ 0 & 0 & 1-\tanh^2\left(r_{13}^{(4)}\right) & 0 \\ 0 & 0 & 0 & 1-\tanh^2\left(r_{14}^{(4)}\right) \end{bmatrix}$$

$$\cdot(\boldsymbol{W}^{(4)})^{\mathrm{T}}\begin{bmatrix} 1 & \eta_1 & 0 & 0 \\ 0 & 0 & 1 & \eta_2 \end{bmatrix} + \left[-\mu_{11}^{(2)}\left(\sigma_{11}^{(2)} - \frac{1}{\sigma_{11}^{(2)}}\right) \;\; -\mu_{12}^{(2)}\left(\sigma_{12}^{(2)} - \frac{1}{\sigma_{12}^{(2)}}\right)\right]\Bigg)$$

$$\cdot\left(1-\tanh^2\left(r_{1k}^{(1)}\right)\right)\cdot\begin{bmatrix} w_{k\to 1}^{(2)} \\ w_{k\to 2}^{(2)} \\ w_{k\to 3}^{(2)} \\ w_{k\to 4}^{(2)} \end{bmatrix},$$

$$\frac{\partial \mathcal{L}_1}{\partial w_{k\to 1}^{(2)}} = \frac{\partial \mathcal{L}_1}{\partial \hat{\boldsymbol{X}}_1}\frac{\partial \hat{\boldsymbol{X}}_1}{\partial z_1^{(4)}}\frac{\partial z_1^{(4)}}{\partial r_1^{(4)}}\frac{\partial r_1^{(4)}}{\partial z_1^{(3)}}\frac{\partial z_1^{(3)}}{\partial \mu_{11}^{(2)}}\frac{\partial \mu_{11}^{(2)}}{\partial w_{k\to 1}^{(2)}}$$

$$= \Bigg((\hat{\boldsymbol{X}}_1 - \boldsymbol{X}_1)\cdot(\boldsymbol{W}^{(5)})^{\mathrm{T}} \cdot \begin{bmatrix} 1-\tanh^2\left(r_{11}^{(4)}\right) & 0 & 0 & 0 \\ 0 & 1-\tanh^2\left(r_{12}^{(4)}\right) & 0 & 0 \\ 0 & 0 & 1-\tanh^2\left(r_{13}^{(4)}\right) & 0 \\ 0 & 0 & 0 & 1-\tanh^2\left(r_{14}^{(4)}\right) \end{bmatrix}$$

$$\cdot(\boldsymbol{W}^{(4)})^{T}\begin{bmatrix} 1 \\ 0 \end{bmatrix} - \mu_{11}^{(2)}\Bigg)\cdot z_{1k}^{(1)},$$

$$\frac{\partial \mathcal{L}_1}{\partial w_{k\to 2}^{(2)}} = \frac{\partial \mathcal{L}_1}{\partial \hat{\boldsymbol{X}}_1}\frac{\partial \hat{\boldsymbol{X}}_1}{\partial z_1^{(4)}}\frac{\partial z_1^{(4)}}{\partial r_1^{(4)}}\frac{\partial r_1^{(4)}}{\partial z_1^{(3)}}\frac{\partial z_1^{(3)}}{\partial \sigma_{11}^{(2)}}\frac{\partial \sigma_{11}^{(2)}}{\partial w_{k\to 2}^{(2)}}$$

$$= \Bigg((\hat{\boldsymbol{X}}_1 - \boldsymbol{X}_1)\cdot(\boldsymbol{W}^{(5)})^{\mathrm{T}} \cdot \begin{pmatrix} 1-\tanh^2\left(r_{11}^{(4)}\right) & 0 & 0 & 0 \\ 0 & 1-\tanh^2\left(r_{12}^{(4)}\right) & 0 & 0 \\ 0 & 0 & 1-\tanh^2\left(r_{13}^{(4)}\right) & 0 \\ 0 & 0 & 0 & 1-\tanh^2\left(r_{14}^{(4)}\right) \end{pmatrix}$$

$$\cdot(\boldsymbol{W}^{(4)})^{T}\begin{bmatrix} \eta_1 \\ 0 \end{bmatrix} + \left(\sigma_{11}^{(2)} - \frac{1}{\sigma_{11}^{(2)}}\right)\Bigg)\cdot z_{1k}^{(2)},$$

$$\frac{\partial \mathcal{L}_1}{\partial w_{k\to 3}^{(2)}} = \frac{\partial \mathcal{L}_1}{\partial \hat{\boldsymbol{X}}_1}\frac{\partial \hat{\boldsymbol{X}}_1}{\partial z_1^{(4)}}\frac{\partial z_1^{(4)}}{\partial r_1^{(4)}}\frac{\partial r_1^{(4)}}{\partial z_1^{(3)}}\frac{\partial z_1^{(3)}}{\partial \mu_{12}^{(2)}}\frac{\partial \mu_{12}^{(2)}}{\partial w_{k\to 3}^{(2)}}$$

$$= \Bigg((\hat{\boldsymbol{X}}_1 - \boldsymbol{X}_1)\cdot(\boldsymbol{W}^{(5)})^{\mathrm{T}} \cdot \begin{pmatrix} 1-\tanh^2\left(r_{11}^{(4)}\right) & 0 & 0 & 0 \\ 0 & 1-\tanh^2\left(r_{12}^{(4)}\right) & 0 & 0 \\ 0 & 0 & 1-\tanh^2\left(r_{13}^{(4)}\right) & 0 \\ 0 & 0 & 0 & 1-\tanh^2\left(r_{14}^{(4)}\right) \end{pmatrix}$$

$$
\cdot (\boldsymbol{W}^{(4)})^T \begin{bmatrix} 0 \\ 1 \end{bmatrix} - \mu_{12}^{(2)} \Bigg) \cdot z_{1k}^{(1)},
$$

$$
\frac{\partial \mathcal{L}_1}{\partial w_{k\to 4}^{(2)}} = \frac{\partial \mathcal{L}_1}{\partial \hat{\boldsymbol{X}}_1} \frac{\partial \hat{\boldsymbol{X}}_1}{\partial z_1^{(4)}} \frac{\partial z_1^{(4)}}{\partial r_1^{(4)}} \frac{\partial r_1^{(4)}}{\partial z_1^{(3)}} \frac{\partial z_1^{(3)}}{\partial \sigma_{12}^{(2)}} \frac{\partial \sigma_{12}^{(2)}}{\partial w_{k\to 4}^{(2)}}
$$

$$
= \Big((\hat{\boldsymbol{X}}_1 - \boldsymbol{X}_1) \cdot (\boldsymbol{W}^{(5)})^{\mathrm{T}}
$$

$$
\cdot \begin{pmatrix} 1-\tanh^2\left(r_{11}^{(4)}\right) & 0 & 0 & 0 \\ 0 & 1-\tanh^2\left(r_{12}^{(4)}\right) & 0 & 0 \\ 0 & 0 & 1-\tanh^2\left(r_{13}^{(4)}\right) & 0 \\ 0 & 0 & 0 & 1-\tanh^2\left(r_{14}^{(4)}\right) \end{pmatrix}
$$

$$
\cdot (\boldsymbol{W}^{(4)})^T \begin{bmatrix} 0 \\ \eta_2 \end{bmatrix} + \left(\sigma_{12}^{(2)} - \frac{1}{\sigma_{12}^{(2)}} \right) \Bigg) \cdot z_{1k}^{(2)} \quad (k = 1, 2, 3, 4),
$$

$$
\frac{\partial \mathcal{L}_1}{\partial w_{p\to q}^{(4)}} = \frac{\partial \mathcal{L}_1}{\partial \hat{\boldsymbol{X}}_1} \frac{\partial \hat{\boldsymbol{X}}_1}{\partial z_{1q}^{(4)}} \frac{\partial z_{1q}^{(4)}}{\partial r_{1q}^{(4)}} \frac{\partial r_{1q}^{(4)}}{\partial w_{p\to q}^{(4)}}
$$

$$
= (\hat{\boldsymbol{X}}_1 - \boldsymbol{X}_1) \cdot \begin{bmatrix} w_{q\to 1}^{(5)} \\ w_{q\to 2}^{(5)} \\ w_{q\to 3}^{(5)} \\ w_{q\to 4}^{(5)} \end{bmatrix}
$$

$$
\cdot \left(1 - \tanh^2\left(r_{1q}^{(4)}\right)\right) \cdot z_{1p}^{(3)} \quad (p = 1, 2; q = 1, 2, 3, 4),
$$

$$
\frac{\partial \mathcal{L}_1}{\partial b_q^{(4)}} = \frac{\partial \mathcal{L}_1}{\partial \hat{\boldsymbol{X}}_1} \frac{\partial \hat{\boldsymbol{X}}_1}{\partial z_{1q}^{(4)}} \frac{\partial z_{1q}^{(4)}}{\partial r_{1q}^{(4)}} \frac{\partial r_{1q}^{(4)}}{\partial b_q^{(4)}}
$$

$$
= (\hat{\boldsymbol{X}}_1 - \boldsymbol{X}_1) \cdot \begin{bmatrix} w_{q\to 1}^{(5)} \\ w_{q\to 2}^{(5)} \\ w_{q\to 3}^{(5)} \\ w_{q\to 4}^{(5)} \end{bmatrix} \cdot \left(1 - \tanh^2\left(r_{1q}^{(4)}\right)\right) \quad (q = 1, 2, 3, 4),
$$

$$
\begin{bmatrix} \frac{\partial \mathcal{L}_1}{\partial w_{q\to 1}^{(5)}} \\ \frac{\partial \mathcal{L}_1}{\partial w_{q\to 2}^{(5)}} \\ \frac{\partial \mathcal{L}_1}{\partial w_{q\to 3}^{(5)}} \\ \frac{\partial \mathcal{L}_1}{\partial w_{q\to 4}^{(5)}} \\ \frac{\partial \mathcal{L}_1}{\partial w_{q\to 5}^{(5)}} \end{bmatrix} = \begin{bmatrix} \frac{\partial \mathcal{L}_1}{\partial \hat{\boldsymbol{X}}_1} \frac{\partial \hat{\boldsymbol{X}}_1}{\partial w_{q\to 1}^{(5)}} \\ \frac{\partial \mathcal{L}_1}{\partial \hat{\boldsymbol{X}}_1} \frac{\partial \hat{\boldsymbol{X}}_1}{\partial w_{q\to 2}^{(5)}} \\ \frac{\partial \mathcal{L}_1}{\partial \hat{\boldsymbol{X}}_1} \frac{\partial \hat{\boldsymbol{X}}_1}{\partial w_{q\to 3}^{(5)}} \\ \frac{\partial \mathcal{L}_1}{\partial \hat{\boldsymbol{X}}_1} \frac{\partial \hat{\boldsymbol{X}}_1}{\partial w_{q\to 4}^{(5)}} \\ \frac{\partial \mathcal{L}_1}{\partial \hat{\boldsymbol{X}}_1} \frac{\partial \hat{\boldsymbol{X}}_1}{\partial w_{q\to 5}^{(5)}} \end{bmatrix}
$$

$$
= (\hat{\boldsymbol{X}}_1 - \boldsymbol{X}_1) \cdot \begin{bmatrix} z_{1q}^{(4)} & 0 & 0 & 0 & 0 \\ 0 & z_{1q}^{(4)} & 0 & 0 & 0 \\ 0 & 0 & z_{1q}^{(4)} & 0 & 0 \\ 0 & 0 & 0 & z_{1q}^{(4)} & 0 \\ 0 & 0 & 0 & 0 & z_{1q}^{(4)} \end{bmatrix} \quad (q = 1, 2, 3, 4),
$$

$$\begin{bmatrix} \frac{\partial \mathcal{L}_1}{\partial b_1^{(5)}} \\ \frac{\partial \mathcal{L}_1}{\partial b_2^{(5)}} \\ \frac{\partial \mathcal{L}_1}{\partial b_3^{(5)}} \\ \frac{\partial \mathcal{L}_1}{\partial b_4^{(5)}} \\ \frac{\partial \mathcal{L}_1}{\partial b_5^{(5)}} \end{bmatrix} = \begin{bmatrix} \frac{\partial \mathcal{L}_1}{\partial \hat{X}_1} \frac{\partial \hat{X}_1}{\partial b_1^{(4)}} \\ \frac{\partial \mathcal{L}_1}{\partial \hat{X}_1} \frac{\partial \hat{X}_1}{\partial b_2^{(4)}} \\ \frac{\partial \mathcal{L}_1}{\partial \hat{X}_1} \frac{\partial \hat{X}_1}{\partial b_3^{(4)}} \\ \frac{\partial \mathcal{L}_1}{\partial \hat{X}_1} \frac{\partial \hat{X}_1}{\partial b_4^{(4)}} \\ \frac{\partial \mathcal{L}_1}{\partial \hat{X}_1} \frac{\partial \hat{X}_1}{\partial b_5^{(4)}} \end{bmatrix} = (\hat{\boldsymbol{X}}_1 - \boldsymbol{X}_1) \cdot \begin{bmatrix} 1 & 0 & 0 & 0 & 0 \\ 0 & 1 & 0 & 0 & 0 \\ 0 & 0 & 1 & 0 & 0 \\ 0 & 0 & 0 & 1 & 0 \\ 0 & 0 & 0 & 0 & 1 \end{bmatrix}.$$

Take $w_{1\to1}^{(1)}$ as an example to illustrate the BP process and the others will be conducted in the same way.

$$\begin{aligned}
\frac{\partial \mathcal{L}_1}{\partial w_{1\to1}^{(1)}} &= \frac{\partial \mathcal{L}_1}{\partial \hat{X}_1} \frac{\partial \hat{X}_1}{\partial z_1^{(4)}} \frac{\partial z_1^{(4)}}{\partial r_1^{(4)}} \frac{\partial r_1^{(4)}}{\partial z_1^{(3)}} \frac{\partial z_1^{(3)}}{\partial z_1^{(2)}} \frac{\partial z_1^{(2)}}{\partial z_{11}^{(1)}} \frac{\partial z_{11}^{(1)}}{\partial r_{11}^{(1)}} \frac{\partial r_{11}^{(1)}}{\partial w_{1\to1}^{(1)}} \\
&= \Bigg((\hat{\boldsymbol{X}}_1 - \boldsymbol{X}_1) \cdot (\boldsymbol{W}^{(5)})^{\mathrm{T}} \\
&\cdot \begin{bmatrix} 1-\tanh^2\left(r_{11}^{(4)}\right) & 0 & 0 & 0 \\ 0 & 1-\tanh^2\left(r_{12}^{(4)}\right) & 0 & 0 \\ 0 & 0 & 1-\tanh^2\left(r_{13}^{(4)}\right) & 0 \\ 0 & 0 & 0 & 1-\tanh^2\left(r_{14}^{(4)}\right) \end{bmatrix} \\
&\cdot (\boldsymbol{W}^{(4)})^T \cdot \begin{bmatrix} 1 & \eta_1 & 0 & 0 \\ 0 & 0 & 1 & \eta_2 \end{bmatrix} + \left[-\mu_{11}^{(2)} \left(\sigma_{11}^{(2)} - \frac{1}{\sigma_{11}^{(2)}} \right) - \mu_{12}^{(2)} \left(\sigma_{12}^{(2)} - \frac{1}{\sigma_{12}^{(2)}} \right) \right] \Bigg) \\
&\cdot \begin{bmatrix} w_{1\to1}^{(2)} \\ w_{1\to2}^{(2)} \\ w_{1\to3}^{(2)} \\ w_{1\to4}^{(2)} \end{bmatrix} \cdot \left(1 - \tanh^2\left(r_{11}^{(1)}\right)\right) \cdot x_{11} \\
&= \left(\begin{bmatrix} -0.1202 & 0.073 \\ 0.6491 & -0.796 \\ 0.0837 & -0.646 \\ -1.9776 & -0.851 \\ -0.3371 & -0.516 \end{bmatrix}^{\mathrm{T}} \cdot \begin{bmatrix} 0.1963 & -0.5255 & 0.2610 & 1.7003 & -0.6352 \\ 0.4165 & -0.0911 & -0.7577 & -1.2179 & -1.7871 \\ -0.7742 & -1.3710 & -0.5602 & -0.6589 & 0.4474 \\ -0.5544 & 0.0430 & 1.2429 & -0.8601 & -0.8536 \end{bmatrix}^{\mathrm{T}} \right. \\
&\cdot \begin{bmatrix} 1-\tanh^2(-0.1526) & 0 & 0 & 0 \\ 0 & 1-\tanh^2(1.5279) & 0 & 0 \\ 0 & 0 & 1-\tanh^2(0.1150) & 0 \\ 0 & 0 & 0 & 1-\tanh^2(-0.1332) \end{bmatrix} \\
&\cdot \begin{bmatrix} -0.7496 & -0.4817 & 2.1771 & 1.8239 \\ 3.1253 & -2.6128 & 1.8581 & -0.6759 \end{bmatrix}^{\mathrm{T}} \cdot \begin{bmatrix} 1 & 0.0425 & 0 & 0 \\ 0 & 0 & 1 & -0.3659 \end{bmatrix}
\end{aligned}$$

$$+\left[0.1821\ 1.1522 - \frac{1}{1.1522}\ 0.0417\ 0.9487 - \frac{1}{0.9487}\right]\right) \cdot \begin{bmatrix} 0.0497 \\ 0.0592 \\ 0.5291 \\ -0.1718 \end{bmatrix}$$

$$\cdot \begin{pmatrix} 1 - \tanh^2(0.0521) \\ 1 - \tanh^2(0.1603) \\ 1 - \tanh^2(-0.0756) \\ 1 - \tanh^2(-0.6363) \end{pmatrix} \cdot (-0.073) = 0.5634.$$

Similarly, for data $\boldsymbol{X}_2$, $\boldsymbol{X}_3$, and $\boldsymbol{X}_4$, we can obtain $\frac{\partial \mathcal{L}_2}{\partial w_{1\to1}^{(1)}} = 0.5183$, $\frac{\partial \mathcal{L}_3}{\partial w_{1\to1}^{(1)}} = -11.1720$, and $\frac{\partial \mathcal{L}_4}{\partial w_{1\to1}^{(1)}} = 0.4260$, and thus

$$\frac{\partial \overline{\mathcal{L}}}{\partial w_{1\to1}^{(1)}} = \frac{\frac{\partial \mathcal{L}_1}{\partial w_{1\to1}^{(1)}} + \frac{\partial \mathcal{L}_2}{\partial w_{1\to1}^{(1)}} + \frac{\partial \mathcal{L}_3}{\partial w_{1\to1}^{(1)}} + \frac{\partial \mathcal{L}_4}{\partial w_{1\to1}^{(1)}}}{4} = -2.4161.$$

Set the learning rate η to be 0.01, and update the value of $w_{1\to1}^{(1)}$, viz.,

$$w_{1\to1(\text{new})}^{(1)} = w_{1\to1(\text{old})}^{(1)} - \eta \frac{\partial \overline{\mathcal{L}}}{\partial w_{1\to1}^{(1)}} = 0.0497 - 0.01 \times (-2.4161) = 0.07386.$$

The other 38 trainable variables are updated in the same way.
After 3000 epochs, the loss $\mathcal{L}$ is 0.57. Table 9.19 shows the R^2 value of the predicted results. Compared with Example 9.8, the prediction accuracy is improved with a more complex model.

Table 9.19 The R^2 of the VAE model

The variable	C	F	T	Fr	H	The average value
R^2	0.7530	0.9165	0.8849	0.9495	0.9476	0.8903

Homework

1.

No.	Cr (x_1)	Mo (x_2)	TT (x_3)	dB (x_4)	S (x_5)	Tensile strength (y)
1	0.08	0.00	0.00	0.00	0.71	0.42
2	0.08	0.00	0.77	0.00	0.81	0.21
3	0.03	0.00	0.77	0.50	1.00	0.22
4	0.94	0.89	0.38	0.00	0.24	0.69

(1) Construct a regression ANN model with the data in the above table, and give the details of updating the parameters with the BP algorithm.
(2) Add L_1 or L_2 regularization terms to the loss function, and provide the details of updating the parameters with the BP algorithm.

2.

Catalysts	GHSV	T	Classes
Mn_5Ce_1/TiO_2	1	0.64	0
Fe_1Mn_8/MnO_x	0.06	0.25	1
Mn_2Ce_1/Graphene	0	0.27	2

(1) Rewrite the class in the above table into one-hot encoding ($[\,c_1\ c_2\ c_3\,]$).
(2) Construct a classification ANN model, calculate the cross-entropy loss, and complete a parameter-update iteration.
(3) Add an activation function to the hidden layer to calculate the cross-entropy loss, and complete a parameter-update iteration.

3. Use the data in Table 9.5 to compare whether the generalization ability of the NN changes after adding the regularization term.
4. Discuss what will happen if the KL divergence term is removed from the loss function in the VAE.

References

Ackley, D. H., Hinton, G. E., & Sejnowski, T. J. (1985). A learning algorithm for Boltzmann machines. *Cognitive science, 9*(1), 147–169.

Andrew, N. (2011). Sparse autoencoder. *CS294A Lecture Notes, 72*, 1–19.

Biron, P. V. (1997). Backpropagation: Theory, architectures, and applications. *Journal of the American Society for Information Science, 48*(1), 88–89.

Blei, D. M., Kucukelbir, A., & McAuliffe, J. D. (2017). Variational inference: A review for statisticians. *Journal of the American Statistical Association, 112*(518), 859–877.

Buscema, M. (1998). Back propagation neural networks. *Substance Use and Misuse, 33*(2), 233–270.

Hopfield, J. J. (1982). Neural networks and physical systems with emergent collective computational abilities. *Proceedings of the national academy of sciences, 79*(8), 2554–2558.

Kingma, D. P., & Welling, M. (2013). Auto-encoding variational Bayes. arXiv preprint arXiv:1312.6114.

McCulloch, W. S., & Pitts, W. (1990). A logical calculus of the ideas immanent in nervous activity. *Bulletin of Mathematical Biology, 52*(1–2), 99–115.

Tickle, A. B., Andrews, R., Golea, M., & Diederich, J. (1998). The truth will come to light: Directions and challenges in extracting the knowledge embedded within trained artificial neural networks. *IEEE Transactions on Neural Networks, 9*(6), 1057–1068.

Vincent, P., Larochelle, H., Bengio, Y., & Manzagol, P. A. (2008). Extracting and composing robust features with denoising autoencoders. In W. Cohen (Ed.), *Proceedings of the 25th International Conference on Machine Learning* (pp. 1096–1103). Association for Computing Machinery.

Chapter 10
Hidden Markov Chains

In the mathematic area of statistics and probability, Markov chains are a fundamental class of stochastic processes. Markov chains, like other mathematic tools, are extremely important to various fields of operational research, economics, finance, social sciences, computer science and engineering, networks, biology, physics, chemistry, and mechanics. The powerful Markov decision process algorithm in reinforcement learning, which will be introduced in Volume II of this book, is developed based on Markov chains. In materials science and engineering, the properties and performance of a material depend on its chemical composition and microstructure, which is in turn formed by the topological arrangement of all involved atoms. Provided that the chemical composition of a material does not change during its service life, the material performance depends only on its microstructure that varies along with service time. The microstructure of a material is named as the state in Markov chains, and the change in the microstructure can be described by the variation in the state. For example, the hot deformation behavior of metals depends on the change in the microstructure, which involves various crystalline defect activities including dislocation multiplication, movement, topological configuration, annihilation, grain boundary movement, phase transformation, dynamic recrystallization, etc. On one side, material researchers are endeavoring to in situ characterize or monitor the change in the microstructure during tests or service, and on the other side, are doing their best to model dynamic behaviors of materials, where Markov chains will find new applications.

10.1 Markov Chain

Markov chains were pioneered by Markov in 1913 (Markov, 1913a, 1913b, 2006) to study a stochastic process that describes a process of events probabilistically. A Markov chain is a sequence of random events expressed by a sequence of variables

T. Zhang, *An Introduction to Materials Informatics*,
https://doi.org/10.1007/978-981-99-7992-9_10

$\boldsymbol{X} = \{x_i\}(i = 0, 1, \cdots, n)$ that possess the Markov properties. As described later, x_0 is preset in a Markov chain and n is denoted as the chain length. The important Markov property is that the subsequent probability of the event is given by the conditional probability

$$P(x_n|x_{n-1}, \cdots, x_0). \tag{10.1a}$$

If the conditional probability of event $P(x_n|x_{n-1}, \cdots, x_0)$ depends only on past n' events, the Markov chain has an order of n' and Eq. (10.1a) is reduced to

$$P(x_n|x_{n-1}, \cdots, x_0) = P(x_n|x_{n-1}, \cdots, x_{n-n'}). \tag{10.1b}$$

Widely studied Markov chains are first-order, i.e., each random event depends only on its predecessor,

$$P(x_n|x_{n-1}, \cdots, x_0) = P(x_n|x_{n-1}). \tag{10.1c}$$

This chapter introduces only Markov chains of first-order.

Baum and Petrie (1966) and Baum and Eagon (1967) developed the hidden Markov model. A hidden Markov chain has a state set, $S = \{s_k\}$ $(k = 1, 2, \cdots, q)$, and in many cases s_k is just a label so that $s_k = k$. The probability for the state s_k occupation at event (i) is denoted by $\pi_i^{(k)} = P(s_k \in S)$ and $\sum_{k=1}^{q} \pi_i^{(k)} = 1$ for $(i = 0, 1, \cdots, n)$. The transition probability matrix $\boldsymbol{T} = \{T_{kl}\}$ is defined by

$$T_{lk} = P\left(\pi_i^{(k)}|\pi_{i-1}^{(l)}\right) \ (l, k = 1, 2, \cdots, q), \quad i \geq 1, \tag{10.2a}$$

with $\sum_{k=1}^{q} T_{lk} = 1$ for $(l = 1, 2, \cdots, q)$. If $T_{ll} = T_{lk} = 1$ for $k = l$ and $T_{lk} = 0$ for $k \neq l$ $(k, l = 1, 2, \cdots, q)$, the state l is called an adsorption state or a terminal state. The initial probability of state occupation $\pi_0^{(l)}$ or π_0 for simplicity is an intrinsic parameter of a hidden Markov chain. The transition probability matrix is another intrinsic parameter of a Markov chain and has the property, being independent of event (i). With the transition probability matrix, the state probability of $\pi_i^{(k)}(k = 1, 2, \cdots, q)$ can be calculated by its preceding state probability via

$$\pi_i^{(k)} = \sum_{l=1}^{q} T_{lk}\pi_{i-1}^{(l)} \ (k = 1, 2, \cdots, q), \quad i \geq 1, \tag{10.2b}$$

or in the matrix and vector form,

$$\boldsymbol{\pi}_i = \boldsymbol{\pi}_{i-1}\boldsymbol{T}. \tag{10.2c}$$

As described above, each row of the transition matrix sums to one and the transition matrix is called a right stochastic matrix with the state transition described by Eq. (10.2c). There is also a left stochastic matrix, with each column summing to one, and a doubly stochastic matrix, in which both of each column and each row sum to one.

Example 10.1 Random walk along a circle

Consider an atom walking randomly along a circle, on which there are $q = 4$ sites for the atom to stay. In this case we have four states, viz., $S = \{s_k\}$ $(k = 1, 2, 3, 4)$. For the random walk along a circle, the atom can move to its nearest neighbors or stays in place without any movement. At each site, the atom has three possibilities, (1) moving clockwise, (2) moving anticlockwise, and (3) staying in place without any movement. If the probabilities for the three movement modes are the same, the transition probability matrix is $\boldsymbol{T} = \begin{bmatrix} \frac{1}{3} & \frac{1}{3} & 0 & \frac{1}{3} \\ \frac{1}{3} & \frac{1}{3} & \frac{1}{3} & 0 \\ 0 & \frac{1}{3} & \frac{1}{3} & \frac{1}{3} \\ \frac{1}{3} & 0 & \frac{1}{3} & \frac{1}{3} \end{bmatrix}$, as shown in Fig. 10.1.

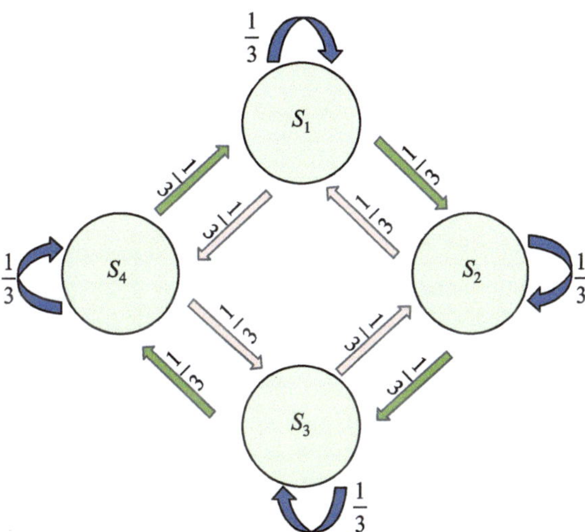

Fig. 10.1 The random walk of an atom along a circle with four staying sites

It is usually assumed that the atom initially stays at one site, say site 1. Thus, the initial state (staying site) occupation probability vector is $\boldsymbol{\pi}_0 = \boldsymbol{\pi}_0^{(1)} = \begin{pmatrix} 1 & 0 & 0 & 0 \end{pmatrix}$. After that, the state occupation probability $\boldsymbol{\pi}_1$, $\boldsymbol{\pi}_2$ and $\boldsymbol{\pi}_3$ are calculated recursively

$$\boldsymbol{\pi}_1 = \boldsymbol{\pi}_0 \boldsymbol{T} = \left(\frac{1}{3} \; \frac{1}{3} \; 0 \; \frac{1}{3} \right), \quad \boldsymbol{\pi}_2 = \boldsymbol{\pi}_0 \boldsymbol{T}^2 = \left(\frac{3}{9} \; \frac{2}{9} \; \frac{2}{9} \; \frac{2}{9} \right), \quad \boldsymbol{\pi}_3 = \boldsymbol{\pi}_0 \boldsymbol{T}^3.$$

A Markov chain is irreducible, if any state can be reached from any other state in a finite number of steps. In the random walk along a circle, the state occupation probability of $\boldsymbol{\pi}_2 = \boldsymbol{\pi}_0 \boldsymbol{T}^2 = \left(\frac{3}{9} \; \frac{2}{9} \; \frac{2}{9} \; \frac{2}{9} \right)$ indicates that every one of the four states (staying sites) already has a certain probability of being taken. If the transition between a state s_k with its nearest neighbor s_l satisfies

$$T_{lk} = P(s_k|s_l) = T_{kl} = P(s_l|s_k), \tag{10.3a}$$

the transition is symmetric. A symmetric transition between two states does not mean an equal state occupation probability, as illustrated in Example 10.1. For the random walk along a circle, although there are only 4 states, the sequence length could be larger or smaller than the number of states. Thus, in a recursive way, we have the probability $\boldsymbol{\pi}_n$,

$$\boldsymbol{\pi}_n = \boldsymbol{\pi}_0 \boldsymbol{T}^n. \tag{10.3b}$$

Using singular value decomposition (SVD), $\boldsymbol{T}$ is decomposed as

$$\boldsymbol{T} = \boldsymbol{U}\boldsymbol{D}\boldsymbol{U}^{-1}, \tag{10.3c}$$

where $\boldsymbol{D}$ is a diagonal eigenvalue matrix with diagonal entries $d_1 \geq d_2 \geq \ldots \geq d_n \geq 0$. Then, the probability $\boldsymbol{\pi}_n$ can be calculated by

$$\boldsymbol{\pi}_n = \boldsymbol{\pi}_0 \boldsymbol{U}\boldsymbol{D}^n \boldsymbol{U}^{-1}. \tag{10.3d}$$

Example 10.2 Two components are used in a structure and the structure can function if one component is working. A Markov chain model is proposed based on domain knowledge: there are 4 possible cases of the structure $S = \{s_k\}(k = 1, 2, 3, 4)$, where s_1 indicates the probability of no damage in the structure, s_2 and s_3 represent the probability of damage in component 1 and component 2, respectively, and s_4 indicates the probability of both components 1 and 2 being damaged. If the structure is examined every day and based on domain knowledge, the failure transition matrix is

$$T = \begin{bmatrix} 0.9 & 0.04 & 0.06 & 0 \\ 0 & 0.9 & 0 & 0.1 \\ 0 & 0 & 0.9 & 0.1 \\ 1 & 0 & 0 & 0 \end{bmatrix}.$$

Calculate the damage probabilities of parts 1, 2, and 12 after the structure service for 1000 days.

We can assume that both components work fine at first. Thus, the initial state probability vector is $\boldsymbol{\pi}_0 = \boldsymbol{\pi}_0^{(1)} = \begin{pmatrix} 1\ 0\ 0\ 0 \end{pmatrix}$. The state probability of the two components after 1000 days is $\boldsymbol{\pi}_{1000} = \boldsymbol{\pi}_0 \boldsymbol{T}^{1000}$. According to Eq. (10.3a), $\boldsymbol{T}$ is decomposed as

$$\boldsymbol{T} = \boldsymbol{U}\boldsymbol{D}\boldsymbol{U}^{-1} = \begin{bmatrix} 0.9 & 0.04 & 0.06 & 0 \\ 0 & 0.9 & 0 & 0.1 \\ 0 & 0 & 0.9 & 0.1 \\ 1 & 0 & 0 & 0 \end{bmatrix}$$

$$= \begin{bmatrix} 0.5 & 0 & 0.440496 & -0.012542 \\ 0.5 & -0.83205 & -0.496447 & 0.111288 \\ 0.5 & 0.5547 & -0.496447 & 0.111288 \\ 0.5 & 0 & 0.559504 & -0.987458 \end{bmatrix}$$

$$\begin{bmatrix} 1 & 0 & 0 & 0 \\ 0 & 0.9 & 0 & 0 \\ 0 & 0 & 0.787298 & 0 \\ 0 & 0 & 0 & 0.012701 \end{bmatrix}$$

$$\cdot \begin{bmatrix} 0.952381 & 0.380952 & 0.571429 & 0.095238 \\ 0 & -0.721110 & 0.721110 & 0 \\ 1.222590 & -0.433921 & -0.650881 & -0.137788 \\ 1.174971 & -0.052968 & -0.079453 & -1.042550 \end{bmatrix}.$$

According to Eq. (10.3b), we can calculate

$$\boldsymbol{\pi}_{1000} = \boldsymbol{\pi}_0 \boldsymbol{U}\boldsymbol{D}^{1000}\boldsymbol{U}^{-1} = \begin{pmatrix} 0.47618958\ 0.19047582\ 0.28571376\ 0.04761896 \end{pmatrix}.$$

Thus, the probabilities of damage in parts 1 and 2, and in both 12 are 0.19047582, 0.28571376, and 0.04761896, respectively.

10.2 Stationary Markov Chain

If a Markov chain is stationary (invariant, steady, or equilibrium), the state distribution satisfies

$$\pi = \pi \boldsymbol{T}. \tag{10.3e}$$

Equation (10.3e) can be rewritten as

$$\pi^{(k)} = \sum_{l=1}^{q} \pi^{(l)} T_{lk} \text{ subject to } \sum_{k=1}^{q} T_{lk} = 1 \text{ and } \sum_{k=1}^{q} \pi^{(k)} = 1. \tag{10.3f}$$

For example, $\left(\frac{1}{2}\ \frac{1}{2}\right) = \left(\frac{1}{2}\ \frac{1}{2}\right)\begin{bmatrix} \frac{1}{2} & \frac{1}{2} \\ \frac{1}{2} & \frac{1}{2} \end{bmatrix}$, $\left(\frac{1}{2}\ \frac{1}{2}\right) = \left(\frac{1}{2}\ \frac{1}{2}\right)\begin{bmatrix} \frac{3}{4} & \frac{1}{4} \\ \frac{1}{4} & \frac{3}{4} \end{bmatrix}$, and $\left(\frac{1}{4}\ \frac{3}{4}\right) = \left(\frac{1}{4}\ \frac{3}{4}\right)\begin{bmatrix} \frac{1}{2} & \frac{1}{2} \\ \frac{1}{6} & \frac{5}{6} \end{bmatrix}$. Clearly, for a given stationary state vector, there are multiple transition matrixes because Eq. (10.3f) has multiple solutions. For a given state vector, we may select a transition matrix to meet Eq. (10.3f). On the other hand, we can select a state vector to meet Eq. (10.3f) for a given transition matrix. For example, if $\boldsymbol{T} = \begin{bmatrix} 1 - T_{12} & T_{12} \\ T_{21} & 1 - T_{21} \end{bmatrix}$, the state vector $\begin{bmatrix} \frac{T_{21}}{T_{12} + T_{21}} & \frac{T_{12}}{T_{12} + T_{21}} \end{bmatrix}$ is stationary. Using SVD, we have $\boldsymbol{\pi}_n = \boldsymbol{\pi}_0 \boldsymbol{U} \boldsymbol{D}^n \boldsymbol{U}^{-1}$. Thus, for a stationary Markov chain and for an arbitrary state vector, the transition matrix has a unity eigenvalue matrix.

10.3 Markov Chain Monte Carlo Methods

Markov chain Monte Carlo (MCMC) methods are a class of Monte Carlo simulations for sampling a given univariate or multivariate probability distribution. The MCMC methods play an extremely important role in Bayesian probability when numerically calculating means, variances, and explorations of the posterior distribution in the Bayesian prior–posterior analysis. Since a Monte Carlo simulation (Metropolis & Ulam, 1949; Metropolis et al., 1953) refers to large numbers of sample drawings by following a distribution of interest, the MCMC methods take Monte Carlo simulations and advantages of Markov chains to calculate means, variances, and explorations of the posterior distribution. As described above, each stochastic event (element, sample, or variable value) in a Markov chain depends only on its predecessor. Using the Markov property in sampling gives the drawing probability $P(x_n|x_{n-1})$. This means that the next element x_n in a Markov chain is drawn by following the probability distribution $P(x_n|x_{n-1})$, where x_{n-1} is the current element in the Markov chain.

As described in Chap. 7, the Bayesian formula is one of the most important equations in statistics and in materials informatics. Suppose that in a given Bayesian model, the prior probability $P(\theta)$ and the likelihood probability $P(x|\theta)$ are known, where x is an observed variable and θ is an unknown parameter in an ML model. Based on the Bayesian formula, the inference about θ is given by the posterior $P(\theta|x)$,

$$P(\theta|x) = \frac{P(x|\theta)P(\theta)}{P(x)}, \tag{10.4a}$$

where $P(x)$ is called the total likelihood. If the total likelihood $P(x) = \int_\theta P(x|\theta)P(\theta)\mathrm{d}\theta$ or the expectation of the posterior distribution, $E(\theta|x) = \int_\theta \theta P(\theta|x)\mathrm{d}\theta$, is of interest, but it is difficult to compute the integrals analytically. In this case, the integral can be calculated by using Monte Carlo simulations. Take the total likelihood as an example. For a given arbitrary x^*, $P(x^*)$ is the expectation $P(x^*) = \int_\theta P(x^*|\theta)P(\theta)\mathrm{d}\theta = E_{P(\theta)}(P(x^*|\theta))$. Following the prior distribution, a sequence of parameter $\boldsymbol{\theta} = (\theta_1, \theta_2, \cdots, \theta_k)$ can be sampled. Subsequently, the likelihood function of $P(x^*|\theta)$ gives the sequence of $P(x^*|\theta_i)(i = 1, 2, \cdots, k)$ for a given x^*. Then, $P(x^*) = E_{P(\theta)}(P(x^*|\theta))$ will be converged from $\frac{1}{k}\sum_{i=1}^{k} P(x^*|\theta_i)$ as $k \to \infty$ with the θ sequence θ_i $(i = 1,2, \cdots, k)$ drawn by following the prior distribution $P(\theta)$.

If $P(\theta)$ is too complex to follow, how to draw samples is an important issue in Monte Carlo simulations. An intuitive method, called importance sampling (Hammersley & Morton, 1954), introduces a normal distribution $q(\theta) \sim N(0, 1)$ or another common distribution $q(\theta)$ so that $P(x^*) = \int_\theta P(x^*|\theta)P(\theta)\mathrm{d}\theta = \int_\theta \frac{P(x^*|\theta)P(\theta)}{q(\theta)} q(\theta)\mathrm{d}\theta = E_{q(\theta)}\left(\frac{P(x^*|\theta)P(\theta)}{q(\theta)}\right)$ is converted to the expectation of function $\frac{P(x^*|\theta)P(\theta)}{q(\theta)}$ over the distribution $q(\theta)$. The drawing from such a normal distribution $q(\theta)$ is much easy. Following the distribution $q(\theta)$, a sequence of parameter $\theta = (\theta_1, \theta_2, \cdots, \theta_k)$ can be sampled, and thereafter, $\frac{P(x^*|\theta)P(\theta)}{q(\theta)}$ gives the sequence of $P(x^*|\theta_i)\frac{P(\theta_i)}{q(\theta_i)}$ $(i = 1, 2, \cdots, k)$ for a given x^* with the fraction $\frac{P(\theta_i)}{q(\theta_i)}$, which is called the weight of samples or importance weight. Then for a given value of x^*, we have the expectation of $P(x^*) = E_{q(\theta)}\left(\frac{P(x^*|\theta)P(\theta)}{q(\theta)}\right)$ estimated by $\frac{1}{k}\sum_{i=1}^{k}\frac{P(x^*|\theta_i)P(\theta_i)}{q(\theta_i)}$ with sufficient samples θ_i $(i = 1, 2, \cdots, k;\ k \to \infty)$ following the distribution $q(\theta)$. However, the weight of samples might not be homogenous, and could depend on the regions of parameter θ, in some regions large and in some regions small. Clearly, the estimated expectation is dominated by the few large values of the weight of samples, even though the probability distribution density is small. On the other hand, a small weight of samples might underestimate the expectation greatly. The weight of samples is an indicator for the goodness of the sampling. The optimal sampling should maintain the weight of samples as close as possible to one. When very little is known about the structural properties of the target distribution, it will be a great challenge to identify a probability distribution $q(\theta)$ that is easy for sampling and its drawing sequence represents such a sequence drawn from a target probability distribution.

The MCMC algorithm provides a strategy to sample a target probability density. In an MCMC simulation, as described above, the sampling process yields a suitably constructed Markov chain. When the Markov chain is longer than a critical length, it reaches the stationary distribution and the stationary Markov chain will represent a target probability distribution of interest. Equations (10.3e) and (10.3f) describe the characteristic of a stationary Markov chain. The segment of the Markov chain before reaching the stationary state is called the burn-in phase. In MCMC simulations, the sampling sequence follows the property of first-order Markov chains so that $q(\theta_{i+1}|\theta_i)$ $(i = 1, 2, \cdots, k)$. An MCMC starts with an arbitrary given initial state θ_0 and draws samples by following the transition rule and a proposed distribution $q(\theta)$. As a result, the samples drawn after the burn-in phase will obey the target probability distribution, e.g., the prior distribution $P(\theta)$. Letting n_0 denote the critical number for the burn-in phase, we have the following samples $\theta_0, \theta_1, \cdots, \theta_{n_0}, \theta_{n_0+1}, \theta_{n_0+2}, \cdots, \theta_{n_0+N}$. The law of large number for a Markov chain indicates that the generated samples $\theta_{n_0+1}, \theta_{n_0+2}, \cdots, \theta_{n_0+N}$ obey the target probability distribution, e.g., the prior distribution $P(\theta)$, thus

$$\lim_{N\to\infty} \frac{1}{N} \sum_{i=1}^{N} P(x^*|\theta_{n_0+i}) \to \int_{\theta} P(x^*|\theta)P(\theta)\mathrm{d}\theta. \tag{10.4b}$$

Some theoretical analysis (Meyn & Tweedie, 1994; Rosenthal, 1995) has suggested how to find the critical value of n_0 for some specific situations. In practice, however, a sufficiently large number, e.g., $\geq 10^5$, of samplings is conducted and the last fifty percent of the samples drawn are used for the target distribution.

10.3.1 Metropolis Hastings (M-H) Algorithm

A Markov chain is constructed by using a proposal distribution $q(\theta_{i+1}|\theta_i)$ $(i = 1, 2, \cdots, k)$, which describes the probability of drawing θ_{i+1} when θ_i is given. The Markov chain is homogeneous, if $q(\theta_{i+1}|\theta_i)$ is independent of the value of θ_i, which means the transition matrix is invariant, as discussed above and later in this chapter. The Metropolis Hastings (M-H) algorithm (Hastings, 1970; Metropolis et al., 1953) is one of the foundational MCMC algorithms. In the M-H algorithm, a random sample θ_0 is initialized as the initial state of a Markov chain. Letting θ' be the current value at time i and θ'' be the candidate value in next time $i + 1$, we select θ'' from $q(\theta''|\theta_i = \theta')$ and the Markov property narrows down the sampling window to a certain extent. The candidate θ'' has only the probability

$$\alpha_{i\to i+1} = \min\left(\frac{P(\theta'')q(\theta_i = \theta'|\theta'')}{P(\theta_i = \theta')q(\theta''|\theta_i = \theta')}, \ 1\right) \tag{10.4c}$$

to be θ_{i+1}, viz., $\theta_{i+1} = \theta''$ has the probability of $\alpha_{i\to i+1}$, where $\alpha_{i\to i+1}$ is called the acceptance probability. Note that $q(\theta_i = \theta'|\theta'')$ in Eq. (10.4c) denotes the probability of the current value $\theta_i = \theta'$ at time i, if its predecessor is θ''. The next value θ_{i+1} has the probability of $(1 - \alpha_{i\to i+1})$ to remain at the current value, viz., $\theta_{i+1} = \theta'$. Clearly, the acceptance probability makes the Markov chain from the proposal transition distribution $q(\theta_{i+1}|\theta_i)$ closer to the target distribution of interest. After a long burn-in period the drawing samples converge to the distribution of $P(\theta)$ gradually. Clearly an appropriate proposal distribution $q(\theta''|\theta')$ plays an important role in MCMC algorithms (Chib, 2001), where θ'' and θ' denote two values of parameter θ. If $q(\theta''|\theta') = q(\theta'|\theta'')$, the proposal distribution q is symmetric. For example, in a multivariate Gaussian distribution, we have

$$q(x|x^*) = \frac{1}{(2\pi)^{h/2}\sigma}\exp\left(\frac{1}{2\sigma^2}(\boldsymbol{x} - \boldsymbol{x}^*)(\boldsymbol{x} - \boldsymbol{x}^*)^{\mathrm{T}}\right) = q(\boldsymbol{x}^*|\boldsymbol{x}) \tag{10.4d}$$

where h is the dimension of multivariate variable x; and x^* and $\sigma^2\boldsymbol{I}$ are the mean and the covariance with $\boldsymbol{I}$ being an $h \times h$ unity matrix, respectively. The acceptance probability for a symmetric proposal distribution q is reduced to

$$\alpha_{i\to i+1} = \min\left(\frac{P(\theta'')}{P(\theta_i = \theta')},\ 1\right). \tag{10.4e}$$

10.3.2 Gibbs Sampling Algorithm

Gibbs sampling (S. Geman & D. Geman, 1984) is another famous MCMC algorithm. The key idea of Gibbs sampling is converting the multi-dimensional sampling problem to a one-dimensional sampling problem by using the conditional distribution of multivariable components. If $\boldsymbol{\theta} \in \mathbf{R}^h$, h components are considered in Gibbs sampling and $\boldsymbol{\theta}' = (\theta_1', \theta_2', \cdots, \theta_h')$ is the current value. Thus the next drawing $\boldsymbol{\theta}'' = (\theta_1'', \theta_2'', \cdots, \theta_h'')$ is produced by a set of conditional distributions $P(\theta_1''|\theta_2 = \theta_2', \theta_3 = \theta_3', \theta_4 = \theta_4', \cdots, \theta_h = \theta_h')$, $P(\theta_1''|\theta_1 = \theta_1'', \theta_3 = \theta_3', \theta_4 = \theta_4', \cdots, \theta_h = \theta_h')$, $\cdots$, $P(\theta_h''|\theta_1 = \theta_1'', \theta_2 = \theta_2'', \theta_3 = \theta_3'', \cdots, \theta_{h-1} = \theta_{h-1}'')$. This means that the Gibbs sampling is one-dimensional conditional sampling and every sampling will calculate h conditional probabilities. Similar to the M-H sampling, there is also a burn-in phase in the Gibbs sampling.

For multiple Gaussian distribution $P(\boldsymbol{\theta}_j) \sim N\left(\begin{pmatrix}\mu_1\\ \mu_2\end{pmatrix}, \begin{pmatrix}\boldsymbol{\Sigma}_{11} & \boldsymbol{\Sigma}_{21}\\ \boldsymbol{\Sigma}_{21} & \boldsymbol{\Sigma}_{22}\end{pmatrix}\right)$, where $\boldsymbol{\theta}_j = \begin{pmatrix}\theta_{j1}\\ \theta_{j2}\end{pmatrix}$ is a (h)-dimensional variable, θ_{j1} is a 1-dimensional variable and θ_{j2} is a $(h\text{–}1)$-dimensional variable, μ_1 is a 1-dimensional mean vector, μ_2 is a $(h\text{–}1)$-dimensional mean vector, $\boldsymbol{\Sigma}_{11}, \boldsymbol{\Sigma}_{12}, \boldsymbol{\Sigma}_{21}, \boldsymbol{\Sigma}_{22}$ are (1×1), $(1 \times h)$,

$(h \times 1)$, and $(h \times h)$ matrixes respectively. There have the condition distributions of $P(\theta_{j1}|\theta_{j2}) \sim N\big(\mu_1 + \boldsymbol{\Sigma}_{12}\boldsymbol{\Sigma}_{22}^{-1}(\theta_{j2} - \mu_{j2}),\ \boldsymbol{\Sigma}_{11} - \boldsymbol{\Sigma}_{12}\boldsymbol{\Sigma}_{22}^{-1}\boldsymbol{\Sigma}_{21}\big)$ and $P(\theta_{j2}|\theta_{j1}) \sim N\big(\mu_2 + \boldsymbol{\Sigma}_{12}\boldsymbol{\Sigma}_{11}^{-1}(\theta_{j1} - \mu_1),\ \boldsymbol{\Sigma}_{22} - \boldsymbol{\Sigma}_{12}\boldsymbol{\Sigma}_{11}^{-1}\boldsymbol{\Sigma}_{12}\big)$ After obtaining the updated value for the first dimension through sampling, the remaining dimensions are subsequently updated, thereby updating the value of all dimensions in the multivariate Gaussian distribution.

Example 10.3

1. Draw samples from $P(\boldsymbol{\theta})$ by using the M-H algorithm for $P(\boldsymbol{\theta}_j) \sim N\left(\begin{pmatrix}1\\4\end{pmatrix}, \begin{bmatrix}3 & 1\\2 & 1\end{bmatrix}\right)$, where $\boldsymbol{\theta}_j = \begin{pmatrix}\theta_{j1}\\\theta_{j2}\end{pmatrix}$ is a two-dimensional variable.

An initial sample $\boldsymbol{\theta}_0 = \begin{pmatrix}0\\0\end{pmatrix}$ is given as the initial drawing sample. According to Eq. (10.4d) all other sample candidates are sampled through a Markov chain of proposal distribution $q(\boldsymbol{\theta}''|\theta_i = \theta') = \frac{1}{(2\pi)^{2/2} \times \sigma}\exp\left(\frac{1}{2\sigma^2}(\boldsymbol{\theta}'' - \boldsymbol{\theta}')(\boldsymbol{\theta}'' - \boldsymbol{\theta}')^{\mathrm{T}}\right)$, $\boldsymbol{\theta}_i = \boldsymbol{\theta}'$ is the current value, and $\boldsymbol{\theta}_0$, $\boldsymbol{\theta}_i$ and $\boldsymbol{\theta}''$ are two-dimensional vectors. The σ^2 is suggested to take the smallest eigenvalue of the covariance matrix $\begin{bmatrix}3 & 1\\2 & 1\end{bmatrix}$ (Faul, 2019). The eigenvalue matrix of $\begin{bmatrix}3 & 1\\2 & 1\end{bmatrix}$ is $\begin{bmatrix}2+\sqrt{3} & 0\\0 & 2-\sqrt{3}\end{bmatrix}$, and we take $\sigma^2 = 2 - \sqrt{3}$, viz., $\sigma = 0.518$. We select the candidates from $q(\theta''|\theta_i = \theta')$, use the acceptance probability of Eq. (10.4c), and generate a sample series of θ_0, $\theta_1, \cdots, \theta_{n_0}, \theta_{n_0+1}, \cdots, \theta_{n_0+N}$, which has a total of $n_0 + 1 + N$ samples with $n_0 + N = 100{,}000$. The first 50,000 samples are deleted, and the last 50,000 samples almost perfectly describe the distribution $P(\boldsymbol{\theta}) \sim N\left(\begin{pmatrix}1\\4\end{pmatrix}, \begin{bmatrix}3 & 1\\2 & 1\end{bmatrix}\right)$, as shown in Fig. 10.2a.

2. Draw samples from $P(\boldsymbol{\theta})$ by using the Gibbs algorithm for $P(\boldsymbol{\theta}_j) \sim N\left(\begin{pmatrix}1\\2\end{pmatrix}, \begin{bmatrix}1 & 1\\1 & 2\end{bmatrix}\right)$, where $\boldsymbol{\theta}_j = \begin{pmatrix}\theta_{j1}\\\theta_{j2}\end{pmatrix}$ is a two-dimensional variable.

The initial sample is given as $\boldsymbol{\theta}_0 = (0,\ 0)^{\mathrm{T}}$. Following $P(\boldsymbol{\theta}) \sim N\left(\begin{pmatrix}1\\2\end{pmatrix}, \begin{bmatrix}1 & 1\\1 & 2\end{bmatrix}\right)$, we have the value of $\boldsymbol{\theta}'' = (\theta_1'', \theta_2'')$ from $\theta_1'' \sim P(\theta_1'|\theta_2') = N(0.5\theta_2', 0.5)$ and $\theta_2'' \sim P(\theta_2|\theta_1'') = N(\theta_1'' + 1, 1)$ for its predecessor $\boldsymbol{\theta}' = (\theta_1', \theta_2')$. Again, the first 50% of samples are discarded after 100,000 times of Gibbs sampling.

Figure 10.2b shows the results of Gibbs sampling, which follow the distribution of $P(\boldsymbol{\theta}) \sim N\left(\begin{pmatrix} 1 \\ 2 \end{pmatrix}, \begin{bmatrix} 1 & 1 \\ 1 & 2 \end{bmatrix}\right)$.

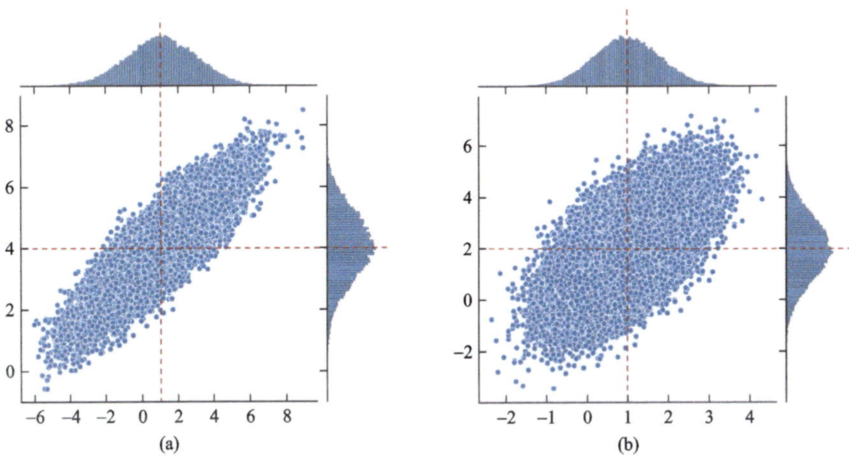

Fig. 10.2 **(a)** M-H algorithm for distribution $P(\theta) \sim N\left(\begin{pmatrix} 1 \\ 4 \end{pmatrix}, \begin{bmatrix} 3 & 1 \\ 2 & 1 \end{bmatrix}\right)$. **(b)** Gibbs sampling for distribution $P(\theta) \sim N\left(\begin{pmatrix} 1 \\ 2 \end{pmatrix}, \begin{bmatrix} 1 & 1 \\ 1 & 2 \end{bmatrix}\right)$. 100,000 times of sampling with a burn-in period of 50,000 times

---

With the state information, a sequence of variables $\{\boldsymbol{X}, \boldsymbol{\pi}\} = \{x_i, \pi_i\}(i = 0, 1, \cdots, n)$ is called a complete dataset for a hidden Markov chain, while the sequence of variables $\boldsymbol{X} = \{x_i\}(i = 0, 1, \cdots, n)$ is an incomplete dataset. In addition to transition probability matrix $\boldsymbol{T}$ and initial state probability π_0, a Markov chain also has an observation (called emission also) set, $\boldsymbol{O} = \{o_j\}$ $(j = 1, 2, \cdots, m)$, and each corresponds to a distinct phenomenon. For example, the observation set (*a*, *b*, *c*, ...) mean $o_1 = a$, $o_2 = b$, etc. Each of the random variables, x_i $(i = 0, 1, \cdots, n)$, including the initial one, is actually an observation set. Thus, the observed variable x at a state s_k relates not only to the state s_k but also to the observation o_j. The observation probability matrix $\boldsymbol{B}$ links the observation o_j to the state s_k, which is defined by

$$\boldsymbol{B} = \{B_{kj}\} = \left\{b_k\left(x_i^{(j)} = o_j\right)\right\} = \left\{P\left(x_i^{(j)} | \pi_i^{(k)}\right)\right\}$$
$$(k = 1, 2, \cdots, q;\ j = 1, 2, \cdots, m). \tag{10.5a}$$

In each state s_k, the probability of the observation o_j is described by B_{kj} and is independent of (i). Clearly, we have $\sum_{j=1}^{m} B_{kj} = 1$. Then, the probability $P\left(x_i^{(j)}\right)$

is

$$P\left(x_i^{(j)}\right) = \pi_i^{(k)} b_k\left(x_i^{(j)} = o_j\right)$$
$$(k = 1, 2, \cdots, q;\ j = 1, 2, \cdots, m;\ i = 0, 1, \cdots, n). \qquad (10.5b)$$

The three intrinsic parameters in a first-order hidden Markov chain are simply expressed by

$$\boldsymbol{\lambda} = (\boldsymbol{T}, \boldsymbol{B}, \boldsymbol{\pi}_0). \qquad (10.5c)$$

With the initial probability $\boldsymbol{\pi}_0$, the probability vector of occupied states is calculated by the transition rule and given by Eq. (10.2b) or Eq. (10.2c). After that, the probability of observed variables can be calculated by Eq. (10.5b).

For a Markov chain with a given length, its structure is like a neuron network, as shown in Fig. 10.3. The number of layers, excluding the input layer, represents the sequence length n. Each layer has the same number of neurons and each neuron represents a state so that the number of neurons is the same as that of states. Each state (neuron) contains the observation basis set and each basis has a fixed fraction (probability) described by the observation probability matrix $\boldsymbol{B}$. The neuron structure is identical in every layer. Obviously, there will be q possible paths from one given neuron in a layer to pass information to the next layer and q^2 possible paths between two nearest layers. Clearly, if there is only one state of $q = 1$, there is a unique path.

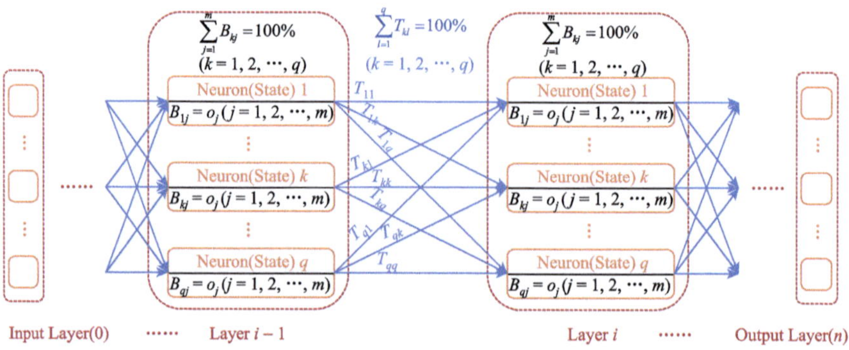

Fig. 10.3 The layer structure of a Markov chain

There are three general tasks in first-order hidden Markov chains (Li, 2018; Rabiner, 1989): (1) With a known Markov chain of the intrinsic parameters $\boldsymbol{\lambda} = (\boldsymbol{T}, \boldsymbol{B}, \boldsymbol{\pi}_0)$ and a observation sequence $\boldsymbol{X} = \{x_i\}(i = 1, 2, \cdots, n)$, how to calculate the probability of the observation sequence $P(\boldsymbol{X}|\boldsymbol{\lambda})$? (2) If a sequence of observations $\boldsymbol{X} = \{x_i\}$ $(i = 0, 1, \cdots, n)$ and the intrinsic parameters, $\boldsymbol{\lambda}$, are all known, how to determine the maximum state chain probability of $P(\pi|\boldsymbol{X}, \boldsymbol{\lambda})$? (3) If many sequences of observations $\boldsymbol{X} = \{x_i\}(i = 0, 1, \cdots, n)$ are known and the sequences follow a

Markov chain of $\boldsymbol{\lambda}$, how to use the maximum likelihood $P(\boldsymbol{X}|\boldsymbol{\lambda})$ to determine the intrinsic parameters $\boldsymbol{\lambda} = (\boldsymbol{T}, \boldsymbol{B}, \boldsymbol{\pi_0})$?

10.4 Calculation Methods for the Probability of Observation Sequence

10.4.1 Direct Method

The probability of the state sequence $P(\tilde{\boldsymbol{\pi}}|\boldsymbol{\lambda})$ for one path from one neuron in the input layer to one neuron in the output layer is given by

$$\begin{aligned} P(\tilde{\boldsymbol{\pi}}|\boldsymbol{\lambda}) &= \pi_0^{(k_0)}\pi_1^{(k_1)}\cdots\pi_i^{(k_i)}\cdots\pi_n^{(k_n)} = \pi_0^{(k_0)}T_{k_0k_1}\cdots T_{k_{i-1}k_i}\cdots T_{k_{n-1}k_n} \\ &= \pi_0^{(k_0)}\prod_{i=1}^{n}T_{k_{i-1}k_i}. \end{aligned} \tag{10.6a}$$

The initial state $\pi_0 = \pi_0^{(k_0)}$ is usually preset at a state with 100% probability, viz., $\pi_0^{(k_0)} = 1$ and $\pi_0^{(k)} = 0$ for $k \neq k_0$ $(k = 1, 2, \cdots, q)$. Of course, the initial state π_0 could also be preset using all the q states with each state having a fixed probability. Hereafter, the initial state means only one certain state unless specially described. Thus, Eq. (10.6a) is simplified to be

$$P(\tilde{\boldsymbol{\pi}}|\boldsymbol{\lambda}) = \prod_{i=1}^{n}T_{k_{i-1}k_i}. \tag{10.6b}$$

The probability of the observation sequence along the path $\tilde{\boldsymbol{\pi}}$ is given by

$$P(\boldsymbol{X}|\tilde{\boldsymbol{\pi}},\boldsymbol{\lambda}) = b_{k_0}(x_0)b_{k_1}(x_1)\cdots b_{k_i}(x_i)\cdots b_{k_n}(x_n) = b_{k_0}(x_0)\prod_{i=1}^{n}b_{k_i}(x_i). \tag{10.6c}$$

The joint probability $P(\boldsymbol{X}, \tilde{\boldsymbol{\pi}}|\boldsymbol{\lambda})$ is calculated by the chain rule,

$$\begin{aligned} P(\boldsymbol{X}, \tilde{\boldsymbol{\pi}}|\boldsymbol{\lambda}) &= P(\boldsymbol{X}|\tilde{\boldsymbol{\pi}},\boldsymbol{\lambda})P(\tilde{\boldsymbol{\pi}}|\boldsymbol{\lambda}) \\ &= b_{k_0}(x_0)T_{k_0k_1}b_{k_1}(x_1)\cdots T_{k_{i-1}k_i}b_{k_i}(x_i)\cdots T_{k_{n-1}k_n}b_{k_n}(x_n) \\ &= b_{k_0}(x_0)\prod_{i=1}^{n}T_{k_{i-1}k_i}\prod_{i=1}^{n}b_{k_i}(x_i). \end{aligned} \tag{10.6d}$$

The probability of $P(\boldsymbol{X}|\boldsymbol{\lambda})$ denotes the probability of all passes, which is the marginal probability of joint probability $P(\boldsymbol{X}, \boldsymbol{\pi}|\boldsymbol{\lambda})$ and is obtained by adding up all states in the state sequence, i.e.,

$$P(\boldsymbol{X}|\boldsymbol{\lambda}) = b_{k_0}(x_0) \prod_{i=1}^{n} \left(\sum_{k_i=1}^{q} \left(T_{k_{i-1}k_i} b_{k_i}(x_i) \right) \right) \tag{10.7a}$$

or equivalently by

$$P(\boldsymbol{X}|\boldsymbol{\lambda}) = b_k(x_0) \sum_{k=1}^{q} \pi_1^{(k)} b_k(x_1) \ldots \sum_{k=1}^{q} \pi_i^{(k)} b_k(x_i) \ldots \sum_{k=1}^{q} \pi_n^{(k)} b_k(x_n). \tag{10.7b}$$

As an example, Figure 10.4 shows a two-layer Markov chain, where in state 1, events a and c will each be observed with a 50% probability, in state 2, events b and c will be observed with 80% and 20% probabilities respectively, and in state 3, events a and b will be observed with 70% and 30% probabilities respectively. When starting from the input layer state 1 to the output layer, the observed sequence "aab" can be achieved via the orange paths.

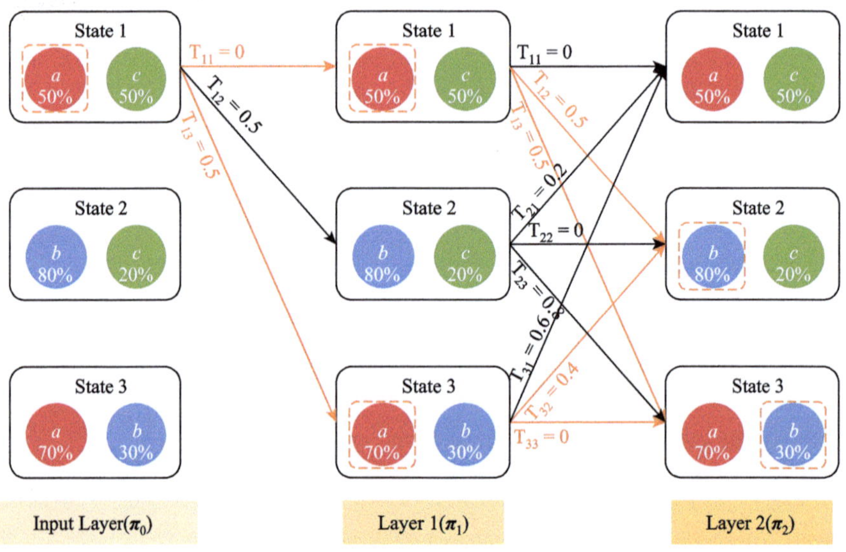

Fig. 10.4 The structure of a two-layer Markov chain, and the observation "*aab*" can be achieved via the orange paths

Example 10.4 The following example shows how to calculate the probability of the observation sequence $P(\boldsymbol{X}|\boldsymbol{\lambda})$ with $\boldsymbol{\lambda} = (\boldsymbol{T}, \boldsymbol{B}, \boldsymbol{\pi}_0)$ and the observed sequence

$\boldsymbol{X} = \{x_i\}(i = 0, 1, \cdots, n)$. Consider a simple Markov chain with $\boldsymbol{X} = \{x_0, x_1\}$, $\boldsymbol{O} = \{a, b, c\}$ and $S = \{1, 2, 3\}$, viz., $n = 1$, $m = 3$ and $q = 3$, meaning that there are only input and output layers with each layer having three neurons. The initial state probability is $\boldsymbol{\pi}_0 = \begin{pmatrix} 1 & 0 & 0 \end{pmatrix}$, the transition matrix is $\boldsymbol{T} = \begin{bmatrix} 1/3 & 1/3 & 1/3 \\ 1/3 & 1/3 & 1/3 \\ 1/3 & 1/3 & 1/3 \end{bmatrix}$, and the observation matrix is $\boldsymbol{B} = \begin{bmatrix} 1 & 0 & 0 \\ 0 & 1 & 0 \\ 0 & 0 & 1 \end{bmatrix}$, meaning that the first, second, and third neurons contain *a*, *b*, and *c*, respectively. With $\boldsymbol{\lambda} = (\boldsymbol{T}, \boldsymbol{B}, \boldsymbol{\pi}_0)$, we have $\boldsymbol{\pi}_1 = \boldsymbol{\pi}_0 \boldsymbol{T} = \frac{1}{3}\begin{pmatrix} 1 & 1 & 1 \end{pmatrix}$, and $x_1^{(1)} = a$, $x_1^{(2)} = b$ and $x_1^{(3)} = c$, which yield three sequences of $\boldsymbol{X}$ of *aa*, *ab*, and *ac*, respectively. Thus, with any $\boldsymbol{X}$ of the three sequences, we have $P(\boldsymbol{X}|\boldsymbol{\lambda}) = 1/3$.

Recursively, we have the sequences of *aaa*, *aba*, *aca*, *aab*, *abb*, *acb*, *aac*, *abc*, and *acc* for $n = 2$; and the sequences of *aaaa*, *abaa*, *acaa*, *aaba*, *abba*, *acba*, *aaca*, *abca*, *acca*, *aaab*, *abab*, *acab*, *aabb*, *abbb*, *acbb*, *aacb*, *abcb*, *accb*, *aaac*, *abac*, *acac*, *aabc*, *abbc*, *acbc*, *aacc*, *abcc*, and *accc* for $n = 3$; etc. With the given $\boldsymbol{\lambda}$, we can calculate the probability of $P(aabcba|\boldsymbol{\lambda}) = \pi_0^{(1)} B_{11} \cdot \pi_1^{(1)} B_{11} \cdot \pi_2^{(2)} B_{22} \cdot \pi_3^{(3)} B_{33} \cdot \pi_4^{(2)} B_{22} \cdot \pi_5^{(1)} B_{11}$ $= \frac{1}{3} \cdot \frac{1}{3} \cdot \frac{1}{3} \cdot \frac{1}{3} \cdot \frac{1}{3} = \left(\frac{1}{3}\right)^5$. The result indicates that if the transition probability is the same for all states and the observation matrix is a unit matrix, the probability of a given sequence will be the reciprocal of the path number.

If $\boldsymbol{\pi}_0 = \begin{pmatrix} 1 & 0 & 0 \end{pmatrix}$, $\boldsymbol{T} = \begin{bmatrix} 0.4 & 0.3 & 0.3 \\ 0.2 & 0.7 & 0.1 \\ 0.1 & 0.3 & 0.6 \end{bmatrix}$, and $\boldsymbol{B} = \begin{bmatrix} 0.5 & 0.2 & 0.3 \\ 0.1 & 0.8 & 0.1 \\ 0.2 & 0.2 & 0.6 \end{bmatrix}$, the first neuron contains ($a = 0.5$, $b = 0.2$, and $c = 0.3$), the second neuron contains ($a = 0.1$, $b = 0.8$, and $c = 0.1$), and the third neuron contains ($a = 0.2$, $b = 0.2$, and $c = 0.6$). The state probabilities are $\boldsymbol{\pi}_1 = \boldsymbol{\pi}_0 \boldsymbol{T} = \begin{pmatrix} 0.4 & 0.3 & 0.3 \end{pmatrix}$ and $\boldsymbol{\pi}_2 = \boldsymbol{\pi}_1 \boldsymbol{T} = \begin{pmatrix} 0.25 & 0.42 & 0.33 \end{pmatrix}$, respectively. The probability of $P(a|\boldsymbol{\lambda}) = \pi_1^{(1)} B_{11} = 0.5$. After that, there are three paths from the starting neuron in the input layer to the three neurons in the first layer, and the sum of three probabilities for the first layer to generate observed event *a* is $P(x_1 = a|\boldsymbol{\lambda}, x_0 = a) = \pi_1^{(1)} B_{11} + \pi_1^{(2)} B_{21} + \pi_1^{(3)} B_{31} = 0.4 \times 0.5 + 0.3 \times 0.1 + 0.3 \times 0.2 = 0.29$. Considering $P(x_0 = a|\boldsymbol{\lambda}) = 0.5$, we have $P(aa|\boldsymbol{\lambda}) = 0.13$. Recursively, we have $P(x_2 = b|\boldsymbol{\lambda}, x_0 = a, x_1 = a) = \pi_2^{(1)} B_{12} + \pi_2^{(2)} B_{22} + \pi_2^{(3)} B_{32} = 0.25 \times 0.2 + 0.42 \times 0.8 + 0.33 \times 0.2 = 0.452$ and $P(aab|\boldsymbol{\lambda}) = 0.05876$.

If $\boldsymbol{\pi}_0 = \begin{pmatrix} 1 & 0 & 0 \end{pmatrix}$, $\boldsymbol{T} = \begin{bmatrix} 0 & 0.7 & 0.3 \\ 0.2 & 0 & 0.8 \\ 0.6 & 0.4 & 0 \end{bmatrix}$, and $\boldsymbol{B} = \begin{bmatrix} 0.5 & 0 & 0.5 \\ 0 & 0.8 & 0.2 \\ 0.7 & 0.3 & 0 \end{bmatrix}$, each of the states does not transition to itself, and the first, second and third states contain ($a = 0.5$, $b = 0$, and $c = 0.5$), ($a = 0$, $b = 0.8$, and $c = 0.2$), and

($a = 0.7$, $b = 0.3$, and $c = 0$), respectively. The state probabilities are $\boldsymbol{\pi}_1 = \boldsymbol{\pi}_0\boldsymbol{T} = \begin{pmatrix}0 & 0.7 & 0.3\end{pmatrix}$ and $\boldsymbol{\pi}_2 = \boldsymbol{\pi}_1\boldsymbol{T} = \begin{pmatrix}0.32 & 0.12 & 0.56\end{pmatrix}$, respectively. The probability of $P(a|\boldsymbol{\lambda}) = \pi_0(1)B_{11} = 0.5$. After that, there are three paths from the starting neuron in the input layer to the three neurons in the first layer, and the three probabilities for the first layer to generate the observed event a are $P\left(x_1^{(1)} = a \mid \boldsymbol{\lambda}, x_0 = a\right) = \pi_1^{(1)}B_{11} = 0 \times 0.5 = 0$, $P(x_1^{(2)} = a \mid \boldsymbol{\lambda}, x_0 = a) = \pi_1^{(2)}B_{21} = 0.7 \times 0 = 0$, and $P\left(x_1^{(2)} = a \mid \boldsymbol{\lambda},\ x_0 = a\right) = 0.3 \times 0.7 = 0.21$. Thus, $P(x_1 = a \mid \boldsymbol{\lambda}, x_0 = a) = \sum_{k=1}^{3} P(x_1^{(k)} = a \mid \boldsymbol{\lambda}, x_0 = a) = 0 + 0 + 0.21 = 0.21$. Considering $P(x_0 = a \mid \boldsymbol{\lambda}) = 0.5$, we have $P(aa\ \ \boldsymbol{\lambda}) = 0.105$. Recursively, we have $P(x_2 = b \mid \boldsymbol{\lambda}, x_0 = a, x_1 = a) = 0.4 \times 0.8 = 0.32$ and $P(aab|\boldsymbol{\lambda}) = 0.0336$.

10.4.2 Forward Method

The direct method is an exhaustive approach to calculating the probability of $P(\boldsymbol{X}|\boldsymbol{\lambda})$ and requires heavy computation. Therefore, forward and backward methods are developed to reduce the computation burden. A forward probability is defined as the sum of probabilities, which a neuron (state) in layer (i) receives from all neurons in layer ($i - 1$), times the observation probability. Due to the first-order property of a Markov chain, the forward probability is probabilistically defined by

$$\alpha_i(k) = P\left(x_0, x_1, \cdots, x_i, \pi_i^{(k)}|\boldsymbol{\lambda}\right) = P\left(x_{i-1}, x_i, \pi_i^{(k)}|\boldsymbol{\lambda}\right). \tag{10.8a}$$

The initial forward probability $\alpha_0(l)$ allows us to delete x_{0-1} from Eq. (10.8a). In this case that only one state in π_0 is non-zero, say $\pi_0^{(l)} = 1$ and $\pi_0^{(k)} = 0$ ($k \neq l; l, k = 1, 2, \cdots, q$), we have the initial forward probability $\alpha_0(l) = P\left(x_0, \pi_0^{(l)}|\boldsymbol{\lambda}\right) = \pi_0^{(l)}b_l\left(x_0^{(j)}\right)$ and $\alpha_0(k \neq l) = 0$, where $b_l\left(x_0^{(j)}\right) = P(x_0|\pi_0, \boldsymbol{\lambda})$, viz., the initial forward probability is the same as the probability of the observed x_0. The initial information propagates forward to all neurons in the first layer and hence the forward probability of neuron (k) in the first layer is calculated by

$$\begin{aligned}\alpha_1(k) &= P\left(x_0, x_1, \pi_1^{(k)}|\boldsymbol{\lambda}\right) = P\left(x_0, x_1|\pi_1^{(k)}, \boldsymbol{\lambda}\right)P\left(\pi_1^{(k)}|\boldsymbol{\lambda}\right)\\ &= \pi_0^{(l)}b_l\left(x_0^{(j)}\right)T_{lk}b_k\left(x_1^{(j)}\right)\\ &= (T_{lk}\alpha_0(l))b_k\left(x_1^{(j)}\right)\ \ (k = 1, 2, \cdots, q),\ \ \text{for } \pi_0^{(l)} = 1 \text{ and } \pi_0^{(k)} = 0\ \ (k \neq l),\end{aligned} \tag{10.8b}$$

The term $(T_{lk}\alpha_0(l))$ or $\left(\sum_{l=1}^{q} T_{lk}\alpha_0(l)\right)$ represents the information received by neuron (k) in the first layer from the initial neuron or all neurons in the input layer. Starting from layer 2, a neuron receives the forward probability information from all its preceding neurons. Recursively the forward probability is calculated by

$$\alpha_{i+1}(k) = \left(\sum_{l=1}^{q} T_{lk}\alpha_i(l)\right) b_k\left(x_{i+1}^{(j)}\right) \quad (k = 1, 2, \cdots, q\ ;\ i = 1, 2, \cdots, n-1). \tag{10.8c}$$

Thus, the probability of $P(\boldsymbol{X}|\boldsymbol{\lambda})$ is calculated by adding up all forward probabilities from all neurons in the output layer, i.e.,

$$P(\boldsymbol{X}|\boldsymbol{\lambda}) = \sum_{k=1}^{q} \alpha_n(k). \tag{10.8d}$$

The forward probability includes the transition probability and the observation probability so that one neuron in a layer receives only one forward probability from all neurons in its preceding layer. Considering Eq. (10.7a) again and taking $(n = 1)$ first, we have

$$P(\boldsymbol{X}|\boldsymbol{\lambda}) = b_l(x_0)\sum_{k=1}^{q} T_{lk}b_k(x_1) = \sum_{k=1}^{q} \alpha_0(l)T_{lk}b_k(x_1) = \sum_{k=1}^{q} \alpha_1(k).$$

Taking $(n = 2)$, we have

$$\begin{aligned} P(\boldsymbol{X}|\boldsymbol{\lambda}) &= b_{k_0}(x_0)\left(\sum_{k_1=1}^{q}\left(T_{k_0k_1}b_{k_1}(x_1)\right)\right)\left(\sum_{k_2=1}^{q}\left(T_{k_1k_2}b_{k_2}(x_2)\right)\right) \\ &= \sum_{k=1}^{q} b_k(x_2)\left(\sum_{l=1}^{q} \alpha_1(l)T_{lk}\right) = \sum_{k=1}^{q} \alpha_2(k). \end{aligned}$$

Recursively, we prove that Eq. (10.8d) is equivalent to Eq. (10.7b).

$$P(\boldsymbol{X}|\boldsymbol{\lambda}) = b_{k_0}(x_0)\prod_{i=1}^{n}\left(\sum_{k_i=1}^{q}\left(T_{k_{i-1}k_i}b_{k_i}(x_i)\right)\right).$$

Example 10.5 Use the forward method to calculate $P(aab|\boldsymbol{\lambda})$ with the same intrinsic parameters of $\boldsymbol{\pi}_0 = \begin{pmatrix}1 & 0 & 0\end{pmatrix}$, $\boldsymbol{T} = \begin{bmatrix} 0 & 0.7 & 0.3 \\ 0.2 & 0 & 0.8 \\ 0.6 & 0.4 & 0 \end{bmatrix}$, and $\boldsymbol{B} = \begin{bmatrix} 0.5 & 0 & 0.5 \\ 0 & 0.8 & 0.2 \\ 0.7 & 0.3 & 0 \end{bmatrix}$. From the observation matrix, we know $B_{12}(b) = 0$, $B_{21}(a) = 0$, and $B_{33}(c) = 0$, which will be excluded in the following calculations. The initial forward probabilities are $\alpha_0(1) = P\left(x_0^{(j)} = a\right) = \pi_0^{(l)} b_1\left(x_0^{(1)} = a\right) = 0.5$, $\alpha_0(2) = \alpha_0(3) = 0$, which give $P(a|\boldsymbol{\lambda}) = \sum_{k=1}^{3} \alpha_0(k) = 0.5 + 0 + 0 = 0.5$. Then we have

$$\alpha_1(k) = (T_{lk}\alpha_0(l))b_k\left(x_1^{(j)} = a\right) = (T_{1k}\alpha_0(1))b_k\left(x_1^{(1)} = a\right) \quad (k = 1, 2, 3),$$

$\alpha_1(k = 1) = (T_{11}\alpha_0(1))b_1\left(x_1^{(1)} = a\right) = 0$ due to $T_{11} = 0$, $\alpha_1(k = 2) = 0$ due to $b_2\left(x_1^{(1)} = a\right) = 0$, and $\alpha_1(k = 3) = (T_{13}\alpha_0(1))b_3\left(x_1^{(1)} = a\right) = 0.3 \times 0.5 \times 0.7 = 0.105$, which give $P(aa|\boldsymbol{\lambda}) = \sum_{k=1}^{3} \alpha_1(k) = 0.105$. After that, using Eq. (10.8c) gives

$$\alpha_2(k) = \left(\sum_{l=1}^{q} T_{lk}\alpha_1(l)\right) b_k\left(x_2^{(j)} = b\right) = T_{3k}\alpha_1(l = 3)b_k\left(x_2^{(2)} = b\right),$$

$\alpha_2(1) = T_{31}\alpha_1(l = 3)b_k\left(x_2^{(2)} = b\right) = 0$ due to $b_1\left(x_2^{(2)} = b\right) = 0$, $\alpha_2(2) = T_{32}\alpha_1(l = 3)B_{22}(b) = 0.4 \times 0.105 \times 0.8 = 0.0336$, and $\alpha_2(3) = T_{33}\alpha_1(l = 3)B_{32}(b) = 0 \times 0.105 \times 0.3 = 0$.

Then using Eq. (10.8d) yields $P(\boldsymbol{X}|\boldsymbol{\lambda}) = \sum_{k=1}^{q} \alpha_2(k) = 0.0336$.

10.4.3 Backward Method

In the reversal manner to the forward method, the backward method considers that a neuron in layer $(i + 1)$ links backward to all neurons in its preceding layer (i). Thus, the backward probability of neuron (k) in layer (i) is defined by

$$\beta_i(k) = P\left(x_n, x_{n+1}, \cdots, x_{i+1}|\boldsymbol{\lambda}, \pi_i^{(k)}\right). \tag{10.9a}$$

To start backward probability calculation, we set a termination layer beyond the output layer, where exists only one neuron so that

$$\beta_n(k) = P\left(x_{n+1}|\boldsymbol{\lambda}, \pi_n^{(k)}\right) = 1 \quad (k = 1, 2, \cdots, q). \tag{10.9b}$$

Then the backward probability of $\beta_{n-1}(k)$ is calculated by

$$\beta_{n-1}(k) = P\left(x_n|\boldsymbol{\lambda}, \pi_{n-1}^{(k)}\right) = \sum_{l=1}^{q} T_{kl} b_l(x_n) \beta_n(l) \quad (k = 1, 2, \cdots, q). \tag{10.9c}$$

Equation (10.9c) indicates that the backward probability $\beta_{n-1}(k)$ adds up the product of three terms of transition matrix component (transition probability), T_{kl}, observation matrix component (observation probability), $b_l(x_n)$, and backward probability, $\beta_n(k)$, and delivers the sum backward. Recursively we have

$$\beta_i(k) = \sum_{l=1}^{q} T_{kl} b_l(x_{i+1}) \beta_{i+1}(l)$$
$$(k = 1, 2, \cdots, q;\ i = 1, 2, \cdots, n-1 \text{ or } i = 0, 1, \cdots, n-1), \tag{10.9d}$$

where $(i = 1, 2, \cdots, n-1)$ is for the case that there is only one state component in π_0, say $\pi_0^{(k)} = 1$ and $\pi_0^{(l)} = 0$ $(l \neq k)$, while $(i = 0, 1, \cdots, n-1)$ is for the case that there are multiple state components in π_0. The two cases yield the probabilities of $P(\boldsymbol{X}|\boldsymbol{\lambda})$,

$$P(\boldsymbol{X}|\boldsymbol{\lambda}) = \pi_0^{(k)} b_k(x_0) \sum_{l=1}^{q} T_{kl} b_l(x_1) \beta_1(l) = b_k(x_0) \sum_{l=1}^{q} T_{kl} b_l(x_1) \beta_1(l). \tag{10.9e}$$

$$P(\boldsymbol{X}|\boldsymbol{\lambda}) = \sum_{l=1}^{q} \pi_0^{(k)} b_k(x_0) \beta_0(k). \tag{10.9f}$$

Considering Eq. (10.7a) again and taking $(n = 1)$ first, we have

$$P(\boldsymbol{X}|\boldsymbol{\lambda}) = b_k(x_0) \sum_{l=1}^{q} T_{kl} b_l(x_1) = b_k(x_0) \sum_{l=1}^{q} T_{kl} b_l(x_1) \beta_1(l)$$

because $\beta_1(l) = 1$. Taking $(n = 2)$, $\beta_2(l) = 1$, and $\beta_1(k) = \sum_{l=1}^{q} T_{kl} b_l(x_{i+1}) \beta_2(l)$, we have

$$\begin{aligned} P(\boldsymbol{X}|\boldsymbol{\lambda}) &= b_{k_0}(x_0) \left(\sum_{k_1=1}^{q} \left(T_{k_0 k_1} b_{k_1}(x_1) \right) \right) \left(\sum_{k_2=1}^{q} \left(T_{k_1 k_2} b_{k_2}(x_2) \right) \right) \\ &= b_l(x_0) \left(\sum_{k=1}^{q} (T_{lk} b_k(x_1)) \right) \left(\sum_{l=1}^{q} (T_{kl} b_l(x_2) \beta_2(l)) \right) \\ &= b_l(x_0) \left(\sum_{k=1}^{q} (T_{lk} b_k(x_1) \beta_1(k)) \right). \end{aligned}$$

Recursively, we prove that Eq. (10.9e) is equivalent to Eq. (10.7a).

Example 10.6 Use the backward method to calculate $P(aab|\boldsymbol{\lambda})$ with the same intrinsic parameters of $\boldsymbol{\pi}_0 = \begin{pmatrix} 1 \, 0 \, 0 \end{pmatrix}$, $\boldsymbol{T} = \begin{bmatrix} 0 & 0.7 & 0.3 \\ 0.2 & 0 & 0.8 \\ 0.6 & 0.4 & 0 \end{bmatrix}$, and $\boldsymbol{B} = \begin{bmatrix} 0.5 & 0 & 0.5 \\ 0 & 0.8 & 0.2 \\ 0.7 & 0.3 & 0 \end{bmatrix}$.
The initial backward probability $\beta_2(k) = 1$. Then,

$$
\begin{aligned}
\beta_1\left(x_1^{(1)} = a\right) &= \sum_{l=1}^{q} T_{kl} b_l(x_2 = b)\beta_2(l) \\
&= 0 \times 0 \times 1 + 0.7 \times 0.8 \times 1 + 0.3 \times 0.3 \times 1 = 0.65, \\
\beta_1\left(x_1^{(2)} = a\right) &= \sum_{l=1}^{q} T_{kl} b_l(x_2 = b)\beta_2(l) \\
&= 0.2 \times 0 \times 1 + 0 \times 0.8 \times 1 + 0.8 \times 0.3 \times 1 = 0.24, \\
\beta_1\left(x_1^{(3)} = a\right) &= \sum_{l=1}^{q} T_{kl} b_l(x_2 = b)\beta_2(l) \\
&= 0.6 \times 0 \times 1 + 0.4 \times 0.8 \times 1 + 0 \times 0.3 \times 1 = 0.32.
\end{aligned}
$$

After that, using Eq. (10.9c) gives

$$
\begin{aligned}
P(\boldsymbol{X}|\boldsymbol{\lambda}) &= \pi_0 B_{kj}(x_0 = a) \sum_{l=1}^{q} T_{kl} b_l(x_1 = a)\beta_1(l) \\
&= 1 \times 0.5 \times (0 + 0 + 0.0672) = 0.0336.
\end{aligned}
$$

10.5 Estimation of Optimal State Sequence

10.5.1 Direct Method

With the sequence $\boldsymbol{X}$ and the intrinsic parameters $\boldsymbol{\lambda}$, the probability of a state sequence, or a path, can be calculated with the forward probability and the backward probability. Using the Bayesian theory, we have the probability of $\pi_i^{(k)}$,

$$\gamma_i(k) = P\left(\pi_i^{(k)}|\boldsymbol{X}, \boldsymbol{\lambda}\right) = \frac{P\left(\pi_i^{(k)}, \boldsymbol{X}|\boldsymbol{\lambda}\right)}{P(\boldsymbol{X}|\boldsymbol{\lambda})} \quad (i = 1, 2, \cdots, n). \tag{10.10a}$$

The joint probability $P\left(\pi_i^{(k)}, \boldsymbol{X}|\boldsymbol{\lambda}\right) = \alpha_i(k)\beta_i(k)$ counts the probability that neuron (k) in layer (i) receives the forward probability of all paths from the input layer and the backward probability that neuron (k) in layer (i) receives the backward probability of all paths from the output layer. Thus, $\sum_{k=1}^{q} P\left(\pi_i^{(k)}, \boldsymbol{X}|\boldsymbol{\lambda}\right) = \sum_{k=1}^{q} \alpha_i(k)\beta_i(k) = P(\boldsymbol{X}|\boldsymbol{\lambda})$. Then the probability of event $x_i^{(j)}$ occurring in state (k) is given by

$$\gamma_i(k) = P\left(\pi_i^{(k)}|\boldsymbol{X}, \boldsymbol{\lambda}\right) = \frac{\alpha_i(k)\beta_i(k)}{\sum_{k=1}^{q} \alpha_i(k)\beta_i(k)} \quad (i = 1, 2, \cdots, n). \tag{10.10b}$$

The optimal path is estimated by the maximum of $\gamma_i(k)$, i.e.,

$$\pi_i^{(k)} = \max(\gamma_i(k)) = \max\left(\frac{\alpha_i(k)\beta_i(k)}{\sum_{k=1}^{q} \alpha_i(k)\beta_i(k)}\right) \quad (i = 1, 2, \cdots, n). \tag{10.10c}$$

In the above analysis, the initial state $(i = 0)$ is excluded if the initial state is preset, otherwise Eqs. (10.10b) and (10.10c) can be applied to the initial state $(i = 0)$.

Example 10.7 Estimate the state sequence with the sequence of $\boldsymbol{X} = (aab)$ and intrinsic parameters of $\boldsymbol{\pi}_0 = \begin{pmatrix} 1\,0\,0 \end{pmatrix}$, $\boldsymbol{T} = \begin{bmatrix} 0 & 0.7 & 0.3 \\ 0.2 & 0 & 0.8 \\ 0.6 & 0.4 & 0 \end{bmatrix}$, and $\boldsymbol{B} = \begin{bmatrix} 0.5 & 0 & 0.5 \\ 0 & 0.8 & 0.2 \\ 0.7 & 0.3 & 0 \end{bmatrix}$.

Since $\boldsymbol{\pi}_0 = \begin{pmatrix} 1\,0\,0 \end{pmatrix}$ is preset, we do not need to calculate $\boldsymbol{\beta}_0$ and $\boldsymbol{\gamma}_0$. From Examples 10.4 and 10.5, we have $\boldsymbol{\alpha}_1 = \begin{pmatrix} 0\,0\,0.105 \end{pmatrix}$ and $\boldsymbol{\alpha}_2 = \begin{pmatrix} 0\,0.0336\,0 \end{pmatrix}$, and $\boldsymbol{\beta}_1 = \begin{pmatrix} 0.0672\,0.2442\,0.195 \end{pmatrix}$ and $\boldsymbol{\beta}_2 = \begin{pmatrix} 0.65\,0.24\,0.32 \end{pmatrix}$. Then $\alpha_1(1)\beta_1(1) = 0$, $\alpha_1(2)\beta_1(2) = 0$, $\alpha_1(3)\beta_1(3) = 0.105 \times 0.195 = 0.020475$ so that $\boldsymbol{\pi}_1 = \begin{pmatrix} 0\,0\,1 \end{pmatrix}$; and $\alpha_2(1)\beta_2(1) = 0$, $\alpha_2(2)\beta_2(2) = 0.0336 \times 0.2442 = 0.008064$, $\alpha_2(3)\beta_2(3) = 0$ so that $\boldsymbol{\pi}_2 = \begin{pmatrix} 0\,1\,0 \end{pmatrix}$. The optimal path for $\boldsymbol{X} = (aab)$ is $\pi_0^{(1)}$, $\pi_1^{(3)}$, and $\pi_2^{(2)}$.

10.5.2 Viterbi Algorithm

The Viterbi algorithm was first proposed by Viterbi in 1967 and applied to various AI areas, (e.g., Forney, 1973; Kruskal, 1983; Vintsyuk, 1968). The Viterbi algorithm is now standard for the application of dynamic programming to any kind of probabilistic maximization problem. The Viterbi algorithm is a dynamic program to find the optimal path for a Markov chain. With given intrinsic parameters $\boldsymbol{\lambda}$ and $\boldsymbol{X}$, the optimal path is obtained from

$$\pi_0^{(k_0)}\pi_1^{(k_1)}\cdots\pi_i^{(k_i)}\cdots\pi_n^{(k_n)} = \underset{\pi}{\operatorname{argmax}}\, P(\boldsymbol{\pi}|\boldsymbol{X}, \boldsymbol{\lambda}). \tag{10.11a}$$

The Viterbi algorithm introduces two probability parameters $v_i^{(l)}$ $(i = 0, 1, \cdots, n)$ and $\theta_i^{(l)}(i = 1, 2, \cdots, n)$ of a state (l) at (i). The probability from $\pi_i^{(l)}$ to $\pi_{i+1}^{(k)}$ $(k = 1, 2, \cdots, q)$ is described by the component T_{lk} in the transition matrix. Each neuron (k) in layer (i) has the probability of

$$\theta_i^{(k)} = \max_{l=1,2,\cdots,q}\left(v_{i-1}^{(l)}T_{lk}\right) \quad (i = 1, 2, \cdots, n) \tag{10.11b}$$

to selectively link to neuron (l) in layer $(i-1)$ in the optimal path and the probability of

$$v_i^{(k)} = b_l(o_j = x_i)\theta_i^{(k)} \quad (i = 1, 2, \cdots, n) \tag{10.11c}$$

to have the observed basis $(o_j = x_i)$. The initial probability parameter $v_0^{(l)}$ is preset and defined as the forward probability, i.e.,

$$v_0^{(l)} = \pi_0^{(l)}b_l(o_j = x_0) = \alpha_0(l) \quad (l = 1, 2, \cdots, q). \tag{10.11d}$$

At the end of $(i = n)$, $v_n^{(k)}$ and $\theta_n^{(k)}$ are calculated and the optimal end state is determined by

$$\hat{\pi}_n = \hat{\pi}_n^{(k)} = \operatorname{argmax}\left(v_n^{(l)}\right) \quad (k, l = 1, 2, \cdots, q), \tag{10.11e}$$

where the number (k) in $\hat{\pi}_n^{(k)}$ takes the number (l) in $\max\left(v_n^{(l)}\right)$ and the number (k) is also the number (k) in $\theta_n^{(k)}$, which in turn determines the number (l) in $v_{n-1}^{(l)}$. Recursively, we can determine the superscript l of $v_i^{(l)}$ through $\theta_{i+1}^{(k)} = \max_{l=1,2,\cdots,q}\left(v_i^{(l)}T_{lk}\right)$, which can be expressed as $v_i^{(l)} = \underset{l=1,2,\cdots,q}{\operatorname{argmax}}\left(v_i^{(l)}T_{lk}\right)$. The superscript of $v_i^{(l)}$ is identical to the superscript of $\hat{\pi}_i^{(l)}$, so we have

$$\hat{\pi}_i^{(l)} = \theta_{i+1}^{(k)}\left(\hat{\pi}_i^{(l)}\right) \quad (i = 1, 2, \cdots, n-1;\ k, l = 1, 2, \cdots, q), \tag{10.11f}$$

and estimate the optimal state sequence.

Example 10.8 Given $\boldsymbol{\pi}_0 = \begin{pmatrix} 1 & 0 & 0 \end{pmatrix}$, $\boldsymbol{T} = \begin{bmatrix} 0.4 & 0.3 & 0.3 \\ 0.2 & 0.7 & 0.1 \\ 0.1 & 0.3 & 0.6 \end{bmatrix}$, $\boldsymbol{B} = \begin{bmatrix} 0.5 & 0.2 & 0.3 \\ 0.1 & 0.8 & 0.1 \\ 0.2 & 0.2 & 0.6 \end{bmatrix}$, and $\boldsymbol{X} = (aabcc)$, estimate the optimal state sequence with the Viterbi algorithm.

As mentioned before, the first neuron contains ($a = 0.5,\ b = 0.2$, and $c = 0.3$), the second neuron contains ($a = 0.1,\ b = 0.8$, and $c = 0.1$) and the third neuron contains ($a = 0.2,\ b = 0.2$, and $c = 0.6$). The state probabilities are $\boldsymbol{\pi}_1 = \begin{pmatrix} 0.4 & 0.3 & 0.3 \end{pmatrix}$, $\boldsymbol{\pi}_2 = \begin{pmatrix} 0.25 & 0.42 & 0.33 \end{pmatrix}$, $\boldsymbol{\pi}_3 = (0.217\ \ 0.468\ \ 0.315)$, and $\boldsymbol{\pi}_4 = (0.2119\ \ 0.4872\ \ 0.3009)$, respectively.

With Eq. (10.11d), we have $\boldsymbol{v}_0 = \begin{pmatrix} 0.5 & 0 & 0 \end{pmatrix}$. With Eq. (10.11b), we have $\theta_1^{(1)} = \max\limits_{l=1,2,\cdots,q} \begin{pmatrix} 0.5 \times 0.4 & 0 \times 0.2 & 0 \times 0.1 \end{pmatrix} = 0.2$, $\theta_1^{(2)} = \max\limits_{l=1,2,\cdots,q} \begin{pmatrix} 0.5 \times 0.3 & 0 \times 0.7 & 0 \times 0.3 \end{pmatrix} = 0.15$, and $\theta_1^{(3)} = \max\limits_{l=1,2,\cdots,q} \begin{pmatrix} 0.5 \times 0.3 & 0 \times 0.1 & 0 \times 0.6 \end{pmatrix} = 0.15$, and hence $\boldsymbol{\theta}_1 = \begin{pmatrix} 0.2 & 0.15 & 0.15 \end{pmatrix}$.

Then with Eq. (10.11c), we have $v_1^{(1)} = b_1(o_j = a)\theta_1^{(1)} = 0.5 \times 0.2 = 0.1$, $v_1^{(2)} = b_2(o_j = a)\theta_1^{(2)} = 0.1 \times 0.15 = 0.015$, and $v_1^{(3)} = b_3(o_j = a)\theta_1^{(3)} = 0.2 \times 0.15 = 0.03$, and hence $\boldsymbol{v}_1 = (0.1\ \ 0.015\ \ 0.03)$.

Recursively, we have $\boldsymbol{\theta}_2 = \begin{pmatrix} 0.04 & 0.03 & 0.03 \end{pmatrix}$ and $\boldsymbol{v}_2 = \begin{pmatrix} 0.008 & 0.024 & 0.006 \end{pmatrix}$, $\boldsymbol{\theta}_3 = (0.0048\ \ 0.0168\ \ 0.0036)$ and $\boldsymbol{v}_3 = \begin{pmatrix} 0.00144 & 0.00168 & 0.00216 \end{pmatrix}$, and $\boldsymbol{\theta}_4 = (0.000576\ 0.001176\ 0.001296)$ and $\boldsymbol{v}_4 = \begin{pmatrix} 0.0001728 & 0.0001176 & 0.0007776 \end{pmatrix}$.

Then, using Eq. (10.11e) gives $v_4^{(3)} = \max\left(v_4^{(l)}\right)(l = 1, 2, 3)$ and $\hat{\pi}_4^{(3)}$. After that, we have, with Eq. (10.11f), $v_3^{(3)} = \left(\text{argmax}\left(v_3^{(l)} T_{l3}\right)\right)(l = 1, 2, 3)$ and $\hat{\pi}_3^{(3)}$, $v_2^{(3)} = \left(\text{argmax}\left(v_2^{(l)} T_{l3}\right)\right)$ $(l = 1, 2, 3)$ and $\hat{\pi}_2^{(3)}$, and $v_1^{(1)} = \left(\text{argmax}\left(v_1^{(l)} T_{l2}\right)\right)(l = 1, 2, 3)$ and $\hat{\pi}_1^{(1)}$. The optimal state sequence is $\left(\hat{\pi}_0^{(1)}, \hat{\pi}_1^{(1)}, \hat{\pi}_2^{(3)}, \hat{\pi}_3^{(3)}, \hat{\pi}_4^{(3)}\right)$.

10.6 Estimation of Intrinsic Parameters—The Baum-Welch Algorithm

The intrinsic parameters of a hidden Markov chain are not obviously shown up in a sequence of variables $\boldsymbol{X} = \{x_i\}(i = 0, 1, \cdots, n)$ or sequences of variables $\boldsymbol{X} = \left\{x_i^{(\tilde{i})}\right\}$ $(i = 0, 1, \cdots, n;\ \tilde{i} = 1, 2, \cdots, \tilde{I})$ with the same sequence length n. One has to estimate the intrinsic parameters from the observed sequence or sequences. In addition to utilizing domain knowledge, the standard ML algorithm to estimate the intrinsic parameters is the forward–backward or Baum-Welch algorithm (Baum, 1972). Once a sequence of variables is treated as a hidden Markov chain, its intrinsic parameters can be estimated by the Baum-Welch algorithm. First, the observation basis set $\boldsymbol{O} = \{o_j\}$ $(j = 1, 2, \cdots, m)$ is directly determined by observing an examined sequence or sequences when the data are large enough. With the state sequence $\boldsymbol{\pi} = \{\pi_i\}$ $(i = 0, 1, \cdots, n)$, a hidden Markov chain can be expressed by $\{x_i, \pi_i\}$ $(i = 0, 1, \cdots, n)$. If all data of $\{x_i, \pi_i\}(i = 0, 1, \cdots, n)$ are known, the data are complete, while $\{x_i\}(i = 0, 1, \cdots, n)$ are incomplete data. The Baum-Welch algorithm is based on the maximum likelihood method. With prior knowledge, the forms of intrinsic parameters must be proposed first and the values of intrinsic parameters are estimated from $\max P(\boldsymbol{X}|\boldsymbol{\lambda})$, which is equivalent to $\max \log(P(\boldsymbol{X}|\boldsymbol{\lambda}))$. Since $P(\boldsymbol{X}|\boldsymbol{\lambda})$ is the marginal distribution of $P(\boldsymbol{X}, \boldsymbol{\pi}|\boldsymbol{\lambda})$, we have, based on Jensen's inequality (Hastie et al., 2009),

$$\begin{aligned}\log(P(\boldsymbol{X}|\boldsymbol{\lambda})) &= \sum_{\tilde{\boldsymbol{\pi}}} \log(P(\boldsymbol{X}, \tilde{\boldsymbol{\pi}}|\boldsymbol{\lambda})) = \sum_{\tilde{\boldsymbol{\pi}}} \frac{\log(P(\boldsymbol{X}, \tilde{\boldsymbol{\pi}}|\boldsymbol{\lambda}))}{P(\boldsymbol{X}, \tilde{\boldsymbol{\pi}}|\boldsymbol{\lambda}^{(t)})} P(\boldsymbol{X}, \tilde{\boldsymbol{\pi}}|\boldsymbol{\lambda}^{(t)}) \\ &\geq \sum_{\tilde{\boldsymbol{\pi}}} \left(\log\left(\frac{P(\boldsymbol{X}, \tilde{\boldsymbol{\pi}}|\boldsymbol{\lambda})}{P(\boldsymbol{X}, \tilde{\boldsymbol{\pi}}|\boldsymbol{\lambda}^{(t)})}\right)\right) P(\boldsymbol{X}, \tilde{\boldsymbol{\pi}}|\boldsymbol{\lambda}^{(t)}),\end{aligned} \tag{10.12a}$$

where $\tilde{\boldsymbol{\pi}}$ denotes a path from one neuron in the input layer to one neuron in the output layer; the summation over $\tilde{\boldsymbol{\pi}}$ denotes all paths; and $\boldsymbol{\lambda}^{(t)}$ denotes temporary values of intrinsic parameters. The term $\sum_{\tilde{\boldsymbol{\pi}}} \left(\log\left(\frac{P(\boldsymbol{X}, \tilde{\boldsymbol{\pi}}|\boldsymbol{\lambda})}{P(\boldsymbol{X}, \tilde{\boldsymbol{\pi}}|\boldsymbol{\lambda}^{(t)})}\right)\right) P(\boldsymbol{X}, \tilde{\boldsymbol{\pi}}|\boldsymbol{\lambda}^{(t)})$ is the lower bound of $\log(P(\boldsymbol{X}|\boldsymbol{\lambda}))$ and the maximum of $\log(P(\boldsymbol{X}|\boldsymbol{\lambda}))$ is converted to the maximum of the lower bound, which can be further decomposed to

$$\begin{aligned}&\sum_{\tilde{\boldsymbol{\pi}}} \left(\log\left(\frac{P(\boldsymbol{X}, \tilde{\boldsymbol{\pi}}|\boldsymbol{\lambda})}{P(\boldsymbol{X}, \tilde{\boldsymbol{\pi}}|\boldsymbol{\lambda}^{(t)})}\right)\right) P(\boldsymbol{X}, \tilde{\boldsymbol{\pi}}|\boldsymbol{\lambda}^{(t)}) \\ &\quad = \sum_{\tilde{\boldsymbol{\pi}}} (\log P(\boldsymbol{X}, \tilde{\boldsymbol{\pi}}|\boldsymbol{\lambda})) P(\boldsymbol{X}, \tilde{\boldsymbol{\pi}}|\boldsymbol{\lambda}^{(t)}) - \sum_{\tilde{\boldsymbol{\pi}}} \log(P(\boldsymbol{X}, \tilde{\boldsymbol{\pi}}|\boldsymbol{\lambda}^{(t)})) P(\boldsymbol{X}, \tilde{\boldsymbol{\pi}}|\boldsymbol{\lambda}^{(t)}).\end{aligned} \tag{10.12b}$$

Since the second term on the right side of Eq. (10.12b) is irrelevant to $\boldsymbol{\lambda}$, the maximum of the lower bound is the maximum of the first term on the right side, which is denoted as the Q function

$$Q(\boldsymbol{\lambda}, \boldsymbol{\lambda}^{(t)}) = \sum_{\tilde{\boldsymbol{\pi}}} (\log P(\boldsymbol{X}, \tilde{\boldsymbol{\pi}}|\boldsymbol{\lambda}))P(\boldsymbol{X}, \tilde{\boldsymbol{\pi}}|\boldsymbol{\lambda}^{(t)}). \tag{10.12c}$$

The maximum of $Q(\boldsymbol{\lambda}, \boldsymbol{\lambda}^{(t)})$ finds the optimal path, which determines the optimal values of intrinsic parameters, viz.,

$$\boldsymbol{\lambda}^{(t+1)} = \underset{\boldsymbol{\lambda}}{\operatorname{argmax}}\, Q(\boldsymbol{\lambda}, \boldsymbol{\lambda}^{(t)}). \tag{10.12d}$$

The iteration goes on until convergence.

Once the $\boldsymbol{\lambda}$ parameters are known, the state joint probability $P\left(\pi_i^{(k)}, \pi_{i+1}^{(l)}|\boldsymbol{X}, \boldsymbol{\lambda}\right)$ $(i = 0, 1, \cdots, n-1)$ is used in the Baum-Welch algorithm, which is denoted by

$$\xi_i(k, l) = P\left(\pi_i^{(k)}, \pi_{i+1}^{(l)}|\boldsymbol{X}, \boldsymbol{\lambda}\right) = \frac{P\left(\pi_i^{(k)}, \pi_{i+1}^{(l)}, \boldsymbol{X}|\boldsymbol{\lambda}\right)}{P(\boldsymbol{X}|\boldsymbol{\lambda})}. \tag{10.13a}$$

The probability of transition component T_{kl} links neuron (k) in layer (i) to neuron (l) in layer $(i+1)$ and the observation component $b_l(x_{i+1})$ represents the probability of the observed x_{i+1} in neuron (l) of layer $(i+1)$ so that $P\left(\pi_i^{(k)}, \pi_{i+1}^{(l)}, \boldsymbol{X}|\boldsymbol{\lambda}\right) = \alpha_i(k)T_{kl}b_l(x_{i+1})\beta_{i+1}(l)$. Therefore, we have

$$\begin{aligned}\xi_i(k, l) &= P\left(\pi_i^{(k)}, \pi_{i+1}^{(l)}|\boldsymbol{X}, \boldsymbol{\lambda}\right) \\ &= \frac{\alpha_i(k)T_{kl}b_l(x_{i+1})\beta_{i+1}(l)}{\sum_{k=1}^{q}\sum_{l=1}^{q}\alpha_i(k)T_{kl}b_l(x_{i+1})\beta_{i+1}(l)} \quad (i = 0, 1, \cdots, n-1).\end{aligned} \tag{10.13b}$$

With a given sequence of $\boldsymbol{X}$, the appearing expectation and transition expectation for neuron (k) are calculated from the $\pi_i^{(k)}$ probability, $\gamma_i(k)$, by $\sum\limits_{i=1}^{n}\gamma_i(k)$ and $\sum\limits_{i=1}^{n-1}\gamma_i(k)$, respectively, and the expectation from neuron (k) to neuron (l) is $\sum\limits_{i=1}^{n-1}\xi_i(k, l)$. The joint probability $P(\boldsymbol{X}, \tilde{\boldsymbol{\pi}}|\boldsymbol{\lambda})$ of Eq. (10.6c) indicates the probability of a path from one neuron in the input layer to one neuron in the output layer, which natural logarithm is

$$\log(P(\boldsymbol{X}, \tilde{\boldsymbol{\pi}}|\boldsymbol{\lambda})) = \log\left(\pi_0^{(k_0)}\right) + \sum_{i=1}^{n}\log\left(T_{k_{i-1}k_i}\right) + \sum_{i=0}^{n}\log\left(b_{k_i}(x_i)\right). \tag{10.14a}$$

Substituting Eq. (10.14a) into Eq. (10.12c) yields

$$Q(\boldsymbol{\lambda}, \boldsymbol{\lambda}^{(t)}) = \sum_{\tilde{\boldsymbol{\pi}}} \log\left(\pi_0^{(k_0)}\right) P(\boldsymbol{X}, \tilde{\boldsymbol{\pi}}|\boldsymbol{\lambda}^{(t)}) + \sum_{\tilde{\boldsymbol{\pi}}} \sum_{i=1}^{n} \log\left(T_{k_{i-1}k_i}\right) P(\boldsymbol{X}, \tilde{\boldsymbol{\pi}}|\boldsymbol{\lambda}^{(t)})$$

$$+ \sum_{\tilde{\boldsymbol{\pi}}} \sum_{i=0}^{n} \log\left(b_{k_i}(x_i)\right) P(\boldsymbol{X}, \tilde{\boldsymbol{\pi}}|\boldsymbol{\lambda}^{(t)}). \quad (10.14b)$$

The three intrinsic parameters form three terms in $Q(\boldsymbol{\lambda}, \boldsymbol{\lambda}^{(t)})$, separately, so that we can update them one by one. The first term can be rewritten as

$$\sum_{\tilde{\boldsymbol{\pi}}} \log\left(\pi_0^{(k_0)}\right) P(\boldsymbol{X}, \tilde{\boldsymbol{\pi}}|\boldsymbol{\lambda}^{(t)}) = \sum_{k=1}^{q} \log\left(\pi_0^{(k)}\right) P\left(\boldsymbol{X}, \pi_0^{(k)}|\boldsymbol{\lambda}^{(t)}\right). \quad (10.15a)$$

The value of π_0 is estimated from

$$\hat{\pi}_0^{(k)} = \underset{\pi_0}{\operatorname{argmax}} \left(\sum_{k=1}^{q} \log\left(\pi_0^{(k)}\right) P\left(\boldsymbol{X}, \pi_0^{(k)}|\boldsymbol{\lambda}^{(t)}\right) \right),$$

$$\text{subject to} \sum_{k=1}^{q} \pi_0^{(k)} = 1. \quad (10.15b)$$

Using the Lagrange method, we estimate the value of π_0 from the Lagrange function

$$\hat{\pi}_0^{(k)} = \underset{\pi_0}{\operatorname{argmax}} \left(\sum_{k=1}^{q} \log\left(\pi_0^{(k)}\right) P\left(\boldsymbol{X}, \pi_0^{(k)}|\boldsymbol{\lambda}^{(t)}\right) + \rho \left(\sum_{k=1}^{q} \pi_0^{(k)} - 1 \right) \right), \quad (10.15c)$$

where ρ is the Lagrange multiplier. The solution of Eq. (10.15c) is

$$\hat{\pi}_0^{(k)} = \frac{P\left(\boldsymbol{X}, \pi_0^{(k)}|\boldsymbol{\lambda}^{(t)}\right)}{P(\boldsymbol{X}|\boldsymbol{\lambda}^{(t)})} = \gamma_0(k) \quad \text{under } \boldsymbol{\lambda}^{(t)}, \quad (10.15d)$$

because of $P\left(\boldsymbol{X}, \pi_0^{(k)}|\boldsymbol{\lambda}^{(t)}\right) = P\left(\pi_0^{(k)}|\boldsymbol{X}, \boldsymbol{\lambda}^{(t)}\right) P(\boldsymbol{X}|\boldsymbol{\lambda}^{(t)}) = \gamma_0(k) P(\boldsymbol{X}|\boldsymbol{\lambda}^{(t)})$. The second term is rewritten as

$$\sum_{\tilde{\boldsymbol{\pi}}} \sum_{i=1}^{n} \log(T_{k_{i-1}k_i}) P(\boldsymbol{X}, \tilde{\boldsymbol{\pi}}|\boldsymbol{\lambda}^{(t)}) = \sum_{k,l=1}^{q} \sum_{i=1}^{n} \log(T_{kl}) P\left(\boldsymbol{X}, \pi_{i-1}^{(k)}, \pi_i^{(l)}|\boldsymbol{\lambda}^{(t)}\right). \quad (10.16a)$$

The value of T_{kl} is estimated from

$$\hat{T}_{kl} = \underset{T_{kl}}{\text{argmax}}\left(\sum_{k,l=1}^{q}\sum_{i=1}^{n}\log(T_{kl})P\left(\boldsymbol{X}, \pi_{i-1}^{(k)}, \pi_i^{(l)}|\boldsymbol{\lambda}^{(t)}\right)\right),$$
$$\text{subject to} \sum_{l=1}^{q} T_{kl} = 1 \quad (k = 1, 2, \cdots, q). \tag{10.16b}$$

Using the Lagrange method, we estimate

$$\hat{T}_{kl} = \frac{\sum_{i=1}^{n} P\left(\boldsymbol{X}, \pi_{i-1}^{(k)}, \pi_i^{(l)}|\boldsymbol{\lambda}^{(t)}\right)}{\sum_{i=1}^{n} P\left(\boldsymbol{X}, \pi_{i-1}^{(k)}|\boldsymbol{\lambda}^{(t)}\right)}. \tag{10.16c}$$

Since $P\left(\boldsymbol{X}, \pi_{i-1}^{(k)}, \pi_i^{(l)}|\boldsymbol{\lambda}^{(t)}\right) = P\left(\pi_{i-1}^{(k)}, \pi_i^{(l)}|\boldsymbol{X}, \boldsymbol{\lambda}^{(t)}\right)P\left(\boldsymbol{X}|\boldsymbol{\lambda}^{(t)}\right) = \xi_{i-1}(k, l)P\left(\boldsymbol{X}|\boldsymbol{\lambda}^{(t)}\right)$ and $P(\boldsymbol{X}, \pi_{i-1}^{(k)}|\boldsymbol{\lambda}^{(t)}) = P\left(\pi_{i-1}^{(k)}|\boldsymbol{X}, \boldsymbol{\lambda}^{(t)}\right)P\left(\boldsymbol{X}|\boldsymbol{\lambda}^{(t)}\right) = \gamma_{i-1}(k)P\left(\boldsymbol{X}|\boldsymbol{\lambda}^{(t)}\right)$, Eq. (10.16c) can be re-written as

$$\hat{T}_{kl} = \frac{\sum_{i=1}^{n} \xi_{i-1}(k, l)}{\sum_{i=1}^{n} \gamma_{i-1}(k)} \quad \text{under } \boldsymbol{\lambda}^{(t)}. \tag{10.16d}$$

The third term can be rewritten as

$$\sum_{\tilde{\boldsymbol{\pi}}}\sum_{i=0}^{n}\log\left(b_{k_i}(x_i)\right)P(\boldsymbol{X}, \tilde{\boldsymbol{\pi}}|\boldsymbol{\lambda}^{(t)}) = \sum_{k=1}^{q}\sum_{i=0}^{n}\log\left(b_k\left(x_i^{(j)}\right)\right)P\left(\boldsymbol{X}, \pi_i^{(k)}|\boldsymbol{\lambda}^{(t)}\right). \tag{10.17a}$$

The value of B_{kj} is estimated from

$$\hat{B}_{kj} = \underset{B_{kj}}{\text{argmax}}\left(\sum_{k=1}^{q}\sum_{i=0}^{n}\log\left(b_k\left(x_i^{(j)}\right)\right)P\left(\boldsymbol{X}, \pi_i^{(k)}|\boldsymbol{\lambda}^{(t)}\right)\right),$$
$$\text{subject to} \sum_{j=1}^{m} B_{kj} = 1 \quad (k = 1, 2, \cdots, q). \tag{10.17b}$$

Again, we use the Lagrange method to estimate $\hat{B}_{kj}$ and have

$$\hat{B}_{kj} = \frac{\sum_{i=0}^{n} P\left(\boldsymbol{X}, \pi_i^{(k)} | \boldsymbol{\lambda}^{(t)}\right) D\left(x_i^{(j)}\right)}{\sum_{i=0}^{n} P\left(\boldsymbol{X}, \pi_i^{(k)} | \boldsymbol{\lambda}^{(t)}\right)}, \tag{10.17c}$$

where $D\left(x_i^{(j)}\right) = \frac{\partial B_{kj}}{\partial b_k\left(x_i^{(j)}\right)}$ and $\frac{\partial B_{kj}}{\partial b_k\left(x_i^{(\bar{j}\neq j)}\right)} = 0$. Similarly, Eq. (10.17c) can be re-expressed as

$$\hat{B}_{kj} = \frac{\sum_{i=0}^{n} \gamma_i(k) D\left(x_i^{(j)}\right)}{\sum_{i=0}^{n} \gamma_i(k)} \quad \text{under } \boldsymbol{\lambda}^{(t)}. \tag{10.17d}$$

Example 10.9 For a sequence of observations, $\boldsymbol{X} = (aabbabaabb)$, we know there are only two observation bases, a and b, and $n = 9$. Suppose there are only two hidden states $S = (s_1, s_2)$ i.e., q = 2, and estimate intrinsic parameters $\boldsymbol{\lambda} = (\boldsymbol{T}, \boldsymbol{B}, \boldsymbol{\pi}_0)$ via the Baum-Welch algorithm.
The initial values of $\boldsymbol{\lambda}^{(0)}$ are guessed as follows:

$$\boldsymbol{T}^{(0)} = \begin{bmatrix} 0.5 \; 0.5 \\ 0.4 \; 0.6 \end{bmatrix}, \ \boldsymbol{B}^{(0)} = \begin{bmatrix} 0.8 \; 0.2 \\ 0.1 \; 0.9 \end{bmatrix}, \text{ and } \boldsymbol{\pi}_0^{(0)} = \left(0.3 \; 0.7\right).$$

With the observed sequence $\boldsymbol{X} = (aabbabaabb)$ and the initial values of $\boldsymbol{\lambda}^{(0)}$, we can calculate $P(aabbabaabb|\boldsymbol{\lambda}^{(0)})$ via the forward method

$$P(aabbabaabb|\boldsymbol{\lambda}^{(0)}) = \sum_{k=1}^{q} \alpha_n(k).$$

According to Eq. (10.8b), we can obtain

$$\alpha_0(1) = 0.8 \times 0.3 = 0.24,$$
$$\alpha_0(2) = 0.7 \times 0.1 = 0.07.$$

Similarly, we can obtain Table 10.1.

Table 10.1 Internal results by backward method

$\alpha_i(k)$	$k = 1$	$k = 2$
$i = 0$	0.24	0.07
$i = 1$	0.1184	0.0162
$i = 2$	0.01314	0.06203
$i = 3$	0.006276	0.03941
$i = 4$	0.01512	0.002678
$i = 5$	0.001726	0.008250
$i = 6$	0.003306	0.0005813
$i = 7$	0.001518	0.0002014
$i = 8$	0.0001679	0.0007920
$i = n = 9$	0.00008015	0.0005032

Thus, we can obtain $P(aabbabaabb|\boldsymbol{\lambda}^{(0)}) = \sum_{k=1}^{q} \alpha_9(k) = 0.00058335$.

Based on Eqs. (10.9b)–(10.9c) of the backward method, we can get the following results at Table 10.2.

Table 10.2 Internal results by forward method

$\beta_i(k)$	$k = 1$	$k = 2$
$i = n = 9$	1	1
$i = 8$	0.55	0.62
$i = 7$	0.3340	0.3788
$i = 6$	0.15254	0.12961
$i = 5$	0.067450	0.056589
$i = 4$	0.032215	0.035958
$i = 3$	0.014684	0.012466
$i = 2$	0.007078	0.007906
$i = 1$	0.004266	0.004836
$i = 0$	0.0019481	0.001655

According to Eq. (10.9d), we can obtain

$$P(aabbabaabb|\boldsymbol{\lambda}^{(0)}) = \pi_0 B_{kj}(x_0) \sum_{l=1}^{q} T_{kl} b_l(x_1) \beta_1(k) = 0.00058335.$$

Therefore, we can calculate $\gamma_i(k)$ via $\gamma_i(k) = \dfrac{\alpha_i(k)\beta_i(k)}{\sum\limits_{k=1}^{q} \alpha_i(k)\beta_i(k)}$ $(i = 1, 2, \cdots, n)$ and $\gamma_0(k) = \dfrac{\sum\limits_{l=1}^{q} \alpha_0(k)T_{kl}b_l(x_1)\beta_1(l)}{\sum\limits_{k=1}^{q}\sum\limits_{l=1}^{q} \alpha_0(k)T_{kl}b_l(x_1)\beta_1(l)}$ according to Eq. (10.10b), as listed in Table 10.3.

Table 10.3 Probability of event $x_i^{(j)}$ occurring in state k

$\gamma_i(k)$	$k = 1$	$k = 2$
$i = 0$	0.801401	0.198599
$i = 1$	0.86572	0.13428
$i = 2$	0.159374	0.840626
$i = 3$	0.157959	0.842041
$i = 4$	0.834931	0.165069
$i = 5$	0.199722	0.800278
$i = 6$	0.870851	0.129149
$i = 7$	0.869224	0.130776
$i = 8$	0.158326	0.841674
$i = 9$	0.13739	0.86261

For $\xi_i(k, l) = \dfrac{\alpha_i(k)T_{kl}b_l(x_{i+1})\beta_{i+1}(l)}{\sum\limits_{k=1}^{q}\sum\limits_{l=1}^{q} \alpha_i(k)T_{kl}b_l(x_{i+1})\beta_{i+1}(l)}$, we can obtain Table 10.4.

Table 10.4 State joint probability

$\xi_i(k, l)$	$(k = 1, l = 1)$	$(k = 1, l = 2)$	$(k = 2, l = 1)$	$(k = 2, l = 2)$
$i = 0$	0.70194	0.09947	0.16378	0.03481
$i = 1$	0.14365	0.72207	0.01572	0.11856
$i = 2$	0.03306	0.12631	0.12490	0.71573
$i = 3$	0.13862	0.01934	0.69631	0.14573
$i = 4$	0.17493	0.66000	0.02479	0.14028
$i = 5$	0.18055	0.01918	0.69030	0.10997
$i = 6$	0.76272	0.10813	0.10650	0.02265
$i = 7$	0.14314	0.72609	0.01519	0.11559
$i = 8$	0.02879	0.12954	0.10860	0.73307

According to Eq. (10.16d), we can update the transition matrix as

$$\hat{T}_{11} = \frac{\sum_{i=0}^{8} \xi_i(1, 1)}{\sum_{i=0}^{8} \gamma_i(1)} = \frac{2.3074}{4.9175} = 0.4692,$$

$$\hat{T}_{12} = \frac{\sum_{i=0}^{8} \xi_i(1, 2)}{\sum_{i=0}^{8} \gamma_i(1)} = \frac{2.6101}{4.9175} = 0.5308,$$

$$\hat{T}_{21} = \frac{\sum_{i=0}^{8} \xi_i(2, 1)}{\sum_{i=0}^{8} \gamma_i(2)} = \frac{1.9461}{4.0825} = 0.4767,$$

$$\hat{T}_{22} = \frac{\sum_{i=0}^{8} \xi_i(2, 2)}{\sum_{i=0}^{8} \gamma_i(2)} = \frac{2.1364}{4.0852} = 0.5233.$$

According to Eq. (10.17d), the observation matrix is estimated as

$$\hat{B}_{1a} = \frac{\sum_{i=0}^{9} \gamma_i(1) D\left(x_i^{(a)}\right)}{\sum_{i=0}^{9} \gamma_i(1)} = 0.8392,$$

$$\hat{B}_{1b} = \frac{\sum_{i=0}^{9} \gamma_i(1) D\left(x_i^{(b)}\right)}{\sum_{i=0}^{9} \gamma_i(1)} = 0.1608,$$

$$\hat{B}_{2a} = \frac{\sum_{i=0}^{9} \gamma_i(2) D\left(x_i^{(a)}\right)}{\sum_{i=0}^{9} \gamma_i(2)} = 0.1533,$$

$$\hat{B}_{2b} = \frac{\sum_{i=0}^{9} \gamma_i(2) D\left(x_i^{(b)}\right)}{\sum_{i=0}^{9} \gamma_i(2)} = 0.8467.$$

According to Table 10.3, the initial states are $\hat{\pi}_0^{(1)}(k=1)=\gamma_0(1)=0.8014$ and $\hat{\pi}_0^{(1)}(k=2)=\gamma_0(2)=0.1986$. For brevity, only the first four significant decimal places are retained here. The intrinsic parameters are updated to $\boldsymbol{T}^{(1)}=\begin{bmatrix}0.4692 & 0.5308\\0.4767 & 0.5233\end{bmatrix}$, $\boldsymbol{B}^{(1)}=\begin{bmatrix}0.8392 & 0.1608\\0.1533 & 0.8467\end{bmatrix}$, and $\pi_0^{(1)}=\begin{pmatrix}0.8014 & 0.1986\end{pmatrix}$.
Based on the updated intrinsic parameters and the forward method, the probability of the observed sequence $\boldsymbol{X}=(aabbabaabb)$ is

$$P(aabbabaabb|\boldsymbol{\lambda}^{(1)})=\sum_{k=1}^{q}\alpha_9(k)=0.00142547>P(aabbabaabb|\boldsymbol{\lambda}^{(0)}).$$

That is, the probability of the observed sequence $\boldsymbol{X}$ increases with updated parameters.
If we repeat the above procedure once more, we can obtain

$$\boldsymbol{T}^{(2)}=\begin{bmatrix}0.4369 & 0.5631\\0.4661 & 0.5339\end{bmatrix},$$

$$\boldsymbol{B}^{(2)}=\begin{bmatrix}0.8549 & 0.1451\\0.1438 & 0.8562\end{bmatrix},$$

$$\boldsymbol{\pi}_0^{(2)}=\begin{pmatrix}0.9563 & 0.0437\end{pmatrix},$$

$$P(aabbabaabb|\boldsymbol{\lambda}^{(2)})=\sum_{k=1}^{q}\alpha_9(k)=0.00026974.$$

Example 10.10 The updated parameters highly depend on the initial guesses. For instance, if we have a Markov chain whose parameters $\boldsymbol{\lambda}$ are

$$\boldsymbol{\pi}_0=\begin{pmatrix}1 & 0 & 0\end{pmatrix},$$

$$\boldsymbol{T}=\begin{bmatrix}0 & 0.7 & 0.3\\0.2 & 0 & 0.8\\0.6 & 0.4 & 0\end{bmatrix},$$

$$\boldsymbol{B}=\begin{bmatrix}3/5 & 0 & 2/5\\1/3 & 1/3 & 1/3\\0 & 1/4 & 3/4\end{bmatrix}.$$

The observed sequence with the highest probability can be $\boldsymbol{X}=aacabcaccaacaccabc$, and the corresponding $P(X|\boldsymbol{\lambda})=9.9263\times10^{-8}$

1. Baum-Welch algorithm without domain knowledge

If we randomly set an initial guess $\boldsymbol{\lambda}^{(0)}$ as

$$\pi_0^{(0)} = \left(0.5\ 0.3\ 0.2\right),$$

$$\boldsymbol{T}^{(0)} = \begin{bmatrix} 0.5\ 0.2\ 0.3 \\ 0.3\ 0.5\ 0.2 \\ 0.2\ 0.3\ 0.5 \end{bmatrix},$$

$$\boldsymbol{B}^{(0)} = \begin{bmatrix} 0.5\ 0.1\ 0.4 \\ 0.1\ 0.7\ 0.2 \\ 0.2\ 0.2\ 0.6 \end{bmatrix},$$

$$P(\boldsymbol{X}|\boldsymbol{\lambda}^{(0)}) = 2.4465 \times 10^{-9}.$$

After applying the Baum-Welch algorithm twice, the updated parameters $\boldsymbol{\lambda}^{(2)}$ are

$$\pi_0^{(2)} = \left(0.9291\ 0.0278\ 0.0431\right),$$

$$\boldsymbol{T}^{(2)} = \begin{bmatrix} 0.5226\ 0.1446\ 0.3328 \\ 0.3767\ 0.3096\ 0.3137 \\ 0.2766\ 0.1757\ 0.5477 \end{bmatrix},$$

$$\boldsymbol{B}^{(2)} = \begin{bmatrix} 0.6261\ 0.0367\ 0.3372 \\ 0.2586\ 0.3363\ 0.4051 \\ 0.3282\ 0.0915\ 0.5803 \end{bmatrix},$$

$$P(\boldsymbol{X}|\boldsymbol{\lambda}^{(2)}) = 3.5312 \times 10^{-8}.$$

After ten times, the updated parameters $\boldsymbol{\lambda}^{(10)}$ are

$$\pi_0^{(10)} = \left(1\ 0\ 0\right),$$

$$\boldsymbol{T}^{(10)} = \begin{bmatrix} 0.3212\ 0.2810\ 0.3978 \\ 0.2712\ 0.2349\ 0.4939 \\ 0.3155\ 0.1414\ 0.5431 \end{bmatrix},$$

$$\boldsymbol{B}^{(10)} = \begin{bmatrix} 0.7744\ 0.0160\ 0.2096 \\ 0.2078\ 0.2910\ 0.5012 \\ 0.2956\ 0.1053\ 0.5991 \end{bmatrix},$$

$$P(\boldsymbol{X}|\boldsymbol{\lambda}^{(10)}) = 5.5584 \times 10^{-8}.$$

2. Baum-Welch algorithm with domain knowledge

The domain knowledge may suggest the following formats of the three intrinsic parameters:

$$\pi_0^{(0)} = \left(1\ 0\ 0\right),$$

$$\boldsymbol{T}^{(0)} = \begin{bmatrix} 0 & T_{12} & T_{13} \\ T_{21} & 0 & T_{23} \\ T_{31} & T_{32} & 0 \end{bmatrix},$$

$$\boldsymbol{B}^{(0)} = \begin{bmatrix} B_{11} & 0 & B_{13} \\ B_{21} & B_{22} & B_{23} \\ 0 & B_{32} & B_{33} \end{bmatrix}.$$

Then, we can guess the initial values of intrinsic parameters $\boldsymbol{\lambda}'^{(0)}$ as

$$\begin{aligned} \boldsymbol{\pi}_0'^{(0)} &= \begin{pmatrix} 1 & 0 & 0 \end{pmatrix}, \\ \boldsymbol{T}'^{(0)} &= \begin{bmatrix} 0 & 0.4 & 0.6 \\ 0.5 & 0 & 0.5 \\ 0.4 & 0.6 & 0 \end{bmatrix}, \\ \boldsymbol{B}'^{(0)} &= \begin{bmatrix} 0.4 & 0 & 0.6 \\ 0.2 & 0.4 & 0.4 \\ 0 & 0.5 & 0.5 \end{bmatrix}, \\ P(\boldsymbol{X}|\boldsymbol{\lambda}'^{(0)}) &= 2.2245 \times 10^{-8}. \end{aligned}$$

The updated parameter after applying Baum-Welch algorithm twice $\boldsymbol{\lambda}'^{(2)}$ will be

$$\begin{aligned} \boldsymbol{\pi}_0'^{(2)} &= \begin{pmatrix} 1 & 0 & 0 \end{pmatrix}, \\ \boldsymbol{T}'^{(2)} &= \begin{bmatrix} 0 & 0.7178 & 0.2822 \\ 0.4546 & 0 & 0.5454 \\ 0.5196 & 0.4804 & 0 \end{bmatrix}, \\ \boldsymbol{B}'^{(2)} &= \begin{bmatrix} 0.6633 & 0 & 0.3367 \\ 0.5742 & 0.0841 & 0.3417 \\ 0 & 0.2861 & 0.7139 \end{bmatrix}, \\ P(\boldsymbol{X}|\boldsymbol{\lambda}'^{(2)}) &= 7.6755 \times 10^{-8}. \end{aligned}$$

As can be seen, the estimated values of parameters $\boldsymbol{\lambda}'^{(2)}$ are closer to the true values $\boldsymbol{\lambda}$ than the initial guesses with domain knowledge $\boldsymbol{\lambda}'^{(0)}$ and estimations without domain knowledge $\boldsymbol{\lambda}^{(2)}$. However, if we repeat the Baum-Welch algorithm more times on the initial guess $\boldsymbol{\lambda}'^{(0)}$, the corresponding parameters $\boldsymbol{\lambda}'^{(10)}$ will be updated to be

$$\begin{aligned} \boldsymbol{\pi}_0'^{(10)} &= \begin{pmatrix} 1 & 0 & 0 \end{pmatrix}, \\ \boldsymbol{T}'^{(10)} &= \begin{bmatrix} 0 & 1 & 0 \\ 0 & 0 & 1 \\ 1 & 0 & 0 \end{bmatrix}, \\ \boldsymbol{B}'^{(10)} &= \begin{bmatrix} 1 & 0 & 0 \\ 1/3 & 1/3 & 1/3 \\ 0 & 0 & 1 \end{bmatrix}, \\ P(\boldsymbol{X}|\boldsymbol{\lambda}'^{(10)}) &= 1.3717 \times 10^{-3}. \end{aligned}$$

The $P(\boldsymbol{X}|\boldsymbol{\lambda}'^{(10)}) = 1.3717 \times 10^{-3}$ is far larger than $P(\boldsymbol{X}|\boldsymbol{\lambda}) = 9.9263 \times 10^{-8}$. Obviously, the hidden Markov chain is overfitted with only one observed sequence, resulting in much simpler transition matrix and observation matrix, which greatly increases the likelihood of the observation sequence. To avoid overfitting, more observation sequences should be introduced and the Baum-Welch algorithm should be constrained using domain knowledge to suggest the format of the intrinsic parameters.

Homework

1. With the initial state of $\boldsymbol{\pi}_0$ =(0.2, 0, 0, 0.2, 0, 0, 0.5, 0, 0.1) and the transition matrix $\boldsymbol{T} = \frac{1}{8}\begin{bmatrix} 0 \cdots 1 \\ \vdots \quad \vdots \\ 1 \cdots 0 \end{bmatrix}_{9\times 9}$, calculate the state vector $\boldsymbol{\pi}_{999}$.
2. The initial state probability is $\boldsymbol{\pi}_0 = \begin{pmatrix} 0.6 \; 0.3 \; 0.1 \end{pmatrix}^{\mathrm{T}}$ of states s_1, s_2, and s_3, the transition matrix is $\boldsymbol{T} = \begin{bmatrix} 0.5 \; 0.2 \; 0.3 \\ 0.1 \; 0.1 \; 0.8 \\ 0.3 \; 0.3 \; 0.4 \end{bmatrix}$, and the observation matrix is $\boldsymbol{B} = \begin{bmatrix} 0.5 \; 0.5 \\ 0.2 \; 0.8 \\ 0.7 \; 0.3 \end{bmatrix}$ for observations a and b. Use the forward and backward methods to calculate the probability of the observation sequence (a, b, a).
3. The initial state probability is $\boldsymbol{\pi}_0 = \begin{pmatrix} 0 \; 0 \; 1 \end{pmatrix}^{\mathrm{T}}$ of states s_1, s_2, and s_3, the transition matrix is $\boldsymbol{T} = \begin{bmatrix} 0.1 \; 0.2 \; 0.7 \\ 0.4 \; 0.3 \; 0.3 \\ 0.2 \; 0.6 \; 0.2 \end{bmatrix}$, and the observation matrix is $\boldsymbol{B} = \begin{bmatrix} 0.5 \; 0.2 \; 0.3 \\ 0.1 \; 0.1 \; 0.8 \\ 0.3 \; 0.3 \; 0.4 \end{bmatrix}$ for observations a, b, and c. Use the forward and backward methods to calculate the probability of the observation sequence (a, c, a, b, c).
4. The initial state probability is $\boldsymbol{\pi}_0 = \begin{pmatrix} 0.2 \; 0.3 \; 0.5 \end{pmatrix}^{\mathrm{T}}$ of states s_1, s_2, and s_3, the transition matrix is $\boldsymbol{T} = \begin{bmatrix} 0.5 \; 0.2 \; 0.3 \\ 0.1 \; 0.1 \; 0.8 \\ 0.3 \; 0.3 \; 0.4 \end{bmatrix}$, and the observation matrix is $\boldsymbol{B} = \begin{bmatrix} 0.5 \; 0.5 \\ 0.2 \; 0.8 \\ 0.7 \; 0.3 \end{bmatrix}$ for observations a and b. Use the forward and backward methods to calculate the probability of s_2 at $\boldsymbol{\pi}_3$, when given the observation sequence (a, a, a, b, a).

5. Conduct Gibbs sampling to draw from $P(\boldsymbol{\theta}) \sim N\left(\begin{pmatrix} 1 \\ 4 \\ 2 \end{pmatrix}, \begin{bmatrix} 2 & 0.2 & 0.4 \\ 0.1 & 1 & 0.2 \\ 0.5 & 0.1 & 5 \end{bmatrix}\right)$.
6. Use the Baum-Welch algorithm to estimate the intrinsic parameters of the hidden Markov chain in Example 10.9.

References

Baum, L. E. (1972). An inequality and associated maximization technique in statistical estimation for probabilistic functions of Markov processes. *Inequalities, 3*(1), 1–8.

Baum, L. E., & Eagon, J. A. (1967). An inequality with applications to statistical estimation for probabilistic functions of Markov processes and to a model for ecology. *Bulletin of the American Mathematical Society, 73*(3), 360–363.

Baum, L. E., & Petrie, T. (1966). Statistical inference for probabilistic functions of finite state Markov chains. *The Annals of Mathematical Statistics, 37*(6), 1554–1563.

Chib, S. (2001). Markov chain Monte Carlo methods: Computation and inference. *Handbook of Econometrics, 1*(5), 3569–3649.

Faul, A. C. (2019). *A concise introduction to machine learning*. CRC Press.

Forney, G. D. (1973). The Viterbi algorithm. *Proceedings of the IEEE, 61*(3), 268–278.

Geman, S., & Geman, D. (1984). Stochastic relaxation, Gibbs distributions, and the Bayesian restoration of images. *IEEE Transactions on Pattern Analysis and Machine Intelligence, 6*, 721–741.

Hastie, T., Tibshirani, R., Friedman, J. H., et al. (2009). The elements of statistical learning: Data mining, inference, and prediction. Springer.

Hammersley, J. M., & Morton, K. W. (1954). Poor man's Monte Carlo. *Journal of the Royal Statistical Society: Series B (Methodological), 16*(1), 23–38.

Hastings, W. K. (1970). Monte Carlo sampling methods using Markov chains and their applications. *Biometrika, 57*(1), 97–109.

Kruskal, J. B. (1983). An overview of sequence comparison: Time warps, string edits, and macromolecules. *SIAM Review, 25*(2), 201–237.

Li, H. (2018). *Statistical learning method* (统计学习方法). Tsinghua University Press.

Markov, A. A. (1913a). Essai d'une recherche statistique sur le texte du roman "Eugene Onegin" illustrant la liaison des epreuve en chain ("Example of a statistical investigation of the text of "Eugene Onegin" illustrating the dependence between samples in chain"). *Izvistia Imperatorskoi Akademii Nauk (Bulletin de l'Académie Impériale des Sciences de St.-Pétersbourg), 7*, 153–162.

Markov, A. A. (1913b). Primer statisticheskogo issledovanija nad tekstom "Evgenija Onegina illjustrirujuschij svjazl" ispytanij v tsep (An example of statistical study on the text of "Eugene Onegin" illustrating the linking of events to a chain). Izvestija Imp. *Akademii Nauk, Serija, 3*, 153–162.

Markov, A. A. (2006). An example of statistical investigation of the text Eugene Onegin concerning the connection of samples in chains. *Science in Context, 19*(4), 591–600.

Metropolis, N., Rosenbluth, A. W., Rosenbluth, M. N., et al. (1953). Equation of state calculations by fast computing machines. *The Journal of Chemical Physics, 21*(6), 1087–1092.

Metropolis, N., & Ulam, S. (1949). The Monte Carlo method. *Journal of the American Statistical Association, 44*(247), 335–341.

Meyn, S. P., & Tweedie, R. L. (1994). Computable bounds for geometric convergence rates of Markov chains. *The Annals of Applied Probability, 1*, 981–1011.

Rabiner, L. R. (1989). A tutorial on hidden Markov models and selected applications in speech recognition. *Proceedings of the IEEE, 77*(2), 257–286.

Rosenthal, J. S. (1995). Rates of convergence for Gibbs sampling for variance component models. *The Annals of Statistics, 23*(3), 740–761.

Vintsyuk, T. K. (1968). Speech discrimination by dynamic programming. *Cybernetics, 4*(1), 52–57.

Viterbi, A. J. (1967). Error bounds for convolutional codes and an asymptotically optimum decoding algorithm. *IEEE Transactions on Information Theory, 2*, 260–269.

Chapter 11
Data Preprocessing and Feature Selection

Materials informatics is data-driven and the goal of materials informatics is to achieve efficient and robust acquisition, management, multi-factor analyses, and dissemination of diverse materials data. As described in Chap. 1, materials data include experimental data, computation data, manufacture data, performance and service data, etc. In addition to one's own data, most data for individual researchers and/or research groups are collected from various sources including scientific literature, patents, and databases. The first step in materials informatics is to collect data. After that, data preprocessing is conducted to make sure all studied data are reliable. This is because most ML algorithms take it for granted that data are reliable. In this sense, reliable data might matter more than ML algorithms, suggesting that more efforts are demanded to cleanse and filter originally collected data. The first part of this chapter introduces some ML methods commonly used in the identification of outliers.

With reliable data, feature selection becomes the essential, necessary, and important step in ML. As described in Chap. 1, bias and variance are the two types of errors in ML and there exists a bias-variance dilemma. This is because it is taken for granted in ML that the true solution can only be approached, but cannot be obtained. The ML goal is to search for the best solution in a feature space so that the sum of bias and variance is minimized. Therefore, features and feature space are two essential and crucial issues in ML. Feature selection selects the most influential features to responses, which can be initial features and/or their combinations. Feature selection is one of the most crucial and challenging steps in ML because the selected features consequently generate a feature space or search space, within which the optimal approximate solution is searched for. The second and major part of this chapter introduces some ML algorithms widely used in feature selection.

T. Zhang, *An Introduction to Materials Informatics*,
https://doi.org/10.1007/978-981-99-7992-9_11

11.1 Reliable Data, Normals and Anomalies

Reliable data mean that the data are reproducible. As described in Chap. 1, initially successful and failed data, with material properties and performance of interest and material composition, processing, testing conditions, or/and service environments, etc. of adjustable parameters, are generated from experiments, calculations, or/and manufacturing. In collecting material data, the conditions to generate these data including data accuracy and noise should be known to ensure the collected data are reproducible. For some unknown reasons, some experimental data may scatter greatly. In this circumstance, one has to increase the number of repeated tests, because only sufficiently repeated tests are able to provide the reliable mean and variance of test results. However, some materials data reported in the literature might be inconsistent and incorrect. The inconsistent data found in the literature refer to the diverse values of some properties for samples with the same chemical composition, microstructures, processing conditions, and characterization conditions. If the data have only small and acceptable inconsistency, their mean might be utilized for the inconsistent data. In the converse case, part of inconsistent data could be presumed based on domain knowledge and/or statistical analysis as the incorrect data, also named outliers or anomalies. Anomaly identification follows the simple principle "obey the majority" and there are statistical criteria/methods like the standard deviation, student's *t*-test, and *F*-test. This chapter introduces ML-based methods including the local outlier factor(LOF), the isolated forest, the one-class support vector machine, and the support vector data description. It should be kept in mind all the time that reliably abnormal data indicate new mechanisms, which deserve more deep investigation, and all advances in science and technology seem to be triggered by experimentally reliably anomaly observations.

11.1.1 Local Outlier Factor

As described in *K*-nearest-neighbors methods, the nearest neighbors of a point are measured by a critical distance to the point, or a critical radius of hypersphere centered at the point in high-dimensional space, within the hypersphere samples are the nearest neighbors of the point. The critical distance, in turn, is determined by a preset critical number of nearest neighbors. Thus, the critical distance determined by the critical number may vary from data point to data point under a preset value of *K*-nearest-neighbors. With a given k, obviously, the larger the critical distance of a point is, the lower the local density of the point will be. The local density of a data point is calculated based on the distances of the point to other points in a dataset (Breunig et al., 2000). Consider a dataset of n data $\boldsymbol{X} = (x_{i1}, x_{i2}, \cdots, x_{im})(i = 1, 2, \cdots, n)$, $\boldsymbol{X} \in \mathbf{R}^m$. A data point p's *K*-nearest-neighbors distance, denoted as $\text{dist}_k(p)$, is defined as the distance from it to its k-th nearest neighbor. Since few data points may have the same distance to point p, the number of data within the *K*-nearest-neighbors distance is $|N_k(p)| \geqslant k$, where $N_k(p)$ denotes the sphere centered at point p in high-dimensional space. For a given k, the reachability distance from point p to point o,

$\text{dist}_k^{\text{reach}}(p, o)$, is defined by

$$\text{dist}_k^{\text{reach}}(p, o) = \max\{\text{dist}_k(p), \text{dist}(p, o)\}. \tag{11.1a}$$

Equation (11.1a) indicates that the lower value of $\text{dist}_k^{\text{reach}}(p, o)$ is $\text{dist}_k(p)$. The reachability distance from point o, one of K-nearest-neighbors of point p, i.e., $o \in N_k(p)$, to point p under the condition of the K-nearest-neighbors distance of point o is

$$\text{dist}_k^{\text{reach}}(o, p) = \max\{\text{dist}_k(o), \text{dist}(o, p)\}. \tag{11.1b}$$

Summing up all reachability distances of K-nearest-neighbors $o \in N_k(p)$ to point p, we have

$$\sum_{o \in N_k(p)} \text{dist}_k^{\text{reach}}(o, p). \tag{11.1c}$$

The local reachability density (lrd) of point p under the K-nearest-neighbors distance is defined as the ratio of the number of data within the K-nearest-neighbors distance over the sum of the reachability distances from all points $o \in N_k(p)$ to point p

$$\text{lrd}_k(p) = \frac{|N_k(p)|}{\sum_{o \in N_k(p)} \text{dist}_k^{\text{reach}}(o, p)} = \left(\frac{\sum_{o \in N_k(p)} \text{dist}_k^{\text{reach}}(o, p)}{|N_k(p)|} \right)^{-1}. \tag{11.2}$$

The term $\frac{\sum_{o \in N_k(p)} \text{dist}_k^{\text{reach}}(o, p)}{|N_k(p)|}$ is the average reachability distance of point p, and $\text{lrd}_k(p)$ is the inverse of the average reachability distance of point p. The local reachability density $\text{lrd}_k(o)$ of every point $o \in N_k(p)$ can be calculated in the same way. After that, the local outlier factor $\text{lof}_k(p)$ is defined by

$$\text{lof}_k(p) = \frac{\sum_{o \in N_k(p)} \frac{\text{lrd}_k(o)}{\text{lrd}_k(p)}}{|N_k(p)|} = \frac{\frac{\sum_{o \in N_k(p)} \text{lrd}_k(o)}{|N_k(p)|}}{\text{lrd}_k(p)}. \tag{11.3}$$

The term $\frac{\sum_{o \in N_k(p)} \text{lrd}_k(o)}{|N_k(p)|}$ is the average of local reachability densities, $\overline{\text{lrd}_k(o)}$, of all points $o \in N_k(p)$. $\text{lof}_k(p)$ is the ratio of the average local reachability density over $\text{lrd}_k(p)$. If $\text{lof}_k(p) = 1$, point p has the same local density as the average of its K-nearest-neighbors. If $\text{lof}_k(p) < 1$, point p has a higher local density than the average of its K-nearest-neighbors, while point p has a lower local density than the average of its K-nearest-neighbors if $\text{lof}_k(p) > 1$. If $\text{lof}_k(p)$ excesses a preset threshold, point p is treated as an outlier.

Example 11.1 Table 11.1 shows four alloys, and their enthalpy of mixing (H_{mix}) and entropy of mixing (S_{mix}) of four alloys. The normalized feature values are considered for calculation, and the Manhattan distance is utilized in the current example. Thus the distances between four data points are

$$\begin{aligned}
\text{dist}(a, b) &= 0.4647, \\
\text{dist}(a, c) &= 0.2192, \\
\text{dist}(a, d) &= 1.6675, \\
\text{dist}(b, c) &= 0.2710, \\
\text{dist}(b, d) &= 1.8678, \\
\text{dist}(c, d) &= 1.5968.
\end{aligned}$$

Table 11.1 Selected data for four alloys for LOF detection

Samples	Alloys	H_{mix} (kJ·mol^{-1})	S_{mix} (J·K^{-1}·mol^{-1})	Normalized H_{mix}	Normalized S_{mix}
a	$Ag_{14}Cu_{78.5}P_{7.5}$	−4.01905	5.483515	0	0.667501
b	$Ag_{10}Mg_{85}Y_5$	−5	4.3082	0.132187	1
c	$Ag_{10.5}Cu_{80.5}P_9$	−5.0946	5.221014	0.144934	0.741764
d	$Ag_{10}Mg_{50}Y_{10}$	−11.44	7.842999	1	0

2-nearest neighbors are considered in this example and the distances of the first and second nearest neighbors can be identified from the calculated distances for each point. To calculate $\text{dist}_k^{\text{reach}}$ with Eq. (11.1b), the values of dist_2 are calculated for all points as

$$\begin{aligned}
\text{dist}_2(a) &= \text{dist}(a, b) = 0.4647, \\
\text{dist}_2(b) &= \text{dist}(b, a) = 0.4647, \\
\text{dist}_2(c) &= \text{dist}(c, b) = 0.2710, \\
\text{dist}_2(d) &= \text{dist}(d, a) = 1.6675.
\end{aligned}$$

The reachability distance and the local reachability density for each point are then calculated. Taking point a as an example, which has points b and c as its 2-nearest neighbors, we have

$$\begin{aligned}
\text{dist}_2^{\text{reach}}(c, a) &= \max\{\text{dist}_2(c), \text{dist}(a, c)\} = 0.2710, \\
\text{dist}_2^{\text{reach}}(b, a) &= \max\{\text{dist}_2(b), \text{dist}(a, b)\} = 0.4647, \\
\text{lrd}_2(a) &= \frac{2}{\text{dist}_2^{\text{reach}}(b, a) + \text{dist}_2^{\text{reach}}(c, a)} = 2.7185.
\end{aligned}$$

Similarly, we obtain

$$\mathrm{lrd}_2(b) = \frac{2}{\mathrm{dist}_2^{\mathrm{reach}}(a, b) + \mathrm{dist}_2^{\mathrm{reach}}(c, b)} = 2.7185,$$
$$\mathrm{lrd}_2(c) = \frac{2}{\mathrm{dist}_2^{\mathrm{reach}}(a, c) + \mathrm{dist}_2^{\mathrm{reach}}(b, c)} = 2.1519,$$
$$\mathrm{lrd}_2(d) = \frac{2}{\mathrm{dist}_2^{\mathrm{reach}}(a, d) + \mathrm{dist}_2^{\mathrm{reach}}(c, d)} = 0.6127.$$

Finally, the local outlier factors are obtained

$$\mathrm{lof}_2(a) = \frac{\mathrm{lrd}_2(b) + \mathrm{lrd}_2(c)}{2\,\mathrm{lrd}_2(a)} = 0.8958,$$
$$\mathrm{lof}_2(b) = \frac{\mathrm{lrd}_2(a) + \mathrm{lrd}_2(c)}{2\mathrm{lrd}_2(b)} = 0.8958,$$
$$\mathrm{lof}_2(c) = \frac{\mathrm{lrd}_2(a) + \mathrm{lrd}_2(b)}{2\mathrm{lrd}_2(c)} = 1.2633,$$
$$\mathrm{lof}_2(d) = \frac{\mathrm{lrd}_2(a) + \mathrm{lrd}_2(c)}{2\mathrm{lrd}_2(d)} = 3.9745.$$

We set a threshold as 1.5. Only $\mathrm{lof}_2(d) > 1.5$ and hence, point d is treated as an outlier. This conclusion matches the visual expression, which can be obtained by plotting all the four points in Fig. 11.1, and point d is obviously an anomaly.

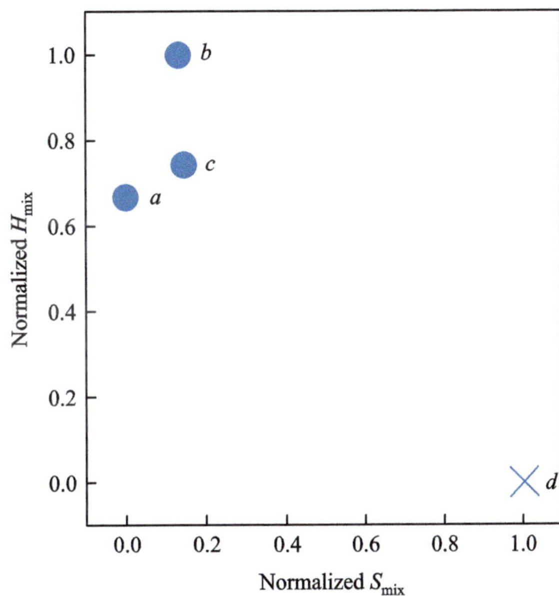

Fig. 11.1 Four samples a, b, c and d in the normalized feature space

---

11.1.2 Isolated Forest

Isolation forest (iForest) is an ensemble algorithm with binary decision trees as base learners, which is developed based on two anomaly characteristics: (1) Anomalies are the minority which are few and different in a dataset; (2) Anomalies have input-feature values that are very different from normal samples. Thus, outliers are more easily to be isolated closer to the root of a decision tree, whereas normal samples are isolated at the deeper end of the tree (Liu et al., 2012).

A data subset with $\hat{n} < n$ data is randomly selected from a dataset of n data $\boldsymbol{X} = (x_{i1}, \cdots, x_{im})(i = 1, 2, \cdots, n)$, $\boldsymbol{X} \in \mathbf{R}^m$. When all samples are distinct, an isolation decision tree (iTree) grows up by randomly and recursively splitting a randomly selected feature to full size that each leaf contains only one sample. A loss function is used in either tree classifier or tree regressor, while no functions are employed in iTrees. Repeating this process t times forms t iTrees. If t iTrees are formed by t-fold division of the original dataset, all data will be used in the training t iTrees.

Once t iTrees are trained, we test every sample x in the dataset $\boldsymbol{X}$ by t iTrees. The path length $h(x)$ of sample x in one iTree is defined as the number of edges (an edge links a parent node to a child node) that the sample traverses from the root node to a terminate node. Then, the mean of $h(x)$ is obtained by (Liu et al., 2012).

$$E(h(x)) = \frac{1}{t}\sum_{i=1}^{t} h_i(x). \tag{11.4}$$

The outlier score s is then defined as

$$s(x, \hat{n}) = 2^{-\frac{E(h(x))}{c(\hat{n})}}, \tag{11.5a}$$

where $c(\hat{n})$ is given by

$$c(\hat{n}) = \begin{cases} 2H(\hat{n}-1) - \frac{2(\hat{n}-1)}{n}, & \text{for } \hat{n} > 2, \\ 1, & \text{for } \hat{n} = 2, \\ 0, & \text{for } n < 2, \end{cases} \tag{11.5b}$$

and $H(i)$ is the harmonic number for natural number i and could be estimated by $\ln(i)+0.5772156649$ (Euler's constant). If the outlier score of a sample is very close to 1, the sample will be an anomaly. Otherwise, if the samples with scores much smaller than 0.5, the samples are normal. A threshold is also needed to determine anomalies.

Example 11.2 The same data given in Example 11.1 are used here. The hyperparameters in this iForest are fixed, three samples are randomly selected to grow an iTree, and four iTrees are trained.

iTree 1: Features H_{mix} and S_{mix}, and samples *a*, *c*, and *d* are selected to grow an iTree, and sample *b* is tested on the iTree. As shown in Fig. 11.2, the path lengths of points *a*, *b*, *c*, and *d* are 2, 2, 2, and 1, respectively. Point *d* has the lowest path length 1.

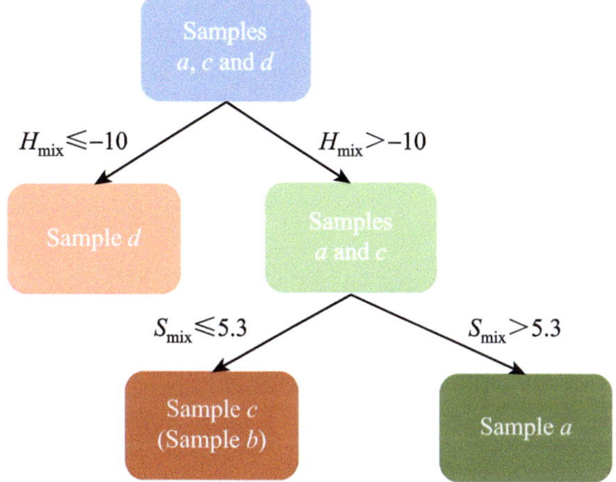

Fig. 11.2 The iTree trained on features H_{mix} and S_{mix}, and samples *a*, *c*, and *d* Point *b* is tested

iTree 2: Feature H_{mix}, and samples *a*, *b*, and *c* are selected to grow an iTree, and sample *d* is tested on the iTree. As shown in Fig. 11.3, the path lengths of points *a*, *b*, *c*, and *d* are 2, 2, 1, and 1, respectively. Points *c* and *d* have the lowest path length 1.

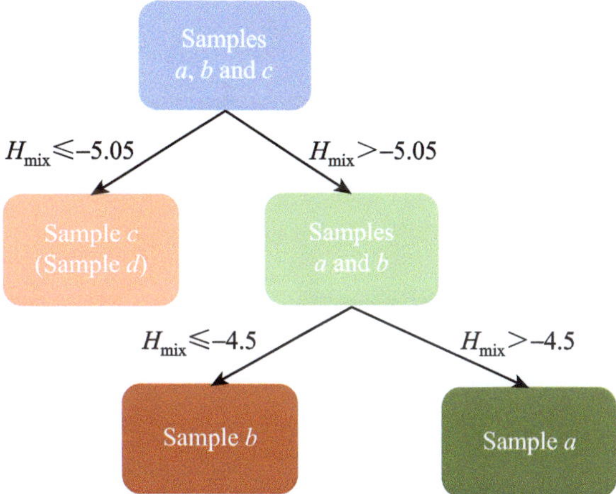

Fig. 11.3 The iTree trained on feature H_{mix} and samples *a*, *b*, and *c*. Point *d* is tested

iTree 3: Feature S_{mix} and H_{mix}, and samples *a*, *b*, and *d* are selected to grow an iTree, and sample *c* is tested on the iTree. As shown in Fig. 11.4, the path lengths of points *a*, *b*, *c*, and *d* are 2, 2, 2, and 1, respectively. Point *d* has the lowest path length 1.

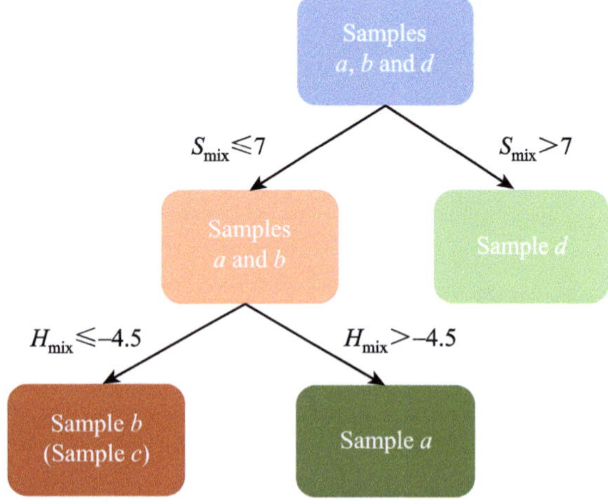

Fig. 11.4 The iTree trained on feature S_{mix} and H_{mix} and samples *a*, *b*, and *d* Point *c* is tested

iTree 4: Feature S_{mix} and samples *b*, *c*, and *d* are selected to grow an iTree, and sample *a* is tested on the iTree. As shown in Fig. 11.5, the path lengths of points *a*, *b*, *c*, and *d* are 2, 1, 2, and 2, respectively. Point *b* has the lowest path length 1.

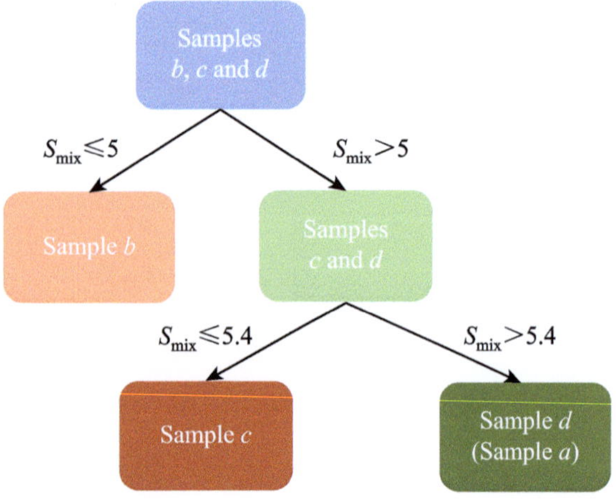

Fig. 11.5 The iTree trained on feature S_{mix} and samples *b*, *c*, and *d*. Point *a* is tested

In summary, the average path length of *a*, *b*, *c*, and *d* are 2.00, 1.75, 1.75, and 1.25, respectively, i.e.,

$$
\begin{aligned}
E(h(a)) &= 2.00,\\
E(h(b)) &= 1.75,\\
E(h(c)) &= 1.75,\\
E(h(d)) &= 1.25.
\end{aligned}
$$

Since

$$c(3) = 2H(2) - \frac{2 \times 2}{4} = 1.541,$$

the outlier scores of four samples are

$$
\begin{aligned}
s(a,3) &= 2^{-\frac{E(h(a))}{c(3)}} = 0.4067,\\
s(b,3) &= 2^{-\frac{E(h(b))}{c(3)}} = 0.4551,\\
s(c,3) &= 2^{-\frac{E(h(c))}{c(3)}} = 0.4551,\\
s(d,3) &= 2^{-\frac{E(h(d))}{c(3)}} = 0.5699.
\end{aligned}
$$

We set a threshold as 0.55. Only $s(d,3) > 0.55$ and hence point d is treated as an outlier.

11.1.3 One-Class Support Vector Machine

One-class support vector machine (one-class SVM) is a special binary support vector classifier (SVC) to discriminate the binary sample classes, namely the normal and outlier samples, while no class labels are provided in advance. Schölkopf et al. (2001) propose to use a hyperplane $(\boldsymbol{W}, -\rho)$, where $\boldsymbol{W}$ and $-\rho$ denote the normal vector and intercept of the hyperplane, respectively, to separate all the data points in a training dataset from the origin and let the training data be on one side of the hyperplane. When a tested sample is located on the same side, it belongs to the same class of training data, otherwise anormaly. The hyperplane $(\boldsymbol{W}, -\rho)$ is determined by maximizing the distance from the origin to this hyperplane. In analogy with what described in Chap. 4 for binary support vector classifier with the soft margin, the hyperplane with slack variables ξ_i is determined from

$$
\begin{gathered}
\min_{\boldsymbol{W},\xi,\rho} \left(\frac{1}{2}\|\boldsymbol{W}\|^2 - \rho + \frac{1}{\nu n}\sum_{i=1}^{n} \xi_i \right),\\
\text{subject to}\, \phi(\boldsymbol{x}_i) W \geq \rho - \xi_i \text{ and } \xi_i \geq 0\ (i = 1, 2, \cdots, n).
\end{gathered}
\tag{11.6a}
$$

Comparing Eq. (11.6a) with Eq. (4.15a) in Chap. 4 indicates that the preset "cost" parameter C is replaced by $\frac{1}{\nu}$. The preset parameter ν is a lower bound of the number of support vectors in training examples. The Lagrange function of Eq. (11.6a) is

$$L_{\mathrm{P}} = \frac{1}{2}\|\boldsymbol{W}\|^2 - \rho + \frac{1}{\nu n}\sum_{i=1}^{n}\xi_i - \sum_{i=1}^{n}\lambda_i(\phi(\boldsymbol{x}_i)\boldsymbol{W} - \rho + \xi_i) - \sum_{i=1}^{n}\mu_i\xi_i, \quad (11.6b)$$

where $\lambda_i \geqslant 0$ and $\mu_i \geqslant 0$ ($i = 1, 2, \ldots, n$) are the Lagrange multipliers. The minimization of L_{P} requires $\frac{\partial L_{\mathrm{P}}}{\partial \boldsymbol{W}} = 0$, $\frac{\partial L_{\mathrm{P}}}{\partial \rho} = 0$, and $\frac{\partial L_{\mathrm{P}}}{\partial \xi_i} = 0$, which yield

$$\boldsymbol{W}^{\mathrm{T}} = \sum_{i=1}^{n}\lambda_i\phi(\boldsymbol{x}_i), \quad (11.7a)$$

$$\sum_{i=1}^{n}\lambda_i = 1, \quad (11.7b)$$

$$\mu_i = \frac{1}{\nu n} - \lambda_i. \quad (11.7c)$$

Substituting Eq. (11.7a)–(11.7c) into Eq. (11.6b) yields the dual objective function

$$L_{\mathrm{D}} = 1 - \frac{1}{2}\sum_{i=1}^{n}\sum_{j=1}^{n}\lambda_i\lambda_j\phi(\boldsymbol{x}_i)(\phi(\boldsymbol{x}_j))^{\mathrm{T}}. \quad (11.8a)$$

The values of λ_i ($i = 1, 2, \cdots, n$) are determined by maximizing the dual objective function, i.e.,

$$\lambda_i = \underset{\lambda}{\operatorname{argmax}}\, L_{\mathrm{D}},$$

$$\text{subject to} \sum_{i=1}^{n}\lambda_i = 1 \quad \text{and} \quad \frac{1}{\nu \times n} \geq \lambda_i \geq 0\,(i = 1, 2, \cdots, n). \quad (11.8b)$$

The KKT conditions must be satisfied in solving the maximization problem.

$$\begin{cases} \lambda_i \geqslant 0, \quad \mu_i \geqslant 0 \text{ (dual feasibility)}, \\ \phi(\boldsymbol{x}_i)W \geqslant \rho - \xi_i, \quad \xi_i \geqslant 0 \text{ (primal feasibility)}, \\ \lambda_i[\phi(\boldsymbol{x}_i)W - \rho + \xi_i] = 0, \quad \mu_i\xi_i = 0 \text{ (complementary slackness)}, \\ W^{\mathrm{T}} = \sum_{i=1}^{n}\lambda_i\phi(\boldsymbol{x}_i), \ \sum_{i=1}^{n}\lambda_i = 1 \text{ (stationary)}. \end{cases}$$

Let S denote the data subset of support vectors and $|S|$ be the number of support vectors, which meet $\lambda_i > 0,\ \xi_i = 0$ and $\phi(\boldsymbol{x}_i)\boldsymbol{W} - \rho + \xi_i = 0 (i \in S)$. Each of the support vectors can be used to determine the ρ value and the average ρ value is given by

$$\rho = \frac{1}{|S|}\sum_{i \in S} \phi(\boldsymbol{x}_i)\boldsymbol{W}. \tag{11.9}$$

The hyperplane $(\boldsymbol{W}, -\rho)$ is determined by Eq. (11.7a) and Eq. (11.9). The decision function is sign(z), which equals 1 for $z \geqslant 0$ and -1 otherwise.

$$\begin{aligned}\operatorname{sign}\left(\sum_{i=1}^{n} \phi(\boldsymbol{x}_i)\boldsymbol{W} - \rho\right) &= \operatorname{sign}\left(\sum_{i,j=1}^{n} \lambda_i \phi(\boldsymbol{x}_i)(\phi(\boldsymbol{x}_j))^{\mathrm{T}} - \rho\right) \\ &= \operatorname{sign}\left(\sum_{i,j=1}^{n} \lambda_i K(\boldsymbol{x}_i, \boldsymbol{x}_j) - \rho\right). \end{aligned} \tag{11.10}$$

Outliers are detected by the sign of Eq. (11.10), which also shows that outliers can be detected via the kernel function $K(\boldsymbol{x}_i, \boldsymbol{x}_j)$.

Example 11.3 Table 11.2 shows 3 high entropy alloys (HEAs) collected from references (Xiong et al., 2020). The average valence electron concentration (VEC) and enthalpy of mixing (H_{mix}) are the two features considered in the example, which definitions are described by Eqs (4.9) and (4.10), respectively, in Chap. 4.

Table 11.2 Three high entropy alloys utilized to train the one-class SVM

Samples	VEC (x_1)	H_{mix} (x_2, kJ·mol^{-1})
$Co_{0.25}Cr_{0.25}FeMn$	7.5	−1.0
NbTaTiVW	5	−3.7
DyGdLuTbTm	3	0

The values of Lagrange multipliers $\hat{\lambda}_i$ are determined by the maximum of the dual objective function L_{D}, as stated by Eq. (11.8a). In this example, Eq. (11.8a) takes the explicit form of

$$L_{\text{D}} = 1 - \frac{1}{2}\left(57.25\lambda_1^2 + 38.69\lambda_2^2 + 9\lambda_3^2 + 82.4\lambda_1\lambda_2 + 45\lambda_1\lambda_3 + 30\lambda_2\lambda_3\right).$$

The values of λ_i must satisfy the condition of $\sum_{i=1}^{n} \lambda_i = 1$, which gives $\lambda_3 = 1 - \lambda_1 - \lambda_2$. With this relationship, the dual objective function L_{D} is rewritten as

$$L_{\text{D}} = -13.5\lambda_1 - 6\lambda_2 - 10.625\lambda_1^2 - 8.845\lambda_2^2 - 12.7\lambda_1\lambda_2 - 3.5.$$

The maximization of L_D requires

$$\begin{cases} \dfrac{\partial L_D}{\partial \lambda_1} = -13.5 - 21.25\lambda_1 - 12.7\lambda_2 = 0, \\ \dfrac{\partial L_D}{\partial \lambda_2} = -6 - 12.7\lambda_1 - 17.69\lambda_2 = 0, \end{cases}$$

which yields

$$\begin{cases} \lambda_1 = -0.7577, \\ \lambda_2 = 0.2048. \end{cases}$$

However, this set of solutions do not satisfy the condition of $\lambda_i \geqslant 0$. Therefore, one Lagrange multiplier must be zero, viz., $\lambda_1 = 0$, $\lambda_2 = 0$, or $\lambda_3 = 0$. If $\lambda_1 = 0$, then we can obtain

$$\begin{cases} \lambda_3 = 1 - \lambda_2, \\ \frac{\partial L_D}{\partial \lambda_2} = -6 - 17.69\lambda_2 = 0. \end{cases}$$

The value of Lagrange multipliers $\lambda_2 = -0.3392$ does not satisfy the condition of $\lambda_i \geqslant 0$. Similarly, $\lambda_2 = 0$ and $\lambda_3 = 0$ lead to negative values of λ_i. Therefore, we should consider the following conditions

$$\begin{cases} \lambda_1 = 1, \\ \lambda_2 = 0, \quad \text{and the maximum } L_D \text{ value } (L_D)_{\max|_{\lambda_2,\lambda_3=0}} = -27.625, \\ \lambda_3 = 0, \end{cases}$$

$$\begin{cases} \lambda_1 = 0, \\ \lambda_2 = 1, \quad \text{and the maximum } L_D \text{ value } (L_D)_{\max|_{\lambda_1,\lambda_3=0}} = -18.345, \\ \lambda_3 = 0, \end{cases}$$

$$\begin{cases} \lambda_1 = 0, \\ \lambda_2 = 0, \quad \text{and the maximum } L_D \text{ value } (L_D)_{\max|_{\lambda_1,\lambda_2=0}} = -3.5. \\ \lambda_3 = 1, \end{cases}$$

Since $(L_D)_{\max|_{\lambda_1,\lambda_2=0}} > (L_D)_{\max|_{\lambda_1,\lambda_3=0}} > (L_D)_{\max|_{\lambda_2,\lambda_3=0}}$, the three values of Lagrange multipliers are determined to be $\lambda_1 = 0$, $\lambda_2 = 0$, and $\lambda_3 = 1$.

Since only $\lambda_3 = 1$ is non-zero, datum $\boldsymbol{X}_3$ is the support vector. The separating plane is determined by $\boldsymbol{W}^{\mathrm{T}} = \sum_{i=1}^{3} \lambda_i \boldsymbol{X}_i = 1 \times (3,\ 0) = (3,\ 0)$ and $\rho = \boldsymbol{X}_3 \boldsymbol{W} = 9$, thereby having the form of $x_1 - 3 = 0$, which is plotted in Fig. 11.6. Take CoCuFeMnNi as a testing sample, whose VEC and H_{mix} values are 9 and 1.8, respectively. The one-class SVM model tests sample CoCuFeMnNi as a normal point due to its VEC value being larger than 3.

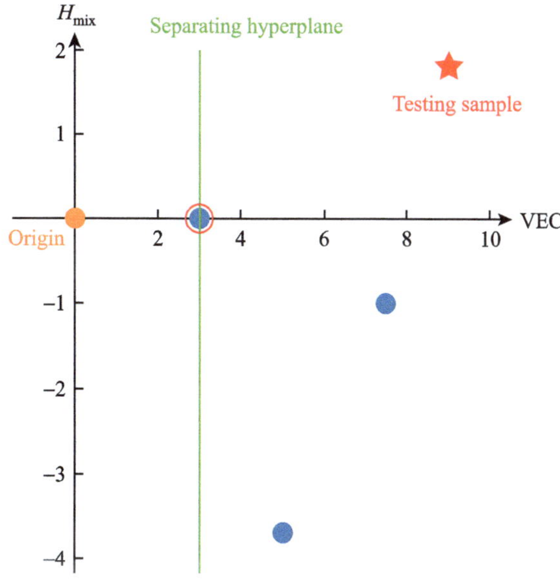

Fig. 11.6 The data in the VEC and H_{mix} feature space, where blue dots represent the training data, the solid line is the optimal separating hyperplane, on which the red circle marks the support vector and the red star highlights the testing sample

---

Example 11.4 Table 11.3 lists the training and testing samples with two features of H_{mix} and S_{mix}. In this example, the one-class SVMs are conducted with the RBF kernel $K(X_i, X_j) = exp\ (-\gamma||X_i, -X_j||^2)$ and a penalty hyperparameter (ν). The hyperparameter values (γ, ν) are (0.01, 0.1), (0.1, 0.1), (1, 0.1), and (1, 0.001) respectively. Figure 11.7 shows the normal half-space and the separating hyperplane in the high-dimensional space appearing as a red closed loop in the two kinds of feature space, where samples on the loop boundary are supporting vectors and inside the loop are normal points. Figure 11.7 indicates that the testing sample is classified as the normal point when $\nu = 0.1$, $\gamma = 0.01$ and $\nu = 0.1$, $\gamma = 0.1$, while the testing sample is classified as the outlier when $\nu = 0.1$, $\gamma = 1$ and $\nu = 0.01$, $\gamma = 1$. The hyperparameters play an important role in the one-class SVM. As described in Chap. 4, the values of hyperparameters are usually optimized by cross-validation. Caution must be used in handling outliers. It is strongly suggested to employ various methods to identify outliers.

Table 11.3 Training samples and the testing sample

Samples	Alloys	H_{mix} (kJ·mol^{-1})	S_{mix} (J·K^{-1}·mol^{-1})
Training samples	$Ag_{14}Cu_{78.5}P_{7.5}$	−4.01905	5.483515
	$Ag_{10}Cu_{85}Ce_5$	−3.49	4.3082
	$Ag_{10.5}Cu_{80.5}P_9$	−5.0946	5.221014
Testing sample	$Ag_{10}Mg_{85}Y_5$	−5	4.3082

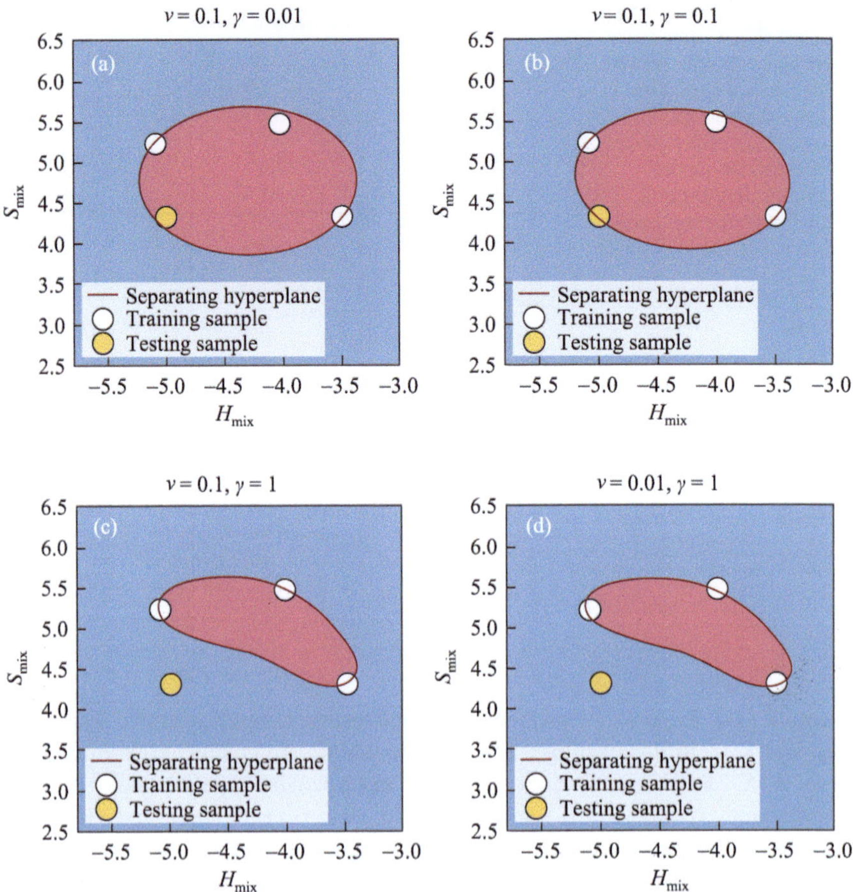

Fig. 11.7 The feature space with white and gold dots stand for the training and testing samples, respectively, and classified by the one-class SVM with the RBF kernel of **(a)** $\gamma = 0.01$ and $\nu = 0.1$, **(b)** $\gamma = 0.1$ and $\nu = 0.1$, **(c)** $\gamma = 1$ and $\nu = 0.1$, and **(d)** $\gamma = 1$ and $\nu = 0.01$. The data located inside and outside the red loop are classified as normal and abnormal points, respectively

---

11.1.4 Support Vector Data Description

Support vector data description (SVDD) (Tax & Duin, 1999, 2004) might be regarded as another type of one-class SVMs because SVDD adopts support vectors. SVDD takes a spherical boundary, instead of a hyperplane, in high-dimensional space to separate the data. The shortest spherical boundary with soft margin in high-dimensional space corresponds to the smallest hypersphere, which should contain most data in a training set within it. The hypersphere has a radius R and is centered at a. Minimizing the radius R is equivalent to minimizing R^2, and hence with a preset "cost" parameter C, minimizing is conducted on

$$\min_{R^2,\xi,a}\left(R^2 + C\sum_{i=1}^{n}\xi_i\right),$$
$$\text{subject to } (\phi(\boldsymbol{x}_i)-a)(\phi(\boldsymbol{x}_i)-a)^{\mathrm{T}} < R^2+\xi_i \text{ and } \xi_i \geq 0\,(i=1,2,\cdots,n). \tag{11.11a}$$

The Lagrange function of Eq. (11.11a) is

$$L_{\mathrm{P}} = R^2 + C\sum_{i=1}^{n}\xi_i - \sum_{i=1}^{n}\lambda_i(R^2+\xi_i-(\phi(\boldsymbol{x}_i)-a)(\phi(\boldsymbol{x}_i)-a)^{\mathrm{T}}) - \sum_{i=1}^{n}\mu_i\xi_i. \tag{11.11b}$$

The minimization of L_{P} requires $\dfrac{\partial L_{\mathrm{P}}}{\partial R^2}=0$, $\dfrac{\partial L_{\mathrm{P}}}{\partial a}=0$, and $\dfrac{\partial L_{\mathrm{P}}}{\partial \xi_i}=0$, which yield

$$\sum_{i=1}^{n}\lambda_i = 1, \tag{11.12a}$$

$$a = \sum_{i=1}^{n}\lambda_i\phi(\boldsymbol{x}_i), \tag{11.12b}$$

$$C = \mu_i + \lambda_i. \tag{11.12c}$$

Substituting Eq. (11.12a) to (11.12c) into Eq. (11.11b) yields the dual objective function

$$L_{\mathrm{D}} = \left[\sum_{i=1}^{n}\lambda_i\phi(\boldsymbol{x}_i)(\phi(\boldsymbol{x}_i))^{\mathrm{T}} - \sum_{i=1}^{n}\sum_{j=1}^{n}\lambda_i\lambda_j\phi(\boldsymbol{x}_i)(\phi(\boldsymbol{x}_j))^{\mathrm{T}}\right]. \tag{11.13a}$$

The values of λ_i $(i=1,2,\cdots,n)$ are determined by maximizing the dual objective function, i.e.,

$$\lambda_i = \underset{\lambda}{\operatorname{argmax}}\, L_{\mathrm{D}}$$
$$\text{subject to } \sum_{i=1}^{n}\lambda_i = 1 \quad \text{and} \quad C \geqslant \lambda_i \geqslant 0\,(i=1,2,\cdots,n). \tag{11.13b}$$

The KKT conditions must be satisfied in solving the maximization problem

$$\begin{cases} \lambda_i \geqslant 0, \mu_i \geqslant 0, \sum\limits_{i=1}^{n} \lambda_i = 1, \\ (\phi(\boldsymbol{x}_i) - a)(\phi(\boldsymbol{x}_i) - a)^{\mathrm{T}} < R^2 + \xi_i, \xi_i \geqslant 0, \\ \lambda_i(R^2 + \xi_i - (\phi(\boldsymbol{x}_i) - a)(\phi(\boldsymbol{x}_i) - a)^{\mathrm{T}}) = 0, \mu_i \xi_i = 0, \\ a = \sum\limits_{i=1}^{n} \lambda_i \phi(\boldsymbol{x}_i), \sum\limits_{i=1}^{n} \lambda_i = 1. \end{cases}$$

Let S denote the data subset of support vectors and $|S|$ be the number of support vectors, which meet $\lambda_i > 0, \xi_i = 0$ and $[R^2 + \xi_i - (\phi(\boldsymbol{x}_i) - a)(\phi(\boldsymbol{x}_i) - a)^{\mathrm{T}}] = 0\,(i \in S)$. Each of the support vectors can be used to determine the R^2 value and the average R^2 value is given by

$$R^2 = \frac{1}{|S|} \sum_{i \in S} (\phi(\boldsymbol{x}_i) - a)(\phi(\boldsymbol{x}_i) - a)^{\mathrm{T}}. \tag{11.14a}$$

The hyperspherical surface (R^2, a) is determined with Eq. (11.12b) and Eq. (11.14a). Substituting Eq. (11.12b) into Eq. (11.14a) and using $\boldsymbol{x}_{\mathrm{v}}$ to denote support vectors, Eq. (11.14a) is re-written in terms of kernel functions as

$$R^2 = K(\boldsymbol{x}_{\mathrm{v}}, \boldsymbol{x}_{\mathrm{v}}) - 2\sum_{i=1}^{n} \lambda_i K(\boldsymbol{x}_{\mathrm{v}}, \boldsymbol{x}_i) + \sum_{i=1}^{n}\sum_{j=1}^{n} \lambda_i \lambda_j K(\boldsymbol{x}_i, \boldsymbol{x}_j). \tag{11.14b}$$

For a tested sample x^*, its distance square $d_{x^*}^2$ is calculated by

$$d_{x^*}^2 = K(x^*, x^*) - 2\sum_{i=1}^{n} \lambda_i K(x^*, \boldsymbol{x}_i) + \sum_{i=1}^{n}\sum_{j=1}^{n} \lambda_i \lambda_j K(\boldsymbol{x}_i, \boldsymbol{x}_j). \tag{11.15}$$

If $d_{x^*}^2 \leqslant R^2$, the tested sample is normal, otherwise abnormal.

Example 11.5 The same data given in Example 11.3 are used here. The values of the Lagrange multiplier $\hat{\lambda}_i$ are determined by the maximum of the dual objective function L_{D}, as stated by Eq. (11.13b). In this case, Eq. (11.13a) can be written as

$$\begin{aligned} L_{\mathrm{D}} &= 57.25\lambda_1 + 38.69\lambda_2 + 9\lambda_3 \\ &\quad - \left(57.25\lambda_1^2 + 38.69\lambda_2^2 + 9\lambda_3^2 + 82.4\lambda_1\lambda_2 + 45\lambda_1\lambda_3 + 30\lambda_2\lambda_3\right). \end{aligned}$$

The values of λ_i must satisfy the condition of $\sum\limits_{i=1}^{n} \lambda_i = 1$, which gives $\lambda_3 = 1 - \lambda_1 - \lambda_2$. With this relationship, the dual objective function L_{D} is rewritten as

$$L_{\mathrm{D}} = 21.25\lambda_1 + 17.69\lambda_2 - 21.25\lambda_1^2 - 17.69\lambda_2^2 - 25.4\lambda_1\lambda_2.$$

The maximization of L_{D} requires

$$\begin{cases} \dfrac{\partial L_{\mathrm{D}}}{\partial \lambda_1} = 21.25 - 42.5\lambda_1 - 25.4\lambda_2 = 0, \\ \dfrac{\partial L_{\mathrm{D}}}{\partial \lambda_2} = 17.69 - 25.4\lambda_1 - 35.38\lambda_2 = 0, \end{cases}$$

which yields

$$\begin{cases} \lambda_1 = 0.35236, \\ \lambda_2 = 0.24703, \\ \lambda_3 = 1 - \lambda_1 - \lambda_2 = 0.40061. \end{cases}$$

The three Lagrange multipliers are all non-zero, meaning that all three data points are the support vectors, as plotted in Fig. 11.8. In this two-dimensional example, the hyperspherical surface is a circle whose center point $\boldsymbol{a}$ and radius squared R^2 are determined by

$$\begin{aligned} \boldsymbol{a} &= \sum_{i=1}^{3} \lambda_i \boldsymbol{X}_i = 0.35236 \times (7.5, -1) + 0.24703 \times (5, -3.7) + 0.40061 \times (3, 0) \\ &= (5.080,\ -1.266), \end{aligned}$$

$$R^2 = (\boldsymbol{x}_{\mathrm{v}}, \boldsymbol{x}_{\mathrm{v}}) - 2\sum_{i=1}^{n} \lambda_i (\boldsymbol{x}_{\mathrm{v}}, \boldsymbol{x}_i) + \sum_{i=1}^{n}\sum_{j=1}^{n} \lambda_i \lambda_j (\boldsymbol{x}_i, \boldsymbol{x}_j) = 5.9287.$$

The circle is described by $(x_1 - 5.080)^2 + (x_2 + 1.266)^2 = 5.9287$.
The testing sample CoCuFeMnNi is located at (9, 1.8), its distance to the circle center is $d_{x^*}^2 = 24.77 > R^2 = 5.9287$, and thereby it is an anomaly point. This result is opposite to that in Example 11.3.

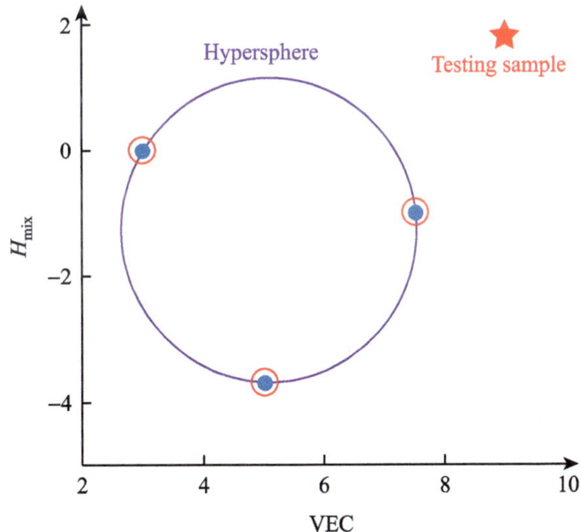

Fig. 11.8 The data in the VEC and H_{mix} feature space, where blue dots represent the training data, the solid circle is the optimal hypersphere, small red circles mark the support vectors, and the red star is the testing sample

---

Example 11.6 The same data given in Example 11.4 are used here to train SVDD models with the RBF kernels of $\gamma = 0.1$ and $\gamma = 1$. The penalty parameter C is set as 1. The trained hypersphere and the distances from data points to the hypersphere center in the feature space are shown by contour maps in Figs. 11.9 and 11.10.

Figure 11.9 shows that when $\gamma = 0.1$, the R^2 value given by the SVDD model is 0.1444. The red star in Fig. 11.9 stands for the testing sample, whose distance $d_{x^*}^2$ is 0.1403, meaning that the testing sample is a normal point. If $\gamma = 1$, as shown in Fig. 11.10, the R^2 value of the hyperspherical surface is 0.5570. However, the distance of the testing sample $d_{x^*}^2$ increases rapidly to 1.0090, indicating that the testing sample will be treated as an outlier.

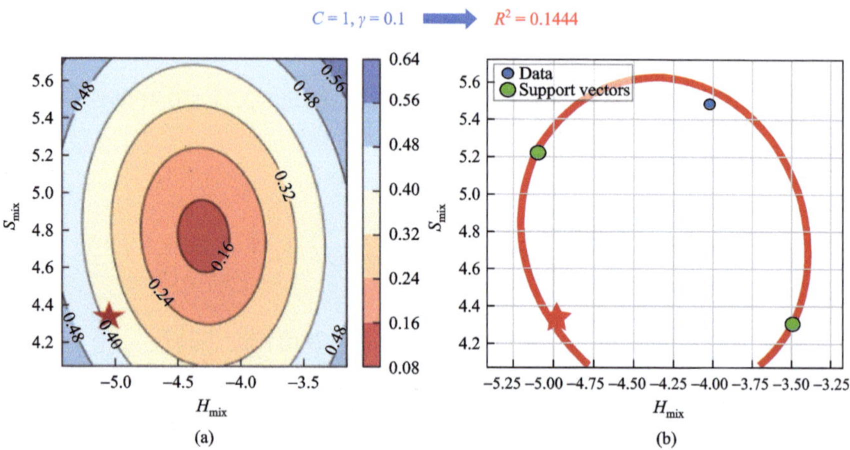

Fig. 11.9 The SVDD model with the RBF kernel ($\gamma = 0.1$) and $C = 1$: **(a)** the distance of data points in the feature space, and **(b)** the hypersphere, where dots denote the training dada, green dots are supporting vectors, and the red star stands for the testing sample

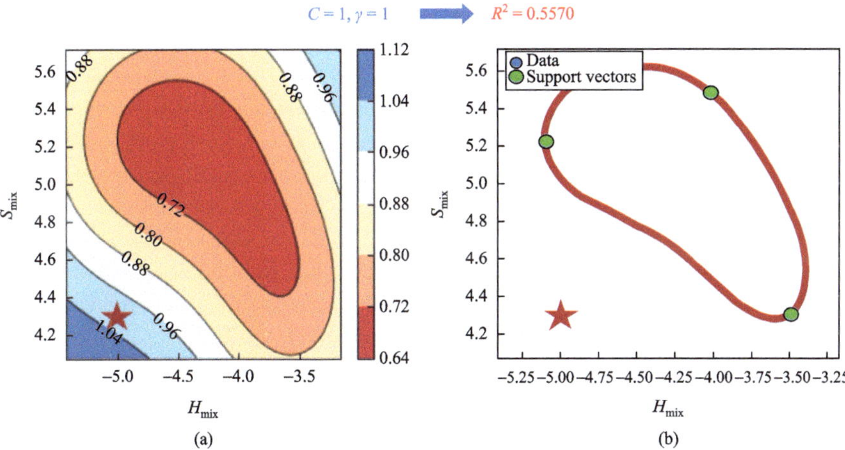

Fig. 11.10 The SVDD model with the RBF kernel ($\gamma = 1$) and $C = 1$: **(a)** the distance of data points in the feature space, and **(b)** the hypersphere, where dots denote the training dada, green dots are supporting vectors, and the red star stands for the testing sample

---

11.2 Feature Selection

In common mathematic analysis, it takes for granted that the function varies with independent variables, meaning that each of the variables can be varied independently and the function will vary accordingly. Besides, a function can be expressed in terms of a basis set of variables, or just a variable, like in polynomial expansion of $f(x)$, where the basis set is $\{1, x, x^2, \cdots, x^m\}$, and in the Fourier transform of a time t dependent function of $f(t)$ into the frequency spectrum, where the basis set is a set of basic, double, triple, $\cdots$, frequencies. In ML, input variables are called features and there are so many features in materials informatics, including chemical compositions, electronic and atomic information, crystalline and phase structures, defects, materials processing conditions, service and application environments and conditions, etc. Usually, as many as possible features should be proposed based on domain knowledge and studied problems in order to avoid any potential missing of important features. The initially proposed features are often correlated, meaning that some of them are superfluous and should be removed. Furthermore, based on the relevancy or irrelevancy to responses of interest, features can be divided into relevant and irrelevant features, and equivalently, features can be ranked by their contributions to the responses. As described in Chap. 2, a good ML model should balance the prediction accuracy, the generalization ability, and the model complexity, meaning that the model performance should be as better as possible and the feature number should be as few as possible. Thus, a penalty term can function as a feature selector. For example, Eq. (2.22) in Chap. 2 shows the regression criterion for LSLR with a general penalty constraint. The L_q regularization or L_q norm with $0\leqslant q\leqslant 1$ has the capacity to rank and select features. In this sense, LASSO might be regarded as the most representative, popular and powerful approach for feature ranking and feature selection. If individuals in evolutionary computation are regarded as features, evolutionary computation will be able to create new features from initially proposed feature blocks and then select features, as described in Chap. 8. In ML, feature selection and feature ranking are essential and crucial, and the first step in feature selection is to remove redundant features. There are various feature selection and feature ranking methods. Very often feature selection and feature ranking can be implemented by the same methods and thus feature selection and feature ranking are not separately discussed for conciseness. In practice, several feature selection methods are often applied sequentially to reduce the feature number in an ML case. Generally speaking, feature selection is to select a feature subset from an originally proposed feature set, and filter, wrapper and embedded feature selections are the widely used approaches for feature selection.

11.2.1 Filter Approach

11.2.1.1 Pearson Correlation Coefficient (*R*)

Filter feature selection is an independent assessment of the general data characteristics and conducted without using any ML models. The filter feature selection sieves the originally proposed feature set and produces the most promising subset before ML commences. The Pearson correlation coefficient (*R*), as described in Chap. 2, is a widely used correlation-based selector to delete redundant features, which is defined as the quotient of the covariance $\text{cov}(x_1, x_2)$ over the variance product of σ_{x_1} and σ_{x_2}, i.e.,

$$R(x_1,\ x_2) = \frac{\text{cov}(x_1,\ x_2)}{\sigma_{x_1}\sigma_{x_2}}, \tag{11.16a}$$

where σ_{x_1} and σ_{x_2} denote the variances of features x_1 and x_2, respectively, and

$$\text{cov}(x_1, x_2) = E[(x_1 - \mu_{x_1})(x_2 - \mu_{x_2})], \tag{11.16b}$$

$$\sigma_{x_j} = E[(x_j - \mu_{x_j})^2]\ \text{and}$$

$$\mu_{x_j} = E[x_j] = \int_{-\infty}^{\infty} f(x_j)x_j \mathrm{d}x_j\ (j = 1, 2) \tag{11.16c}$$

with $\mu_{x_j} = E[x_j]$ being the expectation of feature *j*. For discrete *n* data, the Pearson correlation coefficient is calculated by

$$R = \frac{\sum_{i=1}^{n} (x_{i1} - \mu_{x_1})(x_{i2} - \mu_{x_2})}{\sqrt{\sum_{i=1}^{n} (x_{i1} - \mu_{x_1})^2 \sum_{i=1}^{n} (x_{i2} - \mu_{x_2})^2}} \tag{11.17a}$$

with

$$\mu_{x_j} = \frac{1}{n} \sum_{i=1}^{n} x_{ij} \quad (j = 1,\ 2). \tag{11.17b}$$

If $R = \pm 1$, features x_1 and x_2 are linearly correlated, positively if $R = +1$ or negatively if $R = -1$, meaning that $x_2 = Ax_1 + B$ with A and B being constants. In practice, the Pearson correlation coefficient is calculated for every two features among an initially proposed feature set, and the results are illustrated in a Pearson correlation coefficient map, as shown in Fig. 11.11. A positive threshold is set in the feature selection and when the absolute R value between two features is larger than the threshold, the two features are strongly correlated and only one of the correlated pair will be selected. Besides features, responses may also be taken in the calculation

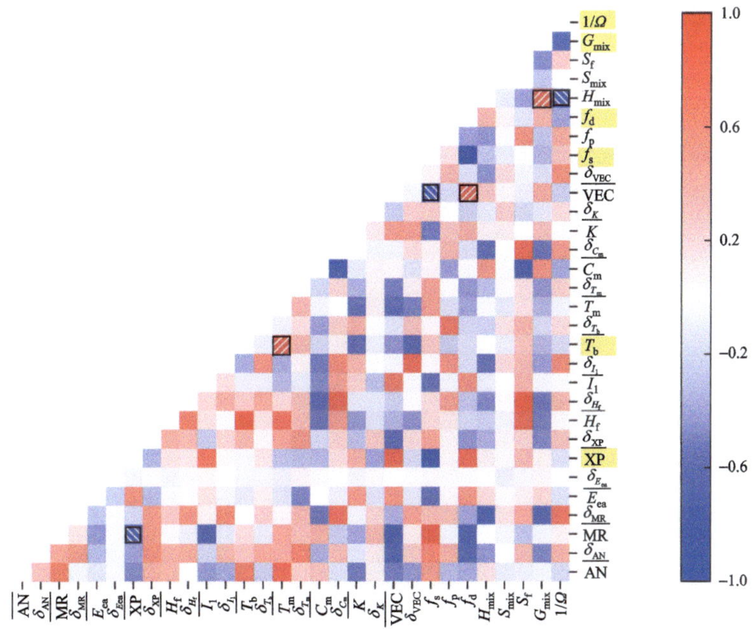

Fig. 11.11 The Pearson correlation coefficient matrix of 30 features in the dataset of 557 alloys. Red represents the positive correlation, and blue represents the negative correlation, respectively, where six strongly-correlated feature pairs (|R| > 0.9) are marked and the six removed features are highlighted in yellow

of the Pearson correlation coefficient (Severson et al., 2019) and hence the Pearson correlation coefficient map is able to measure linear correlation among responses and/or features.

Example 11.7 Xiong et al. (2021) conducted the feature selection by the Pearson correlation coefficient. The initial feature set of complex concentrated alloys (CCAs) includes 22 elemental features (11 elemental property each contributes two features), 5 thermodynamic parameters, and 3 valence electron distributions, and thus in total 30 features, as listed in Table 11.4.

Table 11.4 Feature blocks consisting of elemental parameters, thermodynamic parameters, and valence electron distributions

	Descriptions	Abbreviations	Definitions
Elemental parameters	Atomic number	AN	$\overline{x} = \sum_{i=1}^{N} a_i x_i$
	Metallic radius	MR	
	Melting point	T_m	$\delta_x = \sqrt{\sum_{i=1}^{N} a_i \left(1 - \frac{x_i}{\overline{x}}\right)^2}$
	Boiling point	T_b	
	Pauling electronegativity	XP	a_i: the atomic fraction of element *i*.
	Electron affinity	E_{ea}	N: the number of elements in an alloy
	First ionization potential	I_1	
	Molar heat capacity	C_m	
	Thermal conductivity	K	
	Valence electron concentration	VEC	
	Heat of fusion	H_f	
Thermodynamic parameters	Enthalpy of mixing	H_{mix}	$H_{mix} = 4 \sum_{i=1}^{N} \sum_{j=1}^{N} \Delta H_{ij} a_i a_j$
	Entropy of mixing	S_{mix}	$S_{mix} = -R \sum_{i=1}^{N} a_i \ln a_i$
	Entropy of fusion	S_f	$S_f = \overline{H_f}/\overline{T_m}$
	Gibbs free energy of mixing	G_{mix}	$G_{mix} = H_{mix} - S_{mix} \cdot \overline{T_m}$
	The reciprocal of Ω	$1/\Omega$	$1/\Omega = \lvert H_{mix} \rvert / \overline{T_m} S_{mix}$
VEC distributions	Fraction of the electrons in the s valence orbitals	f_s	$f_s = \overline{\text{sVEC}}/\overline{\text{VEC}}$
	Fraction of the electrons in the p valence orbitals	f_p	$f_p = \overline{\text{pVEC}}/\overline{\text{VEC}}$
	Fraction of the electrons in the d valence orbitals	f_d	$f_d = \overline{\text{dVEC}}/\overline{\text{VEC}}$

The five thermodynamic parameters are the enthalpy of mixing (H_{mix}) (Zhang et al., 2008), the entropy of mixing (S_{mix}) (Zhang et al., 2008), the entropy of fusion

($S_f = \overline{H_f}/\overline{T_m}$) with $\overline{H_f}$ and $\overline{T_m}$ being the average heat of fusion and the average melting temperature, respectively (Zhang et al., 2014), Gibbs free energy of mixing (G_{mix}) (Yeh et al., 2007), and the $1/\Omega = |H_{mix}|/\overline{T_m}S_{mix}$ parameter (Takeuchi et al., 2014; Zhang et al., 2014). The valence electron distributions comprise the fraction (f_k) of the average electrons in the s, p, d valence orbitals, i.e., $f_k = \overline{k\text{VEC}}/\overline{\text{VEC}}(k =$ s, p, d orbitals). The total 30 features are initially proposed and normalized to (0, 1) according to

$$x' = \frac{x - \min x}{\max x - \min x},$$

where x' and x are the normalized and original values, respectively, and $\min x$ and $\max x$ are the original minimum and maximum values of feature x, respectively.

The dataset of 557 alloys is available from Xiong et al. (2021). Based on the threshold of $|R| > 0.9$, there are six strongly-correlated feature pairs, namely, $\overline{\text{MR}}-\overline{\text{XP}}$, $\overline{T_b} - \overline{T_m}$, $\overline{\text{VEC}} - f_s$, $\overline{\text{VEC}} - f_d$, $H_{mix} - G_{mix}$ and $H_{mix} - 1/\Omega$, from which the four features of $\overline{\text{MR}}$, $\overline{T_m}$, $\overline{\text{VEC}}$, and H_{mix} are selected (Hu et al., 2017; Tsai & Yeh, 2014; Ye et al., 2016) and the six features of $\overline{\text{XP}}$, $\overline{T_b}$, f_s, f_d, G_{mix}, and $1/\Omega$ are removed.

11.2.1.2 Maximal Information Coefficient

The maximal information coefficient (MIC) is proposed by Reshef et al. (2011) based on the mutual information $I(x_1, x_2)$, which measures the distance between two probability distributions. For simplicity, consider a dataset S of two features x_1 and x_2 and n unrepeated data (x_{i1}, x_{i2}) $(i = 1, 2, \cdots, n)$. The (x_1, x_2) feature space is partitioned along the x_1 axis into K bins, which is done by $K - 1$ lines perpendicular to the x_1 axis as boundaries. The bin, the data inside the bin, and the data number inside the bin are denoted by a_k, $x_{a_k 1}$, and $|a_k|(k = 1, 2, \cdots, K)$, respectively. Similarly, the (x_1, x_2) feature space is partitioned along the x_2 axis into L bins, which is done by $(L - 1)$ lines perpendicular to the x_2 axis as boundaries. The bin, the data inside the bin, and the data number inside the bin are denoted by b_l, $x_{b_l 2}$, and $|b_l|(l = 1, 2, \cdots, L)$, respectively. Obviously, we have $\sum_{k=1}^{K} |a_k| = n$ and $\sum_{l=1}^{L} |b_l| = n$. The K x_1-bins and L x_2-bins form a grid of $K \times L$ cells in the (x_1, x_2) feature space, which is often denoted by the K-by-L grid. The probability of data in a bin is calculated by the fraction of data located inside the bin, viz.,

$$P(x_{a_k 1}) = \frac{|a_k|}{n} \quad (k = 1, 2, \cdots, K), \tag{11.18a}$$

$$P(x_{b_l 2}) = \frac{|b_l|}{n} \quad (l = 1, 2, \cdots, L). \tag{11.18b}$$

The probability of data in each cell is the joint probability $P(x_{a_k1} \cap x_{b_l2})$ (or expressed by $P(x_{a_k1}, x_{b_l2})$) and calculated by the fraction of data located inside the cell,

$$P(x_{a_k1} \cap x_{b_l2}) = \frac{|a_k \cap b_l|}{n} \quad (k = 1, 2, \cdots, K; \;\; l = 1, 2, \cdots, L). \tag{11.18c}$$

The information entropy is introduced in Chap. 5 to measure the mixture degree of data mixed with different classes. The information entropy is used here to measure the mixture degree of data mixed with different bins and cells, or to characterize the uncertainty of data located in various bins and cells. Take feature x_1 as an example. There are K x_1-bins and the information entropy $H(x_1)$ for such x_1 distribution is defined by

$$H(x_1) = -\sum_{k=1}^{K} P(x_{a_k1}) \ln(P(x_{a_k1})), \tag{11.19a}$$

with $H(0) \triangleq 0$. The term $P(x_{a_k1}) \ln(P(x_{a_k1}))$ represents the information entropy of the data in bin a_k. Similarly, the information entropy $H(x_2)$ for the L x_2-bins is

$$H(x_2) = -\sum_{l=1}^{L} P(x_{b_l2}) \ln(P(x_{b_l2})), \tag{11.19b}$$

and the information entropy $H(x_1 \cap x_2)$ for the grid of $K \times L$ cells is

$$H(x_1 \cap x_2) = -\sum_{k=1}^{K}\sum_{l=1}^{L} P(x_{a_k1} \cap x_{b_l2}) \ln(P(x_{a_k1} \cap x_{b_l2})). \tag{11.19c}$$

Recall that the joint probability $P(x_{a_k1} \cap x_{b_l2})$ can be decomposed as $P(x_{a_k1} \mid x_{b_l2})$ $P(x_{b_l2})$ with the conditional probability of $P(x_{a_k1} \mid x_{b_l2})$ or $P(x_{b_l2} \mid x_{a_k1})P(x_{a_k1})$ with the conditional probability of $P(x_{b_l2} \mid x_{a_k1})$. Under a given x_{b_l2}, the conditional entropy $H\big(x_1 \mid x_2 = x_{b_l2}\big)$ is defined by

$$H(x_1 \mid x_2 = x_{b_l2}) = -\sum_{k=1}^{K} P(x_{a_k1} \mid x_{b_l2}) \ln(P(x_{a_k1} \mid x_{b_l2})). \tag{11.19d}$$

Then, the conditional entropy $H(x_1 \mid x_2)$ is calculated from

$$\begin{aligned} H(x_1|x_2) &= -\sum_{l=1}^{L} P(x_{b_l2}) \sum_{k=1}^{K} P(x_{a_k1}|x_{b_l2}) \ln(P(x_{a_k1}|x_{b_l2})) \\ &= -\sum_{k=1}^{K}\sum_{l=1}^{L} P(x_{a_k1} \cap x_{b_l2}) \ln(P(x_{a_k1}|x_{b_l2})). \end{aligned} \tag{11.19e}$$

In the same way, we have

$$H(x_2|x_1) = -\sum_{k=1}^{K}\sum_{l=1}^{L} P(x_{a_k1} \cap x_{b_l2}) \ln(P(x_{b_l2}|x_{a_k1})). \quad (11.19f)$$

The mutual information $I(x_1, x_2)$ is defined by

$$\begin{aligned} I(x_1, x_2) &= H(x_1) - H(x_1|x_2) = H(x_2) - H(x_2|x_1) \\ &= \sum_{k=1}^{K}\sum_{l=1}^{L} P(x_{a_k1} \cap x_{b_l2}) \ln \frac{P(x_{a_k1} \cap x_{b_l2})}{P(x_{a_k1})P(x_{b_l2})}. \end{aligned} \quad (11.19g)$$

Example 11.8 The left column in Table 11.5 lists 10 data randomly sampled from a zero-mean bivariate Gaussian distribution with the correlation of $\rho = 0.216$ and the right column lists 10 data randomly sampled from a zero-mean bivariate Gaussian distribution with the correlation of $\rho = 0.871$. Calculate the mutual information if the x_1 and x_2 axes are the partition lines.

Table 11.5 Samples from bivariate Gaussian distributions

Dataset 1 ($\rho = 0.216$)		Dataset 2 ($\rho = 0.871$)	
x_1	x_2	x_1	x_2
0.786956	0.03476	0.726466	1.203875
0.206262	1.294678	1.121702	0.727969
−0.03057	1.820368	−1.09712	−0.90634
−0.69502	−1.32109	−0.65383	−0.17763
0.578968	−0.51001	1.366621	0.654354
−1.49513	−0.6379	−1.18755	−1.01806
1.345432	−0.53991	0.055907	0.434227
−0.47557	−0.52533	0.788882	0.593351
−0.36434	−0.33948	0.084153	0.616892
−0.14122	−0.84425	−0.16552	0.44046

The (x_1, x_2) feature space is partitioned into a 2-by-2 grid, as shown in Fig. 11.12. The data numbers inside the cells of dataset 1 are $|x_{a_11} \cap x_{b_12}| = 1$, $|x_{a_11} \cap x_{b_22}| = 5$, $|x_{a_21} \cap x_{b_12}| = 2$, and $|x_{a_21} \cap x_{b_22}| = 2$; and the data numbers inside the cells of dataset 2 are $|x_{a_11} \cap x_{b_12}| = 1$, $|x_{a_11} \cap x_{b_22}| = 3$, $|x_{a_21} \cap x_{b_12}| = 6$, and $|x_{a_21} \cap x_{b_22}| = 0$. The mutual information $I(x_1, x_2)$ between two cells in each dataset is calculated for such 2-by-2 grid.

(1) For data randomly sampled from a zero-mean bivariate Gaussian distribution with the correlation of $\rho = 0.216$, we have $P(x_{a_11}) = \frac{|a_1|}{n} = \frac{6}{10}$, $P(x_{a_21}) = \frac{|a_2|}{n} = \frac{4}{10}$, $P(x_{b_12}) = \frac{|b_1|}{n} = \frac{3}{10}$, and $P(x_{b_22}) = \frac{|b_2|}{n} = \frac{7}{10}$ for the bins,

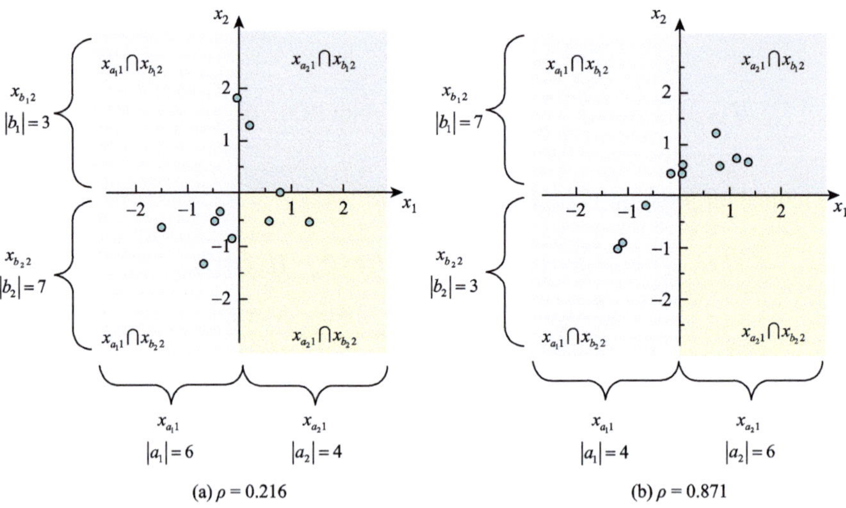

Fig. 11.12 The 2-by-2 grid partitioned by the coordinate axes

and $P(x_{a_1 1} \cap x_{b_1 2}) = \dfrac{|x_{a_1 1} \cap x_{b_1 2}|}{n} = \dfrac{1}{10}$, $P(x_{a_1 1} \cap x_{b_2 2}) = \dfrac{|x_{a_1 1} \cap x_{b_2 2}|}{n} = \dfrac{5}{10}$, $P(x_{a_2 1} \cap x_{b_1 2}) = \dfrac{|x_{a_2 1} \cap x_{b_1 2}|}{n} = \dfrac{2}{10}$, and $P(x_{a_2 1} \cap x_{b_2 2}) = \dfrac{|x_{a_2 1} \cap x_{b_2 2}|}{n} = \dfrac{2}{10}$ for the cells. Then we have

$$
\begin{aligned}
I(x_1, x_2) &= \sum_{k=1}^{K} \sum_{l=1}^{L} p(x_{a_k 1} \cap x_{b_l 2}) \ln \frac{p(x_{a_k 1} \cap x_{b_l 2})}{p(x_{a_k 1}) p(x_{b_l 2})} \\
&= p(x_{a_1 1} \cap x_{b_1 2}) \ln \frac{p(x_{a_1 1} \cap x_{b_1 2})}{p(x_{a_1 1}) p(x_{b_1 2})} \\
&\quad + p(x_{a_1 1} \cap x_{b_2 2}) \ln \frac{p(x_{a_1 1} \cap x_{b_2 2})}{p(x_{a_1 1}) p(x_{b_2 2})} \\
&\quad + p(x_{a_2 1} \cap x_{b_1 2}) \ln \frac{p(x_{a_2 1} \cap x_{b_1 2})}{p(x_{a_2 1}) p(x_{b_1 2})} + p(x_{a_2 1} \cap x_{b_2 2}) \ln \frac{p(x_{a_2 1} \cap x_{b_2 2})}{p(x_{a_2 1}) p(x_{b_2 2})} \\
&= \frac{1}{10} \ln\left(\frac{1/10}{(6/10) \times (3/10)}\right) + \frac{5}{10} \ln\left(\frac{5/10}{(6/10) \times (7/10)}\right) \\
&\quad + \frac{2}{10} \ln\left(\frac{2/10}{(4/10) \times (3/10)}\right) + \frac{2}{10} \ln\left(\frac{2/10}{(4/10) \times (7/10)}\right) = 0.063.
\end{aligned}
$$

(2) For data randomly sampled from a zero-mean bivariate Gaussian distribution with the correlation of $\rho = 0.871$, we have $P(x_{a_1 1}) = \dfrac{|a_1|}{n} = \dfrac{4}{10}$, $P(x_{a_2 1}) = \dfrac{|a_2|}{n} = \dfrac{6}{10}$, $P(x_{b_1 2}) = \dfrac{|b_1|}{n} = \dfrac{7}{10}$, and $P(x_{b_2 2}) = \dfrac{|b_2|}{n} = \dfrac{3}{10}$ for the bins, and $P(x_{a_1 1} \cap x_{b_1 2}) = \dfrac{|x_{a_1 1} \cap x_{b_1 2}|}{n} = \dfrac{1}{10}$, $P(x_{a_1 1} \cap x_{b_2 2}) = \dfrac{|x_{a_1 1} \cap x_{b_2 2}|}{n} = \dfrac{3}{10}$,

$P(x_{a_2 1} \cap x_{b_1 2}) = \frac{|x_{a_2 1} \cap x_{b_1 2}|}{n} = \frac{6}{10}$, and $P(x_{a_2 1} \cap x_{b_2 2}) = \frac{|x_{a_2 1} \cap x_{b_2 2}|}{n} = \frac{0}{10}$ for the cells. Then we have

$$I(x_1, x_2) = \sum_{k=1}^{K} \sum_{l=1}^{L} p(x_{a_k 1} \cap x_{b_l 2}) \ln \frac{p(x_{a_k 1} \cap x_{b_l 2})}{p(x_{a_k 1}) p(x_{b_l 2})}$$

$$= p(x_{a_1 1} \cap x_{b_1 2}) \ln \frac{p(x_{a_1 1} \cap x_{b_1 2})}{p(x_{a_1 1}) p(x_{b_1 2})} + p(x_{a_1 1} \cap x_{b_2 2}) \ln \frac{p(x_{a_1 1} \cap x_{b_2 2})}{p(x_{a_1 1}) p(x_{b_2 2})}$$

$$+ p(x_{a_2 1} \cap x_{b_1 2}) \ln \frac{p(x_{a_2 1} \cap x_{b_1 2})}{p(x_{a_2 1}) p(x_{b_1 2})} + p(x_{a_2 1} \cap x_{b_2 2}) \ln \frac{p(x_{a_2 1} \cap x_{b_2 2})}{p(x_{a_2 1}) p(x_{b_2 2})}$$

$$= \frac{1}{10} \ln\left(\frac{1/10}{(4/10) \times (7/10)}\right) + \frac{3}{10} \ln\left(\frac{3/10}{(4/10) \times (3/10)}\right)$$

$$+ \frac{6}{10} \ln\left(\frac{6/10}{(6/10) \times (7/10)}\right) + \frac{0}{10} \ln\left(\frac{0/10}{(6/10) \times (3/10)}\right) = 0.386.$$

In the example, natural logarithm $\ln(\cdot)$ is used, while $\log_2(\cdot)$ is also often adopted. As expected, the higher the correlation is, the higher the mutual information $I(x_1, x_2)$ will be.

For a given dataset S and bin numbers of K and L, there are many different partitions, viz., there are many different configurations for a K-by-L grid. For example, Fig. 11.13 shows two configurations of a 2-by-2 grid of 18 dimensionless data, where blue lines and green lines are the partition lines in the two configurations, respectively.

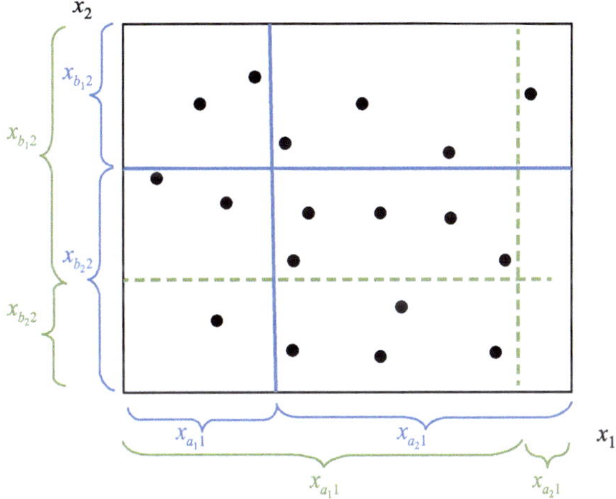

Fig. 11.13 Two different configurations of a 2-by-2 grid

(1) For the blue partition lines, the mutual information $I(x_1, x_2)$ is calculated by $P(x_{a_1}1) = \frac{|a_1|}{n} = \frac{5}{18}$, $P(x_{a_2}1) = \frac{|a_2|}{n} = \frac{13}{18}$, $P(x_{b_1}2) = \frac{|b_1|}{n} = \frac{6}{18}$, $P(x_{b_2}2) = \frac{|b_2|}{n} = \frac{12}{18}$, $P(x_{a_1}1 \cap x_{b_1}2) = \frac{|x_{a_1}1 \cap x_{b_1}2|}{n} = \frac{2}{18}$, $P(x_{a_1}1 \cap x_{b_2}2) = \frac{|x_{a_1}1 \cap x_{b_2}2|}{n} = \frac{3}{18}$, $P(x_{a_2}1 \cap x_{b_1}2) = \frac{|x_{a_2}1 \cap x_{b_1}2|}{n} = \frac{4}{18}$ and $P(x_{a_2}1 \cap x_{b_2}2) = \frac{|x_{a_2}1 \cap x_{b_2}2|}{n} = \frac{9}{18}$.

Subsequently, we have

$$\begin{aligned}
&I(x_1, x_2)\\
&= p(x_{a_1}1 \cap x_{b_1}2) \ln \frac{p(x_{a_1}1 \cap x_{b_1}2)}{p(x_{a_1}1)p(x_{b_1}2)} + p(x_{a_1}1 \cap x_{b_2}2) \ln \frac{p(x_{a_1}1 \cap x_{b_2}2)}{p(x_{a_1}1)p(x_{b_2}2)}\\
&\quad + p(x_{a_2}1 \cap x_{b_1}2) \ln \frac{p(x_{a_2}1 \cap x_{b_1}2)}{p(x_{a_2}1)p(x_{b_1}2)} + p(x_{a_2}1 \cap x_{b_2}2) \ln \frac{p(x_{a_2}1 \cap x_{b_2}2)}{p(x_{a_2}1)p(x_{b_2}2)}\\
&= 0.0038.
\end{aligned}$$

(2) For the green partition lines, the mutual information $I(x_1, x_2)$ is calculated by $P(x_{a_1}1) = \frac{|a_1|}{n} = \frac{17}{18}$, $P(x_{a_2}1) = \frac{|a_2|}{n} = \frac{1}{18}$, $P(x_{b_1}2) = \frac{|b_1|}{n} = \frac{13}{18}$ $P(x_{b_2}2) = \frac{|b_2|}{n} = \frac{5}{18}$, $P(x_{a_1}1 \cap x_{b_1}2) = \frac{|x_{a_1}1 \cap x_{b_1}2|}{n} = \frac{12}{18}$, $P(x_{a_1}1 \cap x_{b_2}2) = \frac{|x_{a_1}1 \cap x_{b_2}2|}{n} = \frac{5}{18}$, $P(x_{a_2}1 \cap x_{b_1}2) = \frac{|x_{a_2}1 \cap x_{b_1}2|}{n} = \frac{1}{18}$ and $P(x_{a_2}1 \cap x_{b_2}2) = \frac{|x_{a_2}1 \cap x_{b_2}2|}{n} = \frac{0}{18}$. Consequently, we have

$$\begin{aligned}
I(x_1, x_2) &= p(x_{a_1}1 \cap x_{b_1}2) \ln \frac{p(x_{a_1}1 \cap x_{b_1}2)}{p(x_{a_1}1)p(x_{b_1}2)}\\
&\quad + p(x_{a_1}1 \cap x_{b_2}2) \ln \frac{p(x_{a_1}1 \cap x_{b_2}2)}{p(x_{a_1}1)p(x_{b_2}2)} + p(x_{a_2}1 \cap x_{b_1}2) \ln \frac{p(x_{a_2}1 \cap x_{b_1}2)}{p(x_{a_2}1)p(x_{b_1}2)}\\
&\quad + p(x_{a_2}1 \cap x_{b_2}2) \ln \frac{p(x_{a_2}1 \cap x_{b_2}2)}{p(x_{a_2}1)p(x_{b_2}2)} = 0.019.
\end{aligned}$$

Obviously, the value of mutual information depends on the configuration, viz., the way to partition the feature space for a certain dataset and values of K and L. Define the maximum of $I(x_1, x_2)$ in a K-by-L grid of dataset S by

$$I^*(x_1, x_2) = \max\{\, I(x_1, x_2)|S, K, L\} \,. \tag{11.20a}$$

The maximum mutual information has the property of $0 \leqslant I^*(x_1, x_2) \leqslant \log(\min\{K, L\})$ (Cover & Thomas, 2003). If the maximum mutual information is normalized by $\log(\min\{K, L\})$, the maximum mutual information will be within the (0, 1) range. The normalized maximum mutual information is expressed by

$$nI^*(K, L) = \frac{I^*(x_1, x_2)}{\ln(\min\{K, L\})}. \tag{11.20b}$$

The normalized maximum mutual information, however, depends on the values of K and L. The maximal information coefficient (MIC) of a dataset S of (x_1, x_2) is defined by

$$\text{MIC}(S) = \max_{K,L}\{nI^*(K, L)\} = \max_{K,L}\left\{\frac{I^*(x_1, x_2)}{\ln(\min\{K, L\})}\right\}, \tag{11.20c}$$

in all possible values of cell numbers. It has been suggested that the size of the (K, L) space is sufficiently large enough for the search for MIC(S) if the cell number $K \times L < B(n)$, where $B(n) = n^{0.6}$ is suggested by Reshef et al. (2011) and $B(n) = n^{0.55}$ is advocated by Kinney and Atwal (2014).

Example 11.9 Filter feature selection of steel's fatigue strength is conducted with the MIC correlation coefficient matrix, on 20 data collected from the database of National Institute for Materials Science, Japan (NIMS). Tables 11.6 and 11.7 list 16 initially proposed features and the actual data of the 20 steel samples, respectively.

Table 11.6 Features of the collected data

Features	Descriptions
NT	Normalizing temperature (°C)
QT	Quenching temperature (°C)
TT	Tempering temperature (°C)
C	wt.% of carbon
Si	wt.% of silicon
Mn	wt.% of manganese
P	wt.% of phosphorus
S	wt.% of sulphur
Ni	wt.% of nickel
Cr	wt.% of chromium
Cu	wt.% of copper
Mo	wt.% of molybdenum
RR	Reduction ratio (%)
dA	Fraction of plastic work inclusions (%)
dB	Fraction of discontinuous array inclusions (%)
dC	Fraction of isolated inclusions (%)

Table 11.7 Twenty collected NIMS data

Sample No.	NT	QT	TT	C	Si	Mn	P	S	Ni	Cr	Cu	Mo	RR	dA	dB	dC	P1
1	900	845	600	0.32	0.24	0.58	0.009	0.01	2.65	0.71	0.06	0	700	0.01	0.01	0.03	510
2	870	855	550	0.41	0.25	0.74	0.013	0.019	0.05	0.99	0.08	0.16	700	0.03	0	0	566
3	870	855	600	0.33	0.27	0.69	0.015	0.008	0.04	0.99	0.05	0.18	640	0.03	0	0	535
4	870	845	630	0.4	0.23	0.7	0.011	0.01	1.82	0.78	0.09	0.18	740	0.05	0	0.01	543
5	865	865	650	0.35	0.25	0.82	0.016	0.023	0.03	0.01	0.01	0	825	0.07	0.04	0	341
6	870	845	550	0.38	0.21	1.59	0.017	0.015	0.03	0.2	0.05	0	480	0.02	0	0	432
7	870	855	650	0.4	0.22	0.7	0.01	0.004	0.04	0.94	0.12	0.17	820	0.02	0	0.02	517
8	845	845	550	0.45	0.24	0.71	0.01	0.009	0.12	0.06	0.13	0	610	0.06	0	0.01	465
9	870	855	550	0.4	0.26	0.77	0.018	0.005	0.05	0.98	0.13	0.17	610	0.05	0	0.01	562
10	870	845	550	0.4	0.22	1.56	0.011	0.019	0.06	0.09	0.05	0	530	0.02	0	0.01	451
11	825	825	650	0.52	0.21	0.8	0.021	0.021	0.02	0.1	0.02	0	1740	0.09	0	0	410
12	825	825	650	0.52	0.29	0.83	0.016	0.025	0.04	0.04	0.02	0	825	0.12	0.02	0	402
13	870	855	600	0.37	0.29	0.76	0.017	0.009	0.12	1.01	0.12	0.16	500	0.05	0	0.02	542
14	900	845	550	0.32	0.3	0.52	0.013	0.009	2.67	0.77	0.06	0	1740	0.03	0	0	570
15	870	855	600	0.41	0.25	0.74	0.013	0.019	0.05	0.99	0.08	0.16	700	0.03	0	0	527
16	870	855	550	0.4	0.25	0.87	0.011	0.015	0.07	1.06	0.11	0.18	700	0.07	0.01	0	603
17	870	845	680	0.4	0.23	0.7	0.011	0.01	1.82	0.78	0.09	0.18	740	0.05	0	0.01	482
18	870	845	550	0.41	0.24	1.51	0.02	0.008	0.04	0.22	0.08	0	610	0.02	0.01	0.02	425
19	870	855	600	0.38	0.27	0.8	0.015	0.012	0.07	1.07	0.12	0.16	1120	0.03	0	0.02	592
20	845	845	550	0.45	0.25	0.79	0.018	0.016	0.02	0.13	0.02	0	1740	0.07	0	0	513

Note P1 denotes the fatigue strength (MPa)

The quantitative feature importance ranking is assessed by the MIC, and with a suitable threshold, the redundant features (strongly-correlated feature pairs) and/or insignificant features (weak correlation with response(s)) can be removed from the original feature set. For the 20 discrete data, the MIC is calculated for every two features among the initially proposed feature set and for each feature with response. Take the variable of the normalizing temperature of steels and the response of fatigue strength as an example to calculate the MIC, and for the sake of simplicity, denote them as x_1 and y, respectively.

We partition the (x_1, y) feature space into a $K \times L$ grid, and the bins and the data number inside the bins along the x_1 axis are denoted by a_k and $|a_k|(k = 1, 2, \cdots, K)$, respectively. Similarly, the (x_1, y) feature space is partitioned along the y axis into L bins, which is done by $L - 1$ lines perpendicular to the y axis as boundaries. The bins and the data number inside the bins are denoted by b_l, and $|b_l|(l = 1, 2, \cdots, L)$, respectively. Following the study of Kinney and Atwal (2014), $B(n) = n^{0.55}$ is adopted here in the binning method, viz., $K \times L < B(n) = 20^{0.55} = 5.195$. As K and L are both integers, $K \times L \leqslant 5$ is adopted here and there are 10 possible combinations of K-by-L, viz., 1-by-L ($L = 1, 2, \cdots, 5$), K-by-1 ($K = 1, 2, \cdots, 5$), and 2-by-2. When K or L equals one, the mutual information $I(x_1, y \mid K = 1, L) = 0$ and $I(x_1, x_2 \mid K, L = 1) = 0$.

Table 11.8 $I(x_1, y|K = 2, L = 2)$ over 76 grid configurations

	B_{x1}^1	B_{x1}^2	B_{x1}^3	B_{x1}^4
B_y^1	0.005412	0.011482	0.073415	0.005412
B_y^2	0.062665	0.025583	0.15683	0.011134
B_y^3	0.229606	0.097046	**0.254456**	0.017204
B_y^4	0.186454	0.060357	0.148452	0.023667
B_y^5	0.15683	0.037644	0.099576	0.030577
B_y^6	0.134129	0.022367	0.06731	0.038002
B_y^7	0.115689	0.085111	0.147044	0.046023
B_y^8	0.100149	0.063676	0.112975	0.054746
B_y^9	0.086715	0.046421	0.085653	0.000559
B_y^{10}	0.074882	0.163897	0.215762	0
B_y^{11}	0.064306	0.139887	0.18338	0.000559
B_y^{12}	0.054746	0.118494	0.154819	0.002273
B_y^{13}	0.046023	0.099195	0.129254	0.005269
B_y^{14}	0.038002	0.081614	0.106106	0.009792
B_y^{15}	0.030577	0.065466	0.08495	0.016285
B_y^{16}	0.023667	0.050534	0.065466	0.025583
B_y^{17}	0.017204	0.036647	0.047407	0.039445
B_y^{18}	0.011134	0.023667	0.030577	0.011134
B_y^{19}	0.005412	0.011482	0.014818	0.005412

Thus, only the values of mutual information $I(x_1, y \mid K = 2,\ L = 2)$ are calculated. Table 11.7 shows that there are 5 different values in the normalizing temperature the and 20 different values in fatigue strength, meaning 4 possible boundaries to partition the and 19 possible boundaries to partition fatigue strength, which yields $p = 4 \times 19 = 76$ configurations of the 2-by-2 grid. The boundaries are denoted by $B_{x_1}^i\ (i = 1,\ 2,\ 3,\ 4)$ and $B_y^j\ (j = 1,\ 2,\ \cdots,\ 19)$, where $B_{x_1}^i$ and B_y^j are the middle of intervals between two adjacent values of x_1 and y, respectively.

Table 11.8 lists 76 values of mutual information $I(x_1, y|K = 2, L = 2)$ for the 76 configurations, showing the maximum $I^*(x_1, y) = 0.254456$. Hereafter, the normalized maximum mutual information is calculated by $nI^*(K = 2, L = 2) = \dfrac{I^*(x_1, y)}{\ln(\min\{K = 2, L = 2\})} = \dfrac{I^*(x_1, y)}{\ln(2)} = 0.3670$. Since $nI^*(K, L) = 0$ for 1-by-L (L =1, 2, $\cdots$, 5) and K-by-1 ($K = 1, 2, \cdots, 5$), the $\mathrm{MIC}(x_1, y) = \max\limits_{K,L}\{nI^*(K, L)\} = 0.367$ for the normalizing temperature and fatigue strength under the condition of $K \times L < 5.195$.

In the same way, for the 20 discrete data, the MIC is calculated for every two variables among the 16 features and 1 response, and the result is illustrated in the MIC matrix of Fig. 11.14.

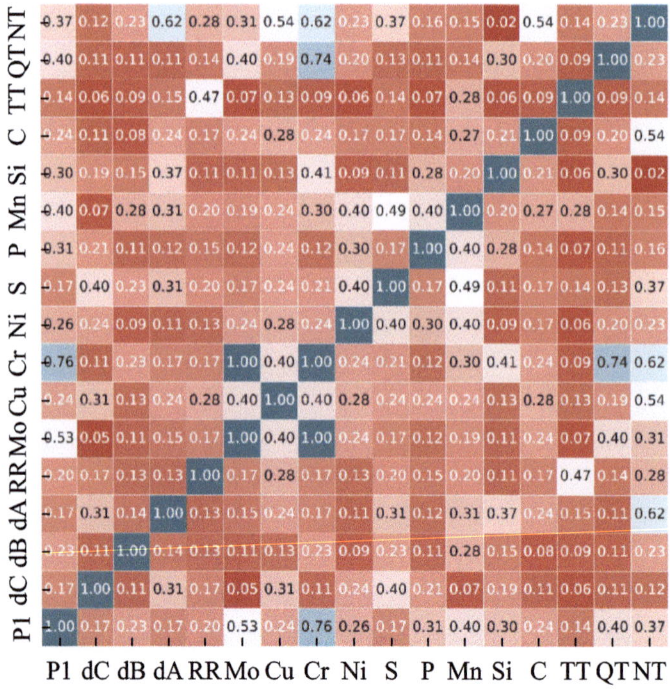

Fig. 11.14 The MIC matrix of the features and the response, where red and blue colors represent the weak and strong correlations, respectively

If a threshold of $(\text{MIC})_c$ is proposed, two features with MIC $\geq (\text{MIC})_c$ are strongly correlated and one of them will be removed. In this example, $(\text{MIC})_c = 0.9$ is adopted and the MIC(Cr, Mo) $= 1 > 0.9$ so that Cr or Mo is redundant. Considering MIC(Cr, y) >MIC(Mo, y), we delete feature Mo from the feature set.

11.2.1.3 The Minimum Redundancy Maximum Relevance Method

The minimum redundancy maximum relevance (mRMR) method is a mutual-information-based feature selection method. Equation (11.20c) gives the mutual information $I(x_1, x_2)$ between features x_1 and x_2 that is defined by the difference between the information entropy and the conditional entropy. Similarly, the mutual information $I(\boldsymbol{x}_j, y)$ between each feature $\boldsymbol{x}_j (j = 1,\ 2,\ \cdots, m)$ and target y is given by

$$I(\boldsymbol{x}_j, y) = \sum_{l=1}^{L} \sum_{k=1}^{K} P(\boldsymbol{x}_j = l, y = k) \ln \frac{P(\boldsymbol{x}_j = l, y = k)}{P(\boldsymbol{x}_j = l) P(y = k)}$$
$$(l = 1,\ 2, \cdots, L; k = 1,\ 2, \cdots, K) \tag{11.21a}$$

for discrete variables with L discrete values of $\boldsymbol{x}_j$ and K discrete values of y, and

$$I(\boldsymbol{x}_j, y) = \int \int p(\boldsymbol{x}_j, y) \ln \frac{p(\boldsymbol{x}_j, y)}{p(\boldsymbol{x}_j) p(y)} \mathrm{d}\boldsymbol{x}_j \mathrm{d}y \tag{11.21b}$$

for continuous data. There are M original features in a studied ML problem and the M features form the original feature set S_M. A feature subset S_m containing m features can be selected from the original feature set $S_M (m \leq M)$ and there are $(2^M - 1)$ feature subsets of S_M. The dependency of feature set S_m on target y is defined by the mutual information $I(S_m, y)$ between the feature set and the target as

$$I(S_m, y) = \int \cdots \int p((\boldsymbol{x}_1, x_2, \cdots, \boldsymbol{x}_m), y)$$
$$\ln \frac{p((\boldsymbol{x}_1, \boldsymbol{x}_2, \cdots, \boldsymbol{x}_m), y)}{p(\boldsymbol{x}_1, \boldsymbol{x}_2, \cdots, \boldsymbol{x}_m) p(y)} \mathrm{d}\boldsymbol{x}_1\ \mathrm{d}\boldsymbol{x}_2\ \cdots\ \mathrm{d}\boldsymbol{x}_m \mathrm{d}y. \tag{11.22a}$$

The maximum dependency is the maximum of all dependencies of the entire feature subsets of the original feature set and thus, the feature subset with the maximum dependency is given by

$$\hat{S}_m = \underset{S_m}{\operatorname{argmax}}\, I(S_m, y). \tag{11.22b}$$

In practice, it is very difficult to calculate the maximum dependency due to high computational burden. Peng et al. (2005) proposed the maximum relevance to replace approximately the maximum dependency. The relevance D of feature subset S_m on target y is defined as the average of the sum of mutual information between each of the features involved in the set and the target, viz.,

$$D = \frac{1}{m}\sum_{j=1}^{m} I(\boldsymbol{x}_j, \boldsymbol{y})\ (\boldsymbol{x}_j \in S_m). \tag{11.23a}$$

Calculating the relevance D of each subset in the entire feature subsets allows one to determine the maximum relevance and the associated feature subset

$$\max D(S_m, \boldsymbol{y}). \tag{11.23b}$$

Similarly, the relevance R between features in a feature subset is defined as the average of the sum of mutual information between every two features involved in the subset

$$R = \frac{1}{m^2}\sum_{i=1,j=1}^{m} I(\boldsymbol{x}_i, \boldsymbol{x}_j)\ (\boldsymbol{x}_i, \boldsymbol{x}_j \in S_m). \tag{11.23c}$$

Since there are m features in the calculated feature set and both $\boldsymbol{x}_i$ and $\boldsymbol{x}_j$ are repeated m times, the average is over m^2. Calculating the relevance R of each subset in the entire feature subsets determines the minimum relevance R

$$\min R(S_m). \tag{11.23d}$$

The mRMR feature selection selects a feature subset from

$$\hat{S}_m = \underset{S_m}{\operatorname{argmax}}(D(S_m, y) - R(S_m)). \tag{11.23e}$$

Example 11.10 Table 11.9 shows some data for lead-free solder alloys data collected from literature, including four chemical elements of Tin (Sn), Silver (Ag), Copper (Cu), and Bismuth (Bi) in wt.% and one property of yield strength in MPa. The mRMR method is conducted to select features (a feature subset).

Table 11.9 Select data for 15 lead-free solder alloys

Sn (wt.%)	Ag (wt.%)	Cu (wt.%)	Bi (wt.%)	Yield Strength (y, MPa)
94.15	3.8	0.7	1.35	34.9
95.6	3.4	0	1	46.28966
94	3.8	0.8	1.4	63
94.85	3.4	0.5	1.25	36.3
95.5	3	0.5	1	47.71034
92	3	3	2	46.09655
93	3	2	2	43.28276
94.25	3.7	0.7	1.35	42.38621
95.375	2.5	0.75	1.375	82.8
96.5	2.5	0	1	48.75862
96	3	0	1	39.64828
96	2.5	0.5	1	45.46207
95.5	3.8	0	0.7	43.36552
96	3.5	0.5	0	46.3
96.21	3.03	0.76	0	46.54

A brute force search is conducted in the mRMR feature selection. The four features and the response are all continuous variables. Thus, the MIC described above is used here with $B(n) = n^{0.6}$ to calculate the normalized mutual information between two continuous variables. The brute force search starts from one feature. There are $C_4^1 = 4$ one-feature subsets. We take $S_1 = \{\text{Sn}\}$ as an example to show the calculation detail for one-feature subsets. The relevance D of feature subset S_1 on target y is $D = \frac{1}{1} \times I(\text{Sn}, y) = 0.17$ according to Eq. (11.23a) and the relevance R between features in one-feature subsets is calculated by Eq. (11.23c) as $R = \frac{1}{1^2} \times I(\text{Sn}, \text{Sn}) = 1$, which yields the mRMR value of $D(S_1, y) - R(S_1) = 0.17 - 1 = -0.83$. Similarly, all mRMR values of the four one-feature subsets are calculated and listed in Table 11.10.

Table 11.10 The mRMR values of one-feature subsets

	$S_1\{\text{Sn}\}$	$S_2\{\text{Ag}\}$	$S_3\{\text{Cu}\}$	$S_4\{\text{Bi}\}$
$D(S_m, \mathbf{y})$	0.17	0.13	0.16	0.17
$R(S_m)$	1	1	1	1
mRMR value	−0.83	−0.87	−0.84	−0.83

There are $C_4^2 = 6$ two-feature subsets. We take $S_5 = \{\text{Sn}, \text{Ag}\}$ as an example to show the calculation detail for the two-feature subsets. The relevance D of feature subset S_5 on target y is $D = \frac{1}{2} \times (I(\text{Sn}, y) + I(\text{Ag}, y)) = 0.15$ and the relevance R between features is $R = \frac{1}{2^2} \times (I(\text{Sn}, \text{Sn}) + I(\text{Sn}, \text{Ag}) + I(\text{Ag}, \text{Sn}) + I(\text{Ag}, \text{Ag})) = 0.58$, which yields the mRMR of $D(S_5, y) - R(S_5) = 0.15 - 0.58 = -0.43$. Table 11.11 lists the mRMR values of the six two-feature subsets.

Table 11.11 The mRMR values of two-feature subsets

	S_5\{Sn, Ag\}	S_6\{Sn,Cu\}	S_7\{Sn, Bi\}
$D(S_m, \boldsymbol{y})$	0.15	0.17	0.17
$R(S_m)$	0.58	0.77	1.00
mRMR value	−0.43	−0.6	−0.83
	S_8\{Ag, Cu\}	S_9\{Ag, Bi\}	S_{10}\{Cu, Bi\}
$D(S_m, \boldsymbol{y})$	0.15	0.15	0.17
$R(S_m)$	0.58	0.61	0.85
mRMR value	−0.43	−0.46	−0.68

There are $C_4^3 = 4$ three-feature subsets. We take $S_{11} = \{\text{Sn}, \text{Ag}, \text{Cu}\}$ as an example to show the details of the calculations mRMR values for the three-feature subsets. The relevance D of feature subset S_{11} on target y is $D = \frac{1}{3} \times (I(\text{Sn}, y) + I(\text{Ag}, y) + I(\text{Cu}, y)) = 0.16$ and the relevance R between features is $R = \frac{1}{3^2} \times (I(\text{Sn}, \text{Sn}) + I(\text{Sn}, \text{Ag}) + I(\text{Sn}, \text{Cu}) + I(\text{Ag}, \text{Sn}) + I(\text{Ag}, \text{Ag}) + I(\text{Ag}, \text{Cu}) + I(\text{Cu}, \text{Sn}) + I(\text{Cu}, \text{Ag}) + I(\text{Cu}, \text{Cu})) = 0.53$, which yields the mRMR of $D(S_{11}, y) - R(S_{11}) = 0.16 - 0.53 = -0.37$. Table 11.12 lists the mRMR values of the four three-feature subsets.

Table 11.12 The mRMR values of three-feature subsets

	S_{11}\{Sn,Ag,Cu\}	S_{12}\{Sn,Ag,Bi\}	S_{13}\{Sn,Cu,Bi\}	S_{14}\{Ag,Cu,Bi\}
$D(S_m, \boldsymbol{y})$	0.16	0.16	0.17	0.16
$R(S_m)$	0.53	0.64	0.83	0.57
mRMR value	−0.37	−0.48	−0.66	−0.41

There is only one four-feature subset, $S_{15} = \{\text{Sn}, \text{Ag}, \text{Cu}, \text{Bi}\}$. The relevance D of feature subset S_{15} on target y is $D = 0.16$ and the relevance R between features is $R = 0.60$, which yields the mRMR of $D(S_{15}, y) - R(S_{15}) = -0.44$. Finally, the maximum mRMR is selected from all mRMR values and the mRMR selects the feature subset $\hat{S}_m = \underset{S_m \in (S_1, S_2, \cdots, S_{15})}{\text{argmax}} (D(S_m, y) - R(S_m)) = S_{11}\{\text{Sn}, \text{Ag}, \text{Cu}\}$, viz., the mRMR selects the three features of Sn, Ag and Cu.

Example 11.11 Table 11.13 shows some selected discrete data for nitrogen oxide (NOx) conversion catalysts. Three features are considered: (1) catalyst support (ST), with 1 and 2 representing TiO2 and glass fiber, respectively; (2) preparation method (PM), with 1, 2, and 3 representing impregnation, sol–gel and jet deposition, respectively; and (3) high temperature (HT) calcination, with 0 and 1 denoting the catalysts have not and have gone through HT calcination, respectively. The target of (NOx) conversion efficiency is ranked at three levels (high, medium, and low), which are denoted by 0, 1, and 2, respectively.

Table 11.13 Selected data for NO_x conversion catalyst data

ST	PM	HT	NO_x conversion
2	1	0	1
2	1	1	1
1	2	1	0
1	1	1	2
1	3	0	2

Note ST:1 represents the TiO2, 2 represents the glass fiber; PM: 1 represents the impregnation method, 2 represents the solgel method, 3 represents the jet deposition

The dataset contains original three features, which yield $2^3 - 1 = 7$ feature subsets, including $S_1\{\text{ST}\}$, $S_2\{\text{PM}\}$, $S_3\{\text{HT}\}$, $S_4\{\text{ST, PM}\}$, $S_5\{\text{ST, HT}\}$, $S_6\{\text{PM, HT}\}$, and $S_7\{\text{ST, PM, HT}\}$. A brute force search is conducted by calculating all mRMR values of the 7 feature subsets and subsequently selecting the maximum mRMR and the associated feature subset. The calculation detail for the feature subset $S_7\{\text{ST, PM, HT}\}$ is illustrated below.

From the data, we have $P(y = 0) = 1/5$, $P(y = 1) = 2/5$, and $P(y = 2) = 2/5$; $P(\text{ST} = 2) = 2/5$ and $P(\text{ST} = 1) = 3/5$; $P(\text{PM} = 1) = 3/5$, $P(\text{PM} = 2) = 1/5$, and $P(\text{PM} = 3) = 1/5$; $P(\text{HT} = 0) = 2/5$ and $P(\text{HT} = 1) = 3/5$.

For feature ST, we have the joint probability $\left\{P(\text{ST} = 1 \cap y = 0) = \frac{1}{5}, P(\text{ST} = 1 \cap y = 1) = 0, P(\text{ST} = 1 \cap y = 2) = \frac{2}{5}, P(\text{ST} = 2 \cap y = 0) = 0, P(\text{ST} = 2 \cap y = 1) = \frac{2}{5}, P(\text{ST} = 2 \cap y = 2) = 0\right\}$.

Thus, with Eq. (11.21a), we have

$$
\begin{aligned}
I(\text{ST}, y) &= \sum_{l=1}^{2}\sum_{k=1}^{3} P(\text{ST} = l, y = k)\ln\frac{P(\text{ST} = l, y = k)}{P(\text{ST} = l)P(y = k)} \\
&= P(\text{ST} = 1 \cap y = 0)\ln\frac{P(\text{ST} = 1 \cap y = 0)}{P(\text{ST} = 1)P(y = 0)} \\
&\quad + P(\text{ST} = 1 \cap y = 1)\ln\frac{P(\text{ST} = 1 \cap y = 1)}{P(\text{ST} = 1)P(y = 1)} \\
&\quad + P(\text{ST} = 1 \cap y = 2)\ln\frac{P(\text{ST} = 1 \cap y = 2)}{P(\text{ST} = 1)P(y = 2)}
\end{aligned}
$$

$$
\begin{aligned}
&+ P(\mathrm{ST}=2\cap y=0)\ln\frac{P(\mathrm{ST}=2\cap y=0)}{P(\mathrm{ST}=2)P(y=0)}\\
&+ P(\mathrm{ST}=2\cap y=1)\ln\frac{P(\mathrm{ST}=2\cap y=1)}{P(\mathrm{ST}=2)P(y=1)}\\
&+ P(\mathrm{ST}=2\cap y=2)\ln\frac{P(\mathrm{ST}=2\cap y=2)}{P(\mathrm{ST}=2)P(y=2)}\\
&= \frac{1}{5}\ln\frac{\frac{1}{5}}{\frac{3}{5}\times\frac{1}{5}} + \frac{2}{5}\ln\frac{\frac{2}{5}}{\frac{3}{5}\times\frac{2}{5}} + \frac{2}{5}\ln\frac{\frac{2}{5}}{\frac{2}{5}\times\frac{2}{5}} = 0.673.
\end{aligned}
$$

Similarly, we have $I(\mathrm{PM}, y) = 0.673$ and $I(\mathrm{HT}, y) = 0.118$. Then, using Eq. (11.23a) gives the relevance D, i.e.,

$$
\begin{aligned}
D(S_7, y) &= \frac{1}{3}(I(\mathrm{ST}, y) + I(\mathrm{PM}, y) + I(\mathrm{HT}, y))\\
&= \frac{1}{3}\times(0.673 + 0.673 + 0.118) = 0.488.
\end{aligned}
$$

In a similar way, the relevance R is calculated with Eq. (11.23c) to be

$$
\begin{aligned}
R(S_7) = &\frac{1}{3^2}(I(\mathrm{ST},\mathrm{ST}) + I(\mathrm{ST},\mathrm{PM}) + I(\mathrm{ST},\mathrm{HT}) + I(\mathrm{PM},\mathrm{ST}) + I(\mathrm{PM},\mathrm{PM})\\
&+ I(\mathrm{PM},\mathrm{HT}) + I(\mathrm{HT},\mathrm{ST}) + I(\mathrm{HT},\mathrm{PM}) + I(\mathrm{HT},\mathrm{HT}))\\
= &\frac{1}{3^2}\times(1 + 0.291 + 0.014 + 0.291 + 1 + 0.291 + 0.014 + 0.291 + 1)\\
= &\,0.466.
\end{aligned}
$$

Thus, the mRMR values of feature subset S_7 is.

$$
D(S_7, y) - R(S_7) = 0.488 - 0.466 = -0.022.
$$

The mRMR $(D - R)$ values are calculated in the same way for all feature subsets and listed in Table 11.14. Obviously, S_4\{ST, PM\} is the best feature subset with the maximal mRMR value.

Table 11.14 The mRMR (D–R) values of all feature subsets

Feature subsets	D–R
S_1\{ST\}	−0.327
S_2\{PM\}	−0.327
S_3\{HT\}	−0.882
S_4\{ST, PM\}	0.0275
S_5\{ST, HT\}	−0.1115
S_6\{PM, HT\}	−0.25
S_7\{ST, PM, HT\}	−0.018

With the increase of the number of original features, the number of feature subsets increases exponentially. It is very hard to conduct a brute force search for the global maximum of mRMR if the number of original features is large. In this case, sequential forward selection (SFS) and sequential backward selection (SBS) are proposed to approximately find the near-maximum mRMR.

11.2.1.4 Sequential Forward Selection (SFS)

SFS starts from one-feature subsets by selecting the one-feature subset with the maximum mRMR. Then, each of the other features is added into the selected one-feature subset, forming the two-feature subsets. The mRMR values are calculated for every two-feature subset and the maximum mRMR is selected associated with a two-feature subset. If the maximum mRMR for two-feature subsets is lower than that for one-feature subsets, the one-feature subset will be selected and the selection process ends. Otherwise, each of the other features is added into the selected two-feature subset, forming the three-feature subsets, and the three-feature subset with the maximum mRMR is selected from the three-feature subsets. Again, if the maximum mRMR for three-feature subsets is lower than that for two-feature subsets, the two-feature subset will be selected and the selection process ends, otherwise SFS continues. SFS is conducted in such a manner and stops until the maximum mRMR found in the current step is lower than the previous step.

In Example 11.11, the selection from one-feature subsets gives two subsets of S_1\{ST\} and S_2\{PM\}, which have the same maximal mRMR value of −0.327, as shown in Table 11.14. It is very rare to have multiple subsets with the same maximal mRMR value. Adding each of the other features into S_1\{ST\} and S_2\{PM\} forms two-feature subsets of S_4\{ST, PM\}, S_5\{ST, HT\}, and S_6\{PM, HT\}. The subset S_4\{ST, PM\} has the maximal mRMR value of 0.0275 among the three two-feature subsets, as shown in Table 11.14, and this value is larger than −0.327, the maximal mRMR value for one-feature subsets. Subsequently, adding feature HT into S_4\{ST, PM\} forms S_7\{ST, PM, HT\}, whose mRMR value is −0.018 which is lower than 0.0275 of S_4\{ST, PM\}. Therefore, the feature subset of S_4\{ST, PM\} is selected by SFS.

Peng et al. (2005) proposed an incremental search method, which starts also from one-feature subsets and selects the one-feature subset with the maximum mRMR. After that, each of the rest features, $\boldsymbol{x}^* \in \{S_M - \boldsymbol{x}_j\}$ with $\boldsymbol{x}_j$ denoting the selected one-feature subset, is added to the selected one-feature subset, forming the two-feature subsets. Instead of selecting the maximum mRMR from the two-feature subsets, the incremental search method selects $\max\limits_{\boldsymbol{x}^* \in \{S_M - \boldsymbol{x}_j\}} \left\{ I(\boldsymbol{x}^*, y) - I\left(\boldsymbol{x}_j, \boldsymbol{x}^*\right) \right\}$, which requires $(M - 1)$ times of calculation, the same computational work as that in SFS. Generally, if $\hat{S}_{m-1}$ is the selected $(m-1)$-feature subset, the m-feature subset is selected from the equation of

$$\hat{S}_m = \underset{x^* \in \{S_M - \hat{S}_{m-1}\}}{\operatorname{argmax}} \left(I(\boldsymbol{x}^*, \boldsymbol{y}) - \frac{1}{m-1} \sum_{j=1}^{m-1} I\left(\boldsymbol{x}_j, \boldsymbol{x}^*\right) \right) \left(\boldsymbol{x}_j \in \hat{S}_{m-1}\right). \quad (11.24)$$

Clearly, this incremental search method is completed in the same approach and with the same computational workload as that of SFS, but with a different selection criterion.

11.2.1.5 Sequential Backward Selection (SBS)

SBS is the typical decremental search method, which starts from the original feature set of M features by calculating its mRMR. Then, one feature is moved out from the original feature set, which forms $M(M-1)$-feature subsets. The mRMR values are calculated for every $(M-1)$-feature subset and the maximum mRMR is selected associated with an $(M-1)$-feature subset. If the maximum mRMR for $(M-1)$-feature subsets is lower than that for the original feature set, the original feature set will be selected and the selection process ends. Otherwise, one feature is moved out from the selected $(M-1)$-feature subset, which forms $(M-1) \times (M-2)$-feature subsets. The $(M-2)$-feature subset is selected with the maximum mRMR among all $(M-2)$-feature subsets. Again, if the maximum mRMR for $(M-2)$-feature subsets is lower than that for $(M-1)$-feature subsets, the $(M-1)$-feature subset will be selected and the selection process ends. SBS is conducted in such a manner until the maximum mRMR is determined.

In Example 11.11, the original feature set is S_7\{ST, PM, HT\} and its mRMR value is −0.018. Removing one feature from the original feature set forms two-feature subsets of S_4\{ST, PM\}, S_5\{ST, HT\}, and S_6\{PM, HT\}. The subset S_4\{ST, PM\} has the maximal mRMR value of 0.0275 among the three two-feature subsets, as shown in Table 11.14, and this value is larger than −0.018, the mRMR value of the original feature set. Removing one feature from the selected two-feature subset forms one-feature subsets of S_1\{ST\} and S_2\{PM\}, both of which have the same mRMR value of −0.327, as shown in Table 11.14, which is lower than 0.0275 of S_4\{ST, PM\}. Therefore, the feature subset of S_4\{ST, PM\} is selected by SBS.

11.2.1.6 ReliefF

The method of Relief relevant features is a distance-based method (Kira & Rendell, 1992) to rank features in classification and the ReliefF algorithm (Kononenko et al., 1994) extends the original Relief algorithm for binary classification to multiple classification. The methodology of ReliefF relevant features is based on the fact that if data are in the same class, the feature values of the data will be clustered together. There are two kinds of features in ML. The first kind of features is called the continuous features. Although randomly sampling a continuous feature gives discrete values, these values vary randomly within a defined range. For example, randomly sampling feature x from the feature x (0, 1) range ten times gives ten random numbers among (0, 1). The other kind of features is named the discrete features, like feature seasons of spring, summer, autumn, and winter. If we label spring, summer, autumn, and winter by 1, 2, 3, and 4, respectively. Feature seasons have only four discrete numbers and randomly sampling feature seasons once give a number of 1, 2, 3, or 4. Consider n data $\boldsymbol{X} = (X_1, X_2, \cdots, X_n)^{\mathrm{T}}$, which belong to K classes expressed by $c_k(k = 1, 2, \cdots, K)$. A sample X_i $(i = 1, 2, \cdots, n) = (x_{i1}, x_{i2}, \cdots, x_{iJ})$ with J initially proposed features is used to evaluate its neighboring records of the same and different classes.

For continuous features, the difference $\mathrm{diff}(x_j, X_i, X_I)$ between data X_i and X_I $(i \neq I$ and $i,\ I = 1, 2, \cdots, n)$ along the feature $x_j(j = 1, 2, \cdots, J)$ direction is calculated by the normalized distance between them, i.e.,

$$\mathrm{diff}(x_j, X_i, X_I) = \frac{|x_{ij} - x_{Ij}|}{\max(x_j) - \min(x_j)}, \tag{11.25a}$$

where $\max(x_j)$ and $\min(x_j)$ are the maximum and the minimum of feature x_j in the dataset, respectively. The normalized distance ensures fairness in the distance measurement for different features.

For discrete features, the difference $\mathrm{diff}(x_j, X_i, X_I)$ between data X_i and X_I $(i \neq I$and $i, I = 1, 2, \cdots, n)$ along the feature $x_j(j = 1, 2, \cdots, J)$ direction is calculated by

$$\mathrm{diff}(x_j, X_i, X_I) = \begin{cases} 0, \text{ if } x_{ij} \text{ and } x_{Ij} \text{ are the same,} \\ 1, \text{ if } x_{ij} \text{ and } x_{Ij} \text{ are different.} \end{cases} \tag{11.25b}$$

In ReliefF, the evaluated samples in the same class are named "near hits", while the evaluated samples in the different classes are called "near misses". If near hits have widely spreading values of a feature, the feature appears to be irrelevant to the label of classes or responses and the feature's weight will be counted less. On the other hand, if near misses have widely spreading values of a feature, the feature appears to be relevant in the sense of separating different classes, and the feature's weight should be counted more. The weight of feature x_j in the initially proposed feature space is calculated by the probability sum of average differences of near misses minus the average difference of near hits. A sample $X_i(i = 1, 2, \cdots, n)$ belonging to c_k is randomly selected first. Then, along the feature x_j direction, the sample's Q $(Q < n)$-nearest-neighbor samples $H_q(q = 1, 2, \cdots, Q)$ in the same class c_k and Q- nearest neighbors $M_q^{c_{\hat{k}}}(q = 1, 2, \cdots, Q)$ in each of the different classes $c_{\hat{k}}$ $(\hat{k} \neq k, \hat{k} = 1, 2, \cdots, K - 1)$ are selected. The weight of feature x_j of sample $X_i(i = 1, 2, \cdots, n)$, $W_{x_j}^i$, is calculated by

$$W_{x_j}^i = \sum_{\hat{k}}^{K-1} \frac{P(c_k)}{1 - P(c_{\hat{k}})} \left(\frac{1}{Q} \sum_{q=1}^{Q} \text{diff}\left(x_j, X_i, M_q^{c_{\hat{k}}}\right) \right) - \frac{1}{Q} \sum_{q=1}^{Q} \text{diff}(x_j, X_i, H_q), \tag{11.26a}$$

where $P(c_k)$ and $P(c_{\hat{k}})$ are the probabilities of sample X_i in class c_k and $c_{\hat{k}}$, respectively. The average weight of feature x_j on n data,

$$W_{x_j}^{\text{AVE}} = \frac{1}{n} \sum_{i=1}^{n} W_{x_j}^i \tag{11.26b}$$

is used to rank the feature importance given in ReliefF. Subsequently, all features are ranked by their average weights, and if a threshold is set, the features with average weights smaller than the threshold will be removed.

When the amount of original data is large, the bootstrap sampling is usually used to generate m $(0 < m \leq n)$ data by randomly sampling m times with replacement (Efron, 1979), and calculating the feature weight based on the bootstrap data (Kira & Rendell, 1992).

Example 11.12 Table 11.15 shows some selected data for 36 high entropy alloys (HEA), including 18 body-centered cubic (BCC) HEAs and 18 face-centered cubic (FCC) HEAs. The VEC and enthalpy of mixing (Hmix, in kJ·mol^{-1}) are the two features considered in the example.

This example is a binary classification problem, and the features are numerical variables. In binary classification, $\dfrac{P(c_k)}{1 - P(c_{\hat{k}})} = 1$, and Eq. (11.26a) is reduced to

$$W_{x_j}^i = \frac{1}{Q} \sum_{q=1}^{Q} \text{diff}\left(x_j, X_i, M_q^{c_{\hat{k}}}\right) - \frac{1}{Q} \sum_{q=1}^{Q} \text{diff}(x_j, X_i, H_q).$$

Let Q = 2, for feature VEC, the maximum and minimum values are 9.50 and 4.17, respectively; for feature Hmix, the maximum and minimum values are 6.25 and −19.84, respectively.

Sample 1 (VEC = 7.27, Hmix = −14.39).

(1) Along the VEC direction, the nearest two samples in the same class (BCC) are No. 3 (VEC = 7.09) and No. 4 (VEC = 7.25), and the nearest two samples in different class (FCC) are No. 24 and No. 31 (VEC = 8.00) and No. 33 (VEC = 7.91). Thus

$$W_{\text{VEC}}^1 = \frac{|7.27 - 7.91| + |7.27 - 8.00|}{2 \times (9.50 - 4.17)} - \frac{|7.27 - 7.09| + |7.27 - 7.25|}{2 \times (9.50 - 4.17)} = 0.1098.$$

Table 11.15 Solid solution structures of 36 high entropy alloys

No.	Alloys	VEC	Hmix	Structures	No.	Alloys	VEC	Hmix	Structures
1	$Al_{0.75}CoCrCu_{0.25}FeNiTi_{0.5}$	7.27	−14.39	BCC	19	$Al_{0.25}CoFeNi$	8.54	−6.06	FCC
2	$Al_{0.8}CrCuFeMn_{1.5}Ni$	7.60	−4.23	BCC	20	$Al_{0.2}CrCuFeNi_2$	8.77	0.12	FCC
3	$AlCoCrCu_{0.25}FeNiTi_{0.5}$	7.09	−15.50	BCC	21	$CoCr_{0.4}Fe_8Mn_{5.4}Ni_{5.2}$	8.26	−3.58	FCC
4	$AlCoCrCu_{0.5}FeNiTi_{0.5}$	7.25	−13.42	BCC	22	CoCrCuFe	8.50	6.25	FCC
5	$AlCoCrFeNiSi_{0.4}$	6.96	−19.84	BCC	23	CoCrFeNi	8.25	−3.75	FCC
6	$AlCoCrFeNiTi_{0.5}$	6.91	−17.92	BCC	24	CoCrMnNi	8.00	−5.50	FCC
7	AlCrCuFeMnNi	7.50	−5.11	BCC	25	CoCrNi	8.33	−4.89	FCC
8	AlHfNbTaTiZr	4.17	−14.78	BCC	26	CoCuFeNi	9.50	5.00	FCC
9	$Cr_2MoNbTaVW$	5.57	−4.82	BCC	27	$CoCuFeNiSn_{0.02}$	9.47	5.02	FCC
10	$Al_{0.3}HfNbTaTiZr$	4.32	−3.99	BCC	28	CoFeMnNi	8.50	−4.00	FCC
11	$Al_{0.5}HfNbTaTiZr$	4.27	−7.67	BCC	29	CoFeNi	9.00	−1.33	FCC
12	$Al_{0.75}HfNbTaTiZr$	4.22	−11.55	BCC	30	$CoFeNiSi_{0.25}$	8.62	−11.83	FCC
13	$Al_{20}(CoCrCuFeMnNiTiV)_{80}$	6.60	−15.44	BCC	31	CoFeNiV	8.00	−10.50	FCC
14	$Cr_{0.5}MoNbTaVW$	5.45	−4.83	BCC	32	CoMnNi	8.67	−5.78	FCC
15	CrMoNbTaVW	5.50	−4.89	BCC	33	$Cr_{0.66}FeMnNi$	7.91	−4.17	FCC
16	$HfMo_{0.5}NbTaTiZr$	4.55	0.60	BCC	34	CrCuFeNi	8.57	3.01	FCC
17	$HfMo_{0.75}NbTaTiZr$	4.61	−0.21	BCC	35	FeMnNi	8.33	−4.44	FCC
18	HfNbTaTiVZr	4.50	0.78	BCC	36	$Co_9Cr_7Cu_{36}Mn_{25}Ni_{23}$	9.24	2.05	FCC

(2) Along the Hmix direction, the nearest two samples in the same class (BCC) are No. 4 (Hmix = −13.42) and No. 8 (Hmix = −14.78), and the nearest two samples in different class (FCC) are No. 31 (Hmix = −10.50) and No. 30 (Hmix = −11.83). Thus

$$W^1_{H_{\text{mix}}} = \frac{|-14.39 + 10.50| + |-14.39 + 11.83|}{2 \times (6.25 + 19.84)} - \frac{|-14.39 + 13.42| + |-14.39 + 14.78|}{2 \times (6.25 + 19.84)} = 0.0975.$$

Sample 2 (VEC = 7.60, Hmix = −4.23).

(1) Along the VEC direction, the nearest two samples in the same class (BCC) are No. 1 (VEC = 7.27) and No. 7 (VEC = 7.50), and the nearest two samples in different class (FCC) are No. 24 (VEC = 8.00) and No. 33 (VEC = 7.91). Thus

$$W^2_{\text{VEC}} = \frac{|7.60 - 7.91| + |7.60 - 8.00|}{2 \times (9.50 - 4.17)} - \frac{|7.60 - 7.27| + |7.60 - 7.50|}{2 \times (9.50 - 4.17)} = 0.02627.$$

(2) Along the Hmix direction, the nearest two samples in the same class (BCC) are No. 9 (Hmix = −4.82) and No. 10 (Hmix = −3.99), and the nearest two samples in different class (FCC) are No. 33 (Hmix = −4.17) and No. 35 (Hmix = −4.44). Thus

$$W^2_{H_{\text{mix}}} = \frac{|-4.23 + 4.17| + |-4.23 + 4.44|}{2 \times (6.25 + 19.84)} - \frac{|-4.23 + 4.82| + |-4.23 + 3.99|}{2 \times (6.25 + 19.84)} = -0.01073.$$

In the same way, the weights of two features are calculated for all $n = 36$ samples. The ReliefF values of two features are given as the average weight

$$W^{\text{AVG}}_{\text{VEC}} = \frac{\sum_{i=1}^{36} W^i_{\text{VEC}}}{36} = 0.288,$$

$$W^{\text{AVG}}_{H_{\text{mix}}} = \frac{\sum_{i=1}^{36} W^i_{H_{\text{mix}}}}{36} = 0.055.$$

If we set a critical value of 0.1 for the ReliefF values, feature Hmix will be removed.

Example 11.13 Table 11.16 shows 6 high entropy alloys collected from Xiong et al. (2021). The 6 alloys belong to single face-centered cubic (FCC), body-centered cubic (BCC) or dual-phase (FCC + BCC) HEAs. Three features are considered here, which include average valance electron concentration (VEC), the enthalpy of mixing (Hmix), and the processing condition (PC) of the as-cast (AC) and the mechanical alloying (MA).

Table 11.16 Crystal structures of 6 high entropy alloys

Samples	Alloy compositions	Processing conditions (x_1)	VEC (x_2)	Hmix (x_3, kJ·mol^{-1})	Phases (Classes)
1	$Al_{0.7}Co_{0.3}CrFeNi$	MA	7.20	−11.40	BCC (1)
2	AlCoCuNiTiZn	MA	8.17	−17.89	BCC (1)
3	CoCrFeMnNi	AC	8.00	−4.16	FCC (2)
4	CoCrFeNi	AC	8.25	−3.75	FCC (2)
5	CoFeMnNi	AC	8.50	−4.00	FCC (2)
6	$CoCr_2FeNi$	AC	7.80	−4.32	FCC + BCC (3)

From Table 11.16, we have the probabilities, $P(1) = \frac{2}{6}$, $P(2) = \frac{3}{6}$, and $P(3) = \frac{1}{6}$. Set $Q = 2$. Feature x_1 is discrete, and features x_2 and x_3 are continuous.

Sample 1 (X_1, y_1) $(\text{MA}, 7.20, \ -11.40, \ 1)$.

(1) Along the x_1 direction, only one sample in the same class is sample 2, the nearest two samples in class 2 are sample 3 and sample 4, and only one sample in class 3 is sample 6. Thus

$$\sum_{q=1}^{Q} \text{diff}(x_1, X_1, H_q) = \text{diff}(x_1, X_1, H_1) = 0,$$

$$\sum_{q=1}^{Q} \text{diff}\left(x_1, X_1, M_q^{c_2}\right) = \text{diff}\left(x_1, X_1, M_1^{c_2}\right) + \text{diff}\left(x_1, X_1, M_2^{c_2}\right) = 1 + 1 = 2,$$

$$\sum_{q=1}^{Q} \text{diff}\left(x_1, X_1, M_q^{c_3}\right) = \text{diff}\left(x_1, X_1, M_1^{c_3}\right) = 1.$$

According to Eq. (11.26a), we have

$$W_{x_1}^1 = \frac{1}{Q} \times \frac{P(c_2)}{1 - P(c_1)} \times \sum_{q=1}^{Q} \text{diff}(x_1,X_1,M_q^{c_2}) + \frac{1}{Q} \times \frac{P(c_3)}{1 - P(c_1)}$$

$$\times \sum_{q=1}^{Q} \text{diff}(x_1,X_1,M_q^{c_3}) - \frac{1}{Q}\sum_{q=1}^{Q} \text{diff}(x_1,X_1,H_q)$$

$$= \frac{1}{2} \times \frac{\frac{3}{6}}{1-\frac{2}{6}} \times 2 + \frac{1}{2} \times \frac{\frac{1}{6}}{1-\frac{2}{6}} \times 1 - \frac{1}{2} \times 0 = 0.875.$$

(2) Along the x_2 direction, only one sample in the same class is sample 2, the nearest two samples in class 2 are sample 3 and sample 4, and only one sample in class 3 is sample 6. Thus

$$\sum_{q=1}^{Q} \text{diff}(x_2, X_1, H_q) = \frac{|x_{2,2} - x_{1,2}|}{\max(x_2) - \min(x_2)} = \frac{|7.20 - 8.17|}{8.50 - 7.20} = 0.746,$$

$$\sum_{q=1}^{Q} \text{diff}(x_2, X_1, M_q^{c_2}) = \frac{|x_{3,2} - x_{1,2}|}{\max(x_2) - \min(x_2)} + \frac{|x_{4,2} - x_{1,2}|}{\max(x_2) - \min(x_2)} = 1.423,$$

$$\sum_{q=1}^{Q} \text{diff}(x_2, X_1, M_q^{c_3}) = \frac{|x_{6,2} - x_{1,2}|}{\max(x_2) - \min(x_2)} = \frac{|7.80 - 7.20|}{8.50 - 7.20} = 0.462.$$

According to Eq. (11.26a), we can obtain that

$$W_{x_2}^1 = \frac{1}{Q} \times \frac{P(c_2)}{1-P(c_1)} \times \sum_{q=1}^{Q} \text{diff}(x_2,X_1,M_q^{c_2}) + \frac{1}{Q} \times \frac{P(c_3)}{1-P(c_1)}$$

$$\times \sum_{q=1}^{Q} \text{diff}(x_2,X_1,M_q^{c_3}) - \frac{1}{Q}\sum_{q=1}^{Q} \text{diff}(x_2,X_1,H_q)$$

$$= \frac{1}{2} \times \frac{\frac{3}{6}}{1-\frac{2}{6}} \times 1.423 + \frac{1}{2} \times \frac{\frac{1}{6}}{1-\frac{2}{6}} \times 0.462 - \frac{1}{2} \times 0.746 = 0.218.$$

(3) Along the x_3 direction, only one sample in the same class is sample 2, the nearest two samples in class 2 are sample 4 and sample 5, and only one sample in class 3 is sample 6. Thus

$$\sum_{q=1}^{Q} \text{diff}(x_3, X_1, H_q) = \frac{|x_{2,3} - x_{1,3}|}{\max(x_3) - \min(x_3)} = \frac{|-17.89 + 11.40|}{-3.75 + 17.89} = 0.459,$$

$$\sum_{q=1}^{Q} \text{diff}(x_3, X_1, M_q^{c_2}) = \frac{|x_{4,3} - x_{1,3}|}{\max(x_3) - \min(x_3)} + \frac{|x_{5,3} - x_{1,3}|}{\max(x_3) - \min(x_3)} = 1.064,$$

$$\sum_{q=1}^{Q} \text{diff}\left(x_3, X_1, M_q^{c_3}\right) = \frac{|x_{6,3} - x_{1,3}|}{\max(x_3) - \min(x_3)} = \frac{|-4.32 + 11.40|}{-3.75 + 17.89} = 0.501.$$

Using Eq. (11.26a), we obtain

$$W_{x_3}^1 = \frac{1}{Q} \times \frac{P(c_2)}{1 - P(c_1)} \times \sum_{q=1}^{Q} \text{diff}\left(x_3,X_1,M_q^{c_2}\right) + \frac{1}{Q} \times \frac{P(c_3)}{1 - P(c_1)} \times \sum_{q=1}^{Q} \text{diff}\left(x_3,X_1,M_q^{c_3}\right)$$
$$- \frac{1}{Q}\sum_{q=1}^{Q} \text{diff}(x_3,X_1,H_q)$$
$$= \frac{1}{2} \times \frac{\frac{3}{6}}{1 - \frac{2}{6}} \times 1.064 + \frac{1}{2} \times \frac{\frac{1}{6}}{1 - \frac{2}{6}} \times 0.501 - \frac{1}{2} \times 0.459 = 0.232.$$

Sample 2 (X_2, y_2)(MA, 8.17, −17.89, 1).

(1) Along the x_1 direction, only one sample in the same class is sample 1, the nearest two samples in class 2 are sample 3 and sample 4, and only one sample in class 3 is sample 6. Thus

$$\sum_{q=1}^{Q} \text{diff}(x_1, X_2, H_q) = \text{diff}(x_1, X_2, H_1) = 0,$$

$$\sum_{q=1}^{Q} \text{diff}\left(x_1, X_2, M_q^{c_2}\right) = \text{diff}\left(x_1, X_2, M_1^{c_2}\right) + \text{diff}\left(x_1, X_2, M_2^{c_2}\right) = 1 + 1 = 2,$$

$$\sum_{q=1}^{Q} \text{diff}\left(x_1, X_2, M_q^{c_3}\right) = \text{diff}\left(x_1, X_2, M_1^{c_3}\right) = 1.$$

Using Eq. (11.26a), we obtain

$$W_{x_1}^2 = \frac{1}{Q} \times \frac{P(c_2)}{1 - P(c_1)} \times \sum_{q=1}^{Q} \text{diff}\left(x_1,X_2,M_q^{c_2}\right) + \frac{1}{Q} \times \frac{P(c_3)}{1 - P(c_1)}$$
$$\times \sum_{q=1}^{Q} \text{diff}\left(x_1,X_2,M_q^{c_3}\right) - \frac{1}{Q}\sum_{q=1}^{Q} \text{diff}(x_1,X_2,H_q)$$
$$= \frac{1}{2} \times \frac{\frac{3}{6}}{1 - \frac{2}{6}} \times 2 + \frac{1}{2} \times \frac{\frac{1}{6}}{1 - \frac{2}{6}} \times 1 - \frac{1}{2} \times 0 = 0.875.$$

(2) Along the x_2 direction, only one sample in the same class is sample 1, the nearest two samples in class 2 are sample 3 and sample 4, and only one sample in class 3 is sample 6. Thus

$$\sum_{q=1}^{Q} \text{diff}(x_2, X_2, H_q) = \frac{|x_{1,2} - x_{2,2}|}{\max(x_2) - \min(x_2)} = \frac{|7.20 - 8.17|}{8.50 - 7.20} = 0.746,$$

$$\sum_{q=1}^{Q} \text{diff}\left(x_2, X_2, M_q^{c_2}\right) = \frac{|x_{3,2} - x_{2,2}|}{\max(x_2) - \min(x_2)} + \frac{|x_{4,2} - x_{2,2}|}{\max(x_2) - \min(x_2)} = 0.192,$$

$$\sum_{q=1}^{Q} \text{diff}\left(x_2, X_2, M_q^{c_3}\right) = \frac{|x_{6,2} - x_{2,2}|}{\max(x_2) - \min(x_2)} = \frac{|7.80 - 8.17|}{8.50 - 7.20} = 0.285.$$

Using Eq. (11.26a), we obtain

$$\begin{aligned} W_{x_2}^2 &= \frac{1}{Q} \times \frac{P(c_2)}{1 - P(c_1)} \times \sum_{q=1}^{Q} \text{diff}\left(x_2, X_2, M_q^{c_2}\right) + \frac{1}{Q} \times \frac{P(c_3)}{1 - P(c_1)} \\ &\quad \times \sum_{q=1}^{Q} \text{diff}\left(x_2, X_2, M_q^{c_3}\right) - \frac{1}{Q} \sum_{q=1}^{Q} \text{diff}(x_2, X_2, H_q) \\ &= \frac{1}{2} \times \frac{\frac{3}{6}}{1 - \frac{2}{6}} \times 0.192 + \frac{1}{2} \times \frac{\frac{1}{6}}{1 - \frac{2}{6}} \times 0.285 - \frac{1}{2} \times 0.746 = -0.265. \end{aligned}$$

(3) Along the x_3 direction, only one sample in the same class is sample 1, the nearest two samples in class 2 are sample 4 and sample 5, and only one sample in class 3 is sample 6. Thus

$$\sum_{q=1}^{Q} \text{diff}(x_3, X_2, H_q) = \frac{|x_{1,3} - x_{2,3}|}{\max(x_3) - \min(x_3)} = \frac{|-11.40 + 17.89|}{-3.75 + 17.89} = 0.459,$$

$$\sum_{q=1}^{Q} \text{diff}\left(x_3, X_2, M_q^{c_2}\right) = \frac{|x_{4,3} - x_{2,3}|}{\max(x_3) - \min(x_3)} + \frac{|x_{5,3} - x_{2,3}|}{\max(x_3) - \min(x_3)} = 1.982,$$

$$\sum_{q=1}^{Q} \text{diff}\left(x_3, X_2, M_q^{c_3}\right) = \frac{|x_{6,3} - x_{2,3}|}{\max(x_3) - \min(x_3)} = \frac{|-4.32 + 17.89|}{-3.75 + 17.89} = 0.960.$$

Using Eq. (11.26a), we obtain

$$W_{x_3}^2 = \frac{1}{Q} \times \frac{P(c_2)}{1 - P(c_1)} \times \sum_{q=1}^{Q} \text{diff}\left(x_3, X_2, M_q^{c_2}\right) + \frac{1}{Q} \times \frac{P(c_3)}{1 - P(c_1)}$$

$$\times \sum_{q=1}^{Q} \text{diff}\left(x_3, X_2, M_q^{c_3}\right) - \frac{1}{Q}\sum_{q=1}^{Q} \text{diff}(x_3, X_2, H_q)$$

$$= \frac{1}{2} \times \frac{\frac{3}{6}}{1-\frac{2}{6}} \times 1.982 + \frac{1}{2} \times \frac{\frac{1}{6}}{1-\frac{2}{6}} \times 0.960 - \frac{1}{2} \times 0.459 = 0.634.$$

In the same way, the weights of three features are calculated for all $n = 6$ samples.

Sample 3 $(X_3, y_3)(\text{AC}, 8.00, -4.16, 2)$, $W_{x_1}^3 = 0.667$, $W_{x_2}^3 = -0.014$, $W_{x_3}^3 = 0.476$;

Sample 4 $(X_4, y_4)(\text{AC}, 8.25, -3.75, 2)$, $W_{x_1}^4 = 0.667$, $W_{x_2}^4 = 0.155$, $W_{x_3}^4 = 0.497$;

Sample 5 $(X_5, y_5)(\text{AC}, 8.50, -4.00, 2)$, $W_{x_1}^5 = 0.667$, $W_{x_2}^5 = 0.219$, $W_{x_3}^5 = 0.491$;

Sample 6 $(X_6, y_6)(\text{AC}, 7.80, -4.32, 3)$, $W_{x_1}^6 = 0.400$, $W_{x_2}^6 = 0.293$, $W_{x_3}^6 = 0.302$.
The ReliefF values of three features are given as the average weight

$$W_{x_1}^{\text{AVG}} = \frac{\sum_{i=1}^{6} W_{x_1}^i}{6} = 0.692,$$

$$W_{x_2}^{\text{AVG}} = \frac{\sum_{i=1}^{6} W_{x_2}^i}{6} = 0.101,$$

$$W_{x_3}^{\text{AVG}} = \frac{\sum_{i=1}^{6} W_{x_3}^i}{6} = 0.409.$$

If we set a critical value of 0.2 for the ReliefF values, the feature VEC will be removed.

11.2.2 Wrapper Approach

The wrapper approach of feature ranking and feature selection is such a method that wraps (utilizes) an ML algorithm to train ML models with different feature subsets selected from an original feature set. From an original feature set of m features $x_i(i = 1, 2, \cdots, m)$, there are m single-feature subsets and m $(m-1)$-feature subsets. The sequential forward selector (SFS) and the sequential backward selector (SBS) are described in the mRMR feature selection, and they are described here again, because both are the most popular wrapper approaches of feature selection.

11.2.2.1 Sequential Forward Selector

As described above SFS firstly selects the most important feature by evaluating the performance of m ML models with the m single-feature subsets, one ML model for a one-feature subset. The selected most important feature is then combined with each of the rest $(m - 1)$ features to forms the $(m - 1)$ bifeature subsets. After that, SFS selects the most important bifeature subset by evaluating the performance of (m−1) ML models with the $(m - 1)$ bifeature subsets, one ML model for a bifeature subset, thereby selecting the second important feature. In the same way, SFS ranks all features. When the performance is assessed by cross-validation, SFS will select features with the best performance subset by adding relevant features one by one.

11.2.2.2 Sequential Backward Selector

SBS firstly selects the most important $(m - 1)$-feature subset by evaluating the performance of m ML models with the m $(m - 1)$-feature subsets, one ML model for an $(m - 1)$-feature subset, thereby selecting the most irrelevant feature. The survived (selected) $(m - 1)$-feature subset can form $(m - 1)$ $(m - 2)$-feature subsets. Then, SBS selects the most important $(m-2)$-feature subset by evaluating the performance of $(m - 1)$ ML models with the $(m - 1)$ $(m - 2)$-feature subsets, one ML model for an $(m - 2)$-feature subset, thereby selecting the second irrelevant feature. In the same way, SBS ranks all features. When the performance is assessed by cross-validation, SBS will select features with the best performance subset by deleting irrelevant features one by one.

As described above, with cross-validation, SFS selects features by adding relevant features one by one, whereas SBS selects features by removing irrelevant features one by one. Both SFS and SBS are heuristic feature selection approaches, which select the local best performance feature subset, rather than the global best performance feature subset. For example, if four features (x_1, x_2, x_3, x_4) are originally proposed and SFS selects the (x_1, x_3) subset, the (x_2, x_3) and (x_3, x_4) subsets will not be evaluated at all. With the four initially proposed features, SBS may select the (x_1, x_2, x_3) subset without evaluating the (x_4, x_2, x_3) and (x_1, x_4, x_3) subsets.

11.2.2.3 Exhaustive Feature Selector

It is possible that the global best performance feature subset exists in the missing evaluation feature subsets. The exhaustive feature selector (EFS) is a brute force approach and evaluates all possible feature subsets to select the global best performance feature subset. If m features are initially proposed, there will be $(2^m - 1)$ feature subsets. EFS utilizes an ML algorithm associated with cross-validation to evaluate the performance of every feature subset.

Example 11.14 The 278 bulk metallic glass data from the collected dataset (Xiong et al, 2020) include five initial features and one response of the shear modulus. The five initial features including the atomic size difference (δ), the average atomic volume (V), the average electronegativity (χ), the electronegativity difference (Δ), and the mixing entropy (S) are defined as $\delta = \sqrt{\sum a_i\left(1 - \frac{r_i}{r}\right)^2}$, $V = \sum a_i \cdot \frac{4}{3}\pi \cdot r_i^3$, $\chi = \sum a_i \chi_i$, $\Delta = \sqrt{\sum_{i=1}^{N} a_i(\chi_i - \chi)^2}$, $S = -R\sum_{i=1}^{N} a_i \ln a_i$, where a_i and x_i (r_i and χ_i) are the atomic fraction and elemental properties (the atomic radius and the electronegativity) of the i-th constituent, respectively, and $r = \sum a_i r_i$ is the average atomic radius. SFS, SBS, and EFS wrap the random forest (RF) algorithm with ten-fold cross-validation (CV) to select features. The coefficient of determination of CV-R^2 values is used to measure the ML performance.

Table 11.17 SFS single-feature subsets

Subsets	CV-R^2
(δ)	0.1148
(V)	0.8874
(χ)	0.8794
(Δ)	−0.0354
(S)	−0.2728

The single-feature subset (V) wins the first step of SFS (Table 11.17).

Table 11.18 SFS bifeature subsets

Subsets	CV-R^2
(δ, V)	0.9509
(V, χ)	0.9308
(V, Δ)	0.9373
(V, S)	0.9055

The bifeature subset (δ, V) wins the second step of SFS (Table 11.18).

Table 11.19 SFS trifeature subsets

Subsets	CV-R^2
(δ, V, χ)	0.9599
(δ, V, Δ)	0.9577
(δ, V, S)	0.9543

The trifeature subset (δ, V, χ) wins the third step (Table 11.19).

The 4-feature subset (δ, V, χ, Δ) wins the fourth step (Table 11.20). The final SFS step gives the full set (δ, V, χ, Δ, S) CV-$R^2 = 0.9588$. Table 11.21 lists all performances of SFS selected subsets, indicating that SFS selects the four features of (δ, V, χ, Δ) with CV-$R^2 = 0.9615$.

Table 11.20 SFS 4-feature subsets

Subsets	CV-R^2
$(\delta, V, \chi, \Delta)$	0.9615
(δ, V, χ, S)	0.9580

Table 11.21 SFS-selected feature subsets

Selected feature subsets	CV-R^2
(V)	0.8874
(δ, V)	0.9509
(δ, V, χ)	0.9599
$(\delta, V, \chi, \Delta)$	0.9615
$(\delta, V, \chi, \Delta, S)$	0.9588

Table 11.22 SBS 4-feature subsets

Subsets	CV-R^2
(V, χ, Δ, S)	0.9433
$(\delta, \chi, \Delta, S)$	0.9709
(δ, V, Δ, S)	0.9564
(δ, V, χ, S)	0.9580
$(\delta, V, \chi, \Delta)$	0.9615

In the first step of SBS, the full feature set $(\delta, V, \chi, \Delta, S)$ has the CV-$R^2 = 0.9588$. The 4-feature subset $(\delta, \chi, \Delta, S)$ wins the second step of SBS by removing feature V (Table 11.22).

Table 11.23 SBS trifeature subsets

Subsets	CV-R^2
(χ, Δ, S)	0.9357
(δ, Δ, S)	0.7012
(δ, χ, S)	0.9684
(δ, χ, Δ)	0.9730

The trifeature subset (δ, χ, Δ) wins the third step of SBS by removing feature S (Table 11.23).

Table 11.24 SBS bifeature subsets

Subsets	CV-R^2
(χ, Δ)	0.9540
(δ, Δ)	0.6293
(δ, χ)	0.9699

All bifeature subsets perform poorly than the trifeature subset (δ, χ, Δ) (Table 11.24). As expected, the best single-feature subset (χ) exhibits even worse performance with CV-$R^2 = 0.7431$. As a result, SBS selects the three-feature subset of (δ, χ, Δ).

Figure 11.15 shows the SFS and SBS results, indicating that SBS selected trifeature subset performs better than the SFS selected 4-feature subset.

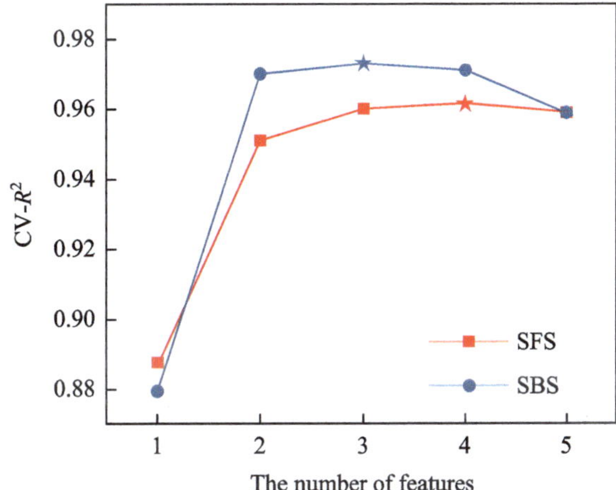

Fig. 11.15 The SFS and SBS feature selection process, where red and blue stars represent the selected subset via SFS and SBS, respectively

Table 11.25 lists the EFS results on the $2^5 - 1 = 31$ feature subsets.

EFS selects three features (δ, χ, Δ) with the CV-R^2 value of 0.9730, which is identical to that selected by SBS. EFS leads to the global maximization with an exhaustive search, while SBS or SFS might obtain a local maximization. The exhaustive search is an NP-hard problem, requires a huge computation source and consumes a lot of computation time, especially when the feature dimensions are high. In practice, SBS and SFS are commonly used in feature selection.

Table 11.25 The performance (CV-R^2) of all 31 possible feature subsets

Feature subsets	CV-R^2	Feature subsets	CV-R^2
(δ)	0.1148	(δ, V, Δ)	0.9577
(V)	0.8874	(δ, V, S)	0.9543
(χ)	0.8794	(δ, χ, Δ)	0.9730
(Δ)	−0.0354	(δ, χ, S)	0.9684
(S)	−0.2728	(δ, Δ, S)	0.7012
(δ, V)	0.9509	(V, χ, Δ)	0.9562
(δ, χ)	0.9699	(V, χ, S)	0.9225
(δ, Δ)	0.6293	(V, Δ, S)	0.9317
(δ, S)	0.4360	(χ, Δ, S)	0.9357
(V, χ)	0.9308	$(\delta, V, \chi, \Delta)$	0.9615
(V, Δ)	0.9373	(δ, V, χ, S)	0.9580
(V, S)	0.9055	(δ, V, Δ, S)	0.9564
(χ, Δ)	0.9540	$(\delta, \chi, \Delta, S)$	0.9709
(χ, S)	0.9267	(V, χ, Δ, S)	0.9433
(Δ, S)	0.4589	$(\delta, V, \chi, \Delta, S)$	0.9588
(δ, V, χ)	0.9599		

11.2.2.4 Permutation Feature Importance

Consider an original dataset of n data, $(x_{i1}, \cdots, x_{ij}, \cdots, x_{im}, y_i)$ $(i = 1, 2, \cdots, n;\ j = 1, 2, \cdots, m)$, which is randomly divided into three subsets, training, validating, and testing data subsets with n_1, n_2 and n_3 data, respectively. An ML model is developed on the training and validating data subsets and the testing subset will assess the performance of the developed model. The permutation feature importance is defined as the change in the prediction score of a developed ML on a testing data subset when the values of a feature in a testing data subset are randomly shuffled, meanwhile the values of other features remain unchanged in the testing data subset (Breiman, 2001). The value shuffling of a feature, say feature j, generates a shuffled dataset $(x_{i1}, \cdots, \hat{x}_{ij}, \cdots, x_{im}, y_i)$, where the values of $\hat{x}_{ij}$ $(i = 1, 2, \cdots, n_3)$ are randomly re-sampled without replacement from the original value set of (x_{ij}), and the shuffled testing dataset is predicted by the developed model. If the joint dataset of training and validating subsets is called the generalized training dataset, the value shuffling of a feature and the permutation feature importance can be calculated on the generalized training set. The permutation feature importance calculated on a testing subset can highlight which features contribute most to the generalization of the developed model, which is emphasized here. Since the value shuffling of a feature provides a new testing dataset, the prediction of the developed model on the shuffled dataset will change. Thus, the change in the ML model prediction indicates how much the model depends on the feature in which values are shuffled. Usually, the change is a drop in prediction and hence a more accurate method to measure the feature importance is mean decrease accuracy. The values of feature j are randomly shuffled many times and the average decrease in accuracy is calculated by

$$P_j = s - \frac{1}{K}\sum_{k=1}^{K} S_{j,k}, \tag{11.27}$$

where P_j represents the permutation feature importance of feature j, s and $S_{j,k}$ are the prediction accuracy of the trained and validated ML model on the original and the k-th shuffled datasets, respectively, and K is the number of shuffling repetitions.

Example 11.15 The efficiency of nitrogen oxide (NOx) conversion of $n = 29$ catalyst data listed in Table 11.26 is employed here to show the calculation detail of the permutation feature importance. About 30% of the data, $n_3 = 9$, are randomly selected as the testing data subset and the rest data form the generalized training subset. A decision tree regression model is developed on the generalized training subset with the leave-one-out cross-validation, which has the coefficient of determination $s = R^2 = 0.886$. The shuffling is done by randomly re-sampling without replacement.

Taking feature SO_2 as an example, shuffling ten times yields ten values of R^2, which are given below:

$$S_{SO_2} = [-1.052, -0.389, -0.739, 0.886, -1.186, -0.739, -1.179, \\ -1.179, -0.739].$$

Eight of nine values of R^2 are negative, meaning that shuffling values of feature SO_2 completely ruins the developed ML model. This result indicates that the developed ML model is very sensitive to feature SO_2, and equivalently, feature SO_2 is important to the developed ML model. Table 11.26 shows there are many same values for different samples, which enhances the possibility that the shuffled values are the same as the original ones, $\hat{x}_{ij} = x_{ij}$ $(i = 1, 2, \cdots, n_3)$. That is why one value of R^2 is identical to the original one. The permutation feature importance of feature SO_2 is calculated to be

$$P_{SO_2} = s - \frac{1}{9}\sum_{k=1}^{9} S_{SO_2,k} = 1.588.$$

Table 11.26 The efficiency of NOx conversion of 29 catalysts

Catalysts	Mn (at.%)	Ce (at.%)	Fe (at.%)	GHSV (h^{-1})	H2O (vol%)	SO_2 (ppm)	T (°C)	NOx conversion
Mn_8Ce_1/MnO_x	0.1407	0.075	0	24,000	0	0	80	0.99
$Mn_{16}Ce_1/MnO_x$	0.1723	0.0693	0	24,000	0	0	100	0.99
Mn_2Ce_1/MnO_x	0.0789	0.1155	0	24,000	10	0	140	0.93
Mn_2Ce_1/MnO_x	0.0789	0.1155	0	24,000	0	200	140	0.73
Mn_2Ce_1/MnO_x	0.0789	0.1155	0	24,000	10	200	140	0.65
Mn_8Ce_1/MnO_x	0.1407	0.075	0	24,000	10	0	140	0.98
Mn_8Ce_1/MnO_x	0.1407	0.075	0	24,000	0	200	140	0.77
Mn/Ce-Acetic acid	0.0391	0	0	60,000	5	100	200	0.67
Mn/Ce-Ethanol	0.0383	0	0	60,000	5	100	200	0.65
$Ce_{0.2}/MnOx$	0.116	0.023	0	40,000	0	100	200	0.70
Mn/Ce-Acetic acid	0.0391	0	0	60,000	0	0	200	0.90
$Fe_{0.2}Mn_{0.8}/TiO_2$	0.139	0	0.023	60,000	0	0	200	0.96
$Fe_{0.3}Mn_{0.8}/TiO_2$	0.132	0	0.033	60,000	0	0	200	0.97
$Fe_{0.1}Mn_{0.8}/TiO_2$	0.153	0	0	60,000	0	0	250	0.90
$Fe_{0.2}Mn_{0.8}/TiO_2$	0.148	0	0.012	60,000	0	0	200	0.92
Mn/Ce-Oxalic acid	0.0496	0	0	60,000	0	0	175	0.93
$Ce_{0.2}/TiO_2$	0	0.046	0	50,000	0	0	300	0.96
$Ce_{0.6}/TiO_2$	0	0.053	0	50,000	0	0	300	0.99
Pure CeO_2	0	0.328	0	50,000	0	0	400	0.85
$Ce_{0.6}/TiO_2$	0	0.053	0	100,000	0	0	350	0.98
$Ce_{0.6}/TiO_2$	0	0.053	0	30,000	0	0	250	0.96
$Ce_{0.6}/TiO_2$	0	0.053	0	12,000	0	0	200	0.99
$Ce_{0.01}/TiO_2$	0	0.0001	0	50,000	0	0	400	0.81

(continued)

Table 11.26 (continued)

Catalysts	Mn (at.%)	Ce (at.%)	Fe (at.%)	GHSV (h^{-1})	H2O (vol%)	SO_2 (ppm)	T (°C)	NOx conversion
$Ce_{0.5}/TiO_2$	0	0.082	0	30,000	0	0	275	1
$Fe_{0.02}Ce/TiO_2$	0	0.079	0.006	30,000	0	0	250	1
$Fe_{0.1}Ce/TiO_2$	0	0.075	0.011	30,000	0	0	250	1
$Fe_{0.2}Ce/TiO_2$	0	0.074	0.015	30,000	0	0	225	1
$Fe_{0.7}Ce/TiO_2$	0	0.068	0.037	30,000	0	0	225	1
$Ce_{0.6}/TiO_2$	0	0.053	0	50,000	0	0	350	0.98

The permutation feature importance scores of other features are calculated in the same way and plotted in Fig. 11.16, indicating that feature SO_2 has the highest permutation feature importance.

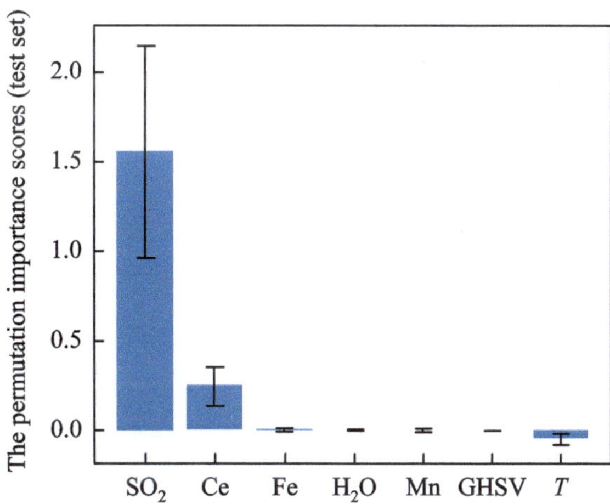

Fig. 11.16 The permutation feature importance scores calculated on the testing dataset ten times, where the blue bars and line segments represent the average values and standard deviations, respectively

---

11.2.3 Embedded Approach

Embedded approaches evaluate feature importance based on the feature behavior in ML models, e.g., in tree-based models. Chapter 5 describes the decision tree in detail, and Chap. 6 introduces the ensemble learning in detail, especially the tree-based ensemble learning algorithms. Figure 11.17 shows a decision tree with splitting by three features of x_1, x_2 and x_3.

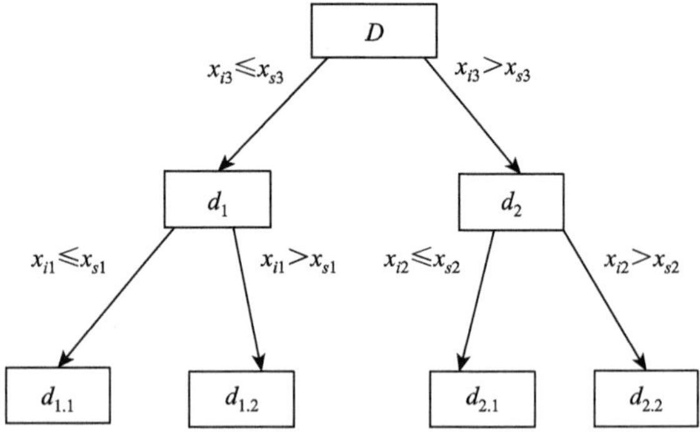

Fig. 11.17 Schematic of a decision tree

The simplest way to measure the feature importance might just be to count the number of times that a feature is used for splitting in an individual tree. For the example shown in Fig. 11.1, each feature is used once for splitting so that each of the three features has the same importance. This approach is adopted in XGBoost as the "weight" method (Chen & Guestrin, 2016); however, it is not widely used in practice.

The impurity used in the evaluation of feature importance usually means the Gini index in classification and the mean squared error (MSE) in regression. When node d is split, by the optimal feature with its optimal value, into child left and right nodes d_{L} and d_{R}, impurity decrease $\Delta i(s_d, d)$ is defined by

$$\Delta i(s_d, d) = \mathrm{Imp}_d - \frac{n_{d_{\mathrm{L}}}}{n_d}\mathrm{Imp}_{d_{\mathrm{L}}} - \frac{n_{d_{\mathrm{R}}}}{n_d}\mathrm{Imp}_{d_{\mathrm{R}}}, \tag{11.28a}$$

where s_d denotes the best splitting, Imp means impurity, n denotes the data number, the subscript denotes the node, and $\frac{n_{d_{\mathrm{L}}}}{n_d}$ and $\frac{n_{d_{\mathrm{R}}}}{n_d}$ are the ratios of data numbers in d_{L} and d_{R} over the data number in node d, respectively. As described in Chap. 5, the best splitting is determined by the maximal impurity decrease, which indicates that the impurity decrease $\Delta i(s_d, d)$ can be easily calculated for all nodes except for terminal nodes in a trained tree. Since there are many trees in a tree-based ensemble learning algorithm, the importance of feature x_j $(j = 1,\ 2,\ \cdots\ ,\ m)$ in decision tree φ is measured by its mean decrease impurity (MDI) (Breiman, 2001), which is defined by

$$\mathrm{Imp}_{\varphi}(x_j) = \sum_{d \in \varphi} I(j_d = j)[p(d)\Delta i(s_d, d)], \tag{11.28b}$$

where $I(j_d = j) = 1$ if feature x_j is used for splitting node d, otherwise $I(j_d = j) = 0$; $p(d) = \frac{n_d}{n}$ with n being the total number of samples. It might occur that some features are never used to split nodes in tree φ and hence their Imp_φ will be zero. This case rarely happens in the random forest thanks to two reasons. The first reason is that the random forest randomly selects a subset f from the entire feature set F at each node so that the possibility of every feature becoming the splitting one increases. The second reason is that a feature that is not used for splitting in a tree may have a high possibility of being used for splitting in other trees due to the change in the data subset.

The mean impurity decrease in the random rorest is named random forest feature importance (Breiman, 2001), and expressed by

$$\text{Imp}_{\text{RF}}(x_j) = \frac{1}{K}\sum_{k=1}^{K}\sum_{d\in\varphi_k} I(j_d = j)[p(d)\Delta i(s_d, d)], \tag{11.28c}$$

where φ_k denotes the tree k ($k = 1, 2, \cdots, K$) and K is the number of trees in the forest.

As described in Chap. 6, the random forest is an ensemble ML algorithm (Breiman, 2001). A bootstrap sample subset (in-bag observations) is randomly selected from a dataset with replacement and the subset is used to build a decision tree. Out-of-bag observations that are not used to build a decision tree can be utilized to estimate the corresponding tree's prediction performance. A powerful RF model is formed by averaging the predictions of all individual trees. A randomly selected feature subset f from the entire feature set F is examined at each node of every decision tree. The node splits on the best feature with the optimal value in subset f according to the maximal impurity decrease criterion. Then, using Eq. (11.28a)–(11.28c), one is able to calculate the feature importance in a trained RF model.

Example 11.16 A dataset consists of data $(\boldsymbol{X}_1, y_1)$, $(\boldsymbol{X}_2, y_2)$, $(\boldsymbol{X}_3, y_3)$, $(\boldsymbol{X}_4, y_4)$, and $(\boldsymbol{X}_5, y_5)$.

$$\boldsymbol{X} = \begin{bmatrix} \boldsymbol{X}_1 \\ \boldsymbol{X}_2 \\ \boldsymbol{X}_3 \\ \boldsymbol{X}_4 \\ \boldsymbol{X}_5 \end{bmatrix} = \begin{bmatrix} x_{11} & x_{12} & x_{13} \\ x_{21} & x_{22} & x_{23} \\ x_{31} & x_{32} & x_{33} \\ x_{41} & x_{42} & x_{43} \\ x_{51} & x_{52} & x_{53} \end{bmatrix} = \begin{bmatrix} 3 & 1 & 2 \\ 7 & 2 & 3 \\ 11 & 3 & 4 \\ 15 & 4 & 5 \\ 19 & 5 & 9 \end{bmatrix} \text{ and y} = \begin{bmatrix} y_1 \\ y_2 \\ y_3 \\ y_4 \\ y_5 \end{bmatrix} = \begin{bmatrix} 50 \\ 170 \\ 290 \\ 362 \\ 485 \end{bmatrix}.$$

An RF model is developed including two trees, each with a data subset built up by bootstrapping, and two features are randomly selected from the original three features for splitting. The two data subsets used are shown below, and the trained trees φ_1 and φ_2 are shown in Figs. 11.18 and 11.19.

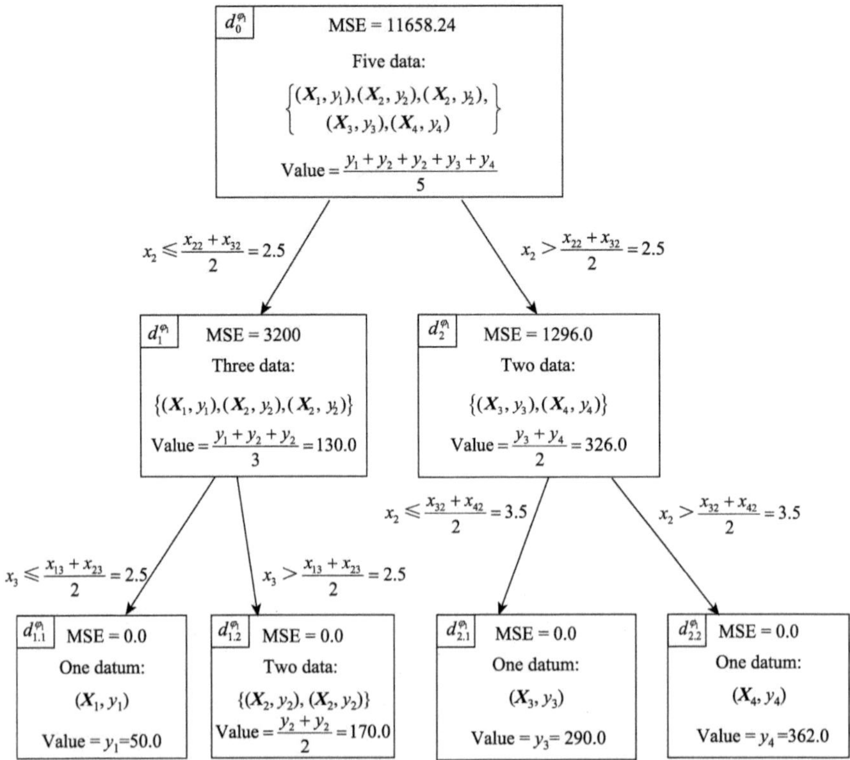

Fig. 11.18 The decision tree φ_1 of the developed RF model is trained on the bootstrap data subset $(X^{\text{Bootstrap\_1}}, y^{\text{Bootstrap\_1}})$ with features x_2 and x_3

$$\boldsymbol{X}^{\text{Bootstrap\_1}} = \begin{bmatrix} \boldsymbol{X}_1 \\ \boldsymbol{X}_2 \\ \boldsymbol{X}_2 \\ \boldsymbol{X}_3 \\ \boldsymbol{X}_4 \end{bmatrix} = \begin{bmatrix} x_{11} & x_{12} & x_{13} \\ x_{21} & x_{22} & x_{23} \\ x_{21} & x_{22} & x_{23} \\ x_{31} & x_{32} & x_{33} \\ x_{41} & x_{42} & x_{43} \end{bmatrix} = \begin{bmatrix} 3 & 1 & 2 \\ 7 & 2 & 3 \\ 7 & 2 & 3 \\ 11 & 3 & 4 \\ 15 & 4 & 5 \end{bmatrix} \text{ and}$$

$$\boldsymbol{y}^{\text{Bootstrap\_1}} = \begin{bmatrix} y_1 \\ y_2 \\ y_2 \\ y_3 \\ y_4 \end{bmatrix} = \begin{bmatrix} 50 \\ 170 \\ 170 \\ 290 \\ 362 \end{bmatrix},$$

$$\boldsymbol{X}^{\text{Bootstrap\_2}} = \begin{bmatrix} \boldsymbol{X}_3 \\ \boldsymbol{X}_4 \\ \boldsymbol{X}_4 \\ \boldsymbol{X}_5 \\ \boldsymbol{X}_5 \end{bmatrix} = \begin{bmatrix} x_{31} & x_{32} & x_{33} \\ x_{41} & x_{42} & x_{43} \\ x_{41} & x_{42} & x_{43} \\ x_{51} & x_{52} & x_{53} \\ x_{51} & x_{52} & x_{53} \end{bmatrix} = \begin{bmatrix} 11 & 3 & 4 \\ 15 & 4 & 5 \\ 15 & 4 & 5 \\ 19 & 5 & 9 \\ 19 & 5 & 9 \end{bmatrix} \text{ and}$$

$$\boldsymbol{y}^{\text{Bootstrap\_2}} = \begin{bmatrix} y_3 \\ y_4 \\ y_4 \\ y_5 \\ y_5 \end{bmatrix} = \begin{bmatrix} 290 \\ 362 \\ 362 \\ 485 \\ 485 \end{bmatrix}.$$

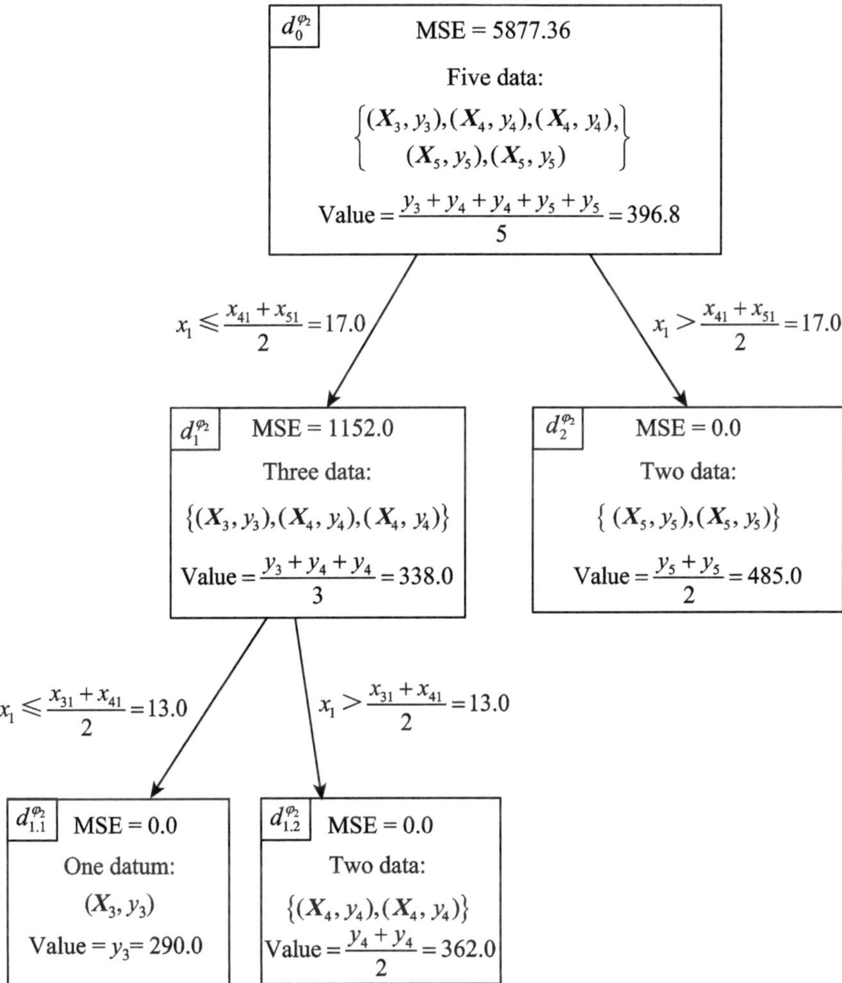

Fig. 11.19 The decision tree φ_2 of the developed RF model is trained on the bootstrap data subset $(X^{\text{Bootstrap\_2}}, y^{\text{Bootstrap\_2}})$ with feature x_1

In decision tree φ_1, feature x_1 is not used for splitting nodes at all, so that

$$\text{Imp}_{\varphi_1}(x_1) = 0.$$

Feature x_2 is used for splitting nodes $d_0^{\varphi_1}$ and $d_2^{\varphi_1}$. The impurity decreases of nodes $d_0^{\varphi_1}$ and $d_2^{\varphi_1}$ are obtained respectively by

$$\Delta i\left(s_{d_0^{\varphi_1}}, d_0^{\varphi_1}\right) = 11658.24 - \frac{3}{5} \times 3200.0 - \frac{2}{5} \times 1296.0 = 9219.84,$$
$$\Delta i\left(s_{d_2^{\varphi_1}}, d_2^{\varphi_1}\right) = 1296.0 - \frac{1}{2} \times 0.0 - \frac{1}{2} \times 0.0 = 1296.0.$$

The feature importance $\text{Imp}_{\varphi_1}(x_2)$ is then given by

$$\begin{aligned}\text{Imp}_{\varphi_1}(x_2) &= p(d_0^{\varphi_1})\Delta i\left(s_{d_0^{\varphi_1}}, d_0^{\varphi_1}\right) + p(d_2^{\varphi_1})\Delta i\left(s_{d_2^{\varphi_1}}, d_2^{\varphi_1}\right)\\ &= \frac{n_{d_0^{\varphi_1}}}{n} \times \Delta i\left(s_{d_0^{\varphi_1}}, d_0^{\varphi_1}\right) + \frac{n_{d_2^{\varphi_1}}}{n} \times \Delta i\left(s_{d_2^{\varphi_1}}, d_2^{\varphi_1}\right)\\ &= \frac{5}{5} \times 9219.84 + \frac{2}{5} \times 1296.0 = 9738.24.\end{aligned}$$

Feature x_3 is only used for splitting node $d_1^{\varphi_1}$, the impurity decrease of node $d_1^{\varphi_1}$ is

$$\Delta i\left(s_{d_1^{\varphi_1}}, d_1^{\varphi_1}\right) = 3200.0 - \frac{1}{3} \times 0.0 - \frac{2}{3} \times 0.0 = 3200.0,$$

and the feature importance $\text{Imp}_{\varphi_1}(x_3)$ is

$$\text{Imp}_{\varphi_1}(x_3) = p(d_1^{\varphi_1})\Delta i\left(s_{d_1^{\varphi_1}}, d_1^{\varphi_1}\right) = \frac{n_{d_1^{\varphi_1}}}{n} \times \Delta i\left(s_{d_1^{\varphi_1}}, d_1^{\varphi_1}\right) = \frac{3}{5} \times 3200.0 = 1920.0.$$

The feature importance of three features in tree φ_1 is finally given in the vector form by

$$\begin{bmatrix}\text{Imp}_{\varphi_1}(x_1)\\ \text{Imp}_{\varphi_1}(x_2)\\ \text{Imp}_{\varphi_1}(x_3)\end{bmatrix} = \begin{bmatrix}0.0\\ 9738.24\\ 1920.0\end{bmatrix}.$$

Similarly, the feature importance of three features in tree φ_1 is calculated and given by

$$\begin{bmatrix}\text{Imp}_{\varphi_2}(x_1)\\ \text{Imp}_{\varphi_2}(x_2)\\ \text{Imp}_{\varphi_2}(x_3)\end{bmatrix} = \begin{bmatrix}5877.36\\ 0.0\\ 0.0\end{bmatrix}.$$

Finally, the mean decrease impurity of feature x_j $(j = 1,\ 2,\ 3)$ on trees φ_1 and φ_2 represents the RF feature importance and is denoted as Imp_{RF}, which is expressed by

$$\begin{bmatrix} \mathrm{Imp}_{\mathrm{RF}}(x_1) \\ \mathrm{Imp}_{\mathrm{RF}}(x_2) \\ \mathrm{Imp}_{\mathrm{RF}}(x_3) \end{bmatrix} = \begin{bmatrix} \frac{\mathrm{Imp}_{\varphi_1}(x_1)+\mathrm{Imp}_{\varphi_2}(x_1)}{2} \\ \frac{\mathrm{Imp}_{\varphi_1}(x_2)+\mathrm{Imp}_{\varphi_2}(x_2)}{2} \\ \frac{\mathrm{Imp}_{\varphi_1}(x_3)+\mathrm{Imp}_{\varphi_2}(x_3)}{2} \end{bmatrix} = \begin{bmatrix} \frac{0.0+5877.36}{2} \\ \frac{9738.24+0.0}{2} \\ \frac{1920.0+0.0}{2} \end{bmatrix} = \begin{bmatrix} 2938.68 \\ 4869.12 \\ 960.0 \end{bmatrix}.$$

Figure 11.20 shows the RF feature importance of the three features.

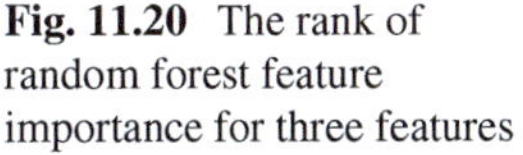

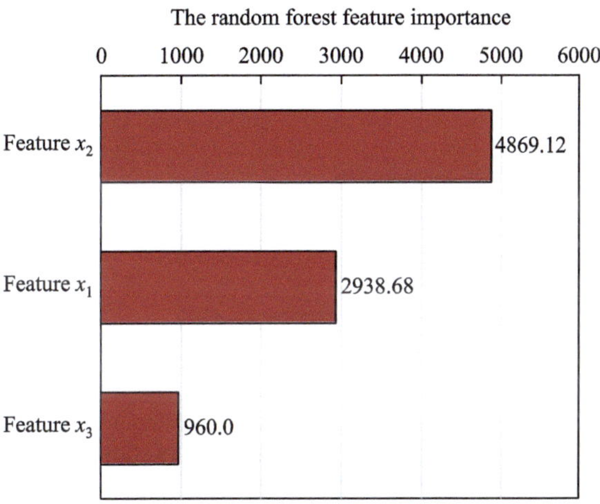

Fig. 11.20 The rank of random forest feature importance for three features

---

Homework

1. Table 11.27 shows four alloys and their enthalpies of mixing (Hmix) and entropies of mixing (Smix). Detect potential outliers among them by using the LOF approach. (Note: the distance utilized in this work should be the Euclidean distance d_{E}.) The distance between point $x = (x_1, x_2, \cdots, x_n)$ and point $y = (y_1, y_2, \cdots, y_n)$ is calculated as $d_{\mathrm{E}}(x, y) = \sqrt{\sum_{i=1}^{n}(x_i - y_i)^2}$.
2. Use isolated forests to find the potential anomalies in the dataset listed in Table 11.27.
3. Please give the Pearson correlation coefficient (R) matrix of features according to the dataset in Example 11.9.
4. Please describe the relationship between MIC and mRMR.
5. Rerun Example 11.13 of ReliefF with setting $Q = 1$.

Table 11.27 Alloys for LOF detection

Samples	Alloys	Smix ($J \cdot K^{-1} \cdot mol^{-1}$)	Hmix ($kJ \cdot mol^{-1}$)
a	$Ag_{20}Mg_{20}Y_{60}$	7.90	−18.40
b	$Ag_{10}Mg_{50}Y_{40}$	7.84	−11.44
c	$Ag_{14}Cu_{78.5}P_{7.5}$	5.48	−4.02
d	$Cu_{34}Zr_{50}Ag_{8}Al_{8}$	9.29	−25.87

6. Please give the merit and demerit of the three wrapper approaches of SFS, SBS and EFS.
7. With the dataset in Table 11.26 and the random forest algorithm, compare the two kinds of impurity-based feature importance, i.e., the permutation importance and the random forest feature importance of MDI.
8. With the dataset from Example 11.16, perform the feature ranking by LASSO and compare the ranking result with Example 11.16.

References

Breiman, L. (2001). Random forests. *Machine Learning, 45*(1), 5–32.

Breunig, M. M., Kriegel, H. P., Ng, R. T., et al. (2000). LOF: Identifying density-based local outliers. In *Proceedings of the 2000 ACM SIGMOD International Conference on Management of Data* (pp. 93–104). New York: Association for Computing Machinery.

Chen, T., & Guestrin, C. (2016). XGBoot: A scalable tree boosting system. In *The 22nd ACM SIGKDD International Conference. ACM*.

Cover, T. M., & Thomas, J. A. (2003). *Elements of information theory*. John Wiley & Sons.

Efron, B. (1979). Bootstrap methods: Another look at the jackknife. *The Annals of Statistics, 7*(1), 1–26.

Hu, Q., Guo, S., Wang, J. M., et al. (2017). Parametric study of amorphous high-entropy alloys formation from two new perspectives: Atomic radius modification and crystalline structure of alloying elements. *Scientific Reports, 7*(1), 1–12.

Kinney, J. B., & Atwal, G. S. (2014). Equitability, mutual information, and the maximal information coefficient. *Proceedings of the National Academy of Sciences, 111*(9), 3354–3359.

Kira, K., & Rendell, L. A. (1992). The Feature selection problem: Traditional methods and a new algorithm. *Proceedings of the Tenth National Conference on Artificial Intelligence, 2*, 129–132.

Kononerko I. (1994). Estimating attributes: Analysis and extension of relief. In *Proceedings of European Conference on Machine Learning* (pp. 171–182). New York: Springer-Verlag.

Liu, F. T., Ting, K. M., & Zhou, Z. H. (2012). Isolation-based anomaly detection. *ACM Transactions on Knowledge Discovery from Data (TKDD), 6*(1), 1–39.

Peng, H., Long, F., & Ding, C. (2005). Feature selection based on mutual information criteria of max-dependency, max-relevance, and min-redundancy. *IEEE Transactions on Pattern Analysis and Machine Intelligence, 27*(8), 1226–1238.

Reshef, D. N., Reshef, Y. A., Finucane, H. K., et al. (2011). Detecting novel associations in large data sets. *Science, 334*, 1518–1524.

Severson, K. A., Attia, P. M., Jin, N., et al. (2019). Data-driven prediction of battery cycle life before capacity degradation. *Nature Energy, 4*(5), 383–391.

Schölkopf, B., Platt, J. C., Shawe-Taylor, J., et al. (2001). Estimating the support of a high-dimensional distribution. *Neural Computation, 13*(7), 1443–1471.

Takeuchi, A., Amiya, K., Wada, T., et al. (2014). High-entropy alloys with a hexagonal close-packed structure designed by equi-atomic alloy strategy and binary phase diagrams. *JOM Journal of the Minerals Metals and Materials Society, 66*(10), 1984–1992.

Tax, D. M., & Duin, R. P. (1999). Support vector domain description. *Pattern Recognition Letters, 20*(11–13), 1191–1199.

Tax, D. M., & Duin, R. P. (2004). Support vector data description. *Machine Learning, 54*(1), 45–66.

Tsai, M. H., & Yeh, J. W. (2014). High-entropy alloys: A critical review. *Materials Research Letters, 2*(3), 107–123.

Xiong, J., Shi, S. Q., & Zhang, T. Y. (2020). A machine-learning approach to predicting and understanding the properties of amorphous metallic alloys. *Materials and Design, 187*, 108378.

Xiong, J., Shi, S. Q., & Zhang, T. Y. (2021). Machine learning of phases and mechanical properties in complex concentrated alloys. *Journal of Materials Science and Technology, 87*, 133–142.

Ye, Y. F., Wang, Q., Lu, J., et al. (2016). High-entropy alloy: Challenges and prospects. *Materials Today, 19*(6), 349–362.

Yeh, J. W., Chen, Y. L., Lin, S. J., et al. (2007). High-entropy alloys—A new era of exploitation. *Materials Science Forum, 560*, 1–9.

Zhang, Y., Zhou, Y. J., Lin, J. P., et al. (2008). Solid-solution phase formation rules for multi-component alloys. *Advanced Engineering Materials, 10*(6), 534–538.

Zhang, Y., Zuo, T. T., Tang, Z., et al. (2014). Microstructures and properties of high-entropy alloys. *Progress Materials Science, 61*, 1–93.

Chapter 12
Interpretative SHAP Value and Partial Dependence Plot

12.1 SHapley Additive ExPlanation Value

A SHapley Additive exPlanation (SHAP) value, whether positive or negative, reflects the contribution of a feature to a predicted response in one datum (Lundberg & Lee, 2017). SHAP values are able to rank the feature importance in an additive manner. Consider a data set with m features of $(x_1,\ x_2,\ \cdots,\ x_m)$ and one response $y = f(x_1,\ x_2,\ \cdots,\ x_m)$.. The game theory (Neumann, 1928a, 1928b) considers all possible contributions a feature x_i $(i = 1, 2, \cdots, m)$ to the response. Based on the game theory, if a data subset contains zero features, it is called an empty subset denoted by S_0 (Lipovetsky & Conklin, 2001). Similarly, if a data subset contains only one feature, two features, three features, and so on, the subset is named correspondingly a one-feature subset, a two-feature subset, a three-feature subset, etc., and donoted by S_1, S_2, S_3, etc. accordingly. One feature can function on its own, bi-function with any other one of the features, tri-function with any other two of the features, and so on. The co-function modes form a feature subset $S_j \cup \{i\}$ $(j = 0,\ 1,\ 2,\ \cdots, m-1)$, where $\{i\}$ means the subset includes feature x_i $(i = 1,\ 2,\ \cdots,\ m)$ only and subset S_j $(j = 0,\ 1,\ 2,\ \cdots, m-1)$ includes S_0 of the empty subset, S_1 of the one-feature subset, S_2 of the two-feature subset, $\cdots$, and S_{m-1} of the $(m-1)$-feature subset, but excludes subset $\{i\}$ $(i = 1,\ 2,\ \cdots, m)$. Obviously, the set of $S_{(m-1)} \cup \{i\}$ is the entire dataset. Thus, the contribution of feature x_i $(i = 1,\ 2,\ \cdots,\ m)$ via one single co-function is given by $W_j[f(S_j \cup \{i\}) - f(S_j)]$, where W_j is called the single contribution weight to $[f(S_j \cup \{i\}) - f(S_j)]$ via the j-feature subset $(j = 0,\ 1, \cdots,$ $m-1)$. In the j-feature subset $(j = 0,\ 1,\ \cdots,\ m-1)$ there are C_{m-1}^{j} co-functions (combinations) with each having the same single contribution weight W_j. Then, the subset weight of $[f(S_j \cup \{i\}) - f(S_j)]$ is $W_j C_{m-1}^{j}$. All contribution weights via all feature subsets are normalized to unit, i.e.,

T. Zhang, *An Introduction to Materials Informatics*,
https://doi.org/10.1007/978-981-99-7992-9_12

$$\sum_{j=0}^{m-1} W_j C_{m-1}^j = 1. \tag{12.1a}$$

Furthermore, the subset weight is treated to be the same, viz.,

$$W_0 \mathrm{C}_{m-1}^0 = W_1 \mathrm{C}_{m-1}^1 = \cdots = W_{m-1} \mathrm{C}_{m-1}^{m-1}. \tag{12.1b}$$

Combining Eqs. (12.1a) and (12.1b) yields

$$W_j = \frac{j!(m-j-1)!}{m!} (j = 0, 1, \cdots, m-1). \tag{12.1c}$$

Finally, the contribution of feature x_i ($i = 1, 2, \cdots, m$) to the response is named the SHAP value ϕ_i (Shapley, 1953) and given by

$$\phi_i = \sum_{S_j \subseteq M \setminus \{i\}} \frac{j!(m-j-1)!}{m!} [f(S_j \cup \{i\}) - f(S_j)], \tag{12.2a}$$

where M is the number of all subsets of co-functions. The response estimator (Lundberg & Lee, 2017) can be expressed by

$$\hat{y} = \phi_0 + \sum_{i=1}^{m} \phi_i, \tag{12.2b}$$

where ϕ_0 denotes the SHAP value of the empty subset, defined by

$$\phi_0 = f(\text{empty}). \tag{12.2c}$$

Figure 12.1 gives a schematic of all single contribution weights for $[f(S_j \cup \{i\}) - f(S_j)]$ ($i = 1, 2, 3$ and $j = 0, 1, 2$). There are three W_0, six W_1, and three W_2, and the four brown lines indicate $[f(S_j \cup \{1\}) - f(S_j)]$ ($j = 0, 1, 2$).

Figure 12.1 takes a three-feature system $y = f(x_1, x_2, x_3)$ to illustrate the above analysis, showing that three $[f(S_0 \cup \{i\}) - f(S_0)]$ with $\{i\}$ ($i = 1, 2, 3$) are with the same weight $W_0 = 1/3$. There are three S_1 and two S_1 exclude $\{i\}$. Thus, combining one S_1 with one $\{i\}$ ($i = 1, 2, 3$) forms one $[f(S_1 \cup \{i\}) - f(S_1)]$ and there are six combinations so that $W_1 = 1/6$. There are also three S_2 and only one S_2 excluding $\{i\}$ so that the weight for $[f(S_2 \cup \{i\}) - f(S_2)]$ is $W_2 = 1/3$. Thus, for three-feature system $y = f(x_1, x_2, x_3)$, we have $W_0 \times \mathrm{C}_2^0 + W_1 \times \mathrm{C}_2^1 + W_2 \times \mathrm{C}_2^2 = 1$ and $W_0 \times \mathrm{C}_2^0 = W_1 \times \mathrm{C}_2^1 = W_2 \times \mathrm{C}_2^2 = 1/3$, which yields $W_0 = 2W_1 = W_2 = 1/3$.

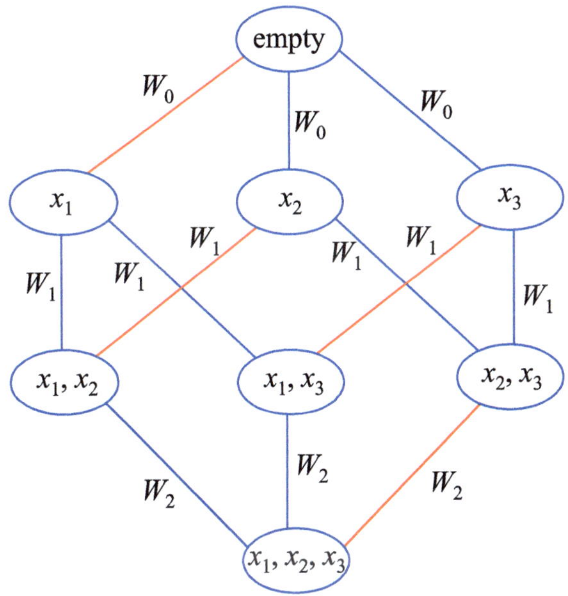

Fig. 12.1 The schematic show of all single contribution weights of $[f(S_j \cup \{i\}) - f(S_j)]$ $(i = 1, 2, 3$ and $j = 0, 1, 2)$. There are three W_0, six W_1, and three W_2, and the four brown lines indicate $[f(S_j \cup \{1\}) - f(S_j)]$ $(j = 0, 1, 2)$

The SHAP interaction value (Fujimoto et al., 2006) is calculated by

$$\phi_{i,j} = \sum_{S_j \subseteq M \backslash \{i,j\}} W_p \nabla_{ij}(S), \tag{12.3a}$$

where W_p is called the single interaction contribution weight between features i and j $(i \neq j; i, j = 1, 2, \cdots, m)$ via the p-feature subset $S(p = 0, 1, \cdots, m-2)$ excluding features i and j, and the contribution from the interaction between features i and j is calculated from

$$\nabla_{ij}(S) = f(S \cup \{i, j\}) - f(S \cup \{i\}) - f(S \cup \{j\}) + f(S). \tag{12.3b}$$

For example, to investigate the interaction between features $i = x_1$ and $j = x_2$ in the three-feature example shown in Fig. 12.1, the p-feature subset S, excluding features x_1 and x_2, includes S_0 and $S_1(x_3)$ only. Thus, for $S = S_0$, $S \cup \{i, j\}$ is $S_2(x_1, x_2)$, $S \cup \{i\}$ is $S_1(x_1)$ and $S \cup \{j\}$ is $S_1(x_2)$; and for $S = S_1(x_3)$, $S \cup \{i, j\}$ is $S_3(x_1, x_2, x_3)$, $S \cup \{i\}$ is $S_2(x_1, x_3)$, and $S \cup \{j\}$ is $S_2(x_2, x_3)$. The definition of Eq. (12.3a) and (12.3b) indicates that $\phi_{i,j} = \phi_{j,i}$ and the total interaction between features i and j is double of $\phi_{i,j}$. The subset weight via the p-feature subset $(p = 0, 1, \cdots, m-2)$ for $\nabla_{ij}(S)$ $(i, j = 1, 2, \cdots, m)$ is $W_p \mathrm{C}_{m-2}^{p}$. All weights via all feature subsets are normalized to unit, i.e.,

$$2 \times \sum_{p=0}^{m-2} W_p \mathrm{C}_{m-2}^{p} = 1, \tag{12.3c}$$

where 2 is added due to $\phi_{i,j} = \phi_{j,i}$. Furthermore, the subset weight is treated to be the same, viz.,

$$W_0 \mathrm{C}_{m-2}^0 = W_1 \mathrm{C}_{m-2}^1 = \cdots = W_{m-2} \mathrm{C}_{m-2}^{m-2}. \tag{12.3d}$$

For the three-feature example shown in Fig. 12.1, we have $W_0 \mathrm{C}_1^0 + W_1 \mathrm{C}_1^1 = 1/2$ and $W_0 = W_1 = 1/4$. Combining Eq. (12.3c) and (12.3d) yields

$$W_p = \frac{p!(m-p-2)!}{2(m-1)!} (p = 0, 1, \cdots, m-2). \tag{12.3e}$$

Substituting Eq. (12.3e) into Eq. (12.3a) gives

$$\phi_{i,j} = \sum_{S \subseteq M \backslash \{i,j\}} \frac{p!(m-p-2)!}{2(m-1)!} \nabla_{ij}(S). \tag{12.3f}$$

A pure contribution of feature *i* to the response is evaluated by

$$\phi_{i,i} = \phi_i - \sum_{j \neq i} \phi_{i,j}. \tag{12.3g}$$

As described above, the co-function number of a feature is $M = \mathrm{C}_{m-1}^0 + \mathrm{C}_{m-1}^1 + \cdots + \mathrm{C}_{m-1}^{m-1} = 2^{m-1}$ and thus the total co-function number is $m \times 2^{m-1}$, which will increase exponentially with the feature number m.

Example 12.1 A dataset includes the original feature dataset $\boldsymbol{X}$ and the response vector $\boldsymbol{y}$.

$$\boldsymbol{X} = \begin{bmatrix} x_{11} & x_{12} & x_{13} \\ x_{21} & x_{22} & x_{23} \\ x_{31} & x_{32} & x_{33} \end{bmatrix} = \begin{bmatrix} 3 & 1 & 2 \\ 7 & 2 & 3 \\ 11 & 3 & 4 \end{bmatrix} \text{ and } \boldsymbol{y} = \begin{pmatrix} 50 \\ 170 \\ 290 \end{pmatrix}. \tag{12.4}$$

A perfect ML model trained from the data is

$$y = (2x_1 - x_2^2 + x_3)^2 + 1. \tag{12.5}$$

With the analytic solution of Eq. (12.5), we shall demonstrate how to calculate the SHAP values explicitly and analytically.

According to Eq. (12.2c), the SHAP value from the empty subset is

$$\phi_0 = y(x_1 = 0, x_2 = 0, x_3 = 0) = 1.$$

The SHAP value of feature x_1 is calculated from Eq. (12.2a). There is only one S_0 subset, and $f(S_0) = 1$, $f(S_0 \cup \{x_1\}) = (2x_1)^2 + 1$, and $f(S_0 \cup \{x_1\}) - f(S_0) = (2x_1)^2$.

There are two S_1 subsets excluding x_1 expressed by $S_1(x_2)$ and $S_1(x_3)$. Thus, we have $f(S_1(x_2)) = (-x_2^2)^2 + 1$, $f(S_1(x_2) \cup \{x_1\}) = (2x_1 - x_2^2)^2 + 1$, and $f(S_1(x_2) \cup \{x_1\}) - f(S_1(x_2)) = (2x_1)^2 - 4x_1x_2^2$; and $f(S_1(x_3)) = (x_3)^2 + 1$, $f(S_1(x_3) \cup \{x_1\}) = (2x_1 + x_3)^2 + 1$ and $f(S_1(x_3) \cup \{x_1\}) - f(S_1(x_2)) = (2x_1)^2 + 4x_1x_3$. There is only one S_2 subset excluding x_1. Thus, we have $f(S_2) = (-x_2^2 + x_3)^2 + 1$, $f(S_2 \cup \{x_1\}) = (2x_1 - x_2^2 + x_3)^2 + 1$, and $f(S_2 \cup \{x_1\}) - f(S_2) = (2x_1)^2 + 4x_1(-x_2^2 + x_3)$. Finally, we have

$$\begin{aligned}\phi_{x_1} &= \frac{1}{3}(2x_1)^2 + \frac{1}{6}\left[(2x_1)^2 - 4x_1x_2^2\right] + \frac{1}{6}[(2x_1)^2 + 4x_1x_3] \\ &+ \frac{1}{3}\left[(2x_1)^2 + 4x_1\left(-x_2^2 + x_3\right)\right] \\ &= 2x_1\left(2x_1 - x_2^2 + x_3\right).\end{aligned} \tag{12.6}$$

Similarly, we have

$$\phi_{x_2} = -x_2^2\left(2x_1 - x_2^2 + x_3\right), \tag{12.7}$$

$$\phi_{x_3} = x_3\left(2x_1 - x_2^2 + x_3\right). \tag{12.8}$$

Then, with Eq. (12.2b), we have

$$y = \phi_0 + \phi_{x_1} + \phi_{x_2} + \phi_{x_3} = \left(2x_1 - x_2^2 + x_3\right)^2 + 1. \tag{12.9}$$

Equation (12.9) is the expected result. Eqs. (12.6)–(12.8) indicate that the SHAP value of feature x_i $(i = 1, 2, 3)$ depends on other features as well, i.e., $\phi_{x_i}(x_1, x_2, x_3)$, which is caused by the interaction among features. The interaction SHAP values are calculated from Eq. (12.3f). The details are described below. For $p = 0$, $W_{p=0} = \dfrac{0!(3-0-2)!}{2(3-1)!} = \dfrac{1}{4}$, $f(S_0 \cup \{x_1, x_2\}) = \left(2x_1 - x_2^2\right)^2 + 1$, $f(S_0 \cup \{x_1\}) = (2x_1)^2 + 1$, $f(S_0 \cup \{x_2\}) = \left(-x_2^2\right)^2 + 1$, and $f(S_0) = 1$. Thus

$$\begin{aligned}\nabla_{x_1,x_2}(S_0) &= f(S_0 \cup \{x_1, x_2\}) - f(S_0 \cup \{x_1\}) - f(S_0 \cup \{x_2\}) + f(S_0) \\ &= \left(2x_1 - x_2^2\right)^2 - (2x_1)^2 - \left(-x_2^2\right)^2 = -4x_1x_2^2.\end{aligned} \tag{12.10a}$$

For $p = 1$, $W_{p=1} = \dfrac{1!(3-1-2)!}{2(3-1)!} = \dfrac{1}{4}$, and there is only one $S_1(x_3)$ excluding features x_1 and x_2, then $f(S_1(x_3) \cup \{x_1, x_2\}) = \left(2x_1 - x_2^2 + x_3\right)^2 + 1$, $f(S_1(x_3) \cup \{x_1\}) = (2x_1 + x_3)^2 + 1$, $f(S_1(x_3) \cup \{x_2\}) = \left(-x_2^2 + x_3\right)^2 + 1$, and $f(S_1(x_3)) = (x_3)^2 + 1$. Thus

$$\begin{aligned}\nabla_{x_1,x_2}(S_1) &= f(S_1(x_3) \cup \{x_1, x_2\}) - f(S_1(x_3) \cup \{x_1\}) - f(S_1(x_3) \cup \{x_2\}) + f(S_1(x_3)) \\ &= \left(2x_1 - x_2^2 + x_3\right)^2 - (2x_1 + x_3)^2 - (-x_2^2 + x_3)^2 + (x_3)^2 = -4x_1x_2^2.\end{aligned} \tag{12.10b}$$

Finally, the interaction SHAP value ϕ_{x_1,x_2} is

$$\phi_{x_1,x_2} = W_{p=0}\nabla_{x_1,x_2}(S_0) + W_{p=1}\nabla_{x_1,x_2}(S_1) = -2x_1x_2^2. \tag{12.11}$$

Similarly, we have

$$\phi_{x_1,x_3} = 2x_1x_3, \tag{12.12}$$

$$\phi_{x_2,x_3} = -x_2^2x_3. \tag{12.13}$$

The pure contribution of each feature is calculated by Eq. (12.3g) to be

$$\phi_{x_1,x_1} = \phi_{x_1} - \phi_{x_1,x_2} - \phi_{x_1,x_3} = 2x_1\left(2x_1 - x_2^2 + x_3\right) + 2x_1x_2^2 - 2x_1x_3 = 4x_1^2. \tag{12.14}$$

Similarly, we have

$$\phi_{x_2,x_2} = \left(-x_2^2\right)^2, \tag{12.15}$$

$$\phi_{x_3,x_3} = x_3^2. \tag{12.16}$$

Equations (12.11)–(12.13) show the interactions between features (x_1, x_2), (x_1, x_3), and (x_2, x_3), respectively, whereas Eqs. (12.14)–(12.16) illustrate the pure contributions from features x_1, x_2, and x_3, respectively. All results of the interactions and the pure contributions are expected. In summary, the SHAP value of a feature counts its interactions with other features, while its pure contribution excludes any interactions with other features.

Lundberg et al. (2018) developed SHAP values for tree ensembles based on the game theory. In decision tree-based algorithms, SHAP values are calculated in a revised way. The following analysis is based on the widely used binary split method, which means that only two child nodes are generated from a parent node. Consider n data of m-dimensional inputs $\{(\boldsymbol{X}_1, y_1), (\boldsymbol{X}_2, y_2), \cdots, (\boldsymbol{X}_n, y_n)\}$, where each datum input $\boldsymbol{X}_i = (x_{i1}, x_{i2}, \cdots, x_{im})(i = 1, 2, \cdots, n) \in \mathbf{R}^m$ is a $1 \times m$ vector and one-dimensional response $y_i \in \mathbf{R}$. A decision tree regressor will split a parent node D based on the optimal feature J with the optimal value of the feature, say x_{sJ}, and in one child node d_1, $x_{iJ} \leq x_{sJ}$, while in the other child node d_2, $x_{iJ} > x_{sJ}$, as described in Chap. 5. In this way, a tree regressor grows up from the root to the leaves. Let P denote the number of all paths from the root to the leaves in a tree and $p(p = 1, 2, \cdots, P)$ be a path from the root to a leaf. Obviously, the number of all paths is identical to the number of leaves. A weight will be assigned to each link between a parent node and a child node by the following definition. The numbers of data in a parent node and its two child nodes are denoted by $|D|$, $|d_1|$ and $|d_2|$, respectively. When studying the SHAP value for a datum of a feature x_k from the

m features ($k = 1, 2, \cdots, m$), the datum is called the studied datum and the other $m-1$ features ($j \neq k, j = 1, 2, \cdots, m$) are called missing features. In the weight assignment, if feature x_k of interest is not the split feature x_J, weights $w_{D\to d_1} = \frac{|d_1|}{|D|}$ and $w_{D\to d_2} = \frac{|d_2|}{|D|}$ are assigned to the links between the parent node and each of the child nodes, respectively, which means that for missing features, the link weight has nothing to do with the studied datum and is calculated by the fraction of the number of data in a child node over the number of data in the parent node, as shown in Fig. 12.2a. If feature x_k is the split feature x_J, i.e., $k = J$, the link weight will be determined by the studied datum. When the value x_{ik} of studied datum i is smaller than or equal to the split value x_{sJ}, i.e., $x_{ik} \leq x_{sJ}$ with $k = J$, the weight of the link between D and d_1 will be one, $w_{D\to d_1} = 1$, and the link weight between D and d_2 will be zero, $w_{D\to d_2} = 0$, as shown in Fig. 12.2b. Otherwise, if $x_{ik} > x_{sJ}$, the weight of the link between D and d_1 will be zero, $w_{D\to d_1} = 0$, and the link weight between between D and d_2 will be one, $w_{D\to d_2} = 1$, as shown in Fig. 12.2c. In this way, all links are assigned weights. Thus, the weight for a path $p(p = 1, 2, \cdots, P)$ from the root to a leaf in a tree will be calculated from

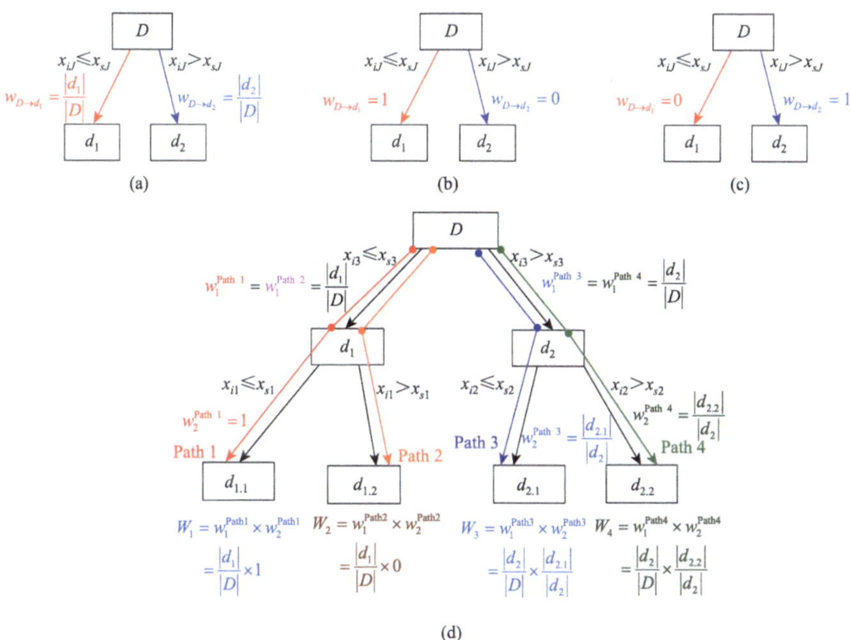

Fig. 12.2 Schematics of the "coverage-split-weighting" assignment. **(a)** Feature x_k of interest is not the split feature x_J. **(b)**, **(c)** Feature x_k of interest is the split feature x_J, **(b)** the x_{ik} value of studied datum i is smaller than or equal to the split value, and **(c)** the x_{ik} value is larger than the split value. **(d)** The calculation of the path weight in a trained tree of four leaves in studying the SHAP value of feature x_1 and the datum studied has $x_{i1} \leq x_{s1}$. In this case,features x_2 and x_3 are missing features

$$W_p = \prod_{i \in p} w_i, \tag{12.17}$$

where $i \in p$ means link i belongs to path p. This weighting method is called the "coverage-split-weighting" (Lundberg et al., 2018). Figure 12.2d shows how to calculate the weight of each link and each path in a trained decision tree regressor on $|D|(|D| = n)$ data with three features of x_1, x_2, and x_3, where the SHAP value of feature x_1 is calculated and thus features x_2 and x_3 are missing features, and the studied datum i has $x_{i1} \le x_{s1}$.

In the tree-based calculation of SHAP values, the root might be regarded as the subset of empty, S_0, and $f(S_0)$ is the average of all data in the root. The combination of $S_0 \cup \{x_k\}(k = 1, 2, \cdots, m)$ forms the subset $S_1(x_k)$. In this case, there is only one feature of interest and the other features are missing features, and $f(S_1(x_k))$ is calculated from

$$f(S_1(x_k)) = \sum_{p=1}^{P} W_p \times \overline{y}_p, \tag{12.18}$$

where $\overline{y}_p$ is the predicted value of leaf p. The combination of $S_0 \cup \{x_k, x_l\}(k \ne l;$ $k, l = 1, 2, \cdots, m)$ or the combination of $S_1\{x_k\} \cup \{x_l\}$ forms the subset $S_2(x_k, x_l)$, where two features are of interest and the other $m-2$ features are missing features. The weight assignment and the model prediction are all the same as described above. In the same way, we can build up subsets $S_3, \cdots, S_m$ and calculate $f(S_3), \cdots, f(S_m)$. Finally, the SHAP value ϕ_i is calculated with Eq. (12.2a), the SHAP interaction value by Eq. (12.3f), and the pure contribution of a feature to the response by Eq. (12.3g).

Example 12.2 A dataset includes the original feature dataset $\boldsymbol{X}$ and the response vector $\boldsymbol{y}$, which consists of data $(\boldsymbol{X}_1, y_1)$, $(\boldsymbol{X}_2, y_2)$, $(\boldsymbol{X}_3, y_3)$, $(\boldsymbol{X}_4, y_4)$ and $(\boldsymbol{X}_5, y_5)$.

$$\boldsymbol{X} = \begin{pmatrix} \boldsymbol{X}_1 \\ \boldsymbol{X}_2 \\ \boldsymbol{X}_3 \\ \boldsymbol{X}_4 \\ \boldsymbol{X}_5 \end{pmatrix} = \begin{bmatrix} x_{11} & x_{12} & x_{13} \\ x_{21} & x_{22} & x_{23} \\ x_{31} & x_{32} & x_{33} \\ x_{41} & x_{42} & x_{43} \\ x_{51} & x_{52} & x_{53} \end{bmatrix} = \begin{bmatrix} 3 & 1 & 2 \\ 7 & 2 & 3 \\ 11 & 3 & 4 \\ 15 & 4 & 5 \\ 19 & 5 & 9 \end{bmatrix} \text{ and } \boldsymbol{y} = \begin{pmatrix} y_1 \\ y_2 \\ y_3 \\ y_4 \\ y_5 \end{pmatrix} = \begin{pmatrix} 50 \\ 170 \\ 290 \\ 362 \\ 485 \end{pmatrix}. \tag{12.19}$$

A decision tree trained on the dataset $(\boldsymbol{X}, \boldsymbol{y})$ is shown in Fig. 12.3 and the associated path weights of the six-feature subsets are shown in Fig. 12.4 in calculating the SHAP values for datum $\boldsymbol{X}_1$.

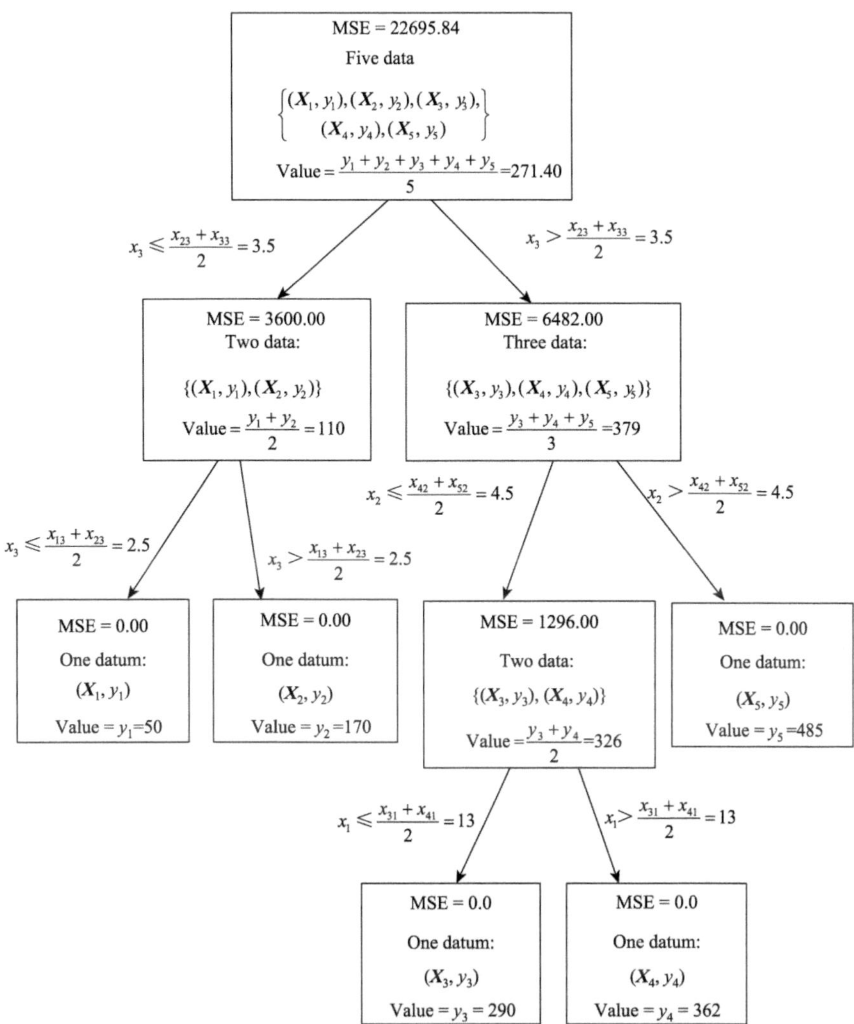

Fig. 12.3 A decision tree trained with the dataset (X, y)

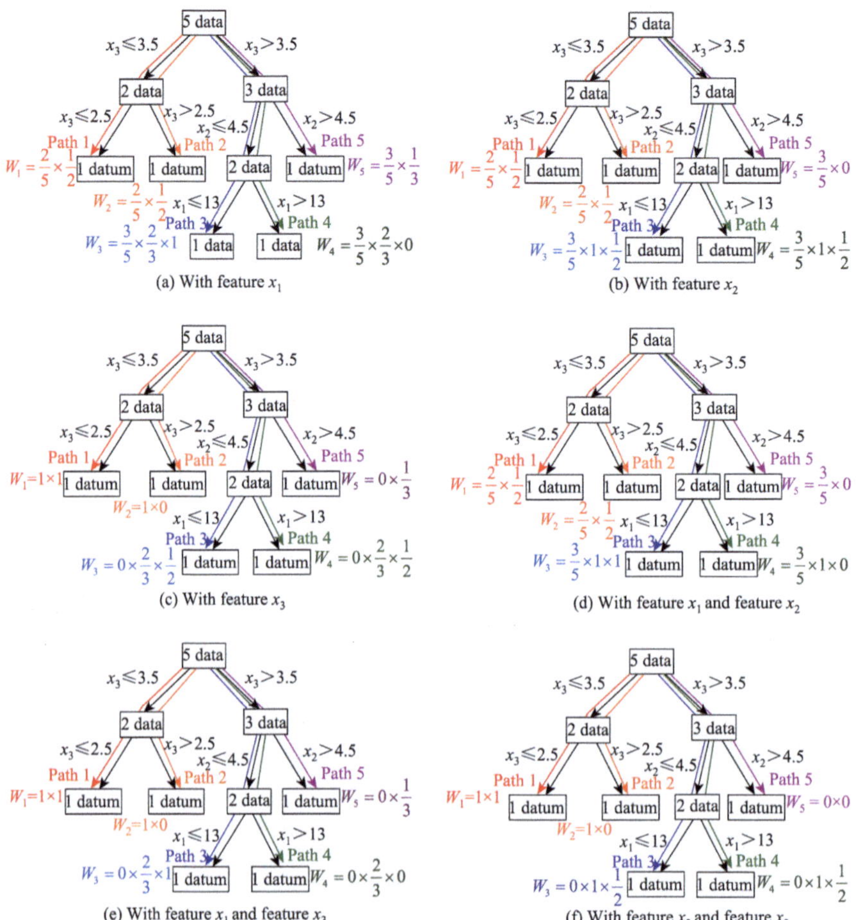

Fig. 12.4 The path weights of datum X_1

In the tree-based calculation of SHAP values, the root might be regarded as the subset of empty, S_0, and $f(S_0)$ is the average of all data in the root. The SHAP value of the empty subset is obtained by

$$\phi_0 = f(S_0) = \frac{y_1 + y_2 + y_3 + y_4 + y_5}{5}. \tag{12.20}$$

The SHAP value of the empty subset is the same for each of the features and to everyone of the data.

As described in Example 12.1, $S_1(x_1) = S_0 \cup \{x_1\}$, $S_2(x_1, x_2) = S_1(x_2) \cup \{x_1\}$, $S_2(x_1, x_3) = S_1(x_3) \cup \{x_1\}$, and $S_3(x_1, x_2, x_3) = S_2(x_2, x_3) \cup \{x_1\}$. Using Eq. (12.2a), we have

$$\begin{aligned}\phi_{x_1} &= W_0[f(S_1(x_1)) - f(S_0)] \\ &\quad + W_1[f(S_2(x_1,x_2)) - f(S_1(x_2))] + W_1[f(S_2(x_1,x_3)) \\ &\quad - f(S_1(x_3))] + W_2[f(S_3(x_1,x_2,x_3)) - f(S_2(x_2,x_3))].\end{aligned} \tag{12.21a}$$

Applying Eq. (12.20) to datum $\boldsymbol{X}_1$ with the path weights gives

$$\begin{aligned}f_{X_1}(S_1(x_1)) &= \frac{2}{5}\times\frac{1}{2}\times y_1 + \frac{2}{5}\times\frac{1}{2}\times y_2 + \frac{3}{5}\times\frac{2}{3}\times 1\times y_3 \\ &\quad + \frac{3}{5}\times\frac{2}{3}\times 0\times y_4 + \frac{3}{5}\times\frac{1}{3}\times y_5 \\ &= \frac{y_1+y_2+2y_3+y_5}{5}\end{aligned} \tag{12.21b}$$

and $f_{X_1}(S_1(x_1)) - f(S_0) = \frac{y_1+y_2+2y_3+y_5}{5} - \frac{y_1+y_2+y_3+y_4+y_5}{5} = \frac{y_3-y_4}{5}$.

There are two S_1 subsets excluding x_1 expressed by $S_1(x_2)$ and $S_1(x_3)$. Thus, we have

$$\begin{aligned}f_{X_1}(S_1(x_2)) &= \frac{2}{5}\times\frac{1}{2}\times y_1 + \frac{2}{5}\times\frac{1}{2}\times y_2 + \frac{3}{5}\times 1\times\frac{1}{2}\times y_3 \\ &\quad + \frac{3}{5}\times 1\times\frac{1}{2}\times y_4 + \frac{3}{5}\times 0\times y_5 \\ &= \frac{2y_1+2y_2+3y_3+3y_4}{10},\end{aligned} \tag{12.21c}$$

$$\begin{aligned}f_{X_1}(S_2(x_1,x_2)) &= \frac{2}{5}\times\frac{1}{2}\times y_1 + \frac{2}{5}\times\frac{1}{2}\times y_2 + \frac{3}{5}\times 1\times 1\times y_3 \\ &\quad + \frac{3}{5}\times 1\times 0\times y_4 + \frac{3}{5}\times 0\times y_5 \\ &= \frac{y_1+y_2+3y_3}{5},\end{aligned} \tag{12.21d}$$

$$\begin{aligned}f_{X_1}(S_2(x_1,x_2)) - f_{X_1}(S_1(x_2)) &= \frac{2y_1+2y_2+6y_3}{10} - \frac{2y_1+2y_2+3y_3+3y_4}{10} \\ &= \frac{3y_3-3y_4}{10}.\end{aligned}$$

Similarly,

$$\begin{aligned}f_{X_1}(S_1(x_3)) &= 1\times 1\times y_1 + 1\times 0\times y_2 + 0\times\frac{2}{3}\times\frac{1}{2}\times y_3 + 0 \\ &\quad \times\frac{2}{3}\times\frac{1}{2}\times y_4 + 0\times\frac{1}{3}\times y_5 = y_1,\end{aligned} \tag{12.21e}$$

$$f_{X_1}(S_2(x_1, x_3)) = 1 \times 1 \times y_1 + 1 \times 0 \times y_2 + 0 \times \frac{2}{3} \times 1 \times y_3 + 0 \times \frac{2}{3} \times 0 \\ \times y_4 + 0 \times \frac{1}{3} \times y_5 = y_1, \quad (12.21f)$$

$$f_{X_1}(S_2(x_1, x_3)) - f_{X_1}(S_1(x_3)) = 0.$$

There is only one $S_2(x_2, x_3)$ subset excluding x_1. Thus, we have

$$f_{X_1}(S_2(x_2, x_3)) = 1 \times 1 \times y_1 + 1 \times 0 \times y_2 + 0 \times 1 \times \frac{1}{2} \times y_3 + 0 \\ \times 1 \times \frac{1}{2} \times y_4 + 0 \times 0 \times y_5 = y_1. \quad (12.21g)$$

When all the three features are of interest, the path weight of $S_3(x_1, x_2, x_3)$ is the same as that of $S_2(x_2, x_3)$ in this example for datum $\boldsymbol{X}_1$ so that

$$f_{X_1}(S_3(x_1, x_2, x_3)) = y_1. \quad (12.21h)$$

Alternatively, the tree-based normal prediction method for datum $\boldsymbol{X}_1$ can be utilized to find the above result when all features are of interest so that there are no missing features. Then, we have

$$f_{X_1}(S_3(x_1, x_2, x_3)) - f_{X_1}(S_2(x_2, x_3)) = 0.$$

Finally, we have

$$\phi_{x_1}(\boldsymbol{X}_1) = \frac{1}{3} \times \left(\frac{y_3 - y_4}{5}\right) + \frac{1}{6} \times \left(\frac{3y_3 - 3y_4}{10}\right) \\ + \frac{1}{6} \times 0 + \frac{1}{3} \times 0 = \frac{7y_3 - 7y_4}{60}. \quad (12.22a)$$

Similarly, we have

$$\phi_{x_2}(\boldsymbol{X}_1) = \frac{2y_3 + y_4 - 3y_5}{30}, \quad (12.22b)$$

$$\phi_{x_3}(\boldsymbol{X}_1) = \frac{48y_1 - 12y_2 - 35y_3 + 5y_4 - 6y_5}{60}. \quad (12.22c)$$

Then, with Eq. (12.2b), we have

$$\hat{y}_1 = \phi_0(\boldsymbol{X}_1) + \phi_{x_1}(\boldsymbol{X}_1) + \phi_{x_2}(\boldsymbol{X}_1) + \phi_{x_3}(\boldsymbol{X}_1)$$
$$= \frac{y_1 + y_2 + 2y_3 + y_5}{5} + \frac{7y_3 - 7y_4}{60} + \frac{2y_3 + y_4 - 3y_5}{30}$$
$$+ \frac{48y_1 - 12y_2 - 35y_3 + 5y_4 - 6y_5}{60} = y_1. \tag{12.22d}$$

Equation (12.21b) is the expected result. Similarly, the SHAP values of features $\boldsymbol{x}_1$, $\boldsymbol{x}_2$ and $\boldsymbol{x}_3$ for data $\boldsymbol{X}_2$, $\boldsymbol{X}_3$, $\boldsymbol{X}_4$ and $\boldsymbol{X}_5$ are calculated and expressed in the vector form of

$$\begin{bmatrix} \hat{y}_1 \\ \hat{y}_2 \\ \hat{y}_3 \\ \hat{y}_4 \\ \hat{y}_5 \end{bmatrix} = \begin{bmatrix} \phi_0(\boldsymbol{X}_1) \\ \phi_0(\boldsymbol{X}_2) \\ \phi_0(\boldsymbol{X}_3) \\ \phi_0(\boldsymbol{X}_4) \\ \phi_0(\boldsymbol{X}_5) \end{bmatrix} + \begin{bmatrix} \phi_{x1}(\boldsymbol{X}_1) \\ \phi_{x1}(\boldsymbol{X}_2) \\ \phi_{x1}(\boldsymbol{X}_3) \\ \phi_{x1}(\boldsymbol{X}_4) \\ \phi_{x1}(\boldsymbol{X}_5) \end{bmatrix} + \begin{bmatrix} \phi_{x2}(\boldsymbol{X}_1) \\ \phi_{x2}(\boldsymbol{X}_2) \\ \phi_{x2}(\boldsymbol{X}_3) \\ \phi_{x2}(\boldsymbol{X}_4) \\ \phi_{x2}(\boldsymbol{X}_5) \end{bmatrix} + \begin{bmatrix} \phi_{x3}(\boldsymbol{X}_1) \\ \phi_{x3}(\boldsymbol{X}_2) \\ \phi_{x3}(\boldsymbol{X}_3) \\ \phi_{x3}(\boldsymbol{X}_4) \\ \phi_{x3}(\boldsymbol{X}_5) \end{bmatrix}$$
$$= \begin{bmatrix} 271.40 \\ 271.40 \\ 271.40 \\ 271.40 \\ 271.40 \end{bmatrix} + \begin{bmatrix} -8.40 \\ -8.40 \\ -24.40 \\ 24.40 \\ 8.80 \end{bmatrix} + \begin{bmatrix} -17.10 \\ -17.10 \\ -47.60 \\ -37.20 \\ 74.40 \end{bmatrix} + \begin{bmatrix} -195.90 \\ -75.90 \\ 90.60 \\ 103.40 \\ 130.40 \end{bmatrix}$$
$$= \begin{bmatrix} 50.00 \\ 170.00 \\ 290.00 \\ 362.00 \\ 485.00 \end{bmatrix}.$$

Usually, the sum of absolute SHAP values of a feature is used to rank the contribution of the feature to the analyzed data (Lundberg et al., 2018). The SHAP values for individual data and the rank of the sum of absolute SHAP values are plotted in Fig. 12.5. The sums of absolute SHAP values are 596.20 for feature x_3, 193.40 for feature x_2, and 74.40 for feature x_1, thereby indicating that feature x_3 might be the most important one to the data.

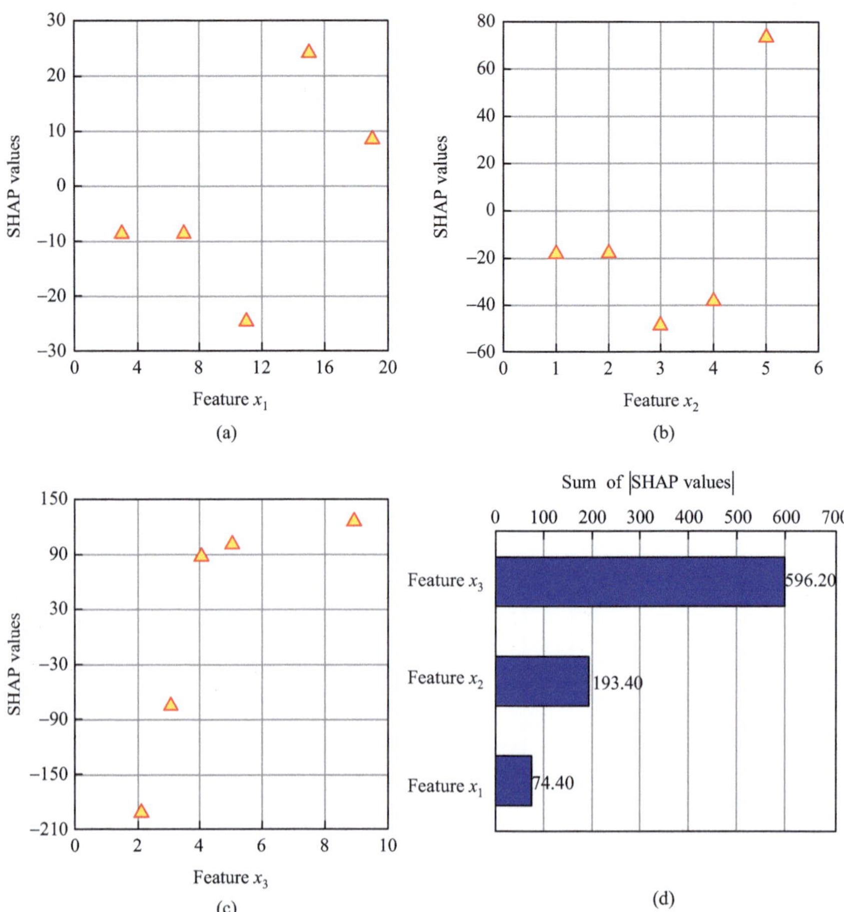

Fig. 12.5 The SHAP values for individual data of **(a)** feature x_1, **(b)** feature x_2, **(c)** feature x_3, and, **(d)** The rank of the sum of absolute SHAP values for three features

With the trained decision tree of Fig. 12.3, the interaction SHAP values between feature x_j and feature $x_k(j \neq k;\ j, k = 1, 2, 3)$ are calculated from Eq. (12.3f) for datum $\boldsymbol{X}_i(i = 1, 2, \cdots, 5)$ and denoted as $\phi_{x_j,x_k}(\boldsymbol{X}_i)$. The following shows how to calculate the interaction SHAP values for datum $\boldsymbol{X}_1$. As described in Example 12.1, we have

$$f_{\boldsymbol{X}_1}(\mathrm{S}_0 \cup \{\mathrm{x}_1, \mathrm{x}_2\}) = \frac{2y_1 + 2y_2 + 6y_3}{10},\ f_{\boldsymbol{X}_1}(\mathrm{S}_0 \cup \{x_1\}) = \frac{y_1 + y_2 + 2y_3 + y_5}{5},$$
$$f_{\boldsymbol{X}_1}(S_0 \cup \{x_2\}) = \frac{2y_1 + 2y_2 + 3y_3 + 3y_4}{10}, \text{ and } f(S_0) = \frac{y_1 + y_2 + y_3 + y_4 + y_5}{5}.$$

Then, using Eq. (12.3b) gives

$$\begin{aligned}\nabla_{x_1,x_2}^{X_1}(S_0) &= f_{X_1}(S_0 \cup \{x_1, x_2\}) - f_{X_1}(S_0 \cup \{x_1\}) - f_{X_1}(S_0 \cup \{x_2\}) + f(S_0) \\ &= \frac{2y_1 + 2y_2 + 6y_3}{10} - \frac{y_1 + y_2 + 2y_3 + y_5}{5} - \frac{2y_1 + 2y_2 + 3y_3 + 3y_4}{10} \\ &\quad + \frac{y_1 + y_2 + y_3 + y_4 + y_5}{5} = \frac{y_3 - y_4}{10}.\end{aligned} \tag{12.23a}$$

Similarly, we have

$$\begin{aligned}\nabla_{x_1,x_2}^{X_1}(S_1) &= f_{X_1}(S_1(x_3) \cup \{x_1, x_2\}) - f_{X_1}(S_1(x_3) \cup \{x_1\}) \\ &\quad - f_{X_1}(S_1(x_3) \cup \{x_2\}) + f_{X_1}(S_1(x_3)) \\ &= y_1 - y_1 - y_1 + y_1 = 0.\end{aligned} \tag{12.23b}$$

Finally, applying Eq. (12.3f) yields the interaction SHAP value $\phi_{x_1,x_2}(X_1)$ of

$$\begin{aligned}\phi_{x_1,x_2}(X_1) &= W_{p=0}\nabla_{x_1,x_2}^{X_1}(S_0) + W_{p=1}\nabla_{x_1,x_2}^{X_1}(S_1) \\ &= \frac{1}{4} \times \frac{y_3 - y_4}{10} + \frac{1}{4} \times 0 = \frac{y_3 - y_4}{40} = -1.8.\end{aligned} \tag{12.23c}$$

In the same way, we calculate the interactions of

$$\phi_{x_1,x_3}(\boldsymbol{X}_1) = 9, \tag{12.23d}$$

$$\phi_{x_2,x_3}(\boldsymbol{X}_1) = 17.7. \tag{12.23e}$$

The non-zero interaction SHAP values of $\phi_{x_1,x_2}(\boldsymbol{X}_1)$, $\phi_{x_1,x_3}(\boldsymbol{X}_1)$ and $\phi_{x_2,x_3}(\boldsymbol{X}_1)$ indicate that there are indeed interactions among any two of the three features.

The pure contribution of a feature is calculated by Eq. (12.3g), which are

$$\phi_{x_1,x_1}(\boldsymbol{X}_1) = \phi_{x_1}(\boldsymbol{X}_1) - \phi_{x_1,x_2}(\boldsymbol{X}_1) - \phi_{x_1,x_3}(\boldsymbol{X}_1) = -8.4 - (-1.8) - 9 = -15.6, \tag{12.24a}$$

$$\phi_{x_2,x_2}(\boldsymbol{X}_1) = -33, \tag{12.24b}$$

$$\phi_{x_3,x_3}(\boldsymbol{X}_1) = -222.6. \tag{12.24c}$$

All results of the interactions and the pure contributions can form the contribution matrix of

$$\begin{bmatrix} \phi_{x_1,x_1}(\boldsymbol{X}_1) & \phi_{x_1,x_2}(\boldsymbol{X}_1) & \phi_{x_1,x_3}(\boldsymbol{X}_1) \\ \phi_{x_2,x_1}(\boldsymbol{X}_1) & \phi_{x_2,x_2}(\boldsymbol{X}_1) & \phi_{x_2,x_3}(\boldsymbol{X}_1) \\ \phi_{x_3,x_1}(\boldsymbol{X}_1) & \phi_{x_3,x_2}(\boldsymbol{X}_1) & \phi_{x_3,x_3}(\boldsymbol{X}_1) \end{bmatrix} = \begin{bmatrix} -15.6 & -1.8 & 9 \\ -1.8 & -33 & 17.7 \\ 9 & 17.7 & -222.6 \end{bmatrix}. \tag{12.25}$$

If there are no interactions, the matrix will be diagonal. The contribution matrixes of the three features to data $\boldsymbol{X}_2$, $\boldsymbol{X}_3$, $\boldsymbol{X}_4$ and $\boldsymbol{X}_5$ can be calculated in the same way. For brevity, only the pure contributions are expressed below in the matrix form of

$$\begin{bmatrix} \phi_{x_1,x_1}(\boldsymbol{X}_1) & \phi_{x_2,x_2}(\boldsymbol{X}_1) & \phi_{x_3,x_3}(\boldsymbol{X}_1) \\ \phi_{x_1,x_1}(\boldsymbol{X}_2) & \phi_{x_2,x_2}(\boldsymbol{X}_2) & \phi_{x_3,x_3}(\boldsymbol{X}_2) \\ \phi_{x_1,x_1}(\boldsymbol{X}_3) & \phi_{x_2,x_2}(\boldsymbol{X}_3) & \phi_{x_3,x_3}(\boldsymbol{X}_3) \\ \phi_{x_1,x_1}(\boldsymbol{X}_4) & \phi_{x_2,x_2}(\boldsymbol{X}_4) & \phi_{x_3,x_3}(\boldsymbol{X}_4) \\ \phi_{x_1,x_1}(\boldsymbol{X}_5) & \phi_{x_2,x_2}(\boldsymbol{X}_5) & \phi_{x_3,x_3}(\boldsymbol{X}_5) \end{bmatrix} = \begin{bmatrix} -15.6 & -33 & -222.6 \\ -15.6 & -33 & -102.6 \\ -13.6 & -31 & 108.4 \\ 13.6 & -32.6 & 106.8 \\ 16.0 & 65.2 & 109.2 \end{bmatrix}. \quad (12.26)$$

To avoid overfitting and enhance the prediction generalization, a decision tree must be pruned. There are many methods to prune a tree, as described in Chap. 5. Will the SHAP values from a pruned tree be different from its counterpart full-size tree? As an example, Fig. 12.6 shows a decision tree obtained by pruning the trained decision tree of Fig. 12.1, where features x_3 and x_2 are used for splitting nodes, while

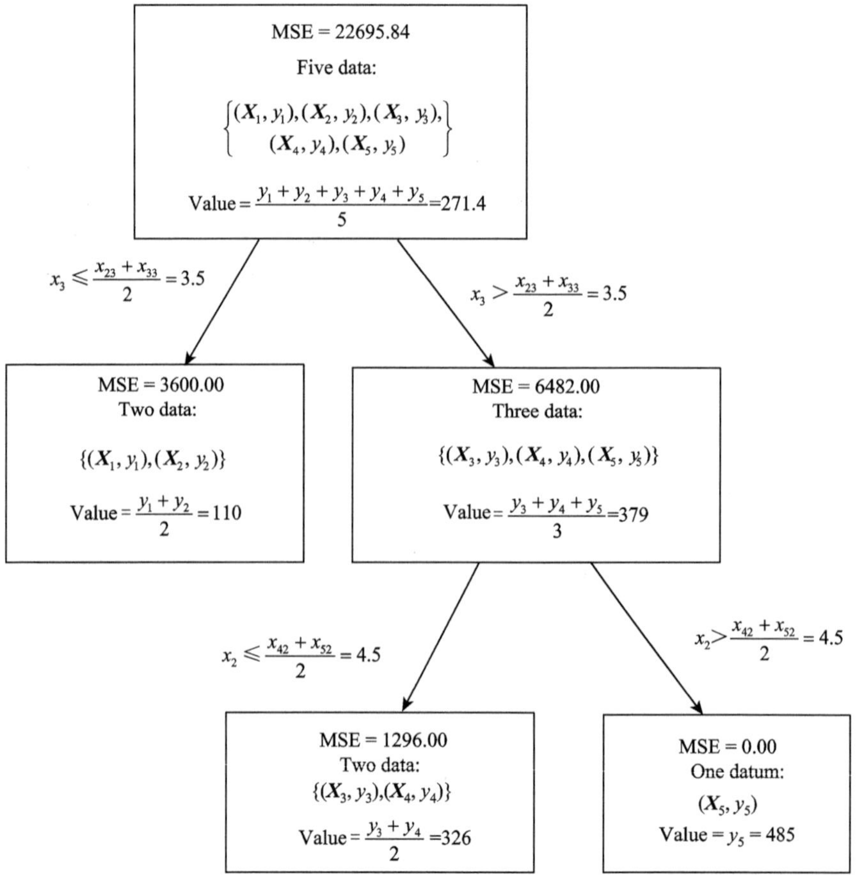

Fig. 12.6 A decision tree obtained by pruning the trained decision tree of Fig. 12.1

feature x_1 is not used for splitting nodes at all. After pruning, each of the two leaves contains two data and one leaf contains only one datum.

Using the same method to calculate the SHAP values of features x_1, x_2 and x_3 for data $\boldsymbol{X}_1$, $\boldsymbol{X}_2$, $\boldsymbol{X}_3$, $\boldsymbol{X}_4$ and $\boldsymbol{X}_5$ based on the pruned decision tree, we have the SHAP values expressed in the vector form of

$$\begin{bmatrix} \hat{y}_1 \\ \hat{y}_2 \\ \hat{y}_3 \\ \hat{y}_4 \\ \hat{y}_5 \end{bmatrix} = \begin{bmatrix} \phi_0(\boldsymbol{X}_1) \\ \phi_0(\boldsymbol{X}_2) \\ \phi_0(\boldsymbol{X}_3) \\ \phi_0(\boldsymbol{X}_4) \\ \phi_0(\boldsymbol{X}_5) \end{bmatrix} + \begin{bmatrix} \phi_{x_1}(\boldsymbol{X}_1) \\ \phi_{x_1}(\boldsymbol{X}_2) \\ \phi_{x_1}(\boldsymbol{X}_3) \\ \phi_{x_1}(\boldsymbol{X}_4) \\ \phi_{x_1}(\boldsymbol{X}_5) \end{bmatrix} + \begin{bmatrix} \phi_{x_2}(\boldsymbol{X}_1) \\ \phi_{x_2}(\boldsymbol{X}_2) \\ \phi_{x_2}(\boldsymbol{X}_3) \\ \phi_{x_2}(\boldsymbol{X}_4) \\ \phi_{x_2}(\boldsymbol{X}_5) \end{bmatrix} + \begin{bmatrix} \phi_{x_3}(\boldsymbol{X}_1) \\ \phi_{x_3}(\boldsymbol{X}_2) \\ \phi_{x_3}(\boldsymbol{X}_3) \\ \phi_{x_3}(\boldsymbol{X}_4) \\ \phi_{x_3}(\boldsymbol{X}_5) \end{bmatrix}$$
$$= \begin{bmatrix} 271.4 \\ 271.4 \\ 271.4 \\ 271.4 \\ 271.4 \end{bmatrix} + \begin{bmatrix} 0 \\ 0 \\ 0 \\ 0 \\ 0 \end{bmatrix} + \begin{bmatrix} -15.9 \\ -15.9 \\ -42.4 \\ -42.4 \\ 84.8 \end{bmatrix} + \begin{bmatrix} -145.5 \\ -145.5 \\ 97 \\ 97 \\ 128.8 \end{bmatrix} = \begin{bmatrix} 110 \\ 110 \\ 326 \\ 326 \\ 485 \end{bmatrix}.$$

The result that the SHAP values of feature x_1 for all the five data are zero indicates that if a feature is not used in splitting at all, the feature will have zero SHAP values. The example also shows that the predicted responses $\hat{y}_i\,(i = 1, 2, 3, 4)$ could deviate from the corresponding real responses $y_i\,(i = 1, 2, 3, 4)$, which might be the expected result, because the average of responses in one leaf is the response for all data in that leaf. The example demonstrates that SHAP values are calculated actually from a trained ML model and hence different SHAP values can be obtained from different ML models trained on the same data.

Using the same method to calculate the interaction SHAP values and pure contributions for data $\boldsymbol{X}_1$, $\boldsymbol{X}_2$, $\boldsymbol{X}_3$, $\boldsymbol{X}_4$ and $\boldsymbol{X}_5$ based on the pruned decision tree, we can obtain the contribution matrix of each datum. For brevity, only the contribution matrix of datum X_1 is shown below, i.e.,

$$\begin{bmatrix} \phi_{x_1,x_1}(\boldsymbol{X}_1) & \phi_{x_1,x_2}(\boldsymbol{X}_1) & \phi_{x_1,x_3}(\boldsymbol{X}_1) \\ \phi_{x_2,x_1}(\boldsymbol{X}_1) & \phi_{x_2,x_2}(\boldsymbol{X}_1) & \phi_{x_2,x_3}(\boldsymbol{X}_1) \\ \phi_{x_3,x_1}(\boldsymbol{X}_1) & \phi_{x_3,x_2}(\boldsymbol{X}_1) & \phi_{x_3,x_3}(\boldsymbol{X}_1) \end{bmatrix} = \begin{bmatrix} 0 & 0 & 0 \\ 0 & -31.8 & 15.9 \\ 0 & 15.9 & 161.4 \end{bmatrix},$$

where the interactions and contribution associated with feature x_1 are all zero, because feature x_1 is not used for splitting at all in the pruned decision tree. The pure contribution matrix is expressed by

$$\begin{bmatrix} \phi_{x_1,x_1}(\boldsymbol{X}_1) & \phi_{x_2,x_2}(\boldsymbol{X}_1) & \phi_{x_3,x_3}(\boldsymbol{X}_1) \\ \phi_{x_1,x_1}(\boldsymbol{X}_2) & \phi_{x_2,x_2}(\boldsymbol{X}_2) & \phi_{x_3,x_3}(\boldsymbol{X}_2) \\ \phi_{x_1,x_1}(\boldsymbol{X}_3) & \phi_{x_2,x_2}(\boldsymbol{X}_3) & \phi_{x_3,x_3}(\boldsymbol{X}_3) \\ \phi_{x_1,x_1}(\boldsymbol{X}_4) & \phi_{x_2,x_2}(\boldsymbol{X}_4) & \phi_{x_3,x_3}(\boldsymbol{X}_4) \\ \phi_{x_1,x_1}(\boldsymbol{X}_5) & \phi_{x_2,x_2}(\boldsymbol{X}_5) & \phi_{x_3,x_3}(\boldsymbol{X}_5) \end{bmatrix} = \begin{bmatrix} 0 & -31.8 & 161.4 \\ 0 & -31.8 & 161.4 \\ 0 & -31.8 & 107.6 \\ 0 & -31.8 & 107.6 \\ 0 & 63.6 & 107.6 \end{bmatrix}$$

12.2 The Joint SHAP Value of Two Features

The SHAP value of a feature can be decomposed into its pure SHAP value and its interaction SHAP values with other features, as stated in Eq. (12.3g). Based on this, Cao et al. (2022) defined the joint SHAP value of two features as

$$\phi_{jk} = \phi_{j,j} + 2\phi_{j,k} + \phi_{k,k} \text{ for } (j, k = 1, 2, \cdots, m), \tag{12.27a}$$

where $\phi_{j,j}$ and $\phi_{k,k}$ are the pure SHAP values of features j and k, respectively, and $\phi_{j,k}$ is the interaction SHAP value between them. In general, ϕ_{jk} is a function of feature x_j and feature x_k.

Cao et al. (2022) proposed a domain knowledge-guided interpretive ML strategy to make machine learning models interpretable. They systematically analyzed the influence of 11 alloying elements on the oxidation behavior of ferritic-martensitic (FM) steels in supercritical water (SCW) and found the five alloying elements of V, Si, Ni, Cr, and Mn play the most important roles. A linear function is proposed to approximately estimate the interaction function, i.e., $\phi_{jk}(x_j, x_k) \approx \phi_{jk0} + a_j^0 x_j + a_k^0 x_k$. The two coefficients a_j^0 and a_k^0 represent the contribution weights of features x_j and x_k, respectively, to their interaction function. The ϕ_{jk0} is intercept. Then letting $a_j^0 \equiv 1$, Cao et al. (2022) introduced an equivalent coefficient $a_k = a_k^0/a_j^0$ for feature x_k and named it the x_j equivalent coefficient of x_k. Letting x_j be the Cr element, they calculated the Cr equivalent coefficient for each element of V, Si, Ni, and Mn, and the sum of the four Cr equivalent coefficients is defined as the oxidation Cr equivalent concentration, denoted by $[\widetilde{\text{Cr}}]$:

$$[\widetilde{\text{Cr}}] = [\text{Cr}] + a_{\text{V}}[\text{V}] + \text{a}_{\text{Si}}[\text{Si}] + \text{a}_{\text{Ni}}[\text{Ni}] + \text{a}_{\text{Mn}}[\text{Mn}], \tag{12.27b}$$

where $[\cdot]$ represents the real concentration in wt.% of an element. Figure 12.7a–d show the joint SHAP values and linear fittings of two features, ϕ_{CrSi}, ϕ_{CrMn}, ϕ_{CrNi}, and ϕ_{CrV}, respectively. After that, the oxidation Cr equivalent concentration is obtained as

$$[\widetilde{\text{Cr}}] = [\text{Cr}] + 40.3[\text{V}] + 2.3[\text{Si}] + 10.7[\text{Ni}] - 1.5[\text{Mn}]. \tag{12.27c}$$

Then the pure SHAP value of the oxidation Cr equivalent concentration is analyzed and plotted in Fig. 12.7e, showing that the pure SHAP value $\phi_{[\widetilde{\text{Cr}}],\,[\widetilde{\text{Cr}}]}$ is almost linearly proportional to $[\widetilde{\text{Cr}}]$ with the Pearson correlation coefficient of $\rho = -0.97$.

The advantage of using the oxidation Cr equivalent concentration lies in the great reduction that one feature replaces five features in the following ML and the final ML model is much more precise, explicit, and interpretive. Figure 12.7e shows that when $[\widetilde{\text{Cr}}] > 19.29$ wt.%, the pure SHAP value $\phi_{[\widetilde{\text{Cr}}],[\widetilde{\text{Cr}}]}$ is all negative, meaning that with such compositions, F-M steels will possess high resistance against oxidation in SCW. The oxidation behaviour of F-M steels in SCW depends not only on the composition, but also on the microstructure of the steels. The oxidation chromium

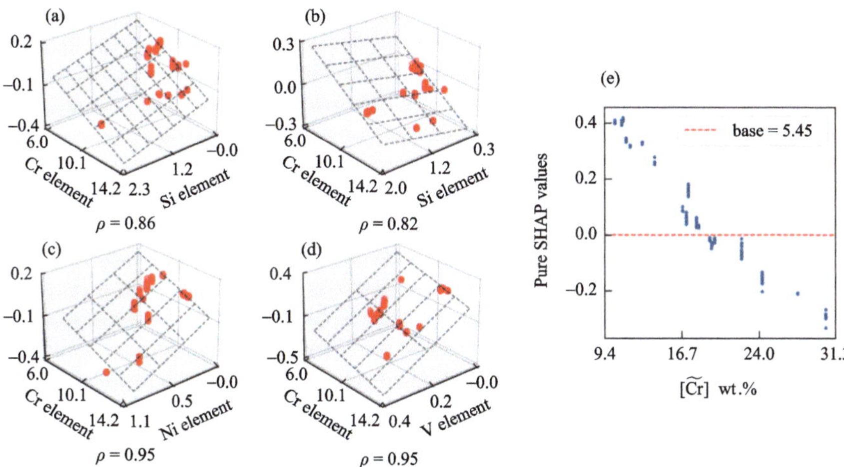

Fig. 12.7 The joint SHAP values of two features and the derived oxidation Cr equivalent concentration. **(a)–(d)** The joint SHAP values of two features, viz., Cr with Si, Mn, Ni, and V respectively. **(e)** The pure SHAP values of the oxidation Cr equivalent concentration feature

(Cr) equivalent concentration is based only on the chemical composition and therefore $[\widetilde{\mathrm{Cr}}]$ is a purely compositional representative. Nevertheless, the oxidation Cr equivalent concentration will provide guidance in the design of anti-oxidation F-M steels in SCW. The application of SHAP values, including pure and interaction SHAP values, will further promote the development of materials informatics.

12.3 Partial Dependence Plot

The approach of partial dependence plot of a feature is similar to that in calculating the marginal probability distribution of that feature from a joint probability distribution (Becker et al., 1996; Friedman, 2001). Consider a discrete bivariate joint probability distribution first. The joint probability distribution function is denoted by $P(x_1, x_2) = P(x_{i1}, x_{j2})(i, j = 1, 2, \cdots, n)$, where n is the amount of data. Then, the marginal probability distribution of variable x_1 is defined by

$$P(x_1) = \sum_{j=1}^{n} P(x_{i1}, x_{j2}). \tag{12.28a}$$

Similarly, the marginal probability distribution of variable x_2 is given by

$$P(x_2) = \sum_{i=1}^{n} P(x_{i1}, x_{j2}). \tag{12.28b}$$

For a continuous bivariate joint probability density distribution $p(x_1, x_2)$, the marginal probability density distributions of variables x_1 and x_2 are respectively defined by

$$p(x_1) = \int_{-\infty}^{+\infty} p(x_1, x_2) \mathrm{d}x_2 \tag{12.28c}$$

and

$$p(x_2) = \int_{-\infty}^{+\infty} p(x_1, x_2) \mathrm{d}x_1. \tag{12.28d}$$

The marginal probability distribution of a variable can be obtained from a joint probability distribution of multiple variables. Taking triple variables for example, we have the discrete joint probability distribution function $P(x_1, x_2, x_3) = P(x_{i1}, x_{j2}, x_{k3})$ $(i, j, k = 1,\ 2,\ \cdots,\ n)$ and the continuous joint probability density distribution $p(x_1, x_2, x_3)$. Then, the marginal discrete probability distribution and the continuous probability distribution density of variable x_1 are given by

$$P(x_1) = \sum_{j=1}^{n} \sum_{k=1}^{n} P\big(x_{i1}, x_{j2}, x_{k3}\big), \tag{12.28e}$$

$$p(x_1) = \int_{-\infty}^{+\infty} p(x_1, x_2, x_3) \mathrm{d}x_2 \mathrm{d}x_3. \tag{12.28f}$$

The marginal discrete probability distributions and the continuous probability distribution densities of variables x_2 and x_3 can be calculated similarly.

12.3.1 Mono-Feature Partial Dependence Plot

A dataset $(x_{i1}, \cdots, x_{im}, y_i)(i = 1, 2, \cdots, n)$ contains n data with m features and one response. A trained ML model gives the prediction of response $\hat{y}(\boldsymbol{X})$ from feature inputs, $\boldsymbol{X} = \{x_1, x_2, \cdots, x_m\}$. To study the partial dependence of a feature, say x_j $(j = 1, 2, \cdots, m)$, the m features are separated into two groups, one feature x_j to be studied and the other features else expressed by $X_{\setminus x_j}$. Similar to the calculation of partial probability distribution, the contributions of other features $X_{\setminus x_j}$ to the response are calculated with the trained ML model $\hat{y}(\boldsymbol{X})$ to obtain the prediction $\hat{y}_{ij}$

$$\hat{y}_{ij}(x_{kj}) = \hat{y}(x_{i1}, \cdots, x_{kj}, \cdots, x_{im}) \quad (i = 1, 2, \cdots, n), \tag{12.29a}$$

where the value of x_{kj} remains unchanged and the values of other features $X_{\backslash x_j}$ change with i. The value of partial dependence of feature x_j for the feature input value of x_{kj}, $P(x_{kj})$, is defined as the averaged $\hat{y}_{ij}(x_{kj})$, i.e.,

$$P(x_{kj}) = \frac{1}{n}\sum_{i=1}^{n} \hat{y}_{ij}(x_{kj}) = \frac{1}{n}\sum_{i=1}^{n} \hat{y}(x_{i1}, \cdots, x_{kj}, \cdots, x_{im}). \tag{12.29b}$$

The number of data n indicates n values of $x_{kj}(k = 1, 2, \cdots, n)$ so that $P(x_j)$ is a function of x_j. Notice that the partial dependence of feature x_j is $\frac{1}{n}$ of the marginal probability distribution of x_j. Equation (12.29b) indicates that a new dataset $\hat{X}_k(x_{i1}, \cdots, x_{kj}, \cdots, x_{im})$ $(i = 1, 2, \cdots, n)$ is constructed with a fixed value of x_{kj}. There will be n new datasets $\hat{X}_k$ because of n values of $x_{kj}(k = 1, 2, \cdots, n)$. In the calculations of partial dependence, the prediction of a trained ML model can be expressed by $\hat{y}_{ij}(\hat{X}_k)$. Averaging the n predictions of $\hat{y}_{ij}(i = 1, 2, \cdots, n)$ for a fixed value of x_{kj} gives a value of $P(x_{kj})$. Then, the partial dependence plots are given by plotting $P(x_{kj})$ versus $x_{kj}(k = 1, 2, \cdots, n)$ for feature $x_j(j = 1, 2, \cdots, m)$.

Example 12.3 A dataset includes the original feature dataset $\boldsymbol{X}$ and the response vector $\boldsymbol{y}$.

$$\boldsymbol{X} = \begin{bmatrix} x_{11} & x_{12} & x_{13} \\ x_{21} & x_{22} & x_{23} \\ x_{31} & x_{32} & x_{33} \end{bmatrix} = \begin{bmatrix} 3 & 1 & 2 \\ 7 & 2 & 3 \\ 11 & 3 & 4 \end{bmatrix} \quad \text{and} \quad \boldsymbol{y} = \begin{pmatrix} 50 \\ 170 \\ 290 \end{pmatrix}. \tag{12.30}$$

A fitness ML model is trained from the data to be

$$y = \left(2x_1 - x_2^2 + x_3\right)^2 + 1. \tag{12.31}$$

New datasets $\hat{\boldsymbol{X}}_k$ $(k = 1, 2, 3)$ for feature x_1 are given by

$$\hat{\boldsymbol{X}}_{x_{11}} = \begin{bmatrix} 3 & 1 & 2 \\ 3 & 2 & 3 \\ 3 & 3 & 4 \end{bmatrix}, \hat{\boldsymbol{X}}_{x_{21}} = \begin{bmatrix} 7 & 1 & 2 \\ 7 & 2 & 3 \\ 7 & 3 & 4 \end{bmatrix}, \quad \text{and} \quad \hat{\boldsymbol{X}}_{x_{31}} = \begin{bmatrix} 11 & 1 & 2 \\ 11 & 2 & 3 \\ 11 & 3 & 4 \end{bmatrix}. \tag{12.32}$$

Substituting (12.32) into Eq. (12.31) gives the predictions of

$$\hat{\boldsymbol{y}}(\hat{\boldsymbol{X}}_{x_{11}}) = \begin{bmatrix} 50 \\ 26 \\ 2 \end{bmatrix}, \hat{\boldsymbol{y}}(\hat{\boldsymbol{X}}_{x_{21}}) = \begin{bmatrix} 226 \\ 170 \\ 82 \end{bmatrix}, \hat{\boldsymbol{y}}(\hat{\boldsymbol{X}}_{x_{31}}) = \begin{bmatrix} 530 \\ 442 \\ 290 \end{bmatrix}. \tag{12.33}$$

The partial dependence of feature x_1 is

$$\boldsymbol{P}(x_{k1}) = \frac{1}{3}\begin{bmatrix} 50+26+2 \\ 226+170+82 \\ 530+442+290 \end{bmatrix} = \begin{bmatrix} 26 \\ 159.3 \\ 420.7 \end{bmatrix}, \tag{12.34}$$

corresponding to $x_1 = [3\ 7\ 11]^{\mathrm{T}}$, and the same expression rule is used hereafter. The partial dependence of feature x_2 and the partial dependence of feature x_3 are calculated in the same way to be respectively

$$\boldsymbol{P}(x_{k2}) = \begin{bmatrix} 311 \\ 224 \\ 119 \end{bmatrix} \text{ and } \boldsymbol{P}(x_{k3}) = \begin{bmatrix} 140.3 \\ 164 \\ 189.6 \end{bmatrix}. \tag{12.35}$$

Example 12.4 Using the same dataset of Eq. (12.16) and the trained decision tree of Fig. 12.3, we calculate the partial dependence of feature $x_j (j = 1, 2, 3)$. New datasets $\hat{\boldsymbol{X}}_k$ $(k = 1, 2, 3, 4, 5)$ for feature x_1 are given by

$$\hat{\boldsymbol{X}}_{x_{11}} = \begin{bmatrix} \hat{\boldsymbol{X}}^1_{x_{11}} \\ \hat{\boldsymbol{X}}^2_{x_{11}} \\ \hat{\boldsymbol{X}}^3_{x_{11}} \\ \hat{\boldsymbol{X}}^4_{x_{11}} \\ \hat{\boldsymbol{X}}^5_{x_{11}} \end{bmatrix} = \begin{bmatrix} 3 & 1 & 2 \\ 3 & 2 & 3 \\ 3 & 3 & 4 \\ 3 & 4 & 5 \\ 3 & 5 & 9 \end{bmatrix}, \hat{\boldsymbol{X}}_{x_{21}} = \begin{bmatrix} \hat{\boldsymbol{X}}^1_{x_{21}} \\ \hat{\boldsymbol{X}}^2_{x_{21}} \\ \hat{\boldsymbol{X}}^3_{x_{21}} \\ \hat{\boldsymbol{X}}^4_{x_{21}} \\ \hat{\boldsymbol{X}}^5_{x_{21}} \end{bmatrix} = \begin{bmatrix} 7 & 1 & 2 \\ 7 & 2 & 3 \\ 7 & 3 & 4 \\ 7 & 4 & 5 \\ 7 & 5 & 9 \end{bmatrix},$$

$$\hat{\boldsymbol{X}}_{x_{31}} = \begin{bmatrix} \hat{\boldsymbol{X}}^1_{x_{31}} \\ \hat{\boldsymbol{X}}^2_{x_{31}} \\ \hat{\boldsymbol{X}}^3_{x_{31}} \\ \hat{\boldsymbol{X}}^4_{x_{31}} \\ \hat{\boldsymbol{X}}^5_{x_{31}} \end{bmatrix} = \begin{bmatrix} 11 & 1 & 2 \\ 11 & 2 & 3 \\ 11 & 3 & 4 \\ 11 & 4 & 5 \\ 11 & 5 & 9 \end{bmatrix}, \hat{\boldsymbol{X}}_{x_{41}} = \begin{bmatrix} \hat{\boldsymbol{X}}^1_{x_{41}} \\ \hat{\boldsymbol{X}}^2_{x_{41}} \\ \hat{\boldsymbol{X}}^3_{x_{41}} \\ \hat{\boldsymbol{X}}^4_{x_{41}} \\ \hat{\boldsymbol{X}}^5_{x_{41}} \end{bmatrix} = \begin{bmatrix} 15 & 1 & 2 \\ 15 & 2 & 3 \\ 15 & 3 & 4 \\ 15 & 4 & 5 \\ 15 & 5 & 9 \end{bmatrix}$$

$$\text{and } \hat{\boldsymbol{X}}_{x_{51}} = \begin{bmatrix} \hat{\boldsymbol{X}}^1_{x_{51}} \\ \hat{\boldsymbol{X}}^2_{x_{51}} \\ \hat{\boldsymbol{X}}^3_{x_{51}} \\ \hat{\boldsymbol{X}}^4_{x_{51}} \\ \hat{\boldsymbol{X}}^5_{x_{51}} \end{bmatrix} = \begin{bmatrix} 19 & 1 & 2 \\ 19 & 2 & 3 \\ 19 & 3 & 4 \\ 19 & 4 & 5 \\ 19 & 5 & 9 \end{bmatrix}. \tag{12.36}$$

Using the trained decision tree of Fig. 12.3, We obtain the predictions of

$$\hat{y}(\hat{\boldsymbol{X}}_{x_{11}}) = \begin{bmatrix} 50 \\ 170 \\ 290 \\ 290 \\ 485 \end{bmatrix}, \hat{y}(\hat{\boldsymbol{X}}_{x_{21}}) = \begin{bmatrix} 50 \\ 170 \\ 290 \\ 290 \\ 485 \end{bmatrix}, \hat{y}(\hat{\boldsymbol{X}}_{x_{31}}) = \begin{bmatrix} 50 \\ 170 \\ 290 \\ 290 \\ 485 \end{bmatrix},$$

$$\hat{y}(\hat{\boldsymbol{X}}_{x_{41}}) = \begin{bmatrix} 50 \\ 170 \\ 362 \\ 362 \\ 485 \end{bmatrix} \text{and} \ \hat{y}(\hat{\boldsymbol{X}}_{x_{51}}) = \begin{bmatrix} 50 \\ 170 \\ 362 \\ 362 \\ 485 \end{bmatrix}. \tag{12.37}$$

The partial dependence of feature x_1 is

$$\boldsymbol{P}(x_{k1}) = \frac{1}{5}\begin{bmatrix} 50+170+290+290+485 \\ 50+170+290+290+485 \\ 50+170+290+290+485 \\ 50+170+362+362+485 \\ 50+170+362+362+485 \end{bmatrix} = \begin{bmatrix} 257 \\ 257 \\ 257 \\ 285.8 \\ 285.8 \end{bmatrix}, \tag{12.38}$$

corresponding to $x_1 = [\ 3\ 7\ 11\ 15\ 19\ ]^{\mathrm{T}}$, and the same expression rule is used hereafter. The partial dependence of feature x_2 and the partial dependence of feature x_3 are calculated in the same way to be respectively

$$\hat{y}(x_{k2}) = \begin{bmatrix} 239.6 \\ 239.6 \\ 239.6 \\ 239.6 \\ 335.0 \end{bmatrix} \quad \text{and} \quad \hat{y}(x_{k3}) = \begin{bmatrix} 50 \\ 170 \\ 379 \\ 379 \\ 379 \end{bmatrix}. \tag{12.39}$$

Then, the partial dependence plots are obtained by plotting $P(x_{kj})$ versus $x_{kj}(k = 1,\ 2,\ 3,\ 4,\ 5;\ j = 1, 2, 3)$ for features x_1, x_2 and x_3, as shown in Fig. 12.8a–c respectively.

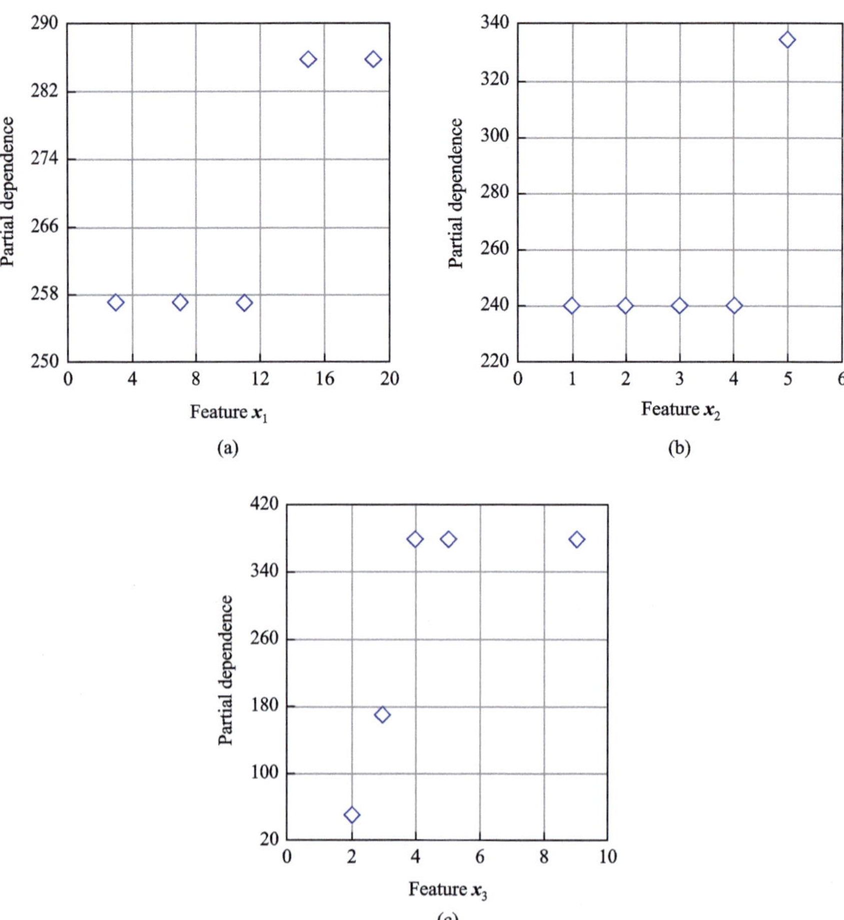

Fig. 12.8 The partial dependence plots of **(a)** feature x_1, **(b)** feature x_2, and **(c)** feature x_3

Based on the pruned decision tree of Fig. 12.6 with the dataset of Eq. (12.16), using the same method to calculate the partial dependence of features x_1, x_2 and x_3, we have the results of

$$\boldsymbol{P}(x_{k1}) = \begin{bmatrix} 271.4 \\ 271.4 \\ 271.4 \\ 271.4 \\ 271.4 \end{bmatrix}; \boldsymbol{P}(x_{k2}) = \begin{bmatrix} 239.6 \\ 239.6 \\ 239.6 \\ 239.6 \\ 335 \end{bmatrix} \text{and } \boldsymbol{P}(x_{k3}) = \begin{bmatrix} 110 \\ 110 \\ 379 \\ 379 \\ 379 \end{bmatrix}. \quad (12.40)$$

An important finding is that the partial dependence of feature x_1 for x_{i1} ($i = 1, 2, \cdots, n$) is the same and equals to the average value $\overline{y}$ of the dataset of Eq. (12.16), indicating that there is no dependence on feature x_1, because feature x_1 is not used for splitting nodes in the pruned decision tree.

12.3.2 Multifeature Partial Dependence Plot

Similarly, in the study of the partial dependence of two features (x_j, x_w) ($j \neq w$, $j = 1,2, \cdots, m$; $w = 1, 2, \cdots, m$), the m features are separated into two groups, the two features (x_j, x_w) form a group and the other features $X_{\boldsymbol{x}_j, x_w}$ form another. With the values of x_{kj} ($k = 1, 2, \cdots, n$) and x_{sw}($s = 1, 2, \cdots, n$), new datasets $\hat{\boldsymbol{X}}_{k,s}$ ($k, s = 1, 2, \cdots, n$) are constructed, i.e.,

$$(x_{i1}, \cdots, x_{kj}, \cdots, x_{sw}, \cdots, x_{im}) \quad (i = 1, 2, \cdots, n; k, s = 1, 2, \cdots, n). \tag{12.41a}$$

A trained ML model gives the prediction $\hat{y}_{iks}(x_{kj}, x_{sw})$ on the newly constructed datasets, i.e.,

$$\hat{y}_{iks}(x_{kj}, x_{sw}) = \hat{y}(x_{i1}, \cdots, x_{kj}, \cdots, x_{sw}, \cdots, x_{im}). \tag{12.41b}$$

The partial dependence of features (x_j, x_w) for the feature input values of (x_{kj}, x_{sw}),$P(x_{kj}, x_{sw})$, is then defined as the averaged $\hat{y}_{iks}(x_{kj}, x_{sw})$, i.e.,

$$P(x_{kj}, x_{sw}) = \frac{1}{n}\sum_{i=1}^{n} \hat{y}_{iks}(x_{kj}, x_{sw}) = \frac{1}{n}\sum_{i=1}^{n} \hat{y}(x_{i1}, \cdots, x_{kj}, \cdots, x_{sw}, \cdots, x_{im}). \tag{12.41c}$$

In the n data, the number of (x_{kj}, x_{sw}) ($k, s = 1, 2, \cdots, n$) is n^2 and thus $P(x_j, x_w)$ is a function of (x_j, x_w). Equation (12.41c) indicates that a new dataset $\hat{\boldsymbol{X}}_{k,s}(x_{i1}, \cdots, x_{kj}, \cdots, x_{sw}, \cdots, x_{im})$ ($i = 1, 2, \cdots, n$) is built up with a fixed value set of (x_{kj}, x_{sw}) and there will be n^2 new datasets $\hat{\boldsymbol{X}}_{k,s}$ because of n^2 values sets of (x_{kj}, x_{sw}) ($k, s = 1, 2, \cdots, n$). The partial dependence plots are given by $P(x_j, x_w)$ versus x_j and x_w for ($j, w = 1, 2, \cdots, m$).

In the same way, the m features are separated into two groups, one group of features is selected from the m features and expressed by $Z_l = \{z_1, z_2, \cdots, z_l\} \subset \{x_1, x_2, \cdots, x_m\}$ and the rest by $Z_{\backslash l}$, $Z_{\backslash l} \cup Z_l = \{x_1, x_2, \cdots, x_m\}$. With the values of $Z_{kl} = \{z_{k_1 1}, z_{k_2 2}, \cdots, z_{k_l l}\}$($k_1, k_2, \cdots, k_l = 1, 2, \cdots, n$), new dataset $\hat{X}_{k_l}$ ($k_1, k_2, \cdots, k_l = 1, 2, \cdots, n$) is constructed, i.e.,

$$\left(x_{i1}, \cdots, z_{k_1 1}, \cdots, z_{k_l}, \cdots, x_{im}\right) \quad (i = 1, 2, \cdots, n; k_1, k_2, \cdots, k_l = 1, 2, \cdots, n). \tag{12.42a}$$

The trained ML model gives the predictions on newly constructed datasets, i.e.,

$$\hat{y}_{iZ_{k_l l}} = \hat{y}(x_{i1}, \cdots, z_{k_1 1}, \cdots, z_{k_l l}, \cdots, x_{im}). \tag{12.42b}$$

The partial dependence of feature $\boldsymbol{Z}_l = \{z_1, z_2, \cdots, z_l\}$ for the feature input value of $(z_{k_1 1}, z_{k_2 2}, \cdots, z_{k_l l})$, $P(z_{k_1 1}, z_{k_2 2}, \cdots, z_{k_l l})$, is defined as the averaged $\hat{y}_i(x_{i1}, \cdots, z_{k_1 1}, \cdots, z_{k_l l}, \cdots, x_{im})$, i.e.,

$$P(z_{k_1 1}, z_{k_2 2}, \cdots, z_{k_l l}) = \frac{1}{n} \sum_{i=1}^{n} \hat{y}_i(x_{i1}, \cdots, z_{k_1 1}, \cdots, z_{k_l l}, \cdots, x_{im}). \tag{12.42c}$$

In the n data, there are n^l values of $(z_{k_1 1}, z_{k_2 2}, \cdots, z_{k_l l})$ $(0 <= l <= m; k_1, k_2, \cdots, k_l = 1, 2, \cdots, n)$, indicating that $P(z_1, z_2, \cdots, z_l)$ is a function of $(z_1, z_2, \cdots, z_l)$. It is more clear to illustrate the partial dependence with an analytic equation. Consider the simple equation of Eq. (12.30), i.e.,

$$y = \left(2x_1 - x_2^2 + x_3\right)^2 + c, \tag{12.43a}$$

where the features range from $\boldsymbol{X}_{\min} = \{x_{\min,1}, x_{\min,2}, x_{\min,3}\} = \{a_1\ a_2\ a_3\}$ to $\boldsymbol{X}_{\max} = \{x_{\max,1}, x_{\max,2}, x_{\max,3}\} = \{b_1\ b_2\ b_3\}$; and c is the constant. Furthermore, (a_i, b_i) $(i = 1,\ 2,\ 3)$ are assumed to be constant, which means that the shape of feature space is tetragonal with plane and line boundaries, which simplifies the integration as described below. For a given value of x_1, we calculate the partial dependence of feature x_1 from the integration, in which x_1 is treated as a constant, to be

$$\begin{aligned} P(x_1) &= \frac{1}{(b_2 - a_2)(b_3 - a_3)} \iint_{a_2, a_3}^{b_2, b_3} [(2x_1 - x_2^2 + x_3)^2 + c] \mathrm{d}x_2 \mathrm{d}x_3 \\ &= 4x_1^2 + 2x_1(b_3 + a_3 - b_2 - a_2) + C_1 \end{aligned} \tag{12.43b}$$

with $C_1 = \dfrac{b_2^2 + a_2^2 + b_2 a_2}{3} + \dfrac{b_3^2 + a_3^2 + b_3 a_3}{3} - \dfrac{1}{2}(b_2 + a_2)(b_3 + a_3) + c$. Similarly, the partial dependence of $P(x_2)$ and the partial dependence $P(x_3)$ are calculated to be

$$\begin{aligned} P(x_2) &= \frac{1}{(b_1 - a_1)(b_3 - a_3)} \iint_{a_1, a_3}^{b_1, b_3} \left[\left(2x_1 - x_2^2 + x_3\right)^2 + c\right] \mathrm{d}x_1 \mathrm{d}x_3 \\ &= x_2^4 - 2x_2^2(b_3 + a_3 - b_2 - a_2) + C_2 \end{aligned} \tag{12.43c}$$

with $C_2 = \dfrac{b_2^2 + a_2^2 + b_2 a_2}{3} + \dfrac{b_3^2 + a_3^2 + b_3 a_3}{3} - \dfrac{1}{2}(b_2 + a_2)(b_3 + a_3) + c$,

$$P(x_3) = \frac{1}{(b_2 - a_2)(b_1 - a_1)} \iint_{a_2,a_1}^{b_2,b_1} \left[\left(2x_1 - x_2^2 + x_3\right)^2 + c \right] \mathrm{d}x_2 \mathrm{d}x_1$$
$$= x_3^2 + 2x_3(b_3 + a_3 - b_2 - a_2) + C_3 \tag{12.43d}$$

with $C_3 = \dfrac{b_2^2 + a_2^2 + b_2 a_2}{3} + \dfrac{b_3^2 + a_3^2 + b_3 a_3}{3} - \dfrac{1}{2}(b_2 + a_2)(b_3 + a_3) + c$.

When calculating the partial dependence of features (x_j, x_w) $(j \neq w, j, w = 1, 2, \cdots, m)$, the values of features (x_j, x_w) will be treated as constants. Thus, the partial dependence of (x_j, x_w) $(j \neq w, j, w = 1, 2, 3)$ with the analytic equation above-mentioned is calculated from

$$P(x_1, x_2) = \frac{1}{b_3 - a_3} \int_{a_3}^{b_3} \left[\left(2x_1 - x_2^2 + x_3\right)^2 + c \right] \mathrm{d}x_3$$
$$= 4x_1^2 + x_2^2 - 4x_1 x_2 + 2x_1(b_3 + a_3) - x_2(b_3 + a_3) + C_{12} \tag{12.44a}$$

with $C_{12} = c + \dfrac{b_3^2 + a_3^2 + b_3 a_3}{3}$;

$$P(x_2, x_3) = \frac{1}{b_1 - a_1} \int_{a_1}^{b_1} \left[\left(2x_1 - x_2^2 + x_3\right)^2 + c \right] \mathrm{d}x_1$$
$$= x_2^2 + x_3^2 - 2x_2(b_1 + a_1) + 2x_3(b_1 + a_1) - x_2 x_3 + C_{23} \tag{12.44b}$$

with $C_{23} = c + \dfrac{4\left(b_1^2 + a_1^2 + b_1 a_1\right)}{3}$;

$$P(x_3, x_1) = \frac{1}{b_2 - a_2} \int_{a_2}^{b_2} \left[\left(2x_1 - x_2^2 + x_3\right)^2 + c \right] \mathrm{d}x_2$$
$$= 4x_1^2 + x_3^2 - 2x_1(b_2 + a_2) + 4x_1 x_3 - x_3(b_2 + a_2) + C_{13} \tag{12.44c}$$

with $C_{13} = c + \dfrac{b_2^2 + a_2^2 + b_2 a_2}{3}$.

Example 12.5 Using the same dataset of Eq. (12.30) and the ML model of Eq. (12.31), we calculate the partial dependence of two features (x_j, x_w) $(j \neq w, j = 1, 2, 3; w = 1, 2, 3)$. New datasets $\hat{X}_{x_{k1}, x_{s2}}$ $(k, s = 1, 2, 3)$ for features (x_1, x_2) are given by

$$\hat{\boldsymbol{X}}_{x_{11}, x_{12}} = \begin{bmatrix} 3 & 1 & 2 \\ 3 & 1 & 3 \\ 3 & 1 & 4 \end{bmatrix}, \quad \hat{\boldsymbol{X}}_{x_{21}, x_{12}} = \begin{bmatrix} 7 & 1 & 2 \\ 7 & 1 & 3 \\ 7 & 1 & 4 \end{bmatrix}, \text{ and } \hat{\boldsymbol{X}}_{x_{31}, x_{12}} = \begin{bmatrix} 11 & 1 & 2 \\ 11 & 1 & 3 \\ 11 & 1 & 4 \end{bmatrix},$$

$$\hat{\boldsymbol{X}}_{x_{11},x_{22}} = \begin{bmatrix} 3 & 2 & 2 \\ 3 & 2 & 3 \\ 3 & 2 & 4 \end{bmatrix}, \quad \hat{\boldsymbol{X}}_{x_{21},x_{22}} = \begin{bmatrix} 7 & 2 & 2 \\ 7 & 2 & 3 \\ 7 & 2 & 4 \end{bmatrix}, \text{ and } \hat{\boldsymbol{X}}_{x_{31},x_{22}} = \begin{bmatrix} 11 & 2 & 2 \\ 11 & 2 & 3 \\ 11 & 2 & 4 \end{bmatrix},$$
$$\hat{\boldsymbol{X}}_{x_{11},x_{32}} = \begin{bmatrix} 3 & 3 & 2 \\ 3 & 3 & 3 \\ 3 & 3 & 4 \end{bmatrix}, \quad \hat{\boldsymbol{X}}_{x_{21},x_{32}} = \begin{bmatrix} 7 & 3 & 2 \\ 7 & 3 & 3 \\ 7 & 3 & 4 \end{bmatrix}, \text{ and } \hat{\boldsymbol{X}}_{x_{31},x_{32}} = \begin{bmatrix} 11 & 3 & 2 \\ 11 & 3 & 3 \\ 11 & 3 & 4 \end{bmatrix}. \tag{12.45}$$

Substituting Eq. (12.45) into Eq. (12.31) gives the predictions of

$$\hat{y}(\hat{\boldsymbol{X}}_{x_{11},x_{12}}) = \begin{bmatrix} 50 \\ 65 \\ 82 \end{bmatrix}, \quad \hat{y}(\hat{\boldsymbol{X}}_{x_{21},x_{12}}) = \begin{bmatrix} 226 \\ 257 \\ 290 \end{bmatrix}, \quad \hat{y}(\hat{\boldsymbol{X}}_{x_{31},x_{12}}) = \begin{bmatrix} 530 \\ 577 \\ 626 \end{bmatrix},$$
$$\hat{y}(\hat{\boldsymbol{X}}_{x_{11},x_{22}}) = \begin{bmatrix} 17 \\ 26 \\ 37 \end{bmatrix}, \quad \hat{y}(\hat{\boldsymbol{X}}_{x_{21},x_{22}}) = \begin{bmatrix} 145 \\ 170 \\ 197 \end{bmatrix}, \quad \hat{y}(\hat{\boldsymbol{X}}_{x_{31},x_{22}}) = \begin{bmatrix} 401 \\ 442 \\ 485 \end{bmatrix},$$
$$\hat{y}(\hat{\boldsymbol{X}}_{x_{11},x_{32}}) = \begin{bmatrix} 2 \\ 1 \\ 2 \end{bmatrix}, \quad \hat{y}(\hat{\boldsymbol{X}}_{x_{21},x_{32}}) = \begin{bmatrix} 50 \\ 65 \\ 82 \end{bmatrix}, \quad \hat{y}(\hat{\boldsymbol{X}}_{x_{31},x_{32}}) = \begin{bmatrix} 226 \\ 257 \\ 290 \end{bmatrix}. \tag{12.46}$$

The partial dependence of features $(\boldsymbol{x}_1, \boldsymbol{x}_2)$ is expressed by

$$\boldsymbol{P}(x_{k1}, x_{s2}) = \begin{bmatrix} 65.7 & 257.7 & 577.7 \\ 26.7 & 170.7 & 442.7 \\ 1.7 & 65.7 & 257.7 \end{bmatrix}, \tag{12.47}$$

where k and s are the row and column numbers of the partial dependence matrix, respectively. The partial dependence of features (x_2, x_3) and the partial dependence of features (x_1, x_3) are calculated in the same way to be respectively

$$\boldsymbol{P}(x_{k2}, x_{s3}) = \begin{bmatrix} 268.7 & 187.7 & 92.7 \\ 299.7 & 212.7 & 107.7 \\ 332.7 & 239.7 & 124.7 \end{bmatrix}, \text{ and } \boldsymbol{P}(x_{k1}, x_{s3}) = \begin{bmatrix} 23 & 140.3 & 385.7 \\ 30.7 & 164 & 425.3 \\ 40.3 & 189.7 & 467 \end{bmatrix}. \tag{12.48}$$

Example 12.6 Using the same dataset of Eq. (12.16) and the trained decision tree of Fig. 12.3, we calculate the partial dependence of two features (x_j, x_w) $(j \neq w,\ j = 1, 2, 3; w = 1, 2, 3)$. New datasets $\hat{\boldsymbol{X}}_{x_{k1},x_{s2}} (k, s = 1, 2, 3, 4, 5)$ for features (x_1, x_2) are given by

$$\hat{\boldsymbol{X}}_{x_{11},x_{12}} = \begin{bmatrix} 3 & 1 & 2 \\ 3 & 1 & 3 \\ 3 & 1 & 4 \\ 3 & 1 & 5 \\ 3 & 1 & 9 \end{bmatrix}, \hat{\boldsymbol{X}}_{x_{21},x_{12}} = \begin{bmatrix} 7 & 1 & 2 \\ 7 & 1 & 3 \\ 7 & 1 & 4 \\ 7 & 1 & 5 \\ 7 & 1 & 9 \end{bmatrix}, \hat{\boldsymbol{X}}_{x_{31},x_{12}} = \begin{bmatrix} 11 & 1 & 2 \\ 11 & 1 & 3 \\ 11 & 1 & 4 \\ 11 & 1 & 5 \\ 11 & 1 & 9 \end{bmatrix},$$

$$\hat{\boldsymbol{X}}_{x_{41},x_{12}} = \begin{bmatrix} 15 & 1 & 2 \\ 15 & 1 & 3 \\ 15 & 1 & 4 \\ 15 & 1 & 5 \\ 15 & 1 & 9 \end{bmatrix}, \hat{\boldsymbol{X}}_{x_{51},x_{12}} = \begin{bmatrix} 19 & 1 & 2 \\ 19 & 1 & 3 \\ 19 & 1 & 4 \\ 19 & 1 & 5 \\ 19 & 1 & 9 \end{bmatrix}, \hat{\boldsymbol{X}}_{x_{11},x_{22}} = \begin{bmatrix} 3 & 2 & 2 \\ 3 & 2 & 3 \\ 3 & 2 & 4 \\ 3 & 2 & 5 \\ 3 & 2 & 9 \end{bmatrix},$$

$$\hat{\boldsymbol{X}}_{x_{21},x_{22}} = \begin{bmatrix} 7 & 2 & 2 \\ 7 & 2 & 3 \\ 7 & 2 & 4 \\ 7 & 2 & 5 \\ 7 & 2 & 9 \end{bmatrix}, \hat{\boldsymbol{X}}_{x_{31},x_{22}} = \begin{bmatrix} 11 & 2 & 2 \\ 11 & 2 & 3 \\ 11 & 2 & 4 \\ 11 & 2 & 5 \\ 11 & 2 & 9 \end{bmatrix}, \hat{\boldsymbol{X}}_{x_{41},x_{22}} = \begin{bmatrix} 15 & 2 & 2 \\ 15 & 2 & 3 \\ 15 & 2 & 4 \\ 15 & 2 & 5 \\ 15 & 2 & 9 \end{bmatrix},$$

$$\hat{\boldsymbol{X}}_{x_{51},x_{22}} = \begin{bmatrix} 19 & 2 & 2 \\ 19 & 2 & 3 \\ 19 & 2 & 4 \\ 19 & 2 & 5 \\ 19 & 2 & 9 \end{bmatrix}, \hat{\boldsymbol{X}}_{x_{11},x_{32}} = \begin{bmatrix} 3 & 3 & 2 \\ 3 & 3 & 3 \\ 3 & 3 & 4 \\ 3 & 3 & 5 \\ 3 & 3 & 9 \end{bmatrix}, \hat{\boldsymbol{X}}_{x_{21},x_{32}} = \begin{bmatrix} 7 & 3 & 2 \\ 7 & 3 & 3 \\ 7 & 3 & 4 \\ 7 & 3 & 5 \\ 7 & 3 & 9 \end{bmatrix},$$

$$\hat{\boldsymbol{X}}_{x_{31},x_{32}} = \begin{bmatrix} 11 & 3 & 2 \\ 11 & 3 & 3 \\ 11 & 3 & 4 \\ 11 & 3 & 5 \\ 11 & 3 & 9 \end{bmatrix}, \hat{\boldsymbol{X}}_{x_{41},x_{32}} = \begin{bmatrix} 15 & 3 & 2 \\ 15 & 3 & 3 \\ 15 & 3 & 4 \\ 15 & 3 & 5 \\ 15 & 3 & 9 \end{bmatrix}, \hat{\boldsymbol{X}}_{x_{51},x_{32}} = \begin{bmatrix} 19 & 3 & 2 \\ 19 & 3 & 3 \\ 19 & 3 & 4 \\ 19 & 3 & 5 \\ 19 & 3 & 9 \end{bmatrix},$$

$$\hat{\boldsymbol{X}}_{x_{11},x_{42}} = \begin{bmatrix} 3 & 4 & 2 \\ 3 & 4 & 3 \\ 3 & 4 & 4 \\ 3 & 4 & 5 \\ 3 & 4 & 9 \end{bmatrix}, \hat{\boldsymbol{X}}_{x_{21},x_{42}} = \begin{bmatrix} 7 & 4 & 2 \\ 7 & 4 & 3 \\ 7 & 4 & 4 \\ 7 & 4 & 5 \\ 7 & 4 & 9 \end{bmatrix}, \hat{\boldsymbol{X}}_{x_{31},x_{42}} = \begin{bmatrix} 11 & 4 & 2 \\ 11 & 4 & 3 \\ 11 & 4 & 4 \\ 11 & 4 & 5 \\ 11 & 4 & 9 \end{bmatrix},$$

$$\hat{\boldsymbol{X}}_{x_{41},x_{42}} = \begin{bmatrix} 15 & 4 & 2 \\ 15 & 4 & 3 \\ 15 & 4 & 4 \\ 15 & 4 & 5 \\ 15 & 4 & 9 \end{bmatrix}, \hat{\boldsymbol{X}}_{x_{51},x_{42}} = \begin{bmatrix} 19 & 4 & 2 \\ 19 & 4 & 3 \\ 19 & 4 & 4 \\ 19 & 4 & 5 \\ 19 & 4 & 9 \end{bmatrix}, \hat{\boldsymbol{X}}_{x_{11},x_{52}} = \begin{bmatrix} 3 & 5 & 2 \\ 3 & 5 & 3 \\ 3 & 5 & 4 \\ 3 & 5 & 5 \\ 3 & 5 & 9 \end{bmatrix},$$

$$\hat{\boldsymbol{X}}_{x_{21},x_{52}} = \begin{bmatrix} 7 & 5 & 2 \\ 7 & 5 & 3 \\ 7 & 5 & 4 \\ 7 & 5 & 5 \\ 7 & 5 & 9 \end{bmatrix}, \hat{\boldsymbol{X}}_{x_{31},x_{52}} = \begin{bmatrix} 11 & 5 & 2 \\ 11 & 5 & 3 \\ 11 & 5 & 4 \\ 11 & 5 & 5 \\ 11 & 5 & 9 \end{bmatrix}, \hat{\boldsymbol{X}}_{x_{41},x_{52}} = \begin{bmatrix} 15 & 5 & 2 \\ 15 & 5 & 3 \\ 15 & 5 & 4 \\ 15 & 5 & 5 \\ 15 & 5 & 9 \end{bmatrix},$$

$$\hat{\boldsymbol{X}}_{x_{51},\,x_{52}} = \begin{bmatrix} 19 & 5 & 2 \\ 19 & 5 & 3 \\ 19 & 5 & 4 \\ 19 & 5 & 5 \\ 19 & 5 & 9 \end{bmatrix}. \tag{12.49}$$

Using the trained decision tree of Fig. 12.3, we obtain the predictions of

$$\begin{aligned}
&\hat{y}\left(\hat{\boldsymbol{X}}_{x_{11},x_{12}}\right) = \begin{bmatrix} 50 \\ 170 \\ 290 \\ 290 \\ 290 \end{bmatrix},\ \hat{y}\left(\hat{\boldsymbol{X}}_{x_{21},x_{12}}\right) = \begin{bmatrix} 50 \\ 170 \\ 290 \\ 290 \\ 290 \end{bmatrix},\ \hat{y}\left(\hat{\boldsymbol{X}}_{x_{31},x_{12}}\right) = \begin{bmatrix} 50 \\ 170 \\ 290 \\ 290 \\ 290 \end{bmatrix},\ \hat{y}\left(\hat{\boldsymbol{X}}_{x_{21},x_{12}}\right) = \begin{bmatrix} 50 \\ 170 \\ 362 \\ 362 \\ 362 \end{bmatrix}, \\
&\hat{y}\left(\hat{\boldsymbol{X}}_{x_{31},x_{12}}\right) = \begin{bmatrix} 50 \\ 170 \\ 362 \\ 362 \\ 362 \end{bmatrix},\ \hat{y}\left(\hat{\boldsymbol{X}}_{x_{11},x_{22}}\right) = \begin{bmatrix} 50 \\ 170 \\ 290 \\ 290 \\ 290 \end{bmatrix},\ \hat{y}\left(\hat{\boldsymbol{X}}_{x_{21},x_{22}}\right) = \begin{bmatrix} 50 \\ 170 \\ 290 \\ 290 \\ 290 \end{bmatrix},\ \hat{y}\left(\hat{\boldsymbol{X}}_{x_{31},x_{22}}\right) = \begin{bmatrix} 50 \\ 170 \\ 290 \\ 290 \\ 290 \end{bmatrix}, \\
&\hat{y}\left(\hat{\boldsymbol{X}}_{x_{21},x_{22}}\right) = \begin{bmatrix} 50 \\ 170 \\ 362 \\ 362 \\ 362 \end{bmatrix},\ \hat{y}\left(\hat{\boldsymbol{X}}_{x_{31},x_{22}}\right) = \begin{bmatrix} 50 \\ 170 \\ 362 \\ 362 \\ 362 \end{bmatrix},\ \hat{y}\left(\hat{\boldsymbol{X}}_{x_{11},x_{32}}\right) = \begin{bmatrix} 50 \\ 170 \\ 290 \\ 290 \\ 290 \end{bmatrix},\ \hat{y}\left(\hat{\boldsymbol{X}}_{x_{21},x_{32}}\right) = \begin{bmatrix} 50 \\ 170 \\ 290 \\ 290 \\ 290 \end{bmatrix}, \\
&\hat{y}\left(\hat{\boldsymbol{X}}_{x_{31},x_{32}}\right) = \begin{bmatrix} 50 \\ 170 \\ 290 \\ 290 \\ 290 \end{bmatrix},\ \hat{y}\left(\hat{\boldsymbol{X}}_{x_{21},x_{32}}\right) = \begin{bmatrix} 50 \\ 170 \\ 362 \\ 362 \\ 362 \end{bmatrix},\ \hat{y}\left(\hat{\boldsymbol{X}}_{x_{31},x_{32}}\right) = \begin{bmatrix} 50 \\ 170 \\ 362 \\ 362 \\ 362 \end{bmatrix},\ \hat{y}\left(\hat{\boldsymbol{X}}_{x_{11},x_{42}}\right) = \begin{bmatrix} 50 \\ 170 \\ 290 \\ 290 \\ 290 \end{bmatrix}, \\
&\hat{y}\left(\hat{\boldsymbol{X}}_{x_{21},x_{42}}\right) = \begin{bmatrix} 50 \\ 170 \\ 290 \\ 290 \\ 290 \end{bmatrix},\ \hat{y}\left(\hat{\boldsymbol{X}}_{x_{31},x_{42}}\right) = \begin{bmatrix} 50 \\ 170 \\ 290 \\ 290 \\ 290 \end{bmatrix},\ \hat{y}\left(\hat{\boldsymbol{X}}_{x_{21},x_{42}}\right) = \begin{bmatrix} 50 \\ 170 \\ 362 \\ 362 \\ 362 \end{bmatrix},\ \hat{y}\left(\hat{\boldsymbol{X}}_{x_{31},x_{42}}\right) = \begin{bmatrix} 50 \\ 170 \\ 362 \\ 362 \\ 362 \end{bmatrix}, \\
&\hat{y}\left(\hat{\boldsymbol{X}}_{x_{11},x_{52}}\right) = \begin{bmatrix} 50 \\ 170 \\ 485 \\ 485 \\ 485 \end{bmatrix},\ \hat{y}\left(\hat{\boldsymbol{X}}_{x_{21},x_{52}}\right) = \begin{bmatrix} 50 \\ 170 \\ 485 \\ 485 \\ 485 \end{bmatrix},\ \hat{y}\left(\hat{\boldsymbol{X}}_{x_{31},x_{52}}\right) = \begin{bmatrix} 50 \\ 170 \\ 485 \\ 485 \\ 485 \end{bmatrix},\ \hat{y}\left(\hat{\boldsymbol{X}}_{x_{21},x_{52}}\right) = \begin{bmatrix} 50 \\ 170 \\ 485 \\ 485 \\ 485 \end{bmatrix}, \\
&\hat{y}\left(\hat{\boldsymbol{X}}_{x_{31},x_{52}}\right) = \begin{bmatrix} 50 \\ 170 \\ 485 \\ 485 \\ 485 \end{bmatrix}.
\end{aligned} \tag{12.50}$$

The partial dependence of features (x_1, x_2) is expressed by

$$P(x_{k1}, x_{s2}) = \frac{1}{5}\begin{bmatrix} 50+170+290+290+290 & 50+170+290+290+290 & 50+170+290+290+290 & 50+170+362+362+362 & 50+170+362+362+362 \\ 50+170+290+290+290 & 50+170+290+290+290 & 50+170+290+290+290 & 50+170+290+290+290 & 50+170+362+362+362 \\ 50+170+290+290+290 & 50+170+290+290+290 & 50+170+290+290+290 & 50+170+290+290+290 & 50+170+362+362+362 \\ 50+170+290+290+290 & 50+170+290+290+290 & 50+170+290+290+290 & 50+170+290+290+290 & 50+170+362+362+362 \\ 50+170+485+485+485 & 50+170+485+485+485 & 50+170+485+485+485 & 50+170+485+485+485 & 50+170+485+485+485 \end{bmatrix}$$

$$= \begin{bmatrix} 218 & 218 & 218 & 261.2 & 261.2 \\ 218 & 218 & 218 & 261.2 & 261.2 \\ 218 & 218 & 218 & 261.2 & 261.2 \\ 218 & 218 & 218 & 261.2 & 261.2 \\ 335 & 335 & 335 & 335 & 335 \end{bmatrix}, \tag{12.51}$$

where k and s are the row and column numbers of the partial dependence matrix, respectively. The partial dependence of features (x_2, x_3) and the partial dependence of features (x_1, x_3) are calculated in the same way to be respectively

$$\boldsymbol{P}(x_{k2}, x_{s3}) = \begin{bmatrix} 50 & 50 & 50 & 50 & 50 \\ 170 & 170 & 170 & 170 & 170 \\ 326 & 326 & 326 & 326 & 485 \\ 326 & 326 & 326 & 326 & 485 \\ 326 & 326 & 326 & 326 & 485 \end{bmatrix}$$

and

$$\boldsymbol{P}(x_{k1}, x_{s3}) = \begin{bmatrix} 50 & 50 & 50 & 50 & 50 \\ 170 & 170 & 170 & 170 & 170 \\ 335 & 335 & 335 & 403 & 403 \\ 335 & 335 & 335 & 403 & 403 \\ 335 & 335 & 335 & 403 & 403 \end{bmatrix}. \tag{12.52}$$

With the above results, the partial dependence of features (x_1, x_2), the partial dependence of features (x_2, x_3) and the partial dependence of features (x_1, x_3) are plotted in Fig. 12.9a–c, respectively.

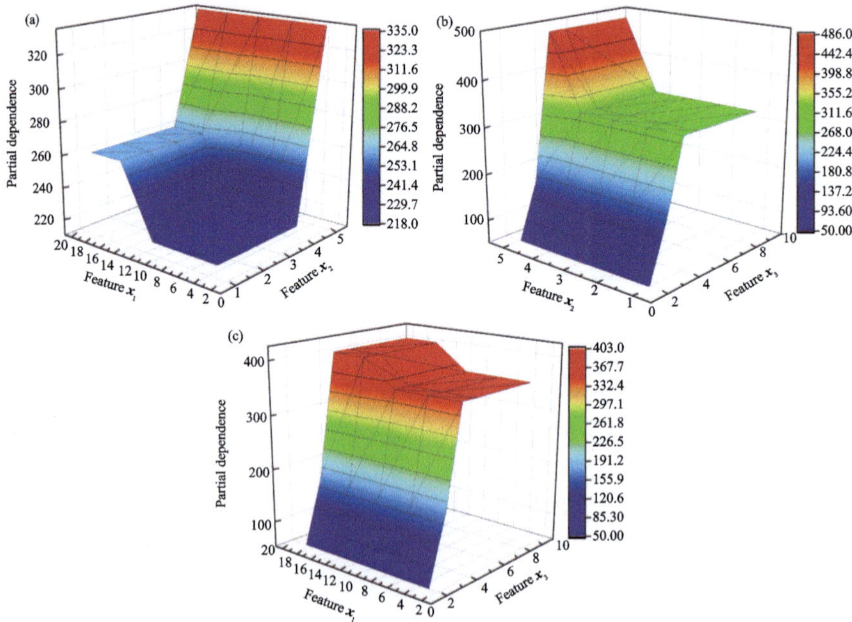

Fig. 12.9 The partial dependence of features **(a)** (x_1, x_2), **(b)** (x_2, x_3), and **(c)** (x_1, x_3)

Based on the pruned decision tree of Fig. 12.6 with the dataset of Eq. (12.16), using the same method to calculate the partial dependence of features (x_1, x_2), (x_2, x_3) and (x_1, x_3), we obtain the results of

$$\hat{\mathbf{y}}(x_{k1}, x_{s2}) = \begin{bmatrix} 239.6 & 239.6 & 239.6 & 239.6 & 239.6 \\ 239.6 & 239.6 & 239.6 & 239.6 & 239.6 \\ 239.6 & 239.6 & 239.6 & 239.6 & 239.6 \\ 239.6 & 239.6 & 239.6 & 239.6 & 239.6 \\ 335 & 335 & 335 & 335 & 335 \end{bmatrix},$$

$$\hat{\mathbf{y}}(x_{k2}, x_{s3}) = \begin{bmatrix} 110 & 110 & 110 & 110 & 110 \\ 110 & 110 & 110 & 110 & 110 \\ 326 & 326 & 326 & 326 & 485 \\ 326 & 326 & 326 & 326 & 485 \\ 326 & 326 & 326 & 326 & 485 \end{bmatrix},$$

and

$$\hat{y}(x_{k1}, x_{s3}) = \begin{bmatrix} 110 & 110 & 110 & 110 & 110 \\ 110 & 110 & 110 & 110 & 110 \\ 379 & 379 & 379 & 379 & 379 \\ 379 & 379 & 379 & 379 & 379 \\ 379 & 379 & 379 & 379 & 379 \end{bmatrix}, \tag{12.53}$$

where the values in each row are the same of the partial dependence matrix for features (x_1, x_2) and (x_1, x_3), because feature x_1 is not used for splitting nodes in the pruned decision tree.

Homework

1. Derive Eqs. (12.1c) and (12.3f).
2. Derive the results of (12.7) and (12.8).
3. Plot the path weights of datum X_2, similar to Fig. 12.4 for datum X_1.
4. Re-calculate the contributions matrix (12.25) with the package of shap (https://pypi.org/project/shap/).
5. Using the dataset of (12.19), calculate the partial dependence of feature x_1.
6. Re-do Example 12.6 in Sect. 12.2 with the package of partial_dependence (https://scikit-learn.org/stable/modules/generated/sklearn.inspection.partial_dependence.html).
7. Derive Eqs. (12.43b) and (12.44a) and discuss their correlation.

References

Becker, R. A., Cleveland, W. S., & Shyu, M. J. (1996). The visual design and control of trellis display. *Journal of Computational and Graphical Statistics, 5*(2), 123–155.

Cao, B., Yang, S., Sun, A., et al. (2022). Domain knowledge-guided interpretive machine learning—formula discovery for the oxidation behaviour of ferritic- martensitic steels in supercritical water. *Journal of Materials Informatics, 2*, 4.

Friedman, J. H. (2001). Greedy function approximation: a gradient boosting machine. *Annals of Statistics, 29*, 1189–1232.

Fujimoto, K., Kojadinovic, I., & Marichal, J. L. (2006). Axiomatic characterizations of probabilistic and cardinal- probabilistic interaction indices. *Games and Economic Behavior, 55*(1), 72–99.

Lipovetsky, S., & Conklin, M. (2001). Analysis of regression in game theory approach. *Applied Stochastic Models in Business and Industry, 17*(4), 319–330.

Lundberg, S. M., & Lee, S. I. (2017). A unified approach to interpreting model predictions. In *Proceedings of the* 31*st International Conference on Neural Information Processing Systems* (pp. 4768–4777). New York: Curran Associates Inc.

Lundberg, S. M., Erion, G. G., & Lee, S. I. (2018). Consistent individualized feature attribution for tree ensembles. arXiv:1802.03888v1.

Shapley, L. S. (1953). A value for *n*-person games. *Contributions to the Theory of Games, 2*, 307–317.

von Neumann, J. (1928a). Sur la théorie des jeux (On game theory). *Comptes Rendus De L'académie Des Sciences, 186*(25), 1689–1691.

von Neumann, J. (1928b). Zur theorie der gesellschaftsspiele (the theory of games). *Mathematische Annalen, 100*, 295–320.

Correction to: An Introduction to Materials Informatics

Correction to:
T. Zhang, *An Introduction to Materials Informatics*, https://doi.org/10.1007/978-981-99-7992-9

The author has provided revised figures, tables, equations, and text. The corrections have been applied to the following chapters:

Symbols and Notations

- The notations of M and SHAP value have been updated.

Chapter 2

- Example 2.3, Equations (2.19a) and (2.19b) and Figure 2.6 have been updated.

The updated version of this book can be found at
https://doi.org/10.1007/978-981-99-7992-9

T. Zhang, *An Introduction to Materials Informatics*,
https://doi.org/10.1007/978-981-99-7992-9_13

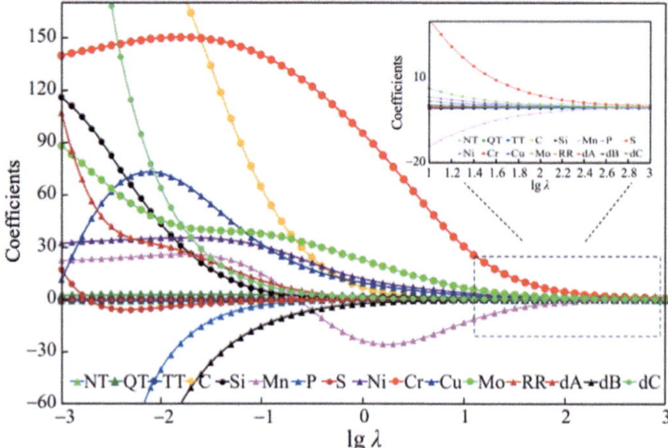

Fig. 2.6 Ridge track plot of ridge regression coefficients versus lg λ

Chapter 3

- The description regarding the recursive update of components has been revised (below Eq. (3.8d)).
- Example 3.2 including Figure 3.4(a) has been updated.

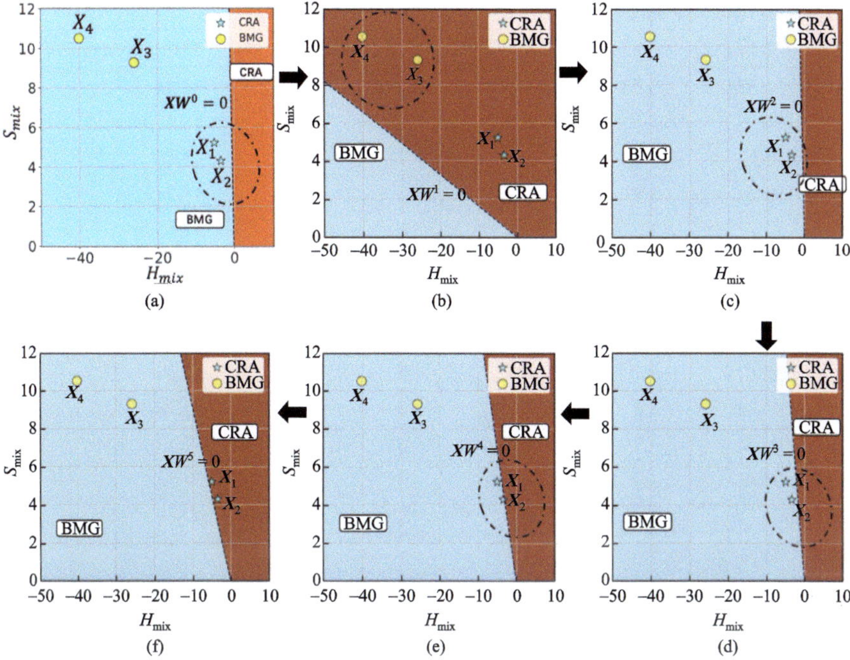

Fig. 3.4 The boundaries calculated by logistic regression in the (H_{mix}, S_{mix}) feature space. *Note* Cyan and yellow dots represent the CRA and BMG data, respectively; the dashed lines are the separating hyperplanes calculated by logistic regression; data inside the dashed ellipses are wrongly classified

Chapter 4

- Equations (4.8a), (4.8b), (4.15a), (4.15b), (4.22), (4.23b), (4.24a), (4.24b), (4.26) and (4.27b) have been corrected.
- The description of KKT conditions has been corrected.
- Tables 4.10 and 4.11 have been updated.

Table 4.10 Training set of 19 data

No.	Sn (wt.%)	Ag (wt.%)	Cu (wt.%)	Bi (wt.%)	Class	Yield Strength(MPa)
1	96.205	3.03	0.76	0.005	1	46.54
2	96	3.5	0.5	0	1	46.3
3	95.5	2.5	0	2	1	45.46
4	95.5	3.8	0.7	0	0	43.37
5	97	3	0	0	0	39.65
6	95	3	0	2	0	37.68
7	95.5	2.5	0	2	1	48.76
8	93.75	2.5	0.75	3	1	82.8
9	95.6	3.7	0.7	0	0	42.39
10	95.48	3.8	0.72	0	0	42.38
11	95.56	3.8	0.64	0	0	37.59
12	95.4	3.9	0.7	0	0	37.58
13	95	3	0	2	1	47.71
14	95	3	2	0	1	46.10
15	93	3	4	0	0	43.28
16	92.8	3.4	0.5	3.3	0	36.3
17	92.4	3.8	0.8	3	1	63
18	91.8	3.4	0	4.8	1	46.29
19	92.5	3.8	0.7	3	0	34.9

Table 4.11 Testing set of 5 data

No.	Sn (wt.%)	Ag (wt.%)	Cu (wt.%)	Bi (wt.%)	Class	Yield Strength(MPa)
1	95.55	3.8	0.65	95.55	0	42.39
2	95.1	2.9	0	95.1	1	47.71
3	95.5	2.5	2	95.5	1	46.10
4	93.1	2.9	4	93.1	0	43.28
5	89.8	3.4	2	89.8	1	46.29

Chapter 6

- Equation (6.8a) has been updated.
- Example 6.6 including Figures 6.5 and 6.6 has been updated.

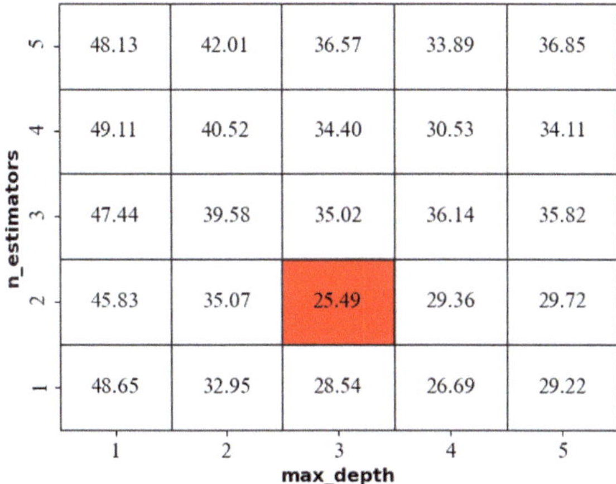

Fig. 6.5 RMSE value of validation set under different parameters

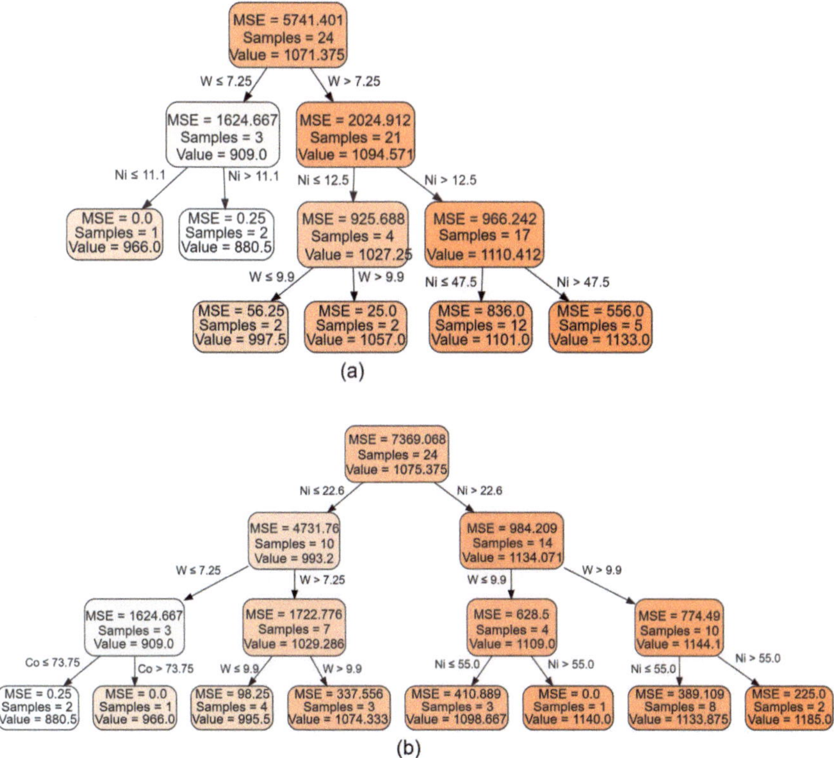

Fig. 6.6 Decision trees in the two-tree random forest. **(a)**Tree one; and **(b)**tree two

Chapter 7

- Equations (7.28b) and (7.29c) have been updated.
- Examples 7.5, 7.7, 7.8 and 7.9 have been updated with minor text revisions.
- Figure 7.11 has been updated.

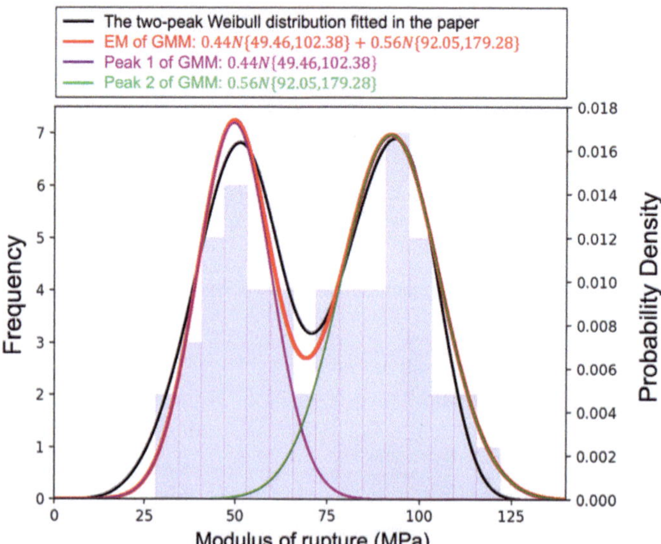

Fig. 7.11 The two-peak Weibull distribution (Fu & Zhang, 1998) and the two-peak Gaussian distribution

Chapter 8

- The description of tournament selection has been updated.

Chapter 10

- All occurrences of λ have been changed to bold $\boldsymbol{\lambda}$ throughout the chapter.
- Equation (10.8c) in the previous version has been deleted.

Chapter 11

- The notation for logarithms has been updated from "log" to "ln".
- Table 11.9 has been updated.

Table 11.9 Select data for 15 lead-free solder alloys

Sn (wt.%)	Ag (wt.%)	Cu (wt.%)	Bi (wt.%)	Yield Strength (y, MPa)
94.15	3.8	0.7	1.35	34.9
95.6	3.4	0	1	46.28966
94	3.8	0.8	1.4	63
94.85	3.4	0.5	1.25	36.3
95.5	3	0.5	1	47.71034
92	3	3	2	46.09655
93	3	2	2	43.28276
94.25	3.7	0.7	1.35	42.38621
95.375	2.5	0.75	1.375	82.8
96.5	2.5	0	1	48.75862
96	3	0	1	39.64828
96	2.5	0.5	1	45.46207
95.5	3.8	0	0.7	43.36552
96	3.5	0.5	0	46.3
96.21	3.03	0.76	0	46.54

The Book and the chapters are updated with the changes.

Appendix A
Vector and Matrix

A.1 Definition

A.1.1 Vector

In an m-dimensional space, a row vector $\boldsymbol{x}$ is expressed by $\boldsymbol{x} = (x_1, x_2, \ldots, x_m)$, where $x_1, x_2, \ldots, x_m$ are the coordinate axes with the basis $\boldsymbol{u}_1 = (1, 0, \ldots, 0)$, $\boldsymbol{u}_2 = (0, 1, \ldots, 0)$, …, $\boldsymbol{u}_m = (0, 0, \ldots, 1)$. If the inner product of two vectors equals zero, $\boldsymbol{x} \cdot \boldsymbol{x}^* = 0$, the two vectors are linearly independent, otherwise linearly dependent.

A.1.2 Matrix

A matrix $\boldsymbol{X}$ that has n rows and m columns is an $n \times m$ matrix, expressed by

$$\boldsymbol{X}_{n\times m} = \begin{bmatrix} x_{11} & \cdots & x_{1m} \\ \vdots & & \vdots \\ x_{n1} & \cdots & x_{nm} \end{bmatrix} \text{ or } \boldsymbol{X} = \left\{x_{ij}\right\}(i = 1, 2, \ldots, n;\ j = 1, 2, \ldots, m),$$

where x_{ij} is an element of matrix $\boldsymbol{X}$.

A.2 Matrix Algebra

The detail of matrix algebra refers to the textbooks of Axler (1997) and Boyd and Vandenberghe (2004), and some are introduced below.

T. Zhang, *An Introduction to Materials Informatics*,
https://doi.org/10.1007/978-981-99-7992-9

A.2.1 Inverse and Transpose

The definitions of inverse $\boldsymbol{A}^{-1}$ and transpose $\boldsymbol{A}^{\mathrm{T}}$ of matrix $\boldsymbol{A}$ are given respectively by

$$\boldsymbol{A} = \{a_{ij}\},\ \boldsymbol{A}^{\mathrm{T}} = \{a_{ji}\},$$

$$\boldsymbol{A}^{-1}\boldsymbol{A} = \boldsymbol{A}\boldsymbol{A}^{-1} = \boldsymbol{I},$$

where $\boldsymbol{I} = \begin{bmatrix} 1 & \cdots & 0 \\ \vdots & & \vdots \\ 0 & \cdots & 1 \end{bmatrix}$ denotes the unit matrix. A square matrix $\boldsymbol{A}$ is invertible only when its determinant is not zero (see Sect. A.2.3). The inverse and transpose of matrices follow the operation rules:

$$\left(\boldsymbol{A}^{\mathrm{T}}\right)^{\mathrm{T}} = \boldsymbol{A},$$

$$(\boldsymbol{A}\boldsymbol{B})^{-1} = \boldsymbol{B}^{-1}\boldsymbol{A}^{-1},$$

$$(\boldsymbol{A}\boldsymbol{B}\boldsymbol{C}\cdots)^{-1} = \cdots\boldsymbol{C}^{-1}\boldsymbol{B}^{-1}\boldsymbol{A}^{-1},$$

$$\left(\boldsymbol{A}^{\mathrm{T}}\right)^{-1} = \left(\boldsymbol{A}^{-1}\right)^{\mathrm{T}},$$

$$(\boldsymbol{A} + \boldsymbol{B})^{\mathrm{T}} = \boldsymbol{A}^{\mathrm{T}} + \boldsymbol{B}^{\mathrm{T}}.$$

A.2.2 Trace

The trace of a square matrix $\boldsymbol{A}$ is defined as

$$\mathrm{tr}(\boldsymbol{A}) = \sum_i a_{ii},$$

$$\mathrm{tr}(\boldsymbol{A}) = \sum_i \lambda_i,\ \lambda_i = \mathrm{eig}(\boldsymbol{A}),$$

where $\mathrm{tr}(\cdot)$ means the trace, a_{ii} is the diagonal elements of matrix $\boldsymbol{A}$ and λ_i is the eigenvalues of matrix $\boldsymbol{A}$ (see Sect. A.2.4).

Trace operation on matrixes $\boldsymbol{A}$, $\boldsymbol{B}$, and $\boldsymbol{C}$ follows the operation rules of

$$\mathrm{tr}(\boldsymbol{A}\boldsymbol{B}) = \mathrm{tr}(\boldsymbol{B}\boldsymbol{A}),$$

$$\mathrm{tr}(\boldsymbol{A}+\boldsymbol{B})=\mathrm{tr}(\boldsymbol{A})+\mathrm{tr}(\boldsymbol{B}),$$

$$\mathrm{tr}(\boldsymbol{ABC})=\mathrm{tr}(\boldsymbol{BCA}).$$

A.2.3 Determinant

The determinant of an $n \times n$ matrix $\boldsymbol{A}$ is defined as

$$\det(\boldsymbol{A})=\sum_{\text{all permutations } \pi_j}\sigma(\pi_j)a_{1j_1}\cdots a_{nj_n},$$

where $\pi_j=(j_1,\ j_2,\ \ldots,\ j_n)$ is one of the $n!$ permutations of the integers for 1 to n, and

$$\sigma(\pi)=\begin{cases}1, & \text{if } \pi \text{ is an even permutation,}\\ -1, & \text{otherwise.}\end{cases}$$

Let both $\boldsymbol{A}$ and $\boldsymbol{B}$ be $n \times n$ matrixes, and the determinant operation has the following rules:

$$\det(\boldsymbol{A})=\prod_i \lambda_i,$$

$$\det(\mathrm{c}\boldsymbol{A})=c^n\det(\boldsymbol{A}),\ \text{when } c \text{ is a constant,}$$

$$\det\left(\boldsymbol{A}^{\mathrm{T}}\right)=\det(\boldsymbol{A}),$$

$$\det(\boldsymbol{AB})=\det(\boldsymbol{A})\det(\boldsymbol{B}),$$

$$\det\left(\boldsymbol{A}^{-1}\right)=\frac{1}{\det(\boldsymbol{A})},$$

$$\det\left(\boldsymbol{A}^n\right)=\det(\boldsymbol{A})^n,$$

$$\det(\boldsymbol{I}+\boldsymbol{A})=1+\det(\boldsymbol{A})+\mathrm{tr}(\boldsymbol{A}).$$

A.2.4 Eigenvalues and Eigenvectors

The eigenvalue equation displays as

$$\boldsymbol{A}\boldsymbol{v} = \lambda\boldsymbol{v},$$

where $\boldsymbol{A}$ is an $n \times n$ matrix, λ is a scalar constant, called an eigenvalue and $\boldsymbol{v}$ is an $n \times 1$ vector, which is usually normalized and called an eigenvector. Solving the equation $\det(\boldsymbol{A} - \lambda\boldsymbol{I}) = 0$ yields n eigenvalues $\lambda_i (i = 1, 2, \ldots, n)$ and then n orthogonal eigenvectors. The n eigenvalues form the diagonal eigenvalue matrix $\boldsymbol{B}$ and the n eigenvectors compose the eigenvector matrix $\boldsymbol{V}$ such that $\boldsymbol{A}\boldsymbol{V} = \boldsymbol{B}\boldsymbol{V}$. The number of non-zero eigenvalues is equal to the rank of matrix $\boldsymbol{A}$.

A.2.5 Singular Value Decomposition (SVD)

SVD is a diagonalization technique to decompose an $n \times m$ matrix $\boldsymbol{T}$ into

$$\boldsymbol{T} = \boldsymbol{U}\boldsymbol{\Sigma}\boldsymbol{V}^{\mathrm{T}},$$

where $\boldsymbol{\Sigma}$ is a $m \times m$ diagonal matrix, $\boldsymbol{\Sigma}^2$ and $\boldsymbol{V}$ are the eigenvalue and eigenvector matrixes of $\boldsymbol{T}^{\mathrm{T}}\boldsymbol{T}$, respectively, and $\boldsymbol{U}$ is a $m \times m$ semi-orthogonal matrix satisfying $\boldsymbol{U}^{\mathrm{T}}\boldsymbol{U} = \boldsymbol{I}$.

A.2.6 Pseudo Inverse

The pseudo inverse of a matrix $\boldsymbol{A}$ is the matrix $\boldsymbol{A}^{+}$ that fulfills: (1) $\boldsymbol{A}\boldsymbol{A}^{+}\boldsymbol{A} = \boldsymbol{A}$, (2) $\boldsymbol{A}^{+}\boldsymbol{A}\boldsymbol{A}^{+} = \boldsymbol{A}^{+}$, (3) $\boldsymbol{A}^{+}\boldsymbol{A}$ and $\boldsymbol{A}\boldsymbol{A}^{+}$ are symmetric. If the SVD of $\boldsymbol{A}$ is $\boldsymbol{A} = \boldsymbol{U}\boldsymbol{D}\boldsymbol{V}^{\mathrm{T}}$, it has

$$\boldsymbol{A}^{+} = \boldsymbol{V}\boldsymbol{D}^{-1}\boldsymbol{U}^{\mathrm{T}}.$$

A.2.7 Some Useful Identities

1 Woodbury Identity

For an invertible $m \times m$ matrix $\boldsymbol{A}$ and another invertible $n \times n$ matrix $\boldsymbol{C}$, the Woodbury identity states

$$(\boldsymbol{A} + \boldsymbol{B}\boldsymbol{C}\boldsymbol{D})^{-1} = \boldsymbol{A}^{-1} - \boldsymbol{A}^{-1}\boldsymbol{B}(\boldsymbol{C}^{-1} - \boldsymbol{D}\boldsymbol{A}^{-1}\boldsymbol{B})^{-1}\boldsymbol{D}\boldsymbol{A}^{-1},$$

where $\boldsymbol{B}$ and $\boldsymbol{D}$ are $m \times n$ and $n \times m$ matrixes, respectively.

2 The Kailath Variant

For an invertible $m \times m$ matrix $\boldsymbol{A}$, the Kailath variant states

$$(\boldsymbol{A} + \boldsymbol{BC})^{-1} = \boldsymbol{A}^{-1} - \boldsymbol{A}^{-1}\boldsymbol{B}\left(\boldsymbol{I} + \boldsymbol{CA}^{-1}\boldsymbol{B}\right)^{-1}\boldsymbol{CA}^{-1},$$

where $\boldsymbol{B}$ and $\boldsymbol{C}$ are $m \times n$ and $n \times m$ matrixes, respectively.

3 Sherman-Morrison Formula

For an invertible $n \times n$ matrix $\boldsymbol{A}$, and $n \times 1$ vectors $\boldsymbol{b}$ and $\boldsymbol{c}$, we have

$$\left(\boldsymbol{A} + \boldsymbol{bc}^{\mathrm{T}}\right)^{-1} = \boldsymbol{A}^{-1} - \frac{\boldsymbol{A}^{-1}\boldsymbol{bc}^{\mathrm{T}}\boldsymbol{A}^{-1}}{1 + \boldsymbol{c}^{\mathrm{T}}\boldsymbol{A}^{-1}\boldsymbol{b}}.$$

4 The Searle Set of Identities

Invertible matrixes $\boldsymbol{A}$ and $\boldsymbol{B}$ follow the identities:

$$\left(\boldsymbol{I} + \boldsymbol{A}^{-1}\right)^{-1} = \boldsymbol{A}(\boldsymbol{A} + \boldsymbol{I})^{-1},$$

$$\left(\boldsymbol{A} + \boldsymbol{BB}^{\mathrm{T}}\right)^{-1}\boldsymbol{B} = \boldsymbol{A}^{-1}\boldsymbol{B}(\boldsymbol{I} + \boldsymbol{B}^{\mathrm{T}}\boldsymbol{A}^{-1}\boldsymbol{B})^{-1},$$

$$\left(\boldsymbol{A}^{-1} + \boldsymbol{B}^{-1}\right)^{-1} = \boldsymbol{A}(\boldsymbol{A} + \boldsymbol{B})^{-1}\boldsymbol{B},$$

$$\boldsymbol{A}^{-1} + \boldsymbol{B}^{-1} = \boldsymbol{A}^{-1}(\boldsymbol{A} + \boldsymbol{B})\boldsymbol{B}^{-1},$$

$$(\boldsymbol{I} + \boldsymbol{AB})^{-1} = \boldsymbol{I} - \boldsymbol{A}(\boldsymbol{I} + \boldsymbol{BA})^{-1}\boldsymbol{B}.$$

A.3 Matrix Analysis

A.3.1 Derivative of Matrix

The detail of matrix derivative refers to the textbook of Horn and Johnson (2012), and some are copied below.

$$\frac{\partial \boldsymbol{A}^{\mathrm{T}}\boldsymbol{X}}{\partial \boldsymbol{X}} = \boldsymbol{A},$$

$$\frac{\partial \boldsymbol{X}^{\mathrm{T}}\boldsymbol{X}}{\partial \boldsymbol{X}} = 2\boldsymbol{X},$$

$$\frac{\partial \boldsymbol{X}^{\mathrm{T}}\boldsymbol{AX}}{\partial \boldsymbol{X}} = \left(\boldsymbol{A} + \boldsymbol{A}^{\mathrm{T}}\right)\boldsymbol{X},$$

$$\frac{\partial \mathrm{tr}(\boldsymbol{XA})}{\partial \boldsymbol{X}} = \boldsymbol{A}^{\mathrm{T}},$$

where $\boldsymbol{X}$ is an $m \times m$ matrix, and $\boldsymbol{A}$ is a constant $m \times m$ matrix.

A.3.2 Derivative of the Determinant of a Matrix

Matrix $\boldsymbol{Y}$ is a function of vector $\boldsymbol{x}$, and $\boldsymbol{A}$ and $\boldsymbol{B}$ are constant matrixes. The following equations give the derivatives of the determinant of matrix $\boldsymbol{Y}$ with respect to vectors $\boldsymbol{x}$ and $\boldsymbol{Y}$.

$$\frac{\partial \det(\boldsymbol{Y})}{\partial \boldsymbol{x}} = \det(\boldsymbol{Y})\mathrm{tr}\left[\boldsymbol{Y}^{-1}\frac{\partial \boldsymbol{Y}}{\partial \boldsymbol{x}}\right],$$

$$\frac{\partial^2 \det(\boldsymbol{Y})}{\partial \boldsymbol{x}^2} = \det(\boldsymbol{Y})\left\{\mathrm{tr}\left[\boldsymbol{Y}^{-1}\frac{\partial^2 \boldsymbol{Y}}{\partial \boldsymbol{x}^2}\right] + \left(\mathrm{tr}\left[\boldsymbol{Y}^{-1}\frac{\partial \boldsymbol{Y}}{\partial \boldsymbol{x}}\right]\right)^2 - \mathrm{tr}\left[\left(\boldsymbol{Y}^{-1}\frac{\partial \boldsymbol{Y}}{\partial \boldsymbol{x}}\right)^2\right]\right\},$$

$$\frac{\partial \det(\boldsymbol{Y})}{\partial \boldsymbol{Y}} = \det(\boldsymbol{Y})\left(\boldsymbol{Y}^{-1}\right)^{\mathrm{T}},$$

$$\frac{\partial \det(\boldsymbol{A}\boldsymbol{Y}\boldsymbol{B})}{\partial \boldsymbol{Y}} = \det(\boldsymbol{A}\boldsymbol{Y}\boldsymbol{B})\left(\boldsymbol{Y}^{-1}\right)^{\mathrm{T}}.$$

A.3.3 Derivative of an Inverse Matrix

For matrix $\boldsymbol{Y}$ and vector $\boldsymbol{x}$, there is

$$\frac{\partial \boldsymbol{Y}^{-1}}{\partial \boldsymbol{x}} = -\boldsymbol{Y}^{-1}\frac{\partial \boldsymbol{Y}}{\partial \boldsymbol{x}}\boldsymbol{Y}^{-1}.$$

A.3.4 Jacobian Matrix and Hessian Matrix

For a scalar-valued matrix function $f = \boldsymbol{X}^{\mathrm{T}}\boldsymbol{A}\boldsymbol{X} + \boldsymbol{b}^{\mathrm{T}}\boldsymbol{X}$, its Jacobian matrix and Hessian matrix are $\nabla f = \frac{\partial f}{\partial \boldsymbol{X}} = \left(\boldsymbol{A} + \boldsymbol{A}^{\mathrm{T}}\right)\boldsymbol{X} + \boldsymbol{b}$ and $\nabla^2 f(x) = \frac{\partial^2 f}{\partial \boldsymbol{X}\partial \boldsymbol{X}^{\mathrm{T}}} = \left(\boldsymbol{A} + \boldsymbol{A}^{\mathrm{T}}\right)$, respectively.

A.3.5 The Chain Rule

Considering a function matrix $\boldsymbol{U} = \{u_{k,l}\}(k = 1, 2, \ldots, K;\ l = 1, 2, \ldots, L)$ and a variable matrix $\boldsymbol{X} = \{x_{i,j}\}(i = 1, 2, \ldots, n;\ j = 1, 2, \ldots, m)$, we have $\boldsymbol{U} = f(\boldsymbol{X})$. The chain rule gives the derivative of a scalar-valued matrix function $g(\boldsymbol{U})$ with respect to $\boldsymbol{X}$

$$\left(\frac{\partial g(\boldsymbol{U})}{\partial \boldsymbol{X}}\right)_{ij} = \frac{\partial g(\boldsymbol{U})}{\partial x_{ij}} = \sum_{k=1}^{K}\sum_{l=1}^{L}\frac{\partial g(\boldsymbol{U})}{\partial u_{kl}}\frac{\partial u_{kl}}{\partial x_{ij}} \quad (i = 1,\ 2,\ \ldots,\ n;\ j = 1, 2, \ldots, m).$$

References

Axler, S. (1997). *Linear algebra done right*. San Francisco: Springer Science & Business Media.

Boyd, S., & Vandenberghe, L. (2004). *Convex optimization*. Cambridge University press.

Horn, R. A., & Johnson, C. R. (2012). *Matrix analysis*. Cambridge University Press.

Appendix B
Basic Statistics

B.1 Probability

B.1.1 Joint Probability

The union of two events A and B is expressed by $A \cup B$, including all occurrences of single event A, single event B and their joint event $A \cap B$. The joint probability $P(A \cap B)$, denoted also by $P(A, B)$ or $P(AB)$, is as the probability that events A and B occur concurrently. The probability of $P(A \cup B)$ is given by $P(A \cup B) = P(A) + P(B) - P(A, B)$.

B.1.2 Bayesian Theorem and Conjugation

A joint distribution probability $P(A, B)$ can be expressed by the chain rule for probability, i.e., $P(A, B) = P(A|B)P(B)$ or $P(A, B) = P(B|A)P(A)$, where $P(A|B)$ and $P(B|A)$ are the conditional probabilities that events A and B occur under the happening of B and A, respectively, and $P(A)$ and $P(B)$ are marginal probabilities of A and B, respectively. Bayesian theorem states

$$P(B|A) = \frac{P(A|B)P(B)}{P(A)},$$

where the conditional probability $P(B|A)$ is also called the posterior probability; The probability $P(B)$ is named the prior probability, which is proposed based on the prior knowledge; The conditional probability $P(A|B)$ is also called the conditional likelihood; And the marginal probability $P(A)$ is termed the total likelihood. In the calculation of the posterior probability under a given value of event $B_k (k = 1, 2, \ldots, K)$, Bayesian theorem states

T. Zhang, *An Introduction to Materials Informatics*,
https://doi.org/10.1007/978-981-99-7992-9

$$P(B_k|A) = \frac{P(A|B_k)P(B_k)}{\sum_k P(A|B_k)P(B_k)} = \frac{P(A|B_k)P(B_k)}{P(A)}.$$

B.1.3 Probability Density of Continuous Variables

For a continuous variable distribution, its probability within an infinitesimal region $\mathrm{d}x$ is given by $p(x)\mathrm{d}x$, where $p(x)$ is called the probability density at point x. Its cumulative distribution function is defined by $F(x) = \int_{-\infty}^{x} p(x)\mathrm{d}x$ so that $p(x) = \frac{\mathrm{d}F(x)}{\mathrm{d}x}$.

B.1.4 Quantile Function

The inverse function of cumulative distribution probability is called the quantile function, defined by $F^{-1}(q) = \inf\{x : F(x) \leq q\}$, for $q \in [0,\ 1]$. The quantile function is an effective tool in sampling.

B.1.5 Expectation, Variance and Covariance of Random Variables

The expected value of the random variable is defined as

$$E[x] = \sum_i x_i P(x_i) \quad \text{for discrete variable}$$

and

$$E[x] = \int_{-\infty}^{\infty} xp(x)\mathrm{d}x, \quad \text{for continuous variable.}$$

The variance of a random variable is defined by

$$\mathrm{var}[x] = E\big[(x - E(x))^2\big] = E\big[x^2\big] - E^2[x].$$

The covariance of two random variables is defined by

$$\mathrm{cov}(X, Y) = E[(X - E(X))(Y - E(Y))] = E[XY] - E[X]E[Y].$$

Two independent random variables lead to $\text{cov}(X, Y) = 0$. If $\text{cov}(X, Y) \approx \pm 1$, the two random variables are strongly positive (+1) or negative (−1) correlated.

B.2 Distributions

The detail of distributions refers to the textbooks of of Rice (2006) and van der Vaart (2000), and some are introduced below.

B.2.1 Bernoulli Distribution

Bernoulli event x has two possible results of success and failure, which are represented by 1 and 0, respectively. A Bernoulli trial is an instantiation of a Bernoulli event, having the probabilities of $P(x = 1) = p$ and $P(x = 0) = 1 - p$, where $p \in [0,\ 1]$. Bernoulli event x has a Bernoulli distribution $f(x) = p^x(1-p)^{1-x}$ for $x = 0$ or 1.

B.2.2 Binomial Distribution

If there are n independent Bernoulli trials, the sequence of success forms binomial distribution

$$P(x) = \begin{cases} \dfrac{n!}{x!(n-x)!} p^x (1-p)^{n-x}, & \text{for } x = 0, 1, \ldots, n, \\ 0, & \text{otherwise,} \end{cases}$$

where x is called the binomial random variable. The binomial distribution has two intrinsic parameters of n and p, and the random variable $x = \{x_1, x_2, \ldots\}$ is drawn from $x \sim \text{Binomial}(n, p)$.

B.2.3 Poisson Distribution

The Poisson distribution with parameter $\lambda(\lambda > 0)$ is given by $P(x) = \dfrac{\lambda^x e^{-\lambda}}{x!}$.

B.2.4 Gaussian Distribution

The normal (or Gaussian) distribution of a continuous variable x has two intrinsic parameters of mean μ and variance σ^2, and is denoted as $N(\mu, \sigma^2)$ with the probability density $p(x) = \frac{1}{\sqrt{2\pi}\sigma}\exp\left[\frac{-(x-\mu)^2}{2\sigma^2}\right]$, $x \in \mathbf{R}$. If $\mu = 0$ and $\sigma^2 = 1$, the normal distribution is standard.

The normal distribution has three properties. (1) If $X \sim N(\mu, \sigma^2)$ and $Z = (X - \mu)/\sigma$, then $Z \sim N(0, 1)$; (2) If $Z \sim N(0, 1)$ and $X = (\sigma Z + \mu)$, then $X \sim N(\mu, \sigma^2)$; (3) If $X_i \sim N(\mu_i, \sigma_i^2)$ and X_i $(i = 1, 2, \ldots, n)$ are independent, then $\sum_{i=1}^{n} X_i \sim N\left(\sum_{i=1}^{n} \mu_i, \sum_{i=1}^{n} \sigma_i^2\right)$.

B.2.5 Weibull Distribution

Without considering the volume effect, the Weibull distribution has two intrinsic parameters of α and β and its cumulative distribution function is $F(x) = 1 - \mathrm{e}^{-\left(\frac{x}{\alpha}\right)^\beta}$. The size-dependent failure probability F_{σ_N} of smooth samples is given by

$$F_{\sigma_N}(V_n, \hat{m}) = 1 - \exp\left[-\frac{V_n}{V_0}\left(\frac{\sigma_N}{\sigma_0}\right)^{\hat{m}}\right],$$

where $\hat{m}$ is the Weibull modulus, σ_0 are the reference strength, and V_n and V_0 are the sample and reference volumes, respectively. The mean and variance values of the Weibull distribution are calculated from

$$\mu^{\mathrm{WB}} = \frac{\sigma_0}{\hat{m}\sqrt{\frac{V_n}{V_0}}}\Gamma\left(1 + \frac{1}{\hat{m}}\right) \quad \text{and}$$

$$\sigma^{\mathrm{WB}} = \left(\frac{\sigma_0}{\hat{m}\sqrt{\frac{V_n}{V_0}}}\right)^2\left[\Gamma\left(1 + \frac{2}{\hat{m}}\right) - \left(\Gamma\left(1 + \frac{1}{\hat{m}}\right)\right)^2\right],$$

respectively, where Γ is the Gamma function.

B.2.6 The Chi-Square (χ^2) Distribution and χ^2-Test

If x is a standard normal random variable, viz., $x \sim N(0, 1)$, the distribution of $Z = x^2$ is called the chi-square (χ^2) distribution with one degree of freedom. If

$Z_1, Z_2, \ldots, Z_n$ are independent one degree of freedom chi-square distributions then $U = \sum_{i=1}^{n} Z_i$ yields a chi-square distribution with n degrees of freedom, denoted as $\chi^2(n)$. For a normal variable $x \sim N(\mu, \sigma^2)$, the sample standard deviation $s^2 = \frac{1}{n-1}\sum_{i=1}^{n}(x_i - \overline{x})^2$ vs. σ^2 follows the χ^2 distribution of an $n-1$ degrees of freedom, viz., $\frac{(n-1)s^2}{\sigma^2} \sim \chi^2(n-1)$ and the value of $\chi^2_{1-\alpha/2}(n-1)$ is available for a given significance α. The χ^2 distribution test evaluates whether s^2 can be accepted under a certain significance.

B.2.7 The Student's t*-Distribution and* t*-Test*

If x is a standard normal random variable, viz., $x \sim N(0, 1)$, the random variable Z obeys an n degrees of freedom χ^2 distribution, then the distribution of $T = \frac{X}{\sqrt{Z/n}}$ obeys the student's t distribution (t distribution) with n degrees of freedom. For a normal variable $x \sim N(\mu, \sigma^2)$, the difference between $\overline{x} = \frac{\sum_{i=1}^{n} x_i}{n}$ and μ follows the student's t distribution of $n-1$ degrees of freedom, viz., $\frac{\overline{x} - \mu}{s/\sqrt{n}} \sim T(n-1)$ and the $\frac{\sigma_0}{\sqrt{n}} T_{\frac{\alpha}{2}}(n-1)$ value is available under for a given significance α. The t-test evaluates whether $\overline{x}$ can be accepted under a certain significance.

References

Rice, J. A. (2006). *Mathematical statistics and data analysis.* Cengage Learning.

van der Vaart, A. W. (2000). *Asymptotic statistics.* Cambridge University.

Index

T. Zhang, *An Introduction to Materials Informatics*,
https://doi.org/10.1007/978-981-99-7992-9

MIX
Papier aus verantwortungsvollen Quellen
Paper from responsible sources
FSC® C105338

If you have any concerns about our products,
you can contact us on
ProductSafety@springernature.com

In case Publisher is established outside the EU,
the EU authorized representative is:
Springer Nature Customer Service Center GmbH
Europaplatz 3, 69115 Heidelberg, Germany

Printed by Libri Plureos GmbH
in Hamburg, Germany